Basic
College
Mathematics

Basic College Mathematics

Jeffrey Slater

John Tobey

North Shore Community College
Beverly, Massachusetts

PRENTICE HALL, Englewood Cliffs, NJ 07632

Library of Congress Cataloging-in-Publication Data

SLATER, JEFFREY, date
   Basic college mathematics/Jeffrey Slater, John Tobey.
      p.   cm.
   Includes index.
   ISBN 0-13-063363-1 :   . — ISBN 0-13-063371-2 (AIE)
   1. Mathematics. I. Tobey, John, date . II. Title.
QA39.2.S558 1991
510—dc20

90-49640
CIP
AC

Editorial/production supervision: Virginia Huebner
Acquisition editor: Priscilla McGeehon
Development editor: Steve Deitmer
Interior design: Judith A. Matz-Coniglio
Cover design: Judith A. Matz-Coniglio
Manufacturing buyer: Lori Bulwin
Prepress buyer: Paula Massenaro
Page layout: Karen Noferi
Photo editor: Lori Morris-Nantz
Photo research: Barbara Scott
Cover photograph: Brooklyn Bridge, © Jeff Perkell/The Stock Market
Photo credits:

CLICK/CHICAGO LTD.
p. 473 Jim Pickerell

THE IMAGE BANK
p. 55 Janeart Ltd.
p. 259 Obremski

PHOTO EDIT
pp. 129, 435, 498, 532 © Tony Freeman
p. 174 © David Conklin

PHOTO RESEARCHERS, INC.
p. 233 Catherine Ursillo
p. 516 Richard Hutchings

THE STOCK MARKET
p. 1 © Paul Barton
p. 202 Michael A. Keller
p. 277 © Jon Feingersh
p. 299 © Blaine Harrington, III
p. 413 © Mugshots/Gabe Palmer

STOCK/BOSTON
p. 75 Stacy Pick
p. 211 Barbara Alper
p. 307 Billy E. Barnes
p. 488, 513 © Robert Rathe

UNIPHOTO PICTURE AGENCY
p. 20 Bob Daemmrich
p. 215 Armed Forces
p. 251 © Arthur Tilley
p. 321 David Stoecklein

WEST LIGHT
p. 151 © Jim Cornfield
p. 253 Walter Hodges

p. 117, Page Poore
p. 351 Beechcraft/PH Photo Archives

Printed in the United States of America
10 9 8 7 6 5 4 3 2 1

ISBN   0-13-063363-1

Prentice-Hall International (UK) Limited, *London*
Prentice-Hall of Australia Pty. Limited, *Sydney*
Prentice-Hall Canada Inc., *Toronto*
Prentice-Hall Hispanoamericana, S.A., *Mexico*
Prentice-Hall of India Private Limited, *New Delhi*
Prentice-Hall of Japan, Inc., *Tokyo*
Simon & Schuster Asia Pte. Ltd., *Singapore*
Editora Prentice-Hall do Brasil, Ltda., *Rio de Janeiro*

This book is dedicated to
Nancy Tobey
Outstanding teacher, friend, mother and wife

Contents

10

Introduction to Algebra *513*

Preface

To the Student

Mathematics becomes more important each year as our society—and the global society of which we are a part—grows and changes with new technologies. Perhaps you do not believe that you need mathematics to handle your everyday life and advance in your job. Farmers, elementary school teachers, bus drivers, laboratory technicians, nurses, telephone operators, photographers, pharmacists, salespeople, doctors, architects and so many others who perhaps believed that they needed little if any mathematics are finding that basic mathematical skills can help them. Mathematics, you will find, can help you too.

People who enter college have a variety of mathematical backgrounds. Many of you may be looking forward to this course. We hope that our book keeps your enthusiasm high. Others of you have never enjoyed math and may never have done too well in it. You may be quite anxious about taking this course. We hope that our book delivers the assistance and support that helps you develop an interest in this most important field.

To show you how mathematics may affect your life and how you can put it to work for yourself, this book draws a great number of examples and problems from everyday situations—maintaining a budget, using a checking account, figuring mileage on a trip, and so on. Also, we explore the mathematical side of situations taken from many different jobs and professions, such as bank loan officer, air traffic controller, environmentalist, physician, police officer, cook, salesperson, and many more. Learning mathematics is more enjoyable if you can see how people put it to use.

Special Features

+ +"To Think About", in-text as well as incorporated into exercises, challenge you. To think in a logical and organized way and expose you to the power of mathematical ideas.

+ +"Developing Your Study Skills", scattered throughout the first few chapters, provide practical advice to you to ensure personal success in the course.

+ +Pretests begin each chapter, diagnosing in advance your strengths and weaknesses in the upcoming material.

+ +Examples and matched Practice Problems give you a chance to solve an exercise similar to the preceding example on your own. Answers and solutions appear a few pages later.

+ +Four-color photos show the mathematics at work in real-life applications and occupations, answering the question "When will I ever need to use this course?"

+ +Chapter Organizers summarize key concepts, provide additional examples for brief review, and give a page reference for further study, in a compact grid format.

++Calculator Problems encourage you to take advantage of this modern technology.

++Practice Quizzes reinforce concepts half-way through each chapter.

++Practice Chapter Tests provide tests that can be used as preparation or as actual tests in a math lab.

++Cumulative Review Problems review the content of previous chapters within exercises sets.

++Cumulative Tests, at the end of each chapter (except the first), cover content from the preceding chapters.

++Extra Practice at the end of every chapter provides the additional drill for those of you who want more than the exercise sets provide.

++Glossary at the end of the book presents both a definition and an example of the term.

This book is the first of a series of three developmental math books. In a tightly co-ordinated sequence of books, *Basic College Mathematics* is followed by *Beginning Algebra* and *Intermediate Algebra*. If you enjoyed using this book, ask your instructor whether he or she will be using another book in the series for your next course.

We would like to thank our editors at Prentice Hall: Priscilla McGeehon, Steve Deitmer, Roberta Lewis, Virginia Huebner and Barbara De Vries; and their designer, Judith Matz-Coniglio, for their advice, support, time and persistence. Without their insight, suggestions, and creative thinking we surely would have had a difficult time finishing this book.

We have been greatly helped by a supportive group of colleagues who not only teach at North Shore Community College but who have provided a number of ideas as well as extensive help on all of our mathematics books. So a special word of thanks to Marie Clark, Hank Harmeling, Tom Rourke, Wally Hersey, Bob McDonald, Judy Carter, Kathy LeBlanc, Bob Oman, Keith Piggott, Russ Sullivan, and Rick Ponticelli who gave help in a variety of ways as this book was being completed. Joan Peabody patiently typed and retyped the entire manuscript. Her cheerful dedication to doing a long job and doing it well have been an invaluable contribution in the task of preparing this book.

We extend our thanks and appreciation to the following reviewers:

Michael A. Contino, *California State University, Hayward*
Deann Christianson, *University of the Pacific*
Jamie King, *Orange Coast College*
Andrea Potts, *Ventura College*
Carol Atnip, *University of Louisville*
H. Allan Edwards, *Parkersburg Community College*
James W. Newsom, *Tidewater Community College*
Gerry C. Vidrine, *Louisiana State University*
Cherie Corr, *Hudson Valley Community College*
Nomiki Shaw, *California State University, Bakersfield*
Gabrielle Andries, *University of Wisconsin, Milwaukee*
Dorothy Pennington, *Tallahassee Community College*
Linda Padilla, *Joliet Junior College*
Marilyn Carlson, *University of Kansas*
William J. Thieman, *Ventura College*
John G. Rose, *Miami Dade Community College*

We also want to thank the following supplements authors:

Lea Pruet Campbell, *Lamar University-Port Arthur*
Cathy Pace, *Louisiana Technical University*
John Garlow, *Tarrant County Junior College*
Bob Martin, *Tarrant County Junior College*

Book writing is impossible for us without the support of our families. Our deepest thanks to Nancy, Johnny, Marcia, Melissa, Shelly, Rusty, and Abby. You have patiently waited, quietly assisted, and given words of encouragement when greatly needed. Finally, we thank God for the strength and energy to write and the opportunity to help others through this textbook.

John Tobey
Jeffrey Slater

Beverly, Massachusetts

Basic
College
Mathematics

Whole Numbers

A realtor makes many mathematical calculations regarding cost, monthly mortgage payments, and interest rates before making a sale.

1

PRETEST CHAPTER 1

If you are familiar with the topics in this chapter, take this test now. Check your answers with those in the back of the book. If an answer was wrong or you couldn't do a problem, study the appropriate section of the chapter.

If you are not familiar with the topics in this chapter, don't take this test now. Instead, study the examples, work the practice problems and then take the test.

This test will help you identify those concepts that you have mastered and those that need more study.

Section 1.1

1. Write in words: 65,287,023.

2. Write in expanded notation: 59,361.

3. Write in standard notation: one million, fifty-eight thousand, four hundred ninety-eight.

Section 1.2 Add.

4.
```
   16
   28
   95
   34
 + 71
```

5.
```
   56,318
    7,193
 + 28,415
```

6.
```
   9873
    400
     28
 + 9134
```

Section 1.3 Subtract.

7.
```
   9865
 − 2734
```

8.
```
   100,980
 −  37,439
```

9.
```
   28,934,125
 − 26,989,372
```

Section 1.4 Multiply.

10. $5 \times 6 \times 2 \times 3 =$

11.
```
    29
 × 48
```

12.
```
   908
 × 375
```

13.
```
   12,658
 ×      4
```

Section 1.5 Divide. If there is a remainder, be sure to state it as part of the answer.

14. $7\overline{)142,107}$

15. $6\overline{)49,520}$

16. $46\overline{)5842}$

Section 1.6

17. Write in exponent form: $7 \times 7 \times 7 \times 7 \times 7$.

18. Evaluate: 4^3.

19. Round to the nearest thousand: 156,503.

20. Round to the nearest hundred: 48,349.

21. Round to the nearest million: 49,830,000.

Section 1.7 Perform the operations in their proper order.

22. $2 \times 7^2 − 3 \times (4 + 2)$

23. $2^4 + 3^2 + 36 \div 2$

24. $5 \times 8 − 3 \times 6 − 2^3 + 20 \div 20$

Section 1.8

25. Emily planned a trip of 1483 miles from her home to Chicago. She traveled 317 miles on her first day of the trip. How many more miles must she travel?

26. Betsey purchased 84 shares of stock at a cost of $1848.00. How much was the cost per share?

27. Dr. Alfonso bought three chairs at $46 each, three lamps at $29 each, and two tables at $37 each for the waiting room of his office. How much did his purchases cost?

1.1 UNDERSTANDING WHOLE NUMBERS

After studying this section, you will be able to:

1 Write numbers in expanded form

2 Write a word name for a number

3 Convert a word name to a number in standard form

Often we learn a new concept in stages. First comes learning the new *terms* and basic assumptions. Then we have to master the *reasoning*, or logic, behind the new concept. This often goes hand in hand with learning a *method* for using the idea. Finally, we can move quickly with a *shortcut*.

For example, in the study of stock investments, you must learn the meaning of such terms as *stock*, *profit*, *loss*, and *commission* before tackling the question "What is my profit from this stock transaction?" After you learn how to answer this question, you can quickly answer many similar ones.

In this book, watch your understanding of mathematics grow through this same process. In the first chapter we consider the whole numbers. The ideas associated with the whole numbers can be so familiar that you have already jumped to the "shortcut" stage. But with a little patience in looking at the terms, reasoning, and step-by-step methods, you'll find your understanding deepens, even with these very familiar numbers, the whole numbers.

To count a number of objects or to answer the question "How many?" we use a set of numbers called **whole numbers**. These whole numbers are as follows:

0, 1, 2, 3, 4, 5, 6, 7, 8, 9, 10, 11, 12, 13, 14, 15, ...

There is no largest whole number. The three dots ... indicate that the set of whole numbers goes on indefinitely. Our number system is based on tens and ones and is called the decimal system (or the base 10 system). The numbers 0, 1, 2, 3, 4, 5, 6, 7, 8, 9 are called **digits**. The position, or placement, of the digits in the number tells the value of the digits. For example, in the number 521, the "5" means 5 hundreds (500). In the number 54, the "5" means 5 tens, or fifty. For this reason, our number system is called a **place-value** system.

Consider the number 5643. The four digits are located in four places that are called, from right to left, the ones, tens, hundreds, and thousands place. By looking at the digits and their places, we see that in the number 5643

- the 5 means "5 thousands."
- the 6 means "6 hundreds."
- the 4 means "4 tens," or forty.
- the 3 means "3 ones."

The *value* of the number is 5 thousand, 6 hundred, 4 tens, 3 ones.

To indicate the value of even greater numbers we can use the following diagram, which shows the names of even more places.

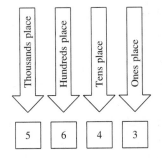

Hundred millions	Ten millions	Millions	Hundred thousands	Ten thousands	Thousands	Hundreds	Tens	Ones

After every group of three digits, moving from right to left, we place a comma to make it easier to read the number. It is usually agreed that a four-digit number does not have a comma, but that numbers with five or more digits do. So 32,000 would be written with a comma but 7000 would not.

A number has the same *value* no matter how we write it. For example, "a million dollars" means the same as "$1,000,000". In fact, any number in our number system can be written in several ways or forms:

- Standard notation   521
- Expanded notation   500 + 20 + 1
- In words   five hundred twenty-one

You may want to write a number in any of these ways. To write a check, you need to use both standard notation and words. To round numbers, which we discuss later, you use expanded notation. Now we show how to rewrite a number in any of these forms.

1 We sometimes write numbers in **expanded notation** to emphasize the place values. In the number 56,327, which is standard notation, we have the digits in the following places:

5 in the ten thousands place
6 in the thousands place
3 in the hundreds place
2 in the tens place
7 in the ones place

So 56,327 in expanded notation is

$$50,000 + 6000 + 300 + 20 + 7$$

(Do you see that in words we would say 56 thousands + 3 hundreds + 2 tens + 7 to describe this?)

Example 1 Write expanded notation for 37,463.

$30,000 + 7000 + 400 + 60 + 3$ ■

Practice Problem 1 Write expanded notation for 89,215. ■

Example 2 Write expanded notation for 1,508,271.

Note that the 0 indicates that there are 0 ten thousands.

$1,000,000 + 500,000 + 8000 + 200 + 70 + 1$ ■

Practice Problem 2 Write expanded notation for 3,520,894. ■

2 We frequently read **word names** for whole numbers. The word name for 56 is fifty-six. The word name for 29 is twenty-nine. The word name for 478 is four hundred seventy-eight.

Example 3 Write the word name for 1695.

One thousand, six hundred ninety-five ■

Practice Problem 3 Write the word name for 2736. ■

Example 4 Write the word name for 200,470.

Two hundred thousand, four hundred seventy ■

Practice Problem 4 Write the word name for 980,306. ■

3 We reverse the process to take a word name and write the corresponding number in standard notation.

Example 5 Write in standard notation: two billion, three hundred eighty-six million, five hundred forty-seven thousand, one hundred ninety.

2,386,547,190 ■

Practice Problem 5   Write in standard notation: eight billion, five hundred seven million, four hundred thirty-six thousand, twenty-four.   ■

Example 6   Last year, the population of Central City was 1,509,637.

(a) In the number 1,509,637, how many ten thousands are there?
(b) How many hundred thousands are there?

(a) There are 0 ten thousands.   **(b)** There are 5 hundred thousands.

■

Practice Problem 6   The campus library has 904,759 books.
(a) What digit tells the number of hundreds?
(b) What digit tells the number of hundred thousands?   ■

To Think About

Not all number systems have had a zero. For example, the system of Roman numerals did not. In a place-value system such as ours, why is a zero so necessary? What would happen if our number system did not have a symbol for zero? If there were no zero, we would have great difficulty writing a number such as 406. This number means four hundreds and six ones. By putting a 0 in the tens place we indicate that the number has four hundreds, six ones, but zero tens. Without a zero symbol there would be no easy way to write 406. Perhaps we would have to write out "four hundreds and six ones"! Zero is a **place holder** in our number system. It holds a position and shows that there is no other digit in that place.   ■

EXERCISES 1.1

Write each number in expanded notation.

1. 6731

2. 5836

3. 108,276

4. 205,793

5. 23,761,345

6. 46,198,253

7. 103,260,768

8. 820,310,574

Write each number in standard notation.

9. 600 + 70 + 1

10. 300 + 80 + 4

11. 9000 + 800 + 60 + 3

12. 7000 + 600 + 50 + 2

13. 50,000 + 4000 + 20 + 7

14. 80,000 + 2000 + 300 + 5

15. 700,000 + 6000 + 200

16. 900,000 + 50,000 + 40 + 7

Write a word name for each number.

17. 53

18. 46

19. 465

20. 278

21. 8936   **22.** 4629   **23.** 105,261   **24.** 370,258

25. 23,561,248   **26.** 19,376,584   **27.** 4,302,156,200   **28.** 7,436,210,400

Write each number in standard notation.

29. Three hundred seventy-five

30. Four hundred twenty-six

31. Fifty-six thousand, two hundred eighty-one

32. Seventy-nine thousand, three hundred forty-six

33. One hundred million, seventy-nine thousand, eight hundred twenty-six

34. Four hundred fifty million, three hundred thousand, two hundred forty-nine

In each sentence write the population number in standard notation.

35. In 1984 the population of New York City was reported to be seven million, one hundred sixty-four thousand, seven hundred forty-two.

36. It has been estimated that in 1991 the population of New York City will be eight million, seventy-one thousand, six hundred thirty-nine.

37. In 1980 the population of Bangalore, India, was estimated to be two million, nine hundred thirteen thousand.

38. In 1981 the population of Bombay, India, was estimated to be eight million, two hundred two thousand.

Specify the digit.

39. The speed of light is approximately 29,979,250,000 centimeters per second.
 (a) What digit tells the number of ten thousands?
 (b) What digit tells the number of ten billions?

40. The value of the radius of the earth is approximately 637,814,000 centimeters.
 (a) What digit tells the number of ten thousands?
 (b) What digit tells the number of ten millions?

41. If you invest $14,000 at a compound interest rate of 12% per year for 57 years, you will have an approximate total of $8,944,485.
 (a) What digit tells the number of hundred thousands?
 (b) What digit tells the number of millions?

42. If you invest $15,000 at a compound interest rate of 13% per year for 58 years, you will have an approximate total of $17,973,845.
 (a) What digit tells the number of ten thousands?
 (b) What digit tells the number of hundred thousands?

? **To Think About**

43. Write in standard notation: five hundred ninety-six trillion, seven hundred thirty-four million, one hundred twenty-nine thousand, twenty-nine.

44. Write a word name for the following number:

2,783,421,632,985,301

(*Hint:* The digit 1 followed by 15 zeros represents the number 1 quadrillion.)

When writing a check, a person must write the word name for the dollar amount of the check.

```
  ┌──────────────────────────────────────────────────────────────┐
  │  Alex J. Writer                                         6848   │
  │  10 Main Street                                                │
  │  Westwood, Texas          April 1    19 91                     │
  │                                                   53-235/113    │
  │  PAY TO THE                                                    │
  │  ORDER OF_____  $ | 1,965 00 |      │
  │                                                                │
  │  _____ DOLLARS     │
  │                                                                │
  │  Last Chance Bank                                              │
  │  Westwood, Texas                                               │
  │  MEMO _____                                          │
  │  ⑆011302357⑈ 011 08115⑈ 6848                                  │
  └──────────────────────────────────────────────────────────────┘
```

45. Alex bought new equipment for his laboratory for $1965. What word name should he write on the check?

46. Alex later bought a new personal computer for $6383. What word name should he write on the check?

For Extra Practice Examples and Exercises, turn to page 58.

Solutions to Odd-Numbered Practice Problems

1. $89{,}215 = 80{,}000 + 9000 + 200 + 10 + 5$ **3.** The word name for 2736 is two thousand, seven hundred thirty-six.
5. 8,507,436,024

Answers to Even-Numbered Practice Problems

2. $3{,}000{,}000 + 500{,}000 + 20{,}000 + 800 + 90 + 4$ **4.** Nine hundred eighty thousand, three hundred six   **6. (a)** 7 **(b)** 9

1.2 ADDITION OF WHOLE NUMBERS

What is addition? We perform addition when we group items together. Consider the manager of the Gourmet Restaurant, who needs five cakes for Saturday and seven cakes for Sunday. When she places her order on Friday for the weekend, she must group together the items needed for each day and come up with a single amount. She adds five cakes and seven cakes and orders twelve cakes:

$$5 + 7 = 12$$

Almost as soon as we learn to count, we join sets of objects together and total the items in the combined grouping. We see the addition process time and again. The carpenter adds to find a total amount of lumber. The auto mechanic adds to get the total parts inventory. The bank teller adds to get a cash total. Of course, we soon learn to add not only by counting the total but by using addition facts, the place-value system, and shortcuts. We learn that we can perform addition not only with concrete objects but also in abstract situations. Geologists add not only five rocks and ten rocks, but also two million years and eight million years. Astronomers add not only two planets and six planets, but also one hundred light-years and two hundred light-years. If you can add well, you can apply addition in any situation.

But to start, we study the basics.

1 Suppose that we have four pencils in the car and we bring three more pencils from home. How many pencils are available to us? Here we added 4 and 3 to obtain a value of 7.

Addition is an operation in which we count or keep track of how many items are in two groups. When numbers are added, the numbers being added are called **addends**. The result is called the **sum**.

□ After studying this section, you will be able to:

1 Add two single-digit numbers

2 Add several single-digit numbers

3 Add numbers of several digits when carrying is not needed

4 Add numbers of several digits when carrying is needed

Let us look at this addition problem again.

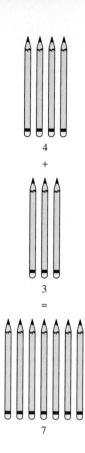

$$\begin{array}{rl} 3 & \text{addend} \\ +\,4 & \text{addend} \\ \hline 7 & \text{sum} \end{array}$$

In this case the numbers 3 and 4 are called addends and the answer 7 is called the sum. Usually when we add numbers, we write them in a column format.

Example 1   Add.

(a) $8 + 5$ **(b)** $3 + 7$ **(c)** $9 + 0$

(a) $\begin{array}{r} 8 \\ +\,5 \\ \hline 13 \end{array}$ **(b)** $\begin{array}{r} 3 \\ +\,7 \\ \hline 10 \end{array}$ **(c)** $\begin{array}{r} 9 \\ +\,0 \\ \hline 9 \end{array}$

Note: When we add zero to any other number, the other number is the sum.

Practice Problem 1   Add.

(a) $\begin{array}{r} 7 \\ +\,5 \\ \hline \end{array}$ **(b)** $\begin{array}{r} 9 \\ +\,4 \\ \hline \end{array}$ **(c)** $\begin{array}{r} 3 \\ +\,0 \\ \hline \end{array}$

The digits 0 to 9 can be recorded in an addition table. If some of these addition facts do not come to mind quickly, a little review is in order. Look over the table, and make sure that you know all the answers. If you are not sure how strong your knowledge is, try doing Exercise Set 1.2, problems 1 and 2.

BASIC ADDITION FACTS

+	0	1	2	3	4	5	6	7	8	9
0	0	1	2	3	4	5	6	7	8	9
1	1	2	3	4	5	6	7	8	9	10
2	2	3	4	5	6	7	8	9	10	11
3	3	4	5	6	7	8	9	10	11	12
4	4	5	6	7	8	9	10	11	12	13
5	5	6	7	8	9	10	11	12	13	14
6	6	7	8	9	10	11	12	13	14	15
7	7	8	9	10	11	12	13	14	15	16
8	8	9	10	11	12	13	14	15	16	17
9	9	10	11	12	13	14	15	16	17	18

Read across the top to the 4 column, and then read down the left to the 7 row. The box where the 4 and 7 meet is 11, which means that $4 + 7 = 11$. Now try reading across the top to the 7 column and down the left to the 4 row. The box where these numbers meet is also 11. We see that the order of numbers in addition does not affect the sum: $4 + 7 = 11$, and $7 + 4 = 11$. We call this fact the *commutative property of addition*.

2 If more than two numbers are to be added, we usually add from the first number to the next number and mentally record the sum. Then we add that sum to the next number, and so on.

Example 2   Add: $3 + 4 + 8 + 2 + 5$.

$$
\begin{array}{r}
3 \\
4 \\
8 \\
2 \\
+\,5 \\
\hline
22
\end{array}
$$

Mentally, we do these steps.

$3 + 4 = 7$
$7 + 8 = 15$
$15 + 2 = 17$
$17 + 5 = 22$

Practice Problem 2   Add: $7 + 6 + 5 + 8 + 2$ ■

Because the order in which we add numbers doesn't matter, we can choose to add from the top down, from the bottom up, or in any other way. One shortcut is to add first any numbers that will give a sum of 10, or 20, or 30, and so on.

Example 3   Add:
$$
\begin{array}{r}
3 \\
4 \\
8 \\
2 \\
+\,6
\end{array}
$$

We mentally group the numbers into tens:

$$
\begin{array}{r}
3 \\
4 \\
8 \\
2 \\
6
\end{array}
$$

$8 + 2 = 10$ $4 + 6 = 10$

The sum is two tens + 3 or 23   ■

Practice Problem 3   Add: $1 + 7 + 2 + 9 + 3$.   ■

3 Of course, many numbers that require addition are more than single-digit numbers. In such cases we must be careful to add separately first the digits in the ones column, then the digits in the tens column, then those in the hundreds column, and so on, moving from right to left.

Example 4   Add: $4304 + 5163$.

$$
\begin{array}{r}
4\ 3\ 0\ 4 \\
+\,5\ 1\ 6\ 3 \\
\hline
9\ 4\ 6\ 7
\end{array}
$$

— sum of 4 ones + 3 ones = 7 ones
— sum of 0 tens + 6 tens = 6 tens
— sum of 3 hundreds + 1 hundred = 4 hundreds
— sum of 4 thousands + 5 thousands = 9 thousands   ■

Practice Problem 4   Add:
$$
\begin{array}{r}
8246 \\
+\,1702
\end{array}
$$ ■

4 When you add several whole numbers, often the sum in a particular column is greater than 9. But in the place-value system we must only use *one* digit in any one place. How are we to write the sum? The following example shows how to do the addition.

Example 5   Add: $45 + 37$.

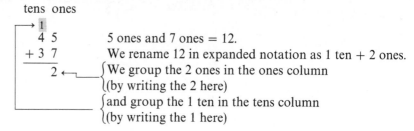

tens ones

$$\begin{array}{r} \overset{1}{4}\;5 \\ +\;3\;7 \\ \hline 2 \end{array}$$

5 ones and 7 ones $= 12$.
We rename 12 in expanded notation as 1 ten $+$ 2 ones.
{ We group the 2 ones in the ones column
(by writing the 2 here)
{ and group the 1 ten in the tens column
(by writing the 1 here)

The placing of the 1 in the tens column is often called "carrying the one."

Now we can add the digits in the tens column,

$$\begin{array}{r} \overset{1}{4}\;5 \\ +\;3\;7 \\ \hline 8\;2 \end{array}$$ ■

Practice Problem 5   Add: $\begin{array}{r} 56 \\ +\;36 \\ \hline \end{array}$ ■

Often you must use carrying several times by bringing the left digit into the next column to the left.

Example 6   Add: $257 + 688 + 94$

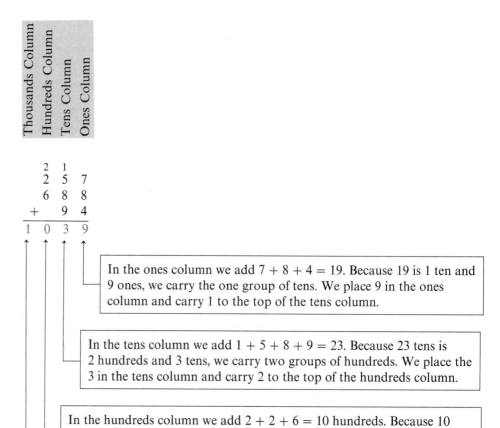

Thousands Column   Hundreds Column   Tens Column   Ones Column

$$\begin{array}{r} \overset{2}{}\;\overset{1}{} \\ 2\;5\;7 \\ 6\;8\;8 \\ +\;\;\;9\;4 \\ \hline 1\;0\;3\;9 \end{array}$$

In the ones column we add $7 + 8 + 4 = 19$. Because 19 is 1 ten and 9 ones, we carry the one group of tens. We place 9 in the ones column and carry 1 to the top of the tens column.

In the tens column we add $1 + 5 + 8 + 9 = 23$. Because 23 tens is 2 hundreds and 3 tens, we carry two groups of hundreds. We place the 3 in the tens column and carry 2 to the top of the hundreds column.

In the hundreds column we add $2 + 2 + 6 = 10$ hundreds. Because 10 hundreds is 1 thousand and zero hundreds, we place the 0 in the hundreds column and place the 1 in the thousands column.

■

Practice Problem 6   Add: $789 + 63 + 297$ ■

Addition may be performed in more than one order. To add $5 + 3 + 7$ we can first add the 5 and 3. We do this by using parentheses to indicate the first operation to be performed.

$$5 + 3 + 7 = (5 + 3) + 7 = 8 + 7 = 15$$

We could also add first the 3 and 7.

$$5 + 3 + 7 = 5 + (3 + 7) = 5 + 10 = 15$$

The sum is the same. This illustrates the *associative law of addition.*

For convenience, we now list three properties of addition we have discussed in this section.

1. **Associative Law of Addition**	$(8 + 2) + 6 = 8 + (2 + 6)$
When we add three numbers, the addition may be grouped in any order.	$10 + 6 = 8 + 8$ $16 = 16$
2. **Commutative Law of Addition**	$5 + 12 = 12 + 5$
Two numbers can be added in either order with the same result.	$17 = 17$
3. **Addition Law of Zero**	$8 + 0 = 8$
When zero is added to a number the sum is that number.	$3 + 0 = 3$

Addition can be checked by adding the numbers in the opposite order. The commutative and associative laws of addition allow us to do this.

Example 7 (a) Add the numbers: $39 + 7284 + 3132$
(b) Check by reversing the order of addition.

$$
\begin{array}{cc}
\textbf{(a)} &
\begin{array}{r}
^{11}\ \ \\
39 \\
7284 \\
+\ 3132 \\
\hline
10455
\end{array}
\end{array}
\qquad
\begin{array}{cc}
\textbf{(b)} &
\begin{array}{r}
^{11}\ \ \\
3132 \\
7284 \\
+\ \ \ 39 \\
\hline
10455
\end{array}
\end{array}
$$

Addition   Check by reversing order.

| The sum is the same in each case. | ■ |

Practice Problem 7

(a) Add: 127
               9876
             + 342

(b) Check by adding: 342
                          9876
                        + 127   ■

To Think About

Is addition always commutative? Can you think of any examples from real life where the order of addition makes the sum different? Consider "adding" footwear to yourself. When you put on your socks and then your shoes, the result isn't the same as if you put on your shoes first and then your socks.   ■

Addition is often used in applied situations. In most real-life situations *you* have to decide if you have enough information to solve the problem. Sometimes you will have *too much* information. Then it is necessary to separate out certain facts that are not part of the problem.

Example 8 The bookkeeper for Smithville Trucking was examining the following data for the company checking account.

Monday: $23,416 was deposited and $17,389 was debited (taken away) from the account.

Tuesday: $44,823 was deposited and $34,089 was debited.

Wednesday: $16,213 was deposited and $20,057 was debited.

What was the total of all *deposits* during this period?

We use *only* the deposit amounts.

Monday: $23,416 was deposited.

Tuesday: $44,823 was deposited.

Wednesday: $16,213 was deposited.

$$\begin{array}{r} \overset{11\ \ 1}{23{,}416} \\ 44{,}823 \\ +\ 16{,}213 \\ \hline \$84{,}452 \end{array}$$

A total of $84,452 was deposited on those three days. ∎

Practice Problem 8 North University has 23,413 men and 18,316 women. South University has 19,316 men and 24,789 women. East University has 20,078 men and 22,965 women. What is the total enrollment of women at the three schools? ∎

Example 9 Mr. Ortiz has a field with the length of each side as labeled on the sketch. What is the total number of feet of fence that would be required to fence in the field?

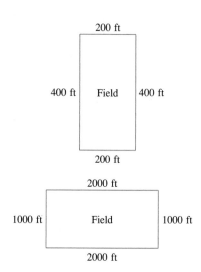

We add the lengths all around the field: 200 feet and 200 feet and 400 feet and 400 feet, for a total of 1200 feet. The amount of fence required is 1200 feet. ∎

The "distance around" an object (such as a field) is called the *perimeter*. Add the lengths all around the object to find the perimeter.

Practice Problem 9 In Vermont, Gretchen fenced the field on which her sheep graze. What is the perimeter of the field? ∎

DEVELOPING YOUR STUDY SKILLS

KEEP TRYING

You may be one of those students who have had much difficulty with mathematics in the past and who are sure that you cannot do well in this course. Perhaps you are thinking, "I have never been any good at mathematics," or "I have always hated mathematics," or "Math always scares me," or "I have not had any math for so long that I have forgotten it all." You may even have picked up on the label "math anxiety" and attached it to yourself. That is most unfortunate, and it is time for you to reprogram your thinking. Replace those negative thoughts with more positive ones. You need to say things like, "I will give this math class my best shot," or "I can learn mathematics if I work at it," or "I will try to do better than I have done in previous math classes." You will be pleasantly surprised at the difference this more positive attitude makes!

We live in a highly technical world, and you cannot afford to give up on the study of mathematics. Dropping mathematics may prevent you from entering certain career fields that you may find interesting. You may not have to take math courses as high-level as calculus, but such courses as intermediate algebra, finite math, college algebra, and trigonometry may be necessary. Learning mathematics can open new doors for you.

Learning mathematics is a process that takes time and effort. You will find that regular study and daily practice are necessary to strengthen your skills and to help you grow academically. This process will lead you toward success in mathematics. Then, as you become more successful, your confidence in your ability to do mathematics will grow.

Complete the addition facts for each table. Strive for total accuracy, but work quickly. Allow a maximum of 3 minutes for each table.

1.

+	3	5	4	8	0	6	7	2	9	1
2										
7										
5										
3										
0										
4										
1										
8										
6										
9										

2.

+	1	6	5	3	0	9	4	7	2	8
3										
9										
4										
0										
2										
7										
8										
1										
6										
5										

Add.

3.
```
  4
  2
  8
+ 9
```

4.
```
  3
  7
  1
+ 6
```

5.
```
  2
  9
  8
+ 7
```

6.
```
  1
  4
  3
+ 8
```

7.
```
   12
   45
+ 30
```

8.
```
   46
   20
+ 33
```

9.
```
   63
   24
+ 12
```

10.
```
   54
   21
+ 23
```

11.
```
   2847
   1634
+ 2098
```

12.
```
   4816
   3015
+ 4798
```

13.
```
   6908
   2173
+ 4255
```

14.
```
   5017
   2984
+ 1328
```

15.
```
   9236
+ 4715
```

16.
```
   3718
+ 2753
```

17.
```
   18,718
+ 24,021
```

18.
```
   99,023
+ 47,325
```

Add from the top. Then check by adding in the reverse order.

19.
```
   47
   31
   26
    8
+ 14
```

20.
```
   24
   39
   16
   14
+  9
```

21.
```
   106
    13
     4
    28
+ 981
```

22.
```
   463
    27
     8
    41
+ 507
```

Add.

23.
```
    126
   8142
     37
 + 9604
```

24.
```
    223
   7021
     89
 + 7634
```

25.
```
  1,362,214
  7,002,316
+ 3,214,896
```

26.
```
  4,002,983
  2,134,702
+ 3,592,001
```

27.
```
  837,241,000
+ 298,039,240
```

28.
```
  982,306,000
+ 583,215,320
```

29.
```
    516,208
     24,317
+ 1,763,295
```

30.
```
     32,500
    763,420
+ 2,837,667
```

31. $12 + 8 + 156 + 72$

32. $85 + 3 + 407 + 26$

33. $15,216 + 485 + 5208$

34. $26,002 + 599 + 3500$

Answer each question.

35. On Monday, Adelaide drove 362 miles, on Tuesday 259 miles and on Wednesday 197 miles. What was the total number of miles she drove?

36. On Wednesday, Tyrone drove 405 miles, on Thursday 273 miles, and on Friday 164 miles. What was the total number of miles he drove?

37. Barbara's taxes on her house two years ago were $2576. Last year they were $2771. This year they are $3428. What is the total amount for three years?

38. Derek's taxes on his house two years ago were $1920. Last year they were $2439. This year they are $2788. What is the total amount for three years?

39. Anthony has a field with the length of each side as labeled on the sketch. What is the total number of feet of fence that would be required to fence in the field? (Find the perimeter of the field.)

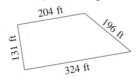

40. Jessica has a field with the length of each side as labeled on the sketch. What is the total number of feet of fence that would be required to fence in the field? (Find the perimeter of the field.)

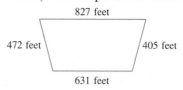

41. In 1984, 2,086,499 people in Illinois, 841,481 people in Indiana, and 1,825,440 people in Ohio voted for Mondale for president. What was the total vote for Mondale in those three states?

42. In 1984, 1,310,936 people in Massachusetts, 890,877 people in Connecticut, and 3,664,763 people in New York voted for Reagan for president. What was the total vote for Reagan in those three states?

43. In 1983, the farm income in Nebraska was $4,217,716,000. In 1984 it was $4,524,309,000. In 1985 it was $4,112,810,000. What was the total farm income in Nebraska for these three years?

44. In 1983, the farm income in Iowa was $5,140,272,000. In 1984 it was $5,014,616,000. In 1985 it was $4,811,147,000. What was the total farm income in Iowa for these three years?

45. The quality control department of a television production center checks television sets produced daily and classifies them as *good* or *defective*. On Monday 473 were good and 2 defective. On Tuesday 495 were good and 8 defective. On Wednesday 506 were good and 14 defective. On Thursday 463 were good and 21 were defective.
   (a) How many good television sets were produced on the four days?
   (b) How many television sets were produced on the four days in all?

46. The quality control division of a car manufacturer classifies the final assembled vehicle as *passing* or *failing* final inspection. In March 36,521 vehicles were classified as passing, whereas 468 failed. In April 20,397 vehicles passed and 989 failed. In May 25,321 vehicles passed and 592 failed.
   (a) How many vehicles passed the inspection during the three months?
   (b) How many vehicles were assembled during the three months in all?

? **To Think About**

47. What would happen to the result of addition problems if addition *were commutative* but *not associative*?

48. What would happen to the result of addition problems if addition *were associative* but *not commutative*?

For problems 49 and 50, consult the following Western University expense chart.

Western University Yearly Expenses	In-State Student	Out-of-State Student, U.S. Citizen	Student from Outside United States
Tuition	$3640	$5276	$8352
Room	1926	2437	2855
Board	1753	1840	1840

49. How much is the total cost for tuition, room, and board for **(a)** an out-of-state U.S. citizen? **(b)** for an in-state student?

50. How much is the total cost for tuition, room, and board for a student from outside the United States?

51. Add: 2,368,521,788 + 5,721,368,701 + 4,027,399,206.

52. Add: 89 + 166 + 23 + 45 + 72 + 190 + 203 + 77 + 18 + 93 + 46 + 73 + 66.

Cumulative Review Problems

Write the word name for each number.

53. 76,208,941

54. 121,000,374

Write each number in standard notation.

55. Eight million, seven hundred twenty-four thousand, three hundred ninety-six

56. Nine million, fifty-one thousand, seven hundred nineteen

Calculator Problems

Add.

57. 99,765,123
    44,669,011
  + 99,288,309
  ‾‾‾‾‾‾‾‾‾‾‾‾

58.   12,309
      9,277
      6,781
    +   999
   ‾‾‾‾‾‾‾‾‾

59. The Conway Construction Company made the following deposits in their checking account:

Monday	$23,277
Tuesday	45,799
Wednesday	71,300
Thursday	32,291
Friday	29,762

What was the total amount of deposits made that week?

For Extra Practice Examples and Exercises, turn to page 58.

Solutions to Odd-Numbered Practice Problems

1. (a)   7   **(b)**   9   **(c)**   3
     + 5   + 4   + 0
    ‾‾‾‾   ‾‾‾‾   ‾‾‾‾
     12   13   3

3.   1
      + 7
      + 2 → 10
      + 9   → 10
      + 3
     ‾‾‾‾
      22

5.   56
       + 36
      ‾‾‾‾
        92

7. (a) $\overset{11}{}$
        127
       9876
      + 342
     ‾‾‾‾‾‾‾‾
      10,345 ← **(b)** Check by adding in opposite order.

        $\overset{111}{}$
         342
        9876
       + 127
      ‾‾‾‾‾‾‾‾
       10,345
    ←—— same ——→

9.   1000
      2000
      1000
    + 2000
   ‾‾‾‾‾‾‾
     6000

Answers to Even-Numbered Practice Problems

2. 28   **4.** 9948   **6.** 1149   **8.** 66,070

☐ After studying this section, you will be able to:

1 *Rapidly use basic subtraction facts*

2 *Subtract whole numbers of several digits*

3 *Subtract whole numbers of several digits when borrowing is necessary*

4 *Check the answer of a subtraction problem*

1.3 SUBTRACTION OF WHOLE NUMBERS

What is subtraction, and how is it related to addition of whole numbers? A story about a restaurant will suggest an answer. Suppose that the Gourmet Restaurant always cuts its Carrot Cake Supreme, hot from the oven, into 12 pieces. At the end of the day, Monsieur Gourmet, the owner, can tell how many pieces were served by counting those that are left. This is the basic idea of subtraction. From a whole, a part is "taken away" or subtracted. When we subtract, we find how many are left.

$12 - 3$ is 9. The answer is called the "difference." Or if 9 remains, how many were "taken away"? $12 - 9 = 3$. Subtraction can be done using a related addition problem:

$$12 - 3 = \boxed{\text{What number}} \overset{\text{implies}}{\Longrightarrow} 3 + \boxed{\text{What number}} = 12$$

1 Subtraction of one whole number from another can be thought of as a "taking away" operation. Suppose that Fred earns $400 per month and has $100 of his pay withheld for taxes. How much does he have left?

$$\underset{\text{salary}}{\$400} \quad \underset{\substack{\text{subtraction} \\ \text{symbol}}}{-} \quad \underset{\substack{\text{amount} \\ \text{withheld}}}{\$100} \quad = \quad \underset{\substack{\text{amount} \\ \text{left}}}{\$300}$$

The answer resulting from subtraction is called the **difference**.

```
  9      8      12      17
- 2    - 3    - 6     - 9
  7      5      6       8
  ↑      ↑      ↑       ↑
```

Each of these is called the difference of the two numbers.

The other two parts of a subtraction problem have labels, although you will not often come across them. The number being subtracted is called the **subtrahend**. The number being subtracted from is called the **minuend**. Let us look at our last subtraction problem again to identify all its parts.

```
  17     minuend
-  9     subtrahend
   8     difference
```

In this case, the number 17 is called the minuend. The number 9 is called the subtrahend. The number 8 is called the difference.

Quick Recall of Subtraction Facts

It is helpful if you can subtract quickly. See if you can do Example 1 correctly in 15 seconds or less. Repeat again with Practice Problem 1. Strive to obtain all answers correctly in 15 seconds or less.

Example 1 Subtract: **(a)** $8 - 2$ **(b)** $13 - 5$ **(c)** $12 - 4$ **(d)** $15 - 8$ **(e)** $16 - 0$

(a)	**(b)**	**(c)**	**(d)**	**(e)**
8	13	12	15	16
− 2	− 5	− 4	− 8	− 0
6	8	8	7	16

Practice Problem 1 Subtract.

(a)	**(b)**	**(c)**	**(d)**	**(e)**
9	12	17	14	18
− 6	− 5	− 8	− 0	− 9

2 If the numbers being subtracted involve more than two digits, line up the ones column, the tens column, the hundreds column, and so on, in order to keep track of your work. Begin with the ones column, and move right to left.

Example 2 Subtract: $9867 - 3725$.

```
  9 8 6 7
- 3 7 2 5
  6 1 4 2
  ↑ ↑ ↑ ↑
```
        └─ 7 ones − 5 ones = 2 ones
      └─ 6 tens − 2 tens = 4 tens
    └─ 8 hundreds − 7 hundreds = 1 hundred
  └─ 9 thousands − 3 thousands = 6 thousands

Practice Problem 2 Subtract: $7695 - 3481$.

3 In the subtraction that we have looked at so far, each digit in the upper number (the minuend) has been greater than the digit in the lower number (the subtrahend) for each place value. Many times, however, a digit in the lower number is greater than the digit in the upper number for that place value.

```
  42
- 28
```

The digit in the one's place in the lower number—the 8 of 28—is greater than the number in the ones place in the upper number—the 2 of 42. To subtract, we must *rename* 42 using place values. This is called **borrowing**.

Example 3   Subtract: 42 − 28.

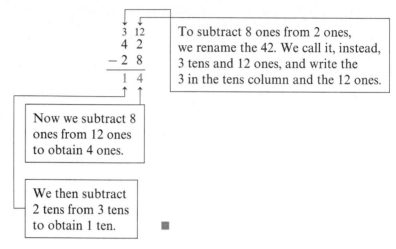

$$\begin{array}{r} 3\ \ 12 \\ 4\ \ 2 \\ -\ 2\ \ 8 \\ \hline 1\ \ 4 \end{array}$$

To subtract 8 ones from 2 ones, we rename the 42. We call it, instead, 3 tens and 12 ones, and write the 3 in the tens column and the 12 ones.

Now we subtract 8 ones from 12 ones to obtain 4 ones.

We then subtract 2 tens from 3 tens to obtain 1 ten.

Practice Problem 3   Subtract: 34 − 16.

Example 4   Subtract: 864 − 548.

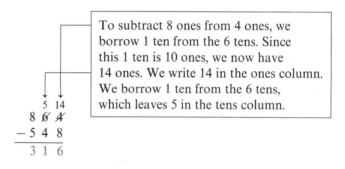

To subtract 8 ones from 4 ones, we borrow 1 ten from the 6 tens. Since this 1 ten is 10 ones, we now have 14 ones. We write 14 in the ones column. We borrow 1 ten from the 6 tens, which leaves 5 in the tens column.

$$\begin{array}{r} 5\ \ 14 \\ 8\ \ 6\ \ 4 \\ -\ 5\ \ 4\ \ 8 \\ \hline 3\ \ 1\ \ 6 \end{array}$$

Now we subtract 8 ones from 14 ones to obtain 6 ones.

Practice Problem 4   Subtract: 693
                                    − 426

Example 5   Subtract: 8040 − 6375.

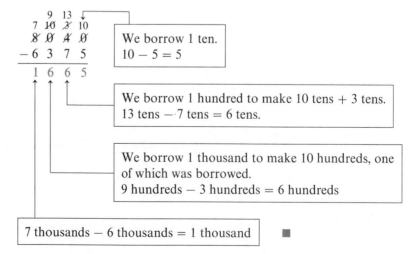

$$\begin{array}{r} 9\ \ 13 \\ 7\ \ 10\ \ \ \ \ 10 \\ 8\ \ 0\ \ 4\ \ 0 \\ -\ 6\ \ 3\ \ 7\ \ 5 \\ \hline 1\ \ 6\ \ 6\ \ 5 \end{array}$$

We borrow 1 ten.
10 − 5 = 5

We borrow 1 hundred to make 10 tens + 3 tens.
13 tens − 7 tens = 6 tens.

We borrow 1 thousand to make 10 hundreds, one of which was borrowed.
9 hundreds − 3 hundreds = 6 hundreds

7 thousands − 6 thousands = 1 thousand

Practice Problem 5   Subtract: 9070
                                       − 5886   ■

Example 6   Subtract.

(a) 9521 − 943   **(b)** 40,000 − 29,056

$$
\begin{array}{r}
\overset{14\ 11}{\overset{8\ \ \cancel{4}\ \ \cancel{1}\ \ 11}{\cancel{9}\ \cancel{5}\ \cancel{2}\ \cancel{1}}} \\
-\ \ \ 9\ 4\ 3 \\ \hline
8\ 5\ 7\ 8
\end{array}
$$

(a)

$$
\begin{array}{r}
\overset{3\ 9\ 9\ 9\ 10}{\cancel{4}\ \cancel{0},\cancel{0}\ \cancel{0}\ \cancel{0}} \\
-\ 2\ 9,0\ 5\ 6 \\ \hline
1\ 0,9\ 4\ 4
\end{array}
$$ ■

(b)

Practice Problem 6   Subtract.

(a)   8964
        − 985

(b)   50,000
        − 32,508   ■

◪ Checking Subtraction Problems

We observe that when $9 - 7 = 2$ it follows that $7 + 2 = 9$. Each subtraction problem is equivalent to a corresponding addition problem. This gives us a convenient way to check our answers to subtraction.

Example 7   Check this subtraction problem.

$$5829 - 3647 = 2182$$

$$
\begin{array}{r}
5\ 8\ 2\ 9 \longleftarrow \\
-3\ 6\ 4\ 7 \\ \hline
2\ 1\ 8\ 2
\end{array}
\quad \text{then} \quad
\begin{array}{r}
3\ 6\ 4\ 7 \\
+2\ 1\ 8\ 2 \\ \hline
5\ 8\ 2\ 9 \longleftarrow
\end{array}
$$

The sum should equal 5829, which it does. We have checked our work, and it is correct.   ■

Practice Problem 7   Check this subtraction problem.

$$9763 - 5732 = 4031$$ ■

Example 8   Subtract and check your answers.

(a) 156,000 − 29,326   **(b)** 1,264,308 − 1,057,612

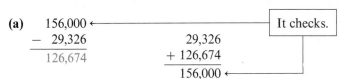

| Subtraction. | Checking by Addition |

(a)
$$
\begin{array}{r}
156,000 \longleftarrow \\
-\ \ 29,326 \\ \hline
126,674
\end{array}
\qquad
\begin{array}{r}
29,326 \\
+126,674 \\ \hline
156,000 \longleftarrow
\end{array}
$$
It checks.

(b)
$$
\begin{array}{r}
1,264,308 \longleftarrow \\
-1,057,612 \\ \hline
206,696
\end{array}
\qquad
\begin{array}{r}
1,057,612 \\
+\ \ 206,696 \\ \hline
1,264,308 \longleftarrow
\end{array}
$$
It checks.   ■

Practice Problem 8   Subtract and check your answers.

(a)   284,000
        − 96,327

(b)   8,526,024
        − 6,397,518   ■

Consider the following population table.

	1960	1970	1980
California	15,717,204	19,971,069	23,667,947
Texas	9,579,677	11,198,655	14,227,799
Arizona	1,302,161	1,775,399	2,716,598
New Mexico	951,023	1,017,055	1,303,302

Example 9

(a) In 1980 how much greater was the population of Texas than the population of Arizona?

(b) How much did the population of California increase from 1960 to 1980?

(a)

14,227,799	1980 population of Texas
− 2,716,598	1980 population of Arizona
11,511,201	difference

The population of Texas was greater by 11,511,201.

(b)

23,667,947	1980 population of California
− 15,717,204	1960 population of California
7,950,743	difference

The population of California increased by 7,950,743 in those 20 years. ■

Practice Problem 9

(a) In 1980 how much greater was the population of California than the population of Texas?

(b) How much did the population of Texas increase from 1960 to 1970? ■

Some problems can be expressed (and solved) with an *equation*. An equation is a number sentence with an equal sign, such as

$$10 = 4 + x$$

Example 10   The librarian knows that he has eight world atlases in all and that five of them are in full color. How many are not in full color?

If we let the number that are *not* in full color be designated as x, the relationship of parts to the whole can be written as a mathematical sentence:

$$x + 5 = 8$$

To *solve* an equation means to find those values that will make the sentence true. We can solve this equation by reasoning and by a knowledge of addition and subtraction: $8 − 5 = 3$, therefore, $x = 3$. We can check if the answer is reasonable by substituting our answer—3—for x in the original problem:

$$x + 5 = 8 \qquad 3 + 5 = 8$$

We see that $x = 3$ checks, so our answer is correct. There are 3 atlases not in full color. ■

Practice Problem 10   Solve.

(a) $x + 12 = 17$

(b) $10 + x = 22$ ■

To Think About

We found that addition is commutative: $5 + 3 = 3 + 5$. However, subtraction is not commutative. For example, $5 − 3 \neq 3 − 5$. (The $\neq$ is read "is not equal to.") Can you think of any numbers that are commutative for subtraction. For example, if a and b represent whole numbers, when does $a − b = b − a$? One example is the number zero. $0 − 0 = 0 − 0$! Can you think of others? See problems 71 and 72. ■

CLASS ATTENDANCE

A student of mathematics needs to get started in the right direction by choosing to attend class every day, beginning with the first day of class. Statistics show that class attendance and good grades go together. Classroom activities are designed to enhance learning, and therefore, you must be in class to benefit from them. Vital information and explanations are given each day that can help you in understanding concepts. Do not be deceived into thinking that you can just find out from a friend what went on in class. There is no good substitute for firsthand experience. Give yourself a push in the right direction by developing to the habit of going to class every day.

CLASS PARTICIPATION

People learn mathematics through active participation, not through observation from the sidelines. If you want to do well in this course, be involved in classroom activities. Sit near the front where you can see and hear well, where your focus is on the instruction process and not on those students around you. Ask questions, be ready to contribute toward solutions, and take part in all classroom activities. Your contributions are valuable to the class and to yourself. Class participation requires an investment of yourself in the learning process, which you will find pays huge dividends.

EXERCISES 1.3

Try to do problems 1–20 in 1 minute or less with no errors.

Subtract.

1. $\begin{array}{r} 8 \\ -2 \end{array}$	**2.** $\begin{array}{r} 7 \\ -5 \end{array}$	**3.** $\begin{array}{r} 13 \\ -8 \end{array}$	**4.** $\begin{array}{r} 9 \\ -0 \end{array}$	**5.** $\begin{array}{r} 6 \\ -5 \end{array}$	**6.** $\begin{array}{r} 13 \\ -8 \end{array}$	**7.** $\begin{array}{r} 15 \\ -6 \end{array}$
8. $\begin{array}{r} 14 \\ -5 \end{array}$	**9.** $\begin{array}{r} 16 \\ -0 \end{array}$	**10.** $\begin{array}{r} 17 \\ -9 \end{array}$	**11.** $\begin{array}{r} 18 \\ -9 \end{array}$	**12.** $\begin{array}{r} 12 \\ -7 \end{array}$	**13.** $\begin{array}{r} 11 \\ -4 \end{array}$	**14.** $\begin{array}{r} 15 \\ -8 \end{array}$
15. $\begin{array}{r} 13 \\ -7 \end{array}$	**16.** $\begin{array}{r} 16 \\ -9 \end{array}$	**17.** $\begin{array}{r} 17 \\ -8 \end{array}$	**18.** $\begin{array}{r} 10 \\ -7 \end{array}$	**19.** $\begin{array}{r} 15 \\ -6 \end{array}$	**20.** $\begin{array}{r} 12 \\ -5 \end{array}$	

Subtract. Check your answers.

21. $\begin{array}{r} 95 \\ -37 \end{array}$	**22.** $\begin{array}{r} 84 \\ -32 \end{array}$	**23.** $\begin{array}{r} 85 \\ -73 \end{array}$	**24.** $\begin{array}{r} 77 \\ -36 \end{array}$	**25.** $\begin{array}{r} 126 \\ -95 \end{array}$
26. $\begin{array}{r} 193 \\ -72 \end{array}$	**27.** $\begin{array}{r} 678 \\ -234 \end{array}$	**28.** $\begin{array}{r} 563 \\ -152 \end{array}$	**29.** $\begin{array}{r} 1763 \\ -422 \end{array}$	**30.** $\begin{array}{r} 1896 \\ -743 \end{array}$
31. $\begin{array}{r} 24{,}396 \\ -13{,}205 \end{array}$	**32.** $\begin{array}{r} 52{,}980 \\ -31{,}660 \end{array}$	**33.** $\begin{array}{r} 986{,}302 \\ -433{,}201 \end{array}$	**34.** $\begin{array}{r} 807{,}965 \\ -304{,}214 \end{array}$	

Check each subtraction. If the problem has not been done correctly, find the correct answer.

35.
$$\begin{array}{r} 127 \\ -\ 19 \\ \hline 108 \end{array}$$

36.
$$\begin{array}{r} 186 \\ -\ 45 \\ \hline 141 \end{array}$$

37.
$$\begin{array}{r} 7293 \\ -6180 \\ \hline 1113 \end{array}$$

38.
$$\begin{array}{r} 5976 \\ -3243 \\ \hline 2733 \end{array}$$

39.
$$\begin{array}{r} 6030 \\ -5020 \\ \hline 1020 \end{array}$$

40.
$$\begin{array}{r} 7890 \\ -3200 \\ \hline 7670 \end{array}$$

41.
$$\begin{array}{r} 98,763 \\ -42,531 \\ \hline 55,232 \end{array}$$

42.
$$\begin{array}{r} 47,969 \\ -33,846 \\ \hline 14,223 \end{array}$$

Subtract. Use borrowing if necessary.

43.
$$\begin{array}{r} 93 \\ -47 \\ \hline \end{array}$$

44.
$$\begin{array}{r} 86 \\ -39 \\ \hline \end{array}$$

45.
$$\begin{array}{r} 125 \\ -\ 88 \\ \hline \end{array}$$

46.
$$\begin{array}{r} 136 \\ -\ 95 \\ \hline \end{array}$$

47.
$$\begin{array}{r} 451 \\ -376 \\ \hline \end{array}$$

48.
$$\begin{array}{r} 706 \\ -435 \\ \hline \end{array}$$

49.
$$\begin{array}{r} 809 \\ -437 \\ \hline \end{array}$$

50.
$$\begin{array}{r} 638 \\ -469 \\ \hline \end{array}$$

51.
$$\begin{array}{r} 10,000 \\ -\ 6,704 \\ \hline \end{array}$$

52.
$$\begin{array}{r} 20,000 \\ -13,120 \\ \hline \end{array}$$

53.
$$\begin{array}{r} 152,000 \\ -117,908 \\ \hline \end{array}$$

54.
$$\begin{array}{r} 361,000 \\ -121,520 \\ \hline \end{array}$$

55.
$$\begin{array}{r} 42,312 \\ -39,998 \\ \hline \end{array}$$

56.
$$\begin{array}{r} 54,913 \\ -29,997 \\ \hline \end{array}$$

57.
$$\begin{array}{r} 760,108 \\ -536,992 \\ \hline \end{array}$$

58.
$$\begin{array}{r} 580,092 \\ -349,905 \\ \hline \end{array}$$

Answer each question.

59. Katherine, Latasha, and Sean ran for class president. Katherine received 365 votes, Latasha 960 votes, and Sean 778 votes. How many more votes did Latasha get than Sean?

60. Thomas drove 982 miles on vacation. Sharon drove 677 miles, and Deidre drove 435 miles. How many more miles did Thomas drive than Sharon?

61. Overland Construction made a bid of $2,364,028 to build the campus cafeteria. Horizon Enterprises made a bid of $1,960,583. How much lower was the Horizon Enterprises bid?

62. The state budget revenue for 1990 was $14,360,560,000. The expenses for the year were $14,267,354,200. How much was left over?

63. Last week Heather earned $398. She had withheld from her pay check $115 for taxes and $39 for retirement. How much money did she actually receive?

64. Last week Won Lin earned $462. He had withheld from his paycheck $163 for taxes and $49 for retirement. How much money did he actually receive?

In answering problems 65–68, consider the following population table.

	1960	1970	1980
Illinois	10,081,158	11,110,285	11,427,409
Michigan	7,823,194	8,881,826	9,262,044
Indiana	4,662,498	5,195,392	5,490,212
Minnesota	3,413,864	3,806,103	4,075,970

65. How much did the population of Minnesota increase from 1960 to 1980?

66. How much did the population of Michigan increase from 1960 to 1980?

67. In 1970, how much greater was the population of Michigan than the population of Indiana?

68. In 1980, how much greater was the population of Indiana than the population of Minnesota?

69. Solve: $x + 12 = 18$.

70. Solve: $x + 20 = 30$.

 To Think About

71. In general, subtraction is not commutative. If a and b are whole numbers, $a - b \neq b - a$. For what types of numbers would it be true that $a - b = b - a$?

72. In general, subtraction is not associative. For example, $8 - (4 - 3) \neq (8 - 4) - 3$. In general, $a - (b - c) \neq (a - b) - c$. Can you find some numbers a, b, c for which $a - (b - c) = (a - b) - c$?

73. Subtract: $28,007,653,121,863 - 27,986,430,705,999$.

74. Subtract: $101,000,101,000,101 - 99,090,989,878,789$.

Cumulative Review Problems

75. Write in standard notation: eight million, four hundred sixty-six thousand, eighty-four.

76. Write a word name for 296,308.

77. Add: $16 + 27 + 82 + 34 + 9$.

78. Add: $156,325$
     $+ 963,209$

 Calculator Problems

Subtract.

79. $100,000$
      $- 97,106$

80. $3,699,000$
      $- 766,827$

81. The Central State University budget was $13,000,120. After $960,396 had been spent. How much was left?

For Extra Practice Examples and Exercises, turn to page 59.

Solutions to Odd-Number Practice Problems

1. (a) 9 **(b)** 12 **(c)** 17 **(d)** 14 **(e)** 18 **3.** $\overset{2\ 14}{3\ \cancel{4}}$ **5.** $\overset{16}{9\ \overset{9}{\cancel{6}}\ \overset{10}{\cancel{7}}\ \cancel{0}}$
      $\underline{-\ 6}$ $\underline{-\ 5}$ $\underline{-\ 8}$ $\underline{-\ 0}$ $\underline{-\ 9}$ $\underline{-1\ 6}$ $\underline{-5\ 8\ 8\ 6}$
       3 7 9 14 9 $1\ 8$ $3\ 1\ 8\ 4$

7. Subtraction   Checking by Addition   **9. (a)** $23,667,947$ **(b)** $11,198,655$
     9763 $\underline{-\ 14,227,799}$ $\underline{-\ 9,579,677}$
    $\underline{-\ 5732}$ ┌─────────────┐ $9,440,148$ $1,618,978$
     4031 │ IT CHECKS │
          └─────────────┘
         5732
        $\underline{+\ 4031}$
         9763

Answers to Even-Numbered Practice Problems

2. 4214   **4.** 267   **6. (a)** 7979 **(b)** 17,492   **8. (a)** 187,673 **(b)** 2,128,506   **10. (a)** $x = 5$ **(b)** $x = 12$

After studying this section, you will be able to:

1 *Master basic multiplication facts*

2 *Multiply a single-digit number by a number with several digits*

3 *Multiply a number by a number that is a multiple of 10.*

4 *Multiply a several-digit number by a several-digit number*

5 *Efficiently multiply a product such as 7 × 20 × 5 × 6*

1.4 MULTIPLICATION OF WHOLE NUMBERS

Like subtraction, multiplication is related to addition. Suppose that the pastry chef at the Gourmet Restaurant bakes croissants on a sheet that holds four croissants across, with room for three rows. How many croissants does the sheet hold?

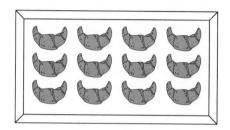

We can add 4 + 4 + 4 to get the total, or we can use a shortcut: 3 rows of 4 is the same as 3 times 4, which equals 12. This is multiplication, a shortcut for repeated addition.

When the numbers are large, multiplication is easier than addition, but for smaller numbers, you can—if you're stuck—do a multiplication problem by working the equivalent addition problem.

1 Multiplication can be thought of as repeated addition. If we had six nickels, each of which is worth 5 cents, we would have 6 × 5 cents, which equals 30 cents. This multiplication is faster than using repeated addition to obtain

$$5¢ + 5¢ + 5¢ + 5¢ + 5¢ + 5¢ = 30¢$$

Your skill in performing multiplication depends on your knowledge of basic multiplication facts. They are listed below.

BASIC MULTIPLICATION FACTS

×	0	1	2	3	4	5	6	7	8	9
0	0	0	0	0	0	0	0	0	0	0
1	0	1	2	3	4	5	6	7	8	9
2	0	2	4	6	8	10	12	14	16	18
3	0	3	6	9	12	15	18	21	24	27
4	0	4	8	12	16	20	24	28	32	36
5	0	5	10	15	20	25	30	35	40	45
6	0	6	12	18	24	30	36	42	48	54
7	0	7	14	21	28	35	42	49	56	63
8	0	8	16	24	32	40	48	56	64	72
9	0	9	18	27	36	45	54	63	72	81

Look them over carefully. You should be able to recall each multiplication fact correctly. If you think you know them well, see if you can do Exercise Set 1.4, problems 1 and 2.

A key fact of multiplication on: When you multiply any number times zero, the result is zero. Notice that

$$2 \times 0 = 0 \qquad 5 \times 0 = 0 \qquad 0 \times 6 = 0 \qquad 0 \times 0 = 0$$

There are several ways of indicating multiplication. Consider 3 times 4:

3×4	$(3)(4)$	$3(4)$	$(3)4$	$3 \cdot 4$	$3 * 4$
with an $\times$	with two sets of parentheses	with a single set of parentheses		with a dot	with a star

For now we will use $\times$ to signal multiplication. It is read as the word "times": 3×4 is read "three times four." If two numbers are denoted by a and b, their multiplication can be indicated by ab, with no symbol between the a and the b.

Multiplication is commutative.

$$3 \times 4 = 4 \times 3$$

The order in which two numbers are multiplied does not change the value of the answer. The numbers that we multiply are called **factors**. The answer is called the **product**.

$$
\begin{array}{rl}
7 & \text{factor} \\
\times\ 8 & \text{factor} \\
\hline
56 & \text{product}
\end{array}
$$

❷ Multiplication of a Single-digit Number by a Number with Several Digits

Example 1 Multiply: 4312×2.

We first multiply the one's column, then the tens column, and so on, moving right to left.

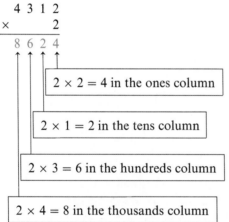

$2 \times 2 = 4$ in the ones column

$2 \times 1 = 2$ in the tens column

$2 \times 3 = 6$ in the hundreds column

$2 \times 4 = 8$ in the thousands column ■

Practice Problem 1 Multiply: 3021×3 ■

Usually, we will have to carry one digit of the result of some of the multiplication into the next left-hand column.

Example 2 Multiply: 36×7.

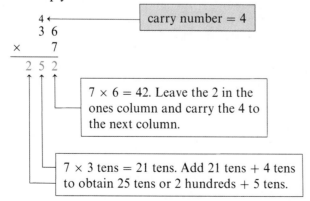

carry number $= 4$

$7 \times 6 = 42$. Leave the 2 in the ones column and carry the 4 to the next column.

7×3 tens $= 21$ tens. Add 21 tens $+$ 4 tens to obtain 25 tens or 2 hundreds $+$ 5 tens. ■

Practice Problem 2 Multiply: 43×8 ■

Example 3   Multiply: 359×9.

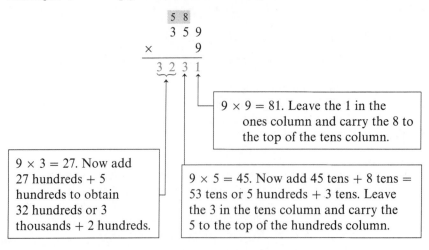

Practice Problem 3   Multiply: 579
$$\times \quad 7 \quad \blacksquare$$

❸ Multiplication by a Multiple of 10

Observe what happens when a number is multiplied by 10, 100, 1000, 10,000, and so on.

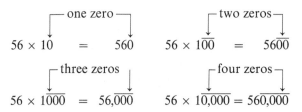

A *power of 10* is a whole number that begins with 1 and ends in one or more zeros. The numbers 10, 100, 1000, 10,000 and so on are powers of 10.

> To multiply a whole number by a power of 10:
> **1.** Count the number of zeros in the power of 10.
> **2.** Add that number of zeros to the right side of the other whole number to obtain the answer.

Example 4   Multiply 358 by each number.
(a) 10   **(b)** 100   **(c)** 1000   **(d)** 100,000

(a) $358 \times 10 = 3580$ (one zero)
(b) $358 \times 100 = 35,800$ (two zeros)
(c) $358 \times 1000 = 358,000$ (three zeros)
(d) $358 \times 100,000 = 35,800,000$ (five zeros)   ∎

Practice Problem 4   Multiply 1267 by each number.
(a) 10   **(b)** 1000   **(c)** 10,000   **(d)** 1,000,000
∎

How can we handle zeros in multiplication involving a number that is not 10, 100, 1000, or any other power of 10? Consider 32×400. We can rewrite 400 as 4×100, which gives us $32 \times 4 \times 100$. We can simply multiply 32×4 and then attach three zeros for the factor 100. We find that $32 \times 4 = 128$. Adding three zeros gives us 128,000, so $32 \times 400 = 128,000$.

4 Multiplication of a Several-Digit Number by a Several-Digit Number

Example 5 Multiply: 234×21.

We can consider 21 as 2 tens (20) and 1 one (1). First we multiply 234 by 1.

$$
\begin{array}{r}
234 \\
\times\ 21 \\
\hline
234
\end{array}
$$

The 234 is called a *partial product*. We next multiply 234 by 20.

$$
\begin{array}{r}
234 \\
\times\ 21 \\
\hline
234 \\
468
\end{array}
$$

Note that 468—also called a partial product—is placed so that its first digit appears in the tens column. This placement is necessary because we are multiplying by the digit in the tens column of the factor 21. We may add a zero to the right of the 468 to make sure that the 8 of 468 lines up in the tens column.

$$
\begin{array}{r}
234 \\
\times\ 21 \\
\hline
234 \\
4680
\end{array}
$$

We now add the two partial products to reach the final product, which is the solution.

$$
\begin{array}{r}
234 \\
\times\ 21 \\
\hline
234 \\
4680 \\
\hline
4914
\end{array}
$$

 234 ⟵ Multiply 234×1.
 4680 ⟵ Multiply 234×20. Write this partial product in the tens column.
 4914 ⟵ Put a zero in the ones column.
   ⎿Add the two partial products. ∎

Practice Problem 5 Multiply:
$$
\begin{array}{r}
323 \\
\times\ 32.
\end{array}
$$
∎

Example 6 Multiply: 671×35.

$$
\begin{array}{r}
671 \\
\times\ 35 \\
\hline
3355 \\
2013\,⓪ \\
\hline
23485
\end{array}
$$

 3 3 5 5 ⟵ First multiply 671×5.
 2 0 1 3 ⓪ ⟵ Now multiply 671×30.
 2 3 4 8 5 ⟵ Now add the two partial products.

Note: The final zero on this line may be omitted. ∎

Practice Problem 6 Multiply:
$$
\begin{array}{r}
385 \\
\times\ 69.
\end{array}
$$
∎

Example 7 Multiply: 14×20.

$$
\begin{array}{r}
14 \\
\times\ 20 \\
\hline
0 \\
280 \\
\hline
280
\end{array}
$$

 0 ⟵ Multiply 14 by 0.
 280 ⟵ Multiply 14 by 2 tens.

Now add the ⟶   280   Place 28 with the 8
partial products.   in the tens column.
         To line up the digits for
         adding, we can insert a 0.

Notice that this result
is the same as $14 \times 2 = 28$
with a zero appended: 280. ∎

Practice Problem 7 Multiply: 34×20. ■

Example 8 Multiply: 120×40.

$$
\begin{array}{r}
120 \\
\times\ 40 \\
\hline
0 \\
4800 \\
\hline
4800
\end{array}
$$

0 ⟵ Multiply 120×0.

4800 ⟵ Multiply 120 by 4 tens.
The answer is 480 tens.
We place the 0 of the 480
in the tens column. To
line up the digits for adding,
we can insert a 0. ■

Now add the ⟶ partial products.

Notice that this result
is the same as $12 \times 4 = 48$
with two zeros appended: 4800.

Practice Problem 8 Multiply: 130×50. ■

Example 9 Multiply: 684×763.

$$
\begin{array}{r}
6\ 8\ 4 \\
\times\ 7\ 6\ 3 \\
\hline
2\ 0\ 5\ 2 \\
4\ 1\ 0\ 4\ \ \\
4\ 7\ 8\ 8\ \ \ \ \\
\hline
5\ 2\ 1\ 8\ 9\ 2
\end{array}
$$

2 0 5 2 ⟵ Multiply 684×3.

4 1 0 4 ⟵ Multiply 684×60. Note that we omit the final zero.

4 7 8 8 ⟵ Multiply 684×700. Note we omit final two zeros.

5 2 1 8 9 2 ■

Practice Problem 9 Multiply: 923×675. ■

5 Multiplication is associative. If we have three numbers to multiply, it does not matter which two numbers we group together first to multiply. For example:

$$2 \times (5 \times 3) = (2 \times 5) \times 3$$

This is obviously true because

$$2 \times (5 \times 3) = 2 \times (15) = 30 \qquad \text{and} \qquad (2 \times 5) \times 3 = (10) \times 3 = 30$$

The final product is the same in either case.

Example 10 Multiply: $14 \times 2 \times 5$.

Since we can group any two numbers together, let's order our steps to take advantage of multiplying by 10.

$$14 \times 2 \times 5 = 14 \times (2 \times 5) = 14 \times 10 = 140 \quad ■$$

Practice Problem 10 Multiply: $25 \times 4 \times 17$. ■

For convenience, we list the properties of multiplication we have discussed in this section.

1. Associative Law of Multiplication. When we multiply three numbers, the multiplication can be grouped in any order.	$(7 \times 3) \times 2 = 7 \times (3 \times 2)$ $21 \times 2 = 7 \times 6$ $42 = 42$
2. Commutative Law of Multiplication. Two numbers can be multiplied in either order with the same result.	$9 \times 8 = 8 \times 9$ $72 = 72$
3. Multiplication Law of Zero. The product of any number and zero yields zero as a result.	$0 \times 14 = 0$ $2 \times 0 = 0$
4. Multiplication Law of One. When one is multiplied by a number the result is that number.	$7 \times 1 = 7$ $1 \times 15 = 15$

Why does our method of grouping numbers in multiplication work? Why can we say in Example 5 that 234×21 is the same as 1×234 and 20×234? What justification do we have? The answer is that multiplication of whole numbers follows the **distributive property of multiplication over addition**.

We know that eight quarters equals two dollars. We can use this fact to illustrate the distributive property.

$$\text{One quarter} \times 8 = \$2.00$$

We use in our example the addition fact that $8 = 5 + 3$, but we could use any other addition fact for 8 ($4 + 4$ or $7 + 1$, for example).

$$\text{One quarter} \times (5 + 3) = \$2.00$$

We can substitute "0.25" for "quarter" and achieve the same results.

$$(\text{One quarter} \times 5) + (\text{one quarter} \times 3)$$

$$\$1.25 + \$0.75 = \$2.00$$

Let's consider an additional example.

$$5 \times 7 = 5 \times (4 + 3) = (5 \times 4) + (5 \times 3) = 20 + 15 = 35$$

We check the multiplication fact that 5×7 does, in fact, equal 35.

In general, for any whole numbers a, b, and c the **distributive property of multiplication over addition** tells us that

$$a \times (b + c) = (a \times b) + (a \times c)$$

The parentheses tell us the order of operations. On the left side b is added to c first. On the right side a is multiplied by b as well as a is multiplied by c first.

Now do you see how we were able to "take apart" the factor 21 in Example 5? We distributed multiplication first over the 1 in 21 and then the 20 in 21. Adding the results gave us the product.

$$234 \times 21 = 234 \times (20 + 1) = (234 \times 20) + (234 \times 1) \quad \blacksquare$$

Example 11   The average annual salary of an employee at Software Associates is \$21,342. There are 38 employees. What is the annual payroll?

$$
\begin{array}{r}
\$21{,}342 \\
\times \qquad 38 \\
\hline
170\,736 \\
640\,26 \\
\hline
810{,}996 \\
\end{array}
$$

The total annual payroll is about \$810,966. $\blacksquare$

Practice Problem 11   The average cost of a new car sold last year at Westover Chevrolet was \$17,348. The dealership sold 378 cars. What were the annual sale of cars at the dealership? $\blacksquare$

DEVELOPING YOUR STUDY SKILLS

WHY STUDY MATHEMATICS?

Students often question the value of mathematics, particularly algebra. They see little real use for algebra in their everyday lives. The manipulation of letters and symbols seems almost meaningless. This is understandable at the beginning or intermediate levels of algebra, because applications of algebra may not be obvious.

The extensive usefulness of mathematics becomes clear as you take higher-level courses, such as college algebra, statistics, trigonometry, and calculus. You may not be planning to take these higher-level courses now, but your college major may require you to do so.

In our present-day, technological world, it is easy to see mathematics at work. Many vocational and professional areas—such as the fields of business, statistics, economics, psychology, finance, computer science, chemistry, physics, engineering, electronics, nuclear energy, banking, quality control, and teaching—require a certain level of expertise in mathematics. Those who want to work in these fields must be able to function at a given mathematical level. Those who cannot will not make it.

So if your field of study requires you to take higher level mathematics courses, be sure to realize the importance of mastering the basics of this course. Then you will be ready for the next one.

EXERCISES 1.4

Complete the multiplication facts for each table. Strive for total accuracy, but work quickly. (For problems 1 and 2, allow a maximum of 3 minutes each.)

Multiply.

1.
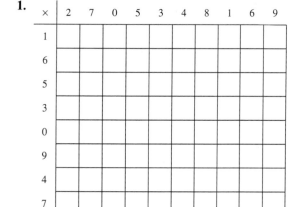

2.
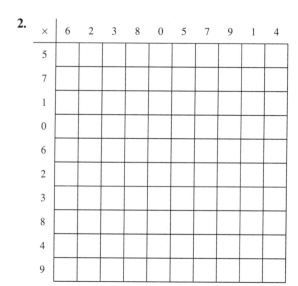

3. 21
   × 4

4. 32
   × 3

5. 104
   × 2

6. 302
   × 3

7. 6102
   × 3

8. 5203
   × 2

9. 101,204
   × 3

10. 301,200
   × 4

11. 12
   × 5

12. 13
   × 4

13. 87
   × 6

14. 95
   × 7

15. 326
   × 5

16. 732
   × 6

17. 1087
   × 7

18. 2605
   × 9

19. 23,695
   × 8

20. 48,761
   × 5

21. 127,054
   × 6

22. 235,702
   × 4

Multiply by multiples of 10.

23. $\begin{array}{r} 156 \\ \times\ 10 \\ \hline \end{array}$ **24.** $\begin{array}{r} 278 \\ \times\ 10 \\ \hline \end{array}$ **25.** $\begin{array}{r} 89{,}361 \\ \times\ \ \ 100 \\ \hline \end{array}$ **26.** $\begin{array}{r} 27{,}158 \\ \times\ \ \ 100 \\ \hline \end{array}$ **27.** $\begin{array}{r} 482 \\ \times\ 1000 \\ \hline \end{array}$

28. $\begin{array}{r} 579 \\ \times\ 1000 \\ \hline \end{array}$ **29.** $\begin{array}{r} 562{,}314 \\ \times\ 100{,}000 \\ \hline \end{array}$ **30.** $\begin{array}{r} 481{,}793 \\ \times\ 100{,}000 \\ \hline \end{array}$ **31.** $\begin{array}{r} 423 \\ \times\ 20 \\ \hline \end{array}$ **32.** $\begin{array}{r} 134 \\ \times\ 20 \\ \hline \end{array}$

33. $\begin{array}{r} 2120 \\ \times\ \ 30 \\ \hline \end{array}$ **34.** $\begin{array}{r} 4230 \\ \times\ \ 20 \\ \hline \end{array}$ **35.** $\begin{array}{r} 62{,}000 \\ \times\ \ 3{,}000 \\ \hline \end{array}$ **36.** $\begin{array}{r} 14{,}000 \\ \times\ \ 4{,}000 \\ \hline \end{array}$

Multiply.

37. $\begin{array}{r} 1022 \\ \times\ \ 31 \\ \hline \end{array}$ **38.** $\begin{array}{r} 3021 \\ \times\ \ 12 \\ \hline \end{array}$ **39.** $\begin{array}{r} 126 \\ \times\ 31 \\ \hline \end{array}$ **40.** $\begin{array}{r} 321 \\ \times\ 24 \\ \hline \end{array}$ **41.** $\begin{array}{r} 89 \\ \times\ 64 \\ \hline \end{array}$

42. $\begin{array}{r} 68 \\ \times\ 49 \\ \hline \end{array}$ **43.** $\begin{array}{r} 768 \\ \times\ 35 \\ \hline \end{array}$ **44.** $\begin{array}{r} 915 \\ \times\ 47 \\ \hline \end{array}$ **45.** $\begin{array}{r} 569 \\ \times\ 73 \\ \hline \end{array}$ **46.** $\begin{array}{r} 817 \\ \times\ 84 \\ \hline \end{array}$

47. $\begin{array}{r} 912 \\ \times\ 76 \\ \hline \end{array}$ **48.** $\begin{array}{r} 498 \\ \times\ 59 \\ \hline \end{array}$ **49.** $\begin{array}{r} 5123 \\ \times\ \ 29 \\ \hline \end{array}$ **50.** $\begin{array}{r} 1268 \\ \times\ \ 38 \\ \hline \end{array}$ **51.** $\begin{array}{r} 6054 \\ \times\ \ 81 \\ \hline \end{array}$

52. $\begin{array}{r} 2098 \\ \times\ \ 62 \\ \hline \end{array}$ **53.** $\begin{array}{r} 178 \\ \times\ 235 \\ \hline \end{array}$ **54.** $\begin{array}{r} 392 \\ \times\ 187 \\ \hline \end{array}$ **55.** $\begin{array}{r} 678 \\ \times\ 132 \\ \hline \end{array}$ **56.** $\begin{array}{r} 493 \\ \times\ 417 \\ \hline \end{array}$

57. $\begin{array}{r} 2076 \\ \times\ \ 105 \\ \hline \end{array}$ **58.** $\begin{array}{r} 5092 \\ \times\ \ 302 \\ \hline \end{array}$ **59.** $\begin{array}{r} 3561 \\ \times\ \ 401 \\ \hline \end{array}$ **60.** $\begin{array}{r} 1298 \\ \times\ \ 304 \\ \hline \end{array}$ **61.** $\begin{array}{r} 1023 \\ \times\ 4005 \\ \hline \end{array}$

62. $\begin{array}{r} 7023 \\ \times\ 2007 \\ \hline \end{array}$ **63.** $\begin{array}{r} 130 \\ \times\ 30 \\ \hline \end{array}$ **64.** $\begin{array}{r} 205 \\ \times\ 200 \\ \hline \end{array}$ **65.** $\begin{array}{r} 307 \\ \times\ 30 \\ \hline \end{array}$ **66.** $\begin{array}{r} 150 \\ \times\ 200 \\ \hline \end{array}$

67. $7 \cdot 2 \cdot 5$ **68.** $8 \cdot 3 \cdot 2$ **69.** $11 \cdot 7 \cdot 4$ **70.** $15 \cdot 3 \cdot 4$ **71.** $50 \cdot 2 \cdot 5$

72. $4 \cdot 25 \cdot 7$ **73.** $12 \cdot 3 \cdot 5 \cdot 2$ **74.** $7 \cdot 6 \cdot 2 \cdot 5$ **75.** What is x if $x = 8 \cdot 7 \cdot 6 \cdot 0$? **76.** What is x if $x = 3 \cdot 12 \cdot 0 \cdot 5$?

Solve each problem.

77. The Nature Club is purchasing 12 bird feeders at $14 per bird feeder. What is the total cost?

78. The reggae band is purchasing 17 shirts at $11 per shirt. What is the total cost?

79. Find the cost of purchasing 326 notebooks at $3 per notebook for the campus store.

80. Find the cost of purchasing 471 bookbags at $4 per bookbag for alumni weekend souvenirs.

81. A student pays $314 per month for his dorm room for nine months. What is his cost for the year?

82. A company rents a car for a salesman at $276 per per month for eight months. What is the cost for the car rental during this time?

83. Marcos has a car that gets 34 miles per gallon on highway driving. Approximately how far can he travel if his gas tank holds 18 gallons?

84. Arlene has a car that gets 27 miles per gallon on highway driving. Approximately how far can she travel if her gas tank holds 19 gallons?

85. There are 84 assembly workers at Ventronics. Each worker assembles 17 VCRs per day. How many VCRs are produced each day?

86. Smithville Services has 96 employees. They all are scheduled to work 38 hours per week. How many working hours should be scheduled in one week, for all employees?

87. If a state has a population of 1,038,420 and an average per capita (per person) income of $12,000 per year, what is the approximate yearly income of people in the state?

88. If a state has a population of 3,021,180 and an average per capita income of $9,000 per year, what is the approximate yearly income in the state?

 To Think About

89. We said that our method of multiplication of several-digit whole numbers requires the use of the property that multiplication is distributive over addition: $a \times (b + c) = (a \times b) + (a \times c)$. Would this be true if we tried our method of multiplication on Roman numerals such as (XII) × (IV)? Why or why not?

90. Is multiplication distributive over subtraction? Why or why not? Give examples.

Multiply (very carefully).

91.
$$\begin{array}{r} 6{,}234{,}189 \\ \times \quad 205{,}307 \\ \hline \end{array}$$

92. $26 \times 32 \times 18 \times 25 \times 7 \times 4$

Cumulative Review Problems

93. Subtract:
$$\begin{array}{r} 34{,}084 \\ -\ 27{,}328 \\ \hline \end{array}$$

94. Add:
$$\begin{array}{r} 263 \\ 27 \\ 891 \\ 5 \\ +\ 63 \\ \hline \end{array}$$

95. Albert Gachet earns $156 per week. He has withheld weekly $12 for taxes, $2 for dues, and $3 for retirement. How much is left?

96. Mrs. May Washington retired at $743 per month. Her cost of living raise increased her payment to $802 per month. What was the increase?

 Calculator Problems

Multiply.

97.
$$\begin{array}{r} 987{,}614 \\ \times\qquad 89 \\ \hline \end{array}$$

98.
$$\begin{array}{r} 2299 \\ \times\ 1123 \\ \hline \end{array}$$

99.
$$\begin{array}{r} 2111 \\ \times\ 1209 \\ \hline \end{array}$$

For Extra Practice Examples and Exercises, turn to page 60.

Solutions to Odd-Numbered Practice Problems

1.
$$\begin{array}{r} 3021 \\ \times\quad 3 \\ \hline 9063 \end{array}$$

3.
$$\begin{array}{r} {}^{56} \\ 579 \\ \times\quad 7 \\ \hline 4053 \end{array}$$

5.
$$\begin{array}{r} 323 \\ \times\ 32 \\ \hline 646 \\ 9\ 69 \\ \hline 10{,}336 \end{array}$$

7.
$$\begin{array}{r} 34 \\ \times\ 20 \\ \hline 680 \end{array}$$

9.
$$\begin{array}{r} 923 \\ \times\ 675 \\ \hline 4\ 615 \\ 64\ 61 \\ 553\ 8 \\ \hline 623{,}025 \end{array}$$

11.
$$\begin{array}{r} \$17{,}348 \\ \times\qquad 378 \\ \hline 138\ 784 \\ 1\ 214\ 36 \\ 5\ 204\ 4 \\ \hline \$6{,}557{,}544 \end{array}$$

Answers to Even-Numbered Practice Problems

2. 344   **4. (a)** 12,670   **(b)** 1,267,000   **(c)** 12,670,000   **(d)** 1,267,000,000   **6.** 26,565   **8.** 6500   **10.** 1700

1.5 DIVISION OF WHOLE NUMBERS

☐ After studying this section, you will be able to:

1 Basic Division Facts

1 *Perform simple division using basic division facts*

Every night 36 carnations are delivered to the Gourmet Restaurant. They are separated into groups of three flowers each, to be placed in vases for the dining tables. We can count out the 36 flowers and repeatedly take out sets of 3 and see how many times this can be done: 12 times. We also could use division: 36 divided by 3 equals 12. Division can be thought of as repeated subtraction.

2 *Perform long division by a one-digit number.*

3 *Perform long division by a two- or three-digit number*

What Does Division Mean?

Suppose that we have eight quarters and want to divide them into two equal piles. We would discover that each pile contains four quarters.

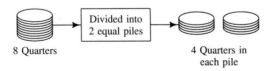

8 Quarters   Divided into 2 equal piles   4 Quarters in each pile

In mathematics we would express this thought by saying that

$$8 \div 2 = 4$$

We are confident that this answer is right because two piles of four quarters is the same dollar amount as eight quarters. In other words, we are confident that $8 \div 2 = 4$ because $2 \times 4 = 8$. These two mathematical sentences are called *related sentences*. The division sentence $8 \div 2 = 4$ is related to the multiplication sentence $2 \times 4 = 8$.

In fact, in mathematics we usually define division in terms of multiplication.

The answer to the division problem $12 \div 3$ is that number which when multiplied by 3 yields 12. Thus $12 \div 3 = 4$ because $3 \times 4 = 12$.

Let us consider one more practical case. Suppose that a surplus of $30 in the French Club budget at the end of the year is to be equally divided among the five club members. We would want to divide the $30 into five equal parts. We would write $30 \div 5$. $30 \div 5 = 6$ because $5 \times 6 = 30$. Thus each of the five people would get $6 in this situation.

$30   5 Piles of $6 each

As a mathematical sentence, $30 \div 5 = 6$.

How Do We Write Division?

The division problem $30 \div 5 = 6$ could also be written $\frac{30}{5} = 6$ or $5\overline{)30}$. When referring to division we sometimes use the words *divisor*, *dividend*, and *quotient* to identify the three parts.

$$\text{divisor} \,\overline{)\, \text{dividend}}^{\text{quotient}}$$

With $30 \div 5 = 6$, 30 is the dividend, 5 is the divisor, and 6 is the quotient.

$$\text{divisor} \longrightarrow 5\overline{)30}\begin{array}{l}\;6 \longleftarrow \text{quotient}\\ \;\;\longleftarrow \text{dividend}\end{array}$$

The answer to a division problem is called the **quotient**. It is important that you can quickly do short problems involving basic division facts.

Example 1 Divide.

(a) $12 \div 4$ (b) $81 \div 9$ (c) $56 \div 8$ (d) $54 \div 6$

(a) $4\overline{)12}^{\,3}$ (b) $9\overline{)81}^{\,9}$ (c) $8\overline{)56}^{\,7}$ (d) $6\overline{)54}^{\,9}$ ∎

Practice Problem 1 Divide.

(a) $36 \div 4$ (b) $25 \div 5$ (c) $72 \div 9$ (d) $30 \div 6$ ∎

Division Problems Involving the Number 1 and The Number 0

It is helpful to remember the following basic concepts:

> 1. Any nonzero number divided by itself is 1 ($7 \div 7 = 1$)
> 2. Any number divided by 1 remains unchanged ($29 \div 1 = 29$).
> 3. Zero may be divided by any nonzero number; the result is always zero ($0 \div 4 = 0$).
> 4. Zero can never be the divisor in a division problem ($3 \div 0$ cannot be done; $0 \div 0$ cannot be done.)

Example 2 Divide if possible. If not possible, state why.

(a) $8 \div 8$ (b) $9 \div 1$ (c) $0 \div 6$ (d) $20 \div 0$ (e) $0 \div 0$

(a) $\frac{8}{8} = 1$ Any number divided by itself is 1.

(b) $\frac{9}{1} = 9$ Any number divided by 1 remains unchanged.

(c) $\dfrac{0}{6} = 0$ Zero divided by any nonzero number is zero.

(d) $\dfrac{20}{0}$ cannot be done   Division by zero is impossible.

(e) $\dfrac{0}{0}$ cannot be done   Division by zero is impossible.   ■

Practice Problem 2   Divide, if possible.

(a) $7 \div 1$ **(b)** $\dfrac{8}{8}$ **(c)** $\dfrac{0}{5}$ **(d)** $12 \div 0$ **(e)** $\dfrac{0}{0}$ ■

To Think About

Why is it that we can never divide by zero, yet zero can be divided by another number? How do we know this? Every division problem has a related multiplication problem.

$$\text{If } 12 \div 6 = 2, \text{ then } 12 = 6 \times 2$$

$$\text{If } 56 \div 7 = 8, \text{ then } 56 = 7 \times 8$$

Suppose that we could divide by zero. Consider the problem $7 \div 0 = $ some number. Let us represent "some number" by the letter a.

$$\text{If } 7 \div 0 = a, \text{ then } 7 = 0 \times a$$

Because zero times any number is zero, $0 \times a = 0$. Thus $7 = 0$, which we know is not true. Therefore, our assumption that $7 \div 0 = a$, where a is a whole number, is wrong. We cannot divide by zero.   ■

◻ Division by a One-Digit Number

Our accuracy with division is improved if we have a checking procedure. For each division fact there is a related multiplication fact.

$$\text{If } 20 \div 4 = 5, \text{ then } 20 = 4 \times 5$$

$$\text{If } 36 \div 9 = 4, \text{ then } 36 = 9 \times 4$$

We will often use multiplication to check our answers.

When two numbers do not divide exactly, a number called the *remainder* is left over. For example, 13 cannot be divided exactly by 2. The number 1 is left over. We call this 1 the remainder.

$$
\begin{array}{r}
6 \\
2\overline{)\,13} \\
\underline{12} \\
1 \longleftarrow \text{remainder}
\end{array}
$$

Thus $13 \div 2 = 6$ with a remainder of 1. We can abbreviate this answer as

$$6\,\text{R}\,1 \quad \text{or} \quad 6\,\text{R} = 1$$

Checking the answer to a division problem uses the property

---

divisor × quotient + remainder = dividend

Example 3   Divide: $33 \div 4$. Check your answer.

$$
\begin{array}{r}
8 \\
4\overline{)\,33} \\
\underline{32} \\
1
\end{array}
$$

$8 \longrightarrow$ How many times can 4 be divided into 33?   8.

$32 \longleftarrow$ What is 8×4?   32.

$1 \longleftarrow$ What is 33 subtract 32?   1.

The answer is 8 with a remainder of 1. We abbreviate 8 R 1. Some people prefer 8 R = 1.

Check:

$$\begin{array}{r} 8 \\ \times 4 \\ \hline 32 \\ 1 \\ \hline 33 \end{array}$$

Multiplying $8 \times 4 = 32$.

Adding the remainder $32 + 1 = 33$.

Because the dividend is 33, the answer is correct. ■

Practice Problem 3 Divide: $45 \div 6$. Check your answer. ■

Example 4 Divide: $158 \div 5$. Check your answer.

$$\begin{array}{r} 31 \\ 5\,)\overline{158} \\ 15 \\ \hline 08 \\ 5 \\ \hline 3 \end{array}$$

5 divided into 15?   3.
What is 3×5?   15.
15 subtract 15?   0.
Bring down 8.
5 divided into 8?   1.
What is 1×5?   The answer is 5.
8 subtract 5?   The answer is 3.

The answer is 31 R 3. Check:

$$\begin{array}{r} 31 \\ \times\ 5 \\ \hline 155 \\ 3 \\ \hline 158 \end{array}$$

Multiplying $31 \times 5 = 155$.

Adding the remainder 3.

Because the dividend is 158, the answer is correct. ■

Practice Problem 4 Divide: $129 \div 6$. Check your answer. ■

Example 5 Divide: $3672 \div 7$.

$$\begin{array}{r} 524 \\ 7\,)\overline{3672} \\ 35 \\ \hline 17 \\ 14 \\ \hline 32 \\ 28 \\ \hline 4 \end{array}$$

How many times can 7 be divided into 36?   5.
What is 5×7?   35.
36 subtract 35?   1.
Bring down 7.
7 divided into 17?   2.
What is 2×7?   14.
17 subtract 14?   3.
Bring down 2.
7 divided into 32?   4.
What is 4×7?   28.
32 subtract 28?   4.

The answer is 524 R 4. ■

Practice Problem 5 Divide: $4237 \div 8$. ■

❸ Division by a Two- or Three-Digit Number

When you divide by a divisor of more than one digit, you may find an estimation technique helpful. Consider only the first digit of the divisor and the first two digits of the dividend. However many times this one-digit divisor goes into this two-digit dividend is a good estimate to use as the first number in the quotient of the larger division problem.

Example 6 Divide: 283 ÷ 41.

First guess:

```
         7
   41 ) 283
       287   too large
```

We estimate the first number of the quotient by asking,
how many times 4 can be divided into 28? The answer is 7.
$7 \times 41 = 287$.
287 is larger than 283!

Second guess:

```
         6
   41 ) 283
       246
        37
```

Because 7 is slightly too large, we try 6.
6×41?   246.
283 subtract 246?   37.

The answer is 6 R 37. (Note that the remainder must always be less than the divisor. Do you see why?)   ■

Practice Problem 6 Divide: 229 ÷ 32.   ■

Example 7 Divide: 33,897 ÷ 56.

First guess:

```
          60
   56 ) 33897
        336
         29
```

How many times can 33 be divided by 5?   6.
What is 6×56?   336.
338 subtract 336?   2.
Bring down 9.
56 cannot be divided into 29.
Write 0 in quotient.

Second set of steps:

```
         605
   56 ) 33897
        336
        297
        280
         17
```

Bring down 7.
How many times can 5 be divided into 29?   5.
What is 5×56?   280.
Subtract $297 - 280$.
Remainder is 17.

The answer is 605 R 17.   ■

Practice Problem 7 Divide: 42,183 ÷ 33.   ■

Example 8 Divide: 5629 ÷ 134.

```
           42
   134 ) 5629
         536
         269
         268
           1
```

How many times does 134 divide into 562?
We guess by saying that 1 divides into 5 five times,
but this is too large. ($5 \times 134 = 670$!)
So we try 4.
What is 4×134?   536.
Subtract $562 - 536$. We obtain 26.
Bring down 9.
How many times does 134 divide into 269?
We guess by saying that 1 divided into 2 goes
2 times
What is 2×134?   268.
Subtract $269 - 268$; the remainder is 1.

The answer is 42 R 1.   ■

Practice Problem 8 Divide: 3227 ÷ 128.   ■

To Think About

Did you realize that division can be thought of as repeated subtraction? Suppose that we want to divide $521 \div 126$ by repeatedly subtracting 126.

$$
\begin{array}{rl}
521 & \\
-126 & \text{first time} \\
\hline
395 & \\
-126 & \text{second time} \\
\hline
269 & \\
-126 & \text{third time} \\
\hline
143 & \\
-126 & \text{fourth time} \\
\hline
17 & \\
\end{array}
$$

We can no longer subtract 126. Thus we find that 521 has four 126's and a remainder of 17. We could say:

$$
\begin{array}{r}
4 \text{ R } 17 \\
126 \overline{)\ 521}
\end{array}
$$

Work $2275 \div 358$ by repeated subtraction. Check your answer by long division. ∎

Example 9 A car traveled 1144 miles in 22 hours. What was the average speed in miles per hour?

$$
\begin{array}{r}
52 \\
22 \overline{)\ 1144} \\
110 \\
\hline
44 \\
44 \\
\hline
0
\end{array}
$$

The car traveled an average of 52 miles per hour. ∎

Practice Problem 9 An airplane traveled 1359 miles in 3 hours. What was the average speed in miles per hour? ∎

EXERCISES 1.5

Divide. See if you can work each of problems 1–30 in 3 minutes or less.

1. $5\overline{)35}$ **2.** $6\overline{)42}$ **3.** $4\overline{)32}$ **4.** $8\overline{)24}$ **5.** $9\overline{)27}$ **6.** $8\overline{)40}$

7. $7\overline{)56}$ **8.** $9\overline{)36}$ **9.** $4\overline{)12}$ **10.** $7\overline{)21}$ **11.** $9\overline{)81}$ **12.** $8\overline{)56}$

13. $7\overline{)63}$ **14.** $6\overline{)54}$ **15.** $4\overline{)28}$ **16.** $8\overline{)16}$ **17.** $8\overline{)40}$ **18.** $9\overline{)63}$

19. $9\overline{)72}$ **20.** $9\overline{)63}$ **21.** $1\overline{)8}$ **22.** $1\overline{)7}$ **23.** $5\overline{)0}$ **24.** $6\overline{)0}$

25. $6\overline{)48}$ **26.** $54 \div 6$ **27.** $42 \div 7$ **28.** $6 \div 6$ **29.** $5 \div 5$ **30.** $0 \div 9$

Divide. Check your answer.

31. $3\overline{)28}$ **32.** $4\overline{)22}$ **33.** $8\overline{)76}$ **34.** $9\overline{)39}$ **35.** $5\overline{)128}$

36. $6\overline{)103}$ **37.** $7\overline{)165}$ **38.** $8\overline{)124}$ **39.** $9\overline{)126}$ **40.** $7\overline{)168}$

41. $5\overline{)185}$ **42.** $8\overline{)224}$ **43.** $4\overline{)1289}$ **44.** $3\overline{)758}$ **45.** $6\overline{)763}$

46. $7\overline{)403}$ **47.** $8\overline{)2896}$ **48.** $9\overline{)2286}$ **49.** $5\overline{)6789}$ **50.** $5\overline{)7324}$

51. $7\overline{)12304}$ **52.** $6\overline{)18127}$ **53.** $9\overline{)22305}$ **54.** $8\overline{)26037}$

Divide.

55. $36\overline{)185}$ **56.** $48\overline{)152}$ **57.** $52\overline{)267}$ **58.** $53\overline{)321}$ **59.** $61\overline{)427}$

60. $72\overline{)432}$ **61.** $12\overline{)1930}$ **62.** $13\overline{)6810}$ **63.** $20\overline{)1156}$ **64.** $20\overline{)1283}$

65. $15\overline{)9236}$ **66.** $16\overline{)8314}$ **67.** $36\overline{)7568}$ **68.** $32\overline{)3527}$ **69.** $18\overline{)3643}$

70. $19\overline{)2174}$ **71.** $124\overline{)620}$ **72.** $132\overline{)792}$ **73.** $241\overline{)1237}$ **74.** $325\overline{)1932}$

75. $3483 \div 129 = x$. What is the value of x?

76. $2214 \div 123 = x$. What is the value of x?

Solve each problems.

77. A company distributes a bonus of $11,102 equally among seven employees. How much does each employee receive?

78. A factory produced 13,412 radios in seven days. What was the average number of radios produced per day?

79. A train traveled 1408 miles in 22 hours. What was the average speed in miles per hour?

80. A sailboat traveled 1656 miles in 276 hours of sailing time. What was the average speed in miles per hour?

81. A car rental company bought 14 cars at exactly the same price. The total bill was $247,128. How much did each car cost?

82. A company bought 12 computers at exactly the same price. The total bill was $214,716. How much did each computer cost?

83. Neil wishes to pay off a loan of $4104 in 24 months. How large will his monthly payments be?

84. A new sorting machine sorts 26 letters per minute. The machine is fed 884 letters. How many minutes will it take to sort the letters?

 To Think About

85. Division is not commutative. For example, $12 \div 4 \neq 4 \div 12$. If $a \div b = b \div a$, what must be true of the numbers a and b besides the fact that $b \neq 0$?

86. Review the "To Think About" question in Section 1.5. Divide $874 \div 138$ by repeated subtraction.

Divide. (Neither problem will have a remainder if you divide correctly.)

87. $2683\overline{)26494625}$

88. $3174\overline{)8436492}$

Cumulative Review Problems

Multiply.

89. $\begin{array}{r} 128 \\ \times\ 43 \\ \hline \end{array}$

90. $\begin{array}{r} 7162 \\ \times\ 145 \\ \hline \end{array}$

Add.

91. $316,214 + 89,981$

Subtract.

92. $1,360,000 - 1,293,156$

Calculator Problems

93. $355\overline{)3126485}$ **94.** $244\overline{)1623088}$

For Extra Practice Examples and Exercises, turn to page 60.

Solutions to Odd-Numbered Practice Problems

1. (a) $4\overline{)36}$ → 9   **(b)** $5\overline{)25}$ → 5   **(c)** $9\overline{)72}$ → 8   **(d)** $6\overline{)30}$ → 5

3. $6\overline{)45}$ → 7
```
 42
 ──
  3
```
Check:
```
   6
 × 7
 ──
  42
 + 3
 ──
  45
```

5. $8\overline{)4237}$ → 529
```
 40
 ──
 23
 16
 ──
 77
 72
 ──
  5
```

7. $33\overline{)42183}$ → 1278
```
 33
 ──
 91
 66
 ──
 258
 231
 ───
 273
 264
 ───
   9
```

9. $3\overline{)1359}$ → 453
```
 12
 ──
 15
 15
 ──
 09
  9
 ──
  0
```
The plane averaged 453 miles per hour.

Answers to Even-Numbered Practice Problems

2. (a) 7   **(b)** 1   **(c)** 0   **(d)** Cannot be done   **(e)** Cannot be done   **4.** 21 R 3   *Check:* $6 \times 21 + 3 = 129$ **6.** 7 R 5
8. 25 R 27

1.6 WHOLE-NUMBER EXPONENTS AND ROUNDING WHOLE NUMBERS

☐ After studying this section, you will be able to:

1 Use and evaluate expressions with whole-number exponents

2 Round whole numbers

Sometimes a simple math idea comes "disguised" in technical language. For example, an *exponent* is just a "shorthand" number that saves writing multiplication of the same numbers:

$$10^3 \quad \text{(the exponent is the 3) means}$$

$$10 \times 10 \times 10 \quad \text{(which takes longer to write)}$$

1 Whole-Number Exponents

The product 5×5 can be written as 5^2. The small number 2 is called the *exponent*. The exponent tells us how many factors are in the multiplication. The number 5 is called the *base*. The base is the number that is multiplied.

$$3 \times 3 \times 3 \times 3 = 3^{4} \quad \text{exponent}$$
$$\text{base}$$

In 3^4 the base is 3 and the exponent is 4. (The 4 is sometimes called the superscript.) 3^4 is read as "3 to the fourth power."

Example 1 Write each product in exponent form.

(a) $15 \times 15 \times 15$ **(b)** $7 \times 7 \times 7 \times 7 \times 7$

(a) $15 \times 15 \times 15 = 15^3$ **(b)** $7 \times 7 \times 7 \times 7 \times 7 = 7^5$ ▪

Practice Problem 1 Write each product in exponent form.
(a) $12 \times 12 \times 12 \times 12$ **(b)** $2 \times 2 \times 2 \times 2 \times 2 \times 2$ ▪

Example 2 Find the value of each expression.

(a) 3^3 **(b)** 7^2 **(c)** 2^5 **(d)** 1^8

(a) $3^3 = 3 \times 3 \times 3 = 27$
(b) $7^2 = 7 \times 7 = 49$
(c) $2^5 = 2 \times 2 \times 2 \times 2 \times 2 = 32$
(d) $1^8 = 1 \times 1 \times 1 \times 1 \times 1 \times 1 \times 1 \times 1 = 1$ ∎

Practice Problem 2 Find the value of each expression.

(a) 12^2 **(b)** 6^3 **(c)** 2^6 **(d)** 1^{10} ∎

If a whole number does not have an exponent visible, the exponent is understood to be 1. Thus,

$$3 = 3^1 \quad \text{or} \quad 10 = 10^1$$

Powers of 10

In today's world, large numbers are often expressed as a power of 10.

$10^1 = 10 = 1$ ten $10^4 = 10,000 = 1$ ten-thousand

$10^2 = 100 = 1$ hundred $10^5 = 100,000 = 1$ hundred-thousand

$10^3 = 1000 = 1$ thousand $10^6 = 1,000,000 = 1$ million

To Think About

What does it mean to have an exponent of zero? What is 10^0? Any whole number that is not zero can be raised to the zero power. The result is 1. Thus $10^0 = 1$, $3^0 = 1$, $5^0 = 1$, and so on. Why is this? What logical reasons would we have for defining $10^0 = 1$? Let's reexamine the powers of 10. As we go down one line at a time, notice the pattern that occurs.

$10^5 = 100,000$
$10^4 = 10,000$
$10^3 = 1000$
$10^2 = 100$
$10^1 = 10$
$10^0 = 1$

As we move down one line, we decrease the exponent by 1.

As we move down one line, we divide the previous number by 10.

Therefore, we present the following definition:

Definition

For any whole number a other than zero, $a^0 = 1$.

∎

2 Rounding Whole Numbers

Large numbers are often expressed to the nearest hundred or to the nearest thousand, because an approximate number is "good enough" for certain uses. When we say that light travels 6,000,000,000,000 miles in a year, the answer is rounded to the nearest trillion. What we are trying to do is pick a certain round-off place (a certain degree of accuracy) and find which of two numbers our given value is closer to. So we say that light travels 6 trillion miles a year because the more complete number—5.865696 trillion—is closer to 6 trillion than to 5 trillion or to 7 trillion miles. To round a number, we first select the round-off place, or it is selected for us.

The numbers on the number line increase from left to right. Whole numbers can be associated with points on a line. We call this line the *number line*. A number is less than a given number if it lies to the left of that number on the number line. 5 is less than 8. It can be written

$$5 < 8$$

5 lies to the left of 8 on a number line.

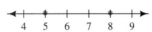

A number is greater than a given number if it lies to the right of the given number. 9 is greater than 1, this is written

$$9 > 1$$

When we round off we are picking the "closest value" on the number line. If we round 368 to the nearest hundred, we get 400. The number 368 is closer to 400 than to 300.

If we round 329 to the nearest hundred, we get 300. The number 329 is closer to 300 than to 400.

Example 3 Round 37,843 to the nearest thousand.

   The thousands place has been selected for us by the directions.

   Locate the thousands round-off place.

The first digit to the right is 5 or more. We will increase the thousands digit by 1.

All the digits to the right of the thousands are replaced by zero.

We have rounded 37,843 to the nearest thousand: 38,000. This means that 37,843 is closer to 38,000 than to 37,000. ■

Practice Problem 3 Round 65,528 to the nearest thousand. ■

Example 4 Round to the nearest hundred thousand: 2,445,360.

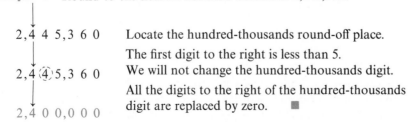

2,4 4 5,3 6 0   Locate the hundred-thousands round-off place.

The first digit to the right is less than 5. We will not change the hundred-thousands digit.

All the digits to the right of the hundred-thousands digit are replaced by zero. ■

Practice Problem 4 Round to the nearest ten thousand: 172,963. ▪

Example 5 Round as shown: **(a)** 561,328 to the nearest ten
(b) 3,798,152 to the nearest hundred **(c)** 51,362,523 to the nearest million

(a) 561,3̌28 = 561,330 to the nearest ten. The digit to the right of tens was 5 or greater.

(b) 3,798,1̌52 = 3,798,200 to the nearest hundred. The digit to the right of hundreds was 5 or greater.

(c) 51,̌362,523 = 51,000,000 to the nearest million. The digit to the right of millions was less than 5. ▪

Practice Problem 5 Round as shown: **(a)** 53,282 to nearest ten
(b) 164,485 to nearest thousand **(c)** 1,365,273 to nearest hundred thousand ▪

Example 6 Round 763,571: **(a)** To nearest thousand **(b)** To nearest ten thousand

(a) 763,̌571 = 764,000 to nearest thousand. The digit to the right of thousands was 5 or greater.

(b) 763,̌571 = 760,000 to the nearest ten thousand. The digit to the right of ten thousand was less than 5. ▪

Practice Problem 6 Round 10,465,121 as indicated.
(a) To nearest hundred thousand **(b)** To nearest million ▪

Example 7 Astronomers use the parsec as a measurement of distance. One **parsec** is approximately 30,800,000,000,000 kilometers. Round 1 parsec to the nearest trillion kilometers.

30,8̌00,000,000,000 km = 31,000,000,000,000 km or 31 trillion km
to the nearest trillion kilometers ▪

Practice Problem 7 One light year is approximately 9,460,000,000,000,000 meters. Round to the nearest hundred trillion meters. ▪

DEVELOPING YOUR STUDY SKILLS

READING THE TEXTBOOK

Homework time each day should begin with the careful reading of the section(s) assigned in your textbook. Much time and effort has gone into the selection of a particular text, and your instructor has chosen a book that will help you to become successful in this mathematics class. Expensive textbooks can be a wise investment if you take advantage of them by reading them.

Reading a mathematics textbook is unlike reading many other types of books that you may find in your literature, history, psychology, or sociology courses. Mathematics texts are technical books that provide you with exercises to practice on. Reading a mathematics text requires slow and careful reading of each word, which takes time and effort.

Begin reading your textbook with a paper and pencil in hand. As you come across a new definition, or concept, underline it in the text and/or write it down in your notebook. Whenever you encounter an unfamiliar term, look it up and make a note of it. When you come to an example, work through it step-by-step. Be sure to read each word and to follow directions carefully.

Notice the helpful hints the author provides to guide you to correct solutions and prevent you from making errors. Take advantage of these pieces of expert advice.

Be sure that you understand what you are reading. Make a note of any of those things that you do not understand and ask your instructor about them. Do not hurry through the material. Learning mathematics takes time.

Write each number in exponent form.

1. $6 \times 6 \times 6 \times 6$

2. $5 \times 5 \times 5$

3. $2 \times 2 \times 2 \times 2 \times 2$

4. $8 \times 8 \times 8 \times 8$

5. $12 \times 12 \times 12$

6. $7 \times 7 \times 7 \times 7 \times 7 \times 7$

7. $1 \times 1 \times 1 \times 1 \times 1 \times 1 \times 1$

8. 26

9. 35

Find the value of each expression.

10. 2^4 **11.** 3^3 **12.** 4^2 **13.** 5^2 **14.** 6^3 **15.** 10^3 **16.** 10^4 **17.** 1^{20}

18. 1^{17} **19.** 2^5 **20.** 2^6 **21.** 4^3 **22.** 3^4 **23.** 13^2 **24.** 15^2 **25.** 8^3

26. 7^3 **27.** 5^4 **28.** 4^4 **29.** 7^0 **30.** 9^0 **31.** 2^7 **32.** 25^2 **33.** 20^3

34. 10^6 **35.** 8^1 **36.** 4^5 **37.** 10^5 **38.** 9^1 **39.** 14^2

Round to the nearest ten.

40. 45   **41.** 83   **42.** 57   **43.** 65   **44.** 92

Round to the nearest hundred.

45. 124   **46.** 826   **47.** 1258   **48.** 2781   **49.** 1643

Round to the nearest thousand.

50. 7621   **51.** 3754   **52.** 18,463   **53.** 14,458   **54.** 27,863   **55.** 94,671

Solve each problem.

56. About 177,000,000 people in the world speak Arabic. Round this figure to the nearest ten million.

57. It is estimated that 788,000,000 people in the world speak Mandarin Chinese. Round this figure to the nearest ten million.

58. In a recent year 38,586 bachelor's degrees were conferred in the field of communications. Round this value to the nearest ten thousand.

59. In a recent year 93,212 bachelor's degrees were conferred in the field of theology. Round this value to the nearest ten thousand.

60. Recently, the total enrollment of children in elementary and secondary schools in Illinois was 349,463. Round this enrollment figure to
(a) The nearest thousand.
(b) The nearest hundred.

61. Recently, the total enrollment of children in elementary and secondary schools in Ohio was 268,357. Round this enrollment figure to
(a) The nearest thousand.
(b) The nearest hundred.

62. Recently, it was estimated that there are 11,182,000 licensed drivers in Texas. Round this figure to
(a) The nearest hundred thousand.
(b) The nearest ten thousand.

63. Recently, it was estimated that there are 17,235,000 licensed drivers in California.
(a) Round this figure to the nearest hundred thousand.
(b) Round this figure to the nearest ten thousand.

 To Think About

64. Evaluate: 2^{20}.

65. Evaluate: 3^{12}.

Cumulative Review Problems

66. Write in words: 1,562,384,000,000.

67. Write in expanded notation: 76,325.

68. Add: $12 + 18 + 64 + 97$.

69. Multiply: $\begin{array}{r} 346 \\ \times\ 28 \\ \hline \end{array}$

70. Divide and check $18\overline{)3764}$. your answer:

 Calculator Problems

Find the value of each expression.

71. 36^3

72. 255^3

For Extra Practice Examples and Exercises, turn to page 61.

Solutions to Odd-Numbered Practice Problems

1. (a) $12 \times 12 \times 12 \times 12 = 12^4$. 12 is the base, 4 is the exponent.
  (b) $2 \times 2 \times 2 \times 2 \times 2 \times 2 = 2^6$. 2 is the base, 6 is the exponent.

3.

6 $\overset{\downarrow}{5}$,5 2 8   Locate the thousands round-off place.

6 $\overset{\downarrow}{5}$,⑤2 8   The first digit to the right is 5 or more. We will increase the thousands digit by 1.

6 6,0 0 0   All digits to the right of thousands are replaced by zero.

5. (a) $53,\overset{\downarrow}{2}82 = 53,280$ to nearest ten. The digit to the right of tens was less than 5.

  (b) $164,\overset{\downarrow}{4}85 = 164,000$ to nearest thousand. The digit to the right of thousands was less than 5.

  (c) $1,\overset{\downarrow}{3}65,273 = 1,400,000$ to nearest hundred thousand. The digit to the right of hundred thousands was greater than 5.
7. $9,460,000,000,000,000 = 9,500,000,000,000,000$ to nearest hundred trillion.

Answer to Even-Numbered Practice Problems

2. (a) 144 **(b)** 216 **(c)** 64 **(d)** 1   **4.** 170,000   **6. (a)** 10,500,000 **(b)** 10,000,000

1.7 ORDER OF OPERATIONS

☐ After studying this section, you will be able to:

1 *Perform in the proper order several arithmetic operations*

If you want to get a certain result and you want to be sure that others who are doing the same task get the same result, you agree on an *order of operations*. Sometimes the order of a set of operations makes little difference. Consider the chefs at the Gourmet Restaurant. The order in which they sharpen their knives and hang their pots and pans probably doesn't matter. But the order in which they add and mix the elements in preparing the food makes all the difference in the world. They may follow a recipe, which is an order of operations for cooking. Following the recipe gives the cooks—and cooks at other restaurants who use the same recipe—consistent results.

In mathematics the order of operations is a list of priorities for working with the numbers in computational problems. This mathematical "recipe" tells how to handle certain indefinite computations. For example, how does a person solve $5 + 3 \times 2$?

1 A problem such as $5 + 3 \times 2$ sometimes causes students difficulty. Some people think $(5 + 3) \times 2 = 8 \times 2 = 16$. Some people think $5 + (3 \times 2) = 5 + 6 = 11$. Only one answer is right—11. To obtain the right answer, we first simplify any expressions with exponents, next multiply and divide from left to right, then add and subtract. The following order is important to learn:

Order of Operations

In the absence of grouping symbols:

Do first   **1.** Perform operations inside parentheses.

↑   **2.** Simplify any expressions with exponents.

↓   **3.** Multiply or divide from left to right.

Do last   **4.** Add or subtract from left to right.

Example 1   Evaluate: $3^2 + 5 - 4$.

$$3^2 + 5 - 4 = 3 \times 3 + 5 - 4 \qquad \text{Simplifying expression with exponents.}$$
$$= 9 + 5 - 4 \qquad \text{Multiplying from left to right.}$$
$$= 14 - 4 \qquad \text{Adding from left to right.}$$
$$= 10 \qquad \text{Subtracting.} \quad \blacksquare$$

Practice Problem 1   Evaluate: $7 \times 3 + 4^3$.   ■

Example 2   Evaluate: $5 + 3 \times 6 - 4 + 12 \div 3$.

There are no numbers to raise to a power, so we first go from *left to right* doing any multiplication or division.

$$5 + 3 \times 6 - 4 + 12 \div 3 = 5 + 18 - 4 + 12 \div 3 \qquad \text{Multiplying.}$$
$$= 5 + 18 - 4 + 4 \qquad \text{Dividing.}$$
$$= 23 - 4 + 4 \qquad \text{Adding.}$$
$$= 19 + 4 \qquad \text{Subtracting.}$$
$$= 23 \qquad \text{Adding.} \quad \blacksquare$$

Practice Problem 2   Evaluate: $7 - 3 \times 4 \div 2 + 5$.   ■

Example 3   Evaluate: $2^3 + 3^2 - 7 \times 2$.

$$= 2 \times 2 \times 2 + 3 \times 3 - 7 \times 2 \qquad \text{Evaluating exponent expressions } 2^3 = 8 \text{ and } 3^2 = 9.$$
$$= 8 + 9 - 7 \times 2$$
$$= 8 + 9 - 14 \qquad \text{Multiplying.}$$
$$= 17 - 14 \qquad \text{Adding.}$$
$$= 3 \qquad \text{Subtracting.} \quad \blacksquare$$

Practice Problem 3   Evaluate: $4^3 - 2 + 3^2$.   ■

Example 4   Evaluate: $2 \times (7 + 5) \div 4 + 3 - 6$.

   First, we combine numbers inside the parentheses by adding the 7 to the 5. Multiplication and division have equal priority, so we work from left to right doing whichever of these operations comes first.

$$= 2 \times 12 \div 4 + 3 - 6$$

$$= 24 \div 4 + 3 - 6 \qquad \text{Multiplying.}$$

$$= 6 + 3 - 6 \qquad \text{Dividing.}$$

$$= 9 - 6 \qquad \text{Adding.}$$

$$= 3 \qquad \text{Subtracting.} \quad ■$$

Practice Problem 4   Evaluate: $(17 + 7) \div 6 \times 2 + 7 \times 3 - 4$.   ■

Example 5   Evaluate: $4^3 + 18 \div 3 - 2^4 - 3 \times (8 - 6)$.

$$= 4^3 + 18 \div 3 - 2^4 - 3 \times 2 \qquad \text{Combining inside the parentheses.}$$

$$= 64 + 18 \div 3 - 16 - 3 \times 2 \qquad \text{Evaluating exponents.}$$

$$= 64 + 6 - 16 - 3 \times 2 \qquad \text{Dividing.}$$

$$= 64 + 6 - 16 - 6 \qquad \text{Multiplying.}$$

$$= 70 - 16 - 6 \qquad \text{Adding.}$$

$$= 54 - 6 \qquad \text{Subtracting.}$$

$$= 48 \qquad \text{Subtracting.} \quad ■$$

Practice Problem 5   Evaluate: $5^2 - 6 \div 2 + 3^4 + 7 \times (12 - 10)$.   ■

To Think About

Why is this rigid order so important? Where in life does this order of operations really make a difference in people's lives? Business, science, and technology applications can often yield problems like $156{,}320 \times 72{,}158 \times 2^3 + 2{,}340{,}561$. Often these are done on a computer or a scientific calculator. In such cases the computer or calculator does these operations in a specific order. The order used is the same order of operations discussed in this section of the book. See problems 31 and 32.   ■

EXERCISES 1.7

Work each problem, using the correct order of operations.

1. $5 \times 6 - 3 + 10$
2. $8 \times 3 - 4 \times 2$
3. $2 + 6 \times 12 \div 2$
4. $6 \times 7 - 4 \div 2$

5. $100 \div 5^2 + 3$
6. $24 \div 2^2 + 4$
7. $7 \times 3^2 + 4 - 8$
8. $2^3 \times 4 + 6 - 9$

9. $3 \times 12^2 - 4 \times (8 - 5)$
10. $5 \times 2^6 - 2 \times (12 - 7)$

11. $20 \div 4 \times 3 - 6 + 2 \times (1 + 4)$

12. $50 \div 10 \times 2 - 8 + 3 \times (17 - 11)$

13. $2^3 + 3^4 + 5^2$

14. $4^3 + 3^2 + 2^5$

15. $100 \div 5 \times 3 \times 2 \div 2$

16. $36 \div 4 \times 5 \times 6 \div 6$

17. $12^2 - 6 \times 3 \times 4 \times 0$

18. $14^2 - 5 \times 2 \times 3 \times 0$

19. $8^2 \times 6 \div 4$

20. $5^3 \times 2 \div 10$

21. $5 + 6 \times 2 \div 4 - 1$

22. $7 + 5 \times 4 \div 10 - 2$

23. $8 + 2^3 \times 5 + 3$

24. $7 \times 4^3 \times 6 + 1$

25. $3 \times 5 + 7 \times (8 - 5) - 5 \times 4$

26. $(5 + 3) \times 9 + 6 \times 5 - 6 \times 6$

27. $12 \div 2 \times 3 - 2^4$

28. $30 \div 10 \times 5 - 3^2$

29. $3^2 \times 6 \div 9 + 4 \times 3$

30. $5^2 \times 3 \div 25 + 7 \times 6$

 To Think About

31. Suppose that a computer operator wants to solve $2000 \times 7400 \times 6^3 + 160,000$, but she does not follow our order of operations. She first multiplies 6^3 to get 216, then she multiplies 216×7400 to get, 1,598,400. Then she multiplies $1,598,400 \times 2000$ to get 3,196,800,000. She then adds 160,000. Would she get the right answer? Why or why not?

32. Suppose that a math student wants to solve $12 \times 5 + 3 \times 5 + 7 \times 5$, and she adds $12 + 3 + 7$ to get 22 first. She then multiplies 22×5 to get an answer. Is her answer correct? Why or why not? What happens if you follow the order of operations of this section in working this problem?

 Calculator Problems

33. $2^6 + 4^5 + 3^4 + 12^2 \div 4^2 \div 9 \times 5^2 \times 4^3$

34. $16 \times 15 \times 14 \times 13 \times 12 \times 11 \times 10 \times 9 \times 8$
$\qquad - 15 \times 14 \times 13 \times 12 \times 11 \times 10 \times 9 \times 8$
(*Hint:* Look for a pattern in each part of problem 34.)

Cumulative Review Problems

35. Write in expanded notation: 156,312.

36. Write in symbols: two hundred million, seven hundred sixty-five thousand, nine hundred nine.

37. Write in words: 261,763,002.

38. Add: 1563
2736
381
25
+ 9823

39. Subtract: 16,093
− 14,937

40. Multiply: 5003
            $\times$ 126

41. Divide: $18\overline{)12510}$.

42. Evaluate: 4^4.

43. The earth actually rotates once every 86,164 seconds. Round this figure to the nearest ten thousand seconds.

 Calculator Problems

44. $(135)^2 - 2525 \times 5 + 25{,}876$

45. $(66)^3 \div 6 + 132{,}684 \times 2^2$

For Extra Practice Examples and Exercises, turn to page 61.

Solutions to Odd-Numbered Practice Problems

1. $7 \times 3 + 4^3$

$\begin{aligned}
&= 7 \times 3 + 4 \times 4 \times 4 && \text{Evaluating exponents.}\\
&= 7 \times 3 + 64 && 4^3 = 64.\\
&= 21 + 64 && \text{Multiplying.}\\
&= 85 && \text{Adding.}
\end{aligned}$

3. $4^3 - 2 + 3^2$

$\begin{aligned}
&= 4 \times 4 \times 4 - 2 + 3 \times 3 && \text{Evaluating exponents.}\\
&= 64 - 2 + 9 && 4^3 = 64 \text{ and } 3^2 = 9.\\
&= 62 + 9 && \text{Subtracting.}\\
&= 71 && \text{Adding.}
\end{aligned}$

5. $5^2 - 6 \div 2 + 3^4 + 7 \times 2$

$\begin{aligned}
&= 25 - 6 \div 2 + 81 + 7 \times 2 && 5^2 = 25 \text{ and } 3^4 = 81.\\
&= 25 - 3 + 81 + 14 && 6 \div 2 = 3 \text{ and } 7 \times 2 = 14.\\
&= 22 + 81 + 14 && \text{Subtracting.}\\
&= 117 && \text{Adding.}
\end{aligned}$

Combining inside the parentheses.

Answers to Even-Numbered Practice Problems

2. 6   **4.** 25

☐ After studying this section, you will be able to:

1 *Solve applied problems involving one type of operation*

2 *Solve applied problems in volving more than one type of operation*

1.8 APPLIED PROBLEMS

1 When solving an applied problem, the following steps are helpful.

> **Steps to Solving an Applied Problem**
>
> **1.** Read over the problem carefully. Find out what is asked for. Draw a picture if it will help you to visualize the situation.
> **2.** Write down which whole numbers are to be used in solving the problem.
> **3.** Estimate the answer.
> **4.** Determine what operation needs to be done: addition, subtraction, multiplication, division, or a combination of these.
> **5.** Perform the necessary computations. State the answer, including the unit of measure.
> **6.** Compare your answer with the estimate. See if it is reasonable.

Example 1 Gerald made deposits of $317, $512, $84, and $161 into his checking account. What was the total of his deposits?

1. We are trying to find the total of Gerald's four deposits.
2. The deposits are $317 $512 $84 $161 .
3. Estimate. Here we make a reasonable guess. We try to make a simpler problem. Here we estimate by rounding.

$$
\begin{array}{cccc}
317 \longrightarrow & \text{Round each} & \longrightarrow 300 & \text{We add} \\
512 \longrightarrow & \text{deposit to} & \longrightarrow 500 & \text{up our} \\
84 \longrightarrow & \text{nearest} & \longrightarrow 100 & \text{rounded} \\
161 \longrightarrow & \text{hundred.} & \longrightarrow 200 & \text{values.} \\
& & \overline{1100} &
\end{array}
$$

Our estimate is $1100.

4. We need to *add* to find that sum of the deposits.

5.
$$
\begin{array}{r}
\overset{11}{317} \\
512 \\
84 \\
+\,161 \\
\hline
1074
\end{array}
$$

The total of the four deposits is $1074.

6. $1074 is close to our estimated answer, $1100. Our answer is reasonable. ✓ ■

Practice Problem 1 The hospital distributes aspirin to patients in pain from several aspirin bottles. One bottle contains 340 tablets, another 161 tablets, another 87 tablets, another 477 tablets. If these were combined into one bottle, how many tablets would it contain? ■

An odometer in a car records its mileage. It is usually located on the car's speedometer.

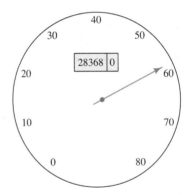

Example 2 Theofilos began a trip with an odometer reading of 28,368 miles. When he ended the trip the odometer read 30,162 miles. How many miles did he travel?

1. We want to find the number of miles traveled.
2. We have his final odometer reading → 30,162 and his initial odometer reading → 28,368.
3. We estimate the number of miles traveled. We round the values: 30 thousand minus 28 thousand equals 2 thousand miles.

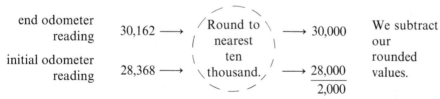

Our estimate is 2000 miles.

4. We need to subtract them to find the difference in the number of miles.

5.
$$
\begin{array}{r}
3\,0\,,1\;6\;2 \\
-\,2\;8\,,3\;6\;8 \\
\hline
1\,,7\;9\;4
\end{array}
$$

The trip totaled 1794 miles.

6. We compare this answer with our estimate. Our answer in reasonable. ✓ ■

Practice Problem 2 In 1984, 2,678,559 people in Ohio voted for Reagan. The same year 1,825,440 voted for Mondale. By how many votes did the Republican candidate beat the Democratic candidate that year in Ohio? ■

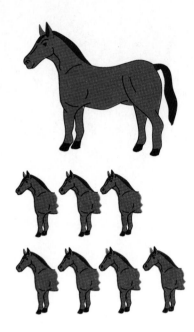

Example 3 One horsepower is the power to lift 550 pounds a distance of a foot in 1 second. How many pounds can be lifted 1 foot in 1 second by 7 horsepower?

1. We want to find the number of pounds lifted 1 foot in 1 second by 7 horsepower.
2. Each horsepower can lift 550 pounds.
3. We estimate our answer. We round 550 to 600 pounds: $600 \times 7 = 4200$ pounds. Our estimate is 4200 pounds.
4. To solve the problem, we *multiply* the 7 horsepower by 550 pounds for each horsepower.

$$550 \qquad 550 \qquad 550 \qquad 550 \qquad 550 \qquad 550 \qquad 550$$

5. $$\begin{array}{r} \overset{3}{550} \\ \times \quad 7 \\ \hline 3850 \end{array}$$

7 horsepower moves 3850 pounds 1 foot in 1 second.

6. Is this reasonable? This answer is close to our estimate. Our answer is reasonable. ✓ ■

Practice Problem 3 In a measure of liquid capacity, 1 gallon = 1024 fluid drams. How many fluid drams would be in 9 gallons? ■

Example 4 Laura can type 35 words per minute. She has to type an English theme that has 5180 words. How many minutes will it take her to type the theme?

1. We want to find how long it will take Laura to type her theme.
2. The theme has *5180 words*; she can type 35 words per minute.
3. Estimate: 5180 words is approximately 5000 words.

5180 words $\longrightarrow$ 5000 words rounded to nearest thousand

35 words per minute $\longrightarrow$ 40 words per minute rounded to nearest ten

$$\begin{array}{r} 125 \\ 40 \overline{)\,5000} \end{array}$$
We divide our estimated values

Our estimate is 125 minutes.

4. A picture may help. If we *divide* the total number of words by units of 35 words, we can get the number of minutes needed. Each "package of 1 minute is 35 words." We want to know how many packages make up 5180 words.

35 words in 1 minute	35 words in 1 minute	35 words in 1 minute	
5180 words			

5. Division

$$\begin{array}{r} 148 \\ 35 \overline{)\,5180} \\ \underline{35} \\ 168 \\ \underline{149} \\ 280 \\ \underline{280} \end{array}$$

She can type the paper in 148 minutes. Since 60 minutes = 1 hour and 120 min = 2 hours,

$$\begin{array}{r} 148 \\ -120 \\ \hline 28 \end{array}\quad\begin{array}{l} 148\ \text{minutes} \\ 2\ \text{hours} \\ 28\ \text{minutes left} \end{array}$$

An alternative answer is 2 hours, 28 minutes.

6. This is close to our estimated answer. Our answer is reasonable. ✓ ■

Practice Problem 4 Donna bought 45 shares of stock for $1620. How much did it cost her per share? ■

2 Applied Problems That Involve More Than One Calculation

Example 5 Cleanway Rent-A-Car bought four luxury sedans at $21,000, three compact sedans at $14,000, and seven subcompact sedans at $8000. What was the total cost of the purchase?

1. We need to add up the total cost of buying luxury sedans, compact sedans, and subcompact sedans.

2. We organize the data by groups.

Quantity	Type of Car	Cost per Car
4	Luxury sedans	$21,000
3	Compact sedans	14,000
7	Subcompact sedans	8,000

3. Estimate: Round car cost to nearest $10,000.

$$\begin{array}{rcl} 4 \times 20,000 & = & 80,000 \\ 3 \times 10,000 & = & 30,000 \\ 7 \times 10,000 & = & +\ 70,000 \\ \hline & & \$180,000 \end{array}$$

4. First we need to *multiply* to get the cost per car.

$$4 \times 21,000 = \text{cost of luxury sedans}$$

$$3 \times 14,000 = \text{cost of compact sedans}$$

$$7 \times 8000\ \ = \text{cost of subcompact sedans}$$

Then we need to add to get the total cost.

5. It may be helpful to visualize an imaginary bill of sale.

Car Fleet Sales, Inc. Hamilton, Massachusetts			
Customer: Cleanway Rent-A-Car			Amount for This Type of Car
Quantity	Type of Car	Cost per Car	
4	Luxury Sedans	$21,000	$ 84,000 (4 x $21,000 = $84,000)
3	Compact Sedans	$14,000	$ 42,000 (3 x $14,000 = $42,000)
7	Subcompact Sedans	$ 8,000	$ 56,000 (7 x $8,000 = $56,000)
		TOTAL	$182,000 (Sum of the three Amounts)

Thus the total is $182,000.

6. Our calculated sum is close to the estimated answer. The answer is reasonable. ✓ ■

Practice Problem 5 Anderson Dining Commons purchased 50 tables at $200 each, 180 chairs at $40 each, and six moving carts at $65 each. What was the cost of the total purchase? ■

Example 6 Dawn had a balance of $410 in her checking account last month. She made deposits of $46, $18, $150, $379, and $22. She made out checks for $316, $400, and $89. What is her new balance?

1. We want to obtain the number of dollars in the new balance.

2.

Old Balance	Deposits	Checks
$410	$ 46	$316
	18	400
	150	89
	379	
	22	

3. We estimate. If we round all values to nearest hundred:

Old Balance	Deposits	Checks
410 ⟶ 400	$ 46 ⟶ 0	$316 ⟶ 300
	18 ⟶ 0	400 ⟶ 400
	150 ⟶ 200	89 ⟶ 100
	379 ⟶ 400	
	22 ⟶ 0	
400	600	800

= $1000 − $800 = $200   estimated

4. We want to *add* to get a total of all deposits and *add* to get a total of all checks.

5. ┌──────────────┐   ┌──────────────────┐   ┌────────────────┐   ┌──────────────┐
   │ Old balance │ + │ total of deposits │ − │ total of checks │ = │ new balance │
   └──────────────┘   └──────────────────┘   └────────────────┘   └──────────────┘

Total sum of deposits   46   Total sum of checks   316
                         18   400
                        150   + 89
                        379                              $805
                     + 22
                      $615

Finally,

Old balance	410
+ total deposits	+ 615
	1025
− total checks	− 805
New balance	220

The new balance of checking account is $220.00.

6. This is close to our estimated answer. Our answer is reasonable. ✓   ■

Practice Problem 6 Last month Bridget had $498 in a savings account. She made two deposits: one for $607 and one for $163. The bank credited her with $36 interest. Since last month she made four withdrawals: $19, $158, $582, and $74. What is her balance this month? ■

Example 7   When Lorenzo took a car trip, his gas tank was full when the odometer read 76,358 miles. He ended his trip at 76,668 miles and filled the gas tank with 10 gallons of gas. How many miles per gallon does he get with his car?

1. We need to find the number of miles driven and then calculate how many miles per gallon he obtains.

2. Basic facts:

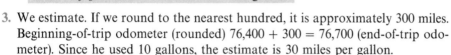

End-of-trip odometer reading:	76,668
Beginning-of-trip odometer reading:	76,358
Gallons of gas used:	10 gallons

3. We estimate. If we round to the nearest hundred, it is approximately 300 miles. Beginning-of-trip odometer (rounded) 76,400 + 300 = 76,700 (end-of-trip odometer). Since he used 10 gallons, the estimate is 30 miles per gallon.

4. We need to subtract odometer readings to obtain miles driven and then divide by the number of gallons used.

5.
$$\begin{array}{r} 76,668 \\ -\ 76,358 \\ \hline 310 \end{array}$$

$$\begin{array}{r} 31 \\ 10\overline{)310} \end{array}$$

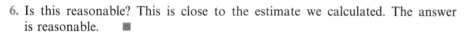

$$\frac{310 \text{ miles}}{10 \text{ gallons}} = 31 \text{ miles per gallon}$$

6. Is this reasonable? This is close to the estimate we calculated. The answer is reasonable. ■

Practice Problem 7   Deidre took a car trip with a full tank of gas. Her trip began with the odometer at 50,698 and ended at 51,118 miles. She then filled the tank with 12 gallons of gas. How many miles per gallon did her car get on the trip? ■

DEVELOPING YOUR STUDY SKILLS

EXAM TIME: HOW TO REVIEW

Reviewing adequately for an exam enables you to bring together the concepts you have learned over several sections. For your review, you will need to:

1. Reread your textbook. Make a list of any terms, rules, or formulas you need to know for the exam. Be sure you understand them all.
2. Reread your notes. Go over returned homework and quizzes. Redo the problems you missed.
3. Practice some of each type of problem covered in the chapter(s) you are to be tested on.
4. Use the end-of-chapter materials provided in your textbook. Read carefully through the Chapter Organizer. Do the Extra Practice Sections. Take the Practice Quizzes. When you are finished, check your answers. Redo any problems you missed.
5. Get help if any concepts give you difficulty.

EXERCISES 1.8

Solve each applied problem. In each case, only one type of operation is required.

1. The university has 12,845 students, 624 faculty, and 826 staff members. What is the total of these three groups?

2. This semester Robert Chu made deposits of $412, $363, $1451, and $86 into his checking account. What were his total deposits this semester?

3. When Melissa Sullivan began her trip the car odometer read 54,360 miles. When she finished the trip it read 54,932 miles. How many miles did she travel?

4. An airplane was flying at an altitude of 4760 feet. It had to clear a mountain of 4356 feet. By how many feet did it clear the mountain?

5. There are 500 sheets in a ream of paper. Dr. Elizabeth Jones ordered 12 reams of paper. How many sheets did she order?

6. There are 144 pencils in a gross. Mr. Jim Weston ordered 14 gross of pencils for the office. How many pencils did he order?

7. A 16-ounce can of beets cost 96¢. What is the unit cost of the beets? (How much did they cost per ounce?)

8. A 14-ounce can of soup cost 98¢. What is the unit cost of the soup? (How much does the soup cost per ounce?)

9. Shirley planned to bicycle a total distance of 1645 miles this summer. The first day Shirley bicycled from Overtown to Lost Woods, a distance of 126 miles. How many miles does she yet have to bicycle?

10. Dr. Smithson bought 425 shares of stock for $6800. What was her cost per share?

11. Carlos can type 28 words per minute. His history essay is 4480 words long. How many minutes will it take him to type the essay?

12. There are 44,010 square feet in 1 acre. A man owns 7 acres of woodland. How many square feet of land does he own?

13. Michelle's new watch has a crystal that vibrates exactly 32,768 times in 1 second. How many times does it vibrate in 2 minutes? (60 seconds = 1 minute.)

14. Roberto had $2158 in his savings account six months ago. In the last six months he made four deposits: $156, $238, $1119, and $866. The bank deposited $136 in interest over the six-month period. How much does he have in his savings account at present?

15. A Minnesota homeowner is determining how to decrease electricity use in heating his home this winter. He has estimated that new storm windows would save 3487 kilowatt-hours, attic insulation would save 9342 kilowatt-hours, a more efficient hot-water heater would save 4612 killowatt-hours, and lower-wattage light bulbs would save 161 kilowatt-hours. How many kilowatt-hours will he save in total?

16. When Jason began a trip his car odometer read 38,125 miles. When he finished the trip it read 39,602 miles. How many miles did he travel?

Solve each applied problem. In each case, more than one type of operation is required.

17. Lydia bought three blouses at $34 each, two pairs of slacks at $46 each, and three pairs of shoes at $41 each. What was the total cost of her shopping spree?

18. Hector bought three shirts at $26 each, two pairs of pants at $45 each, and three pairs of shoes at $48 each. What was the total cost of his shopping spree?

19. Angela had a balance in her checking account of $212. She then put in deposits of $128, $100, and $67. She also made out four checks, for $29, $74, $16, and $203. When all the deposits are recorded and all the checks clear, what balance will she have in her checking account?

20. Yoshiko had a balance in her checking account of $387. She then put in deposits of $116, $34, $20, and $215. She also made out three checks, for $22, $112, and $277. When all the deposits are recorded and all the checks clear, what balance will she have in her checking account?

21. Christina owns 120 acres of farmland. She rents it to a farmer for $160 per acre per year. Her property taxes are $43 per acre per year. How much profit does she make on the land each year?

22. Todd owns 13 acres of commercially zoned land in the city. He rents it to a construction company for $12,350 per acre per year. His property taxes to the city are $7362 per acre per year. How much profit does he make on the land each year?

23. Nancy wants to determine the miles per gallon rating of her car. She filled the tank when the odometer read 14,926 miles. She then drove the car for a trip. At the end of the trip the odometer read 15,276 miles. It took 14 gallons to fill the tank. How many miles per gallon does her car deliver?

24. Garcia wants to determine the miles per gallon rating of his car. He filled the tank when the odometer read 36,339 miles. He drove the car for a week. The odometer then read 36,781 miles and the tank required 17 gallons to be filled. How many miles per gallon did Garcia's car achieve?

 To Think About

In solving problems 25 and 26, consider the following price list.

Eleazor's Electronics Discount Prices

21-in. color TV sets	$300	Compact disc players	$264
16-in. color TV sets	160	Stereo radio/tape players	89
9-in. color TV sets	115	Portable refrigerators	78

25. Old South Dorm bought from Eleazor's Electronics three 21-inch color television sets, eight 9-inch color television sets, four compact disc players, and 10 portable refrigerators. The cost of these purchases was divided equally among the 457 students in Old South Dorm. How much did each student contribute?

26. Old North Dorm bought from Eleazor's Electronics five 16-inch color television sets, three 9-inch color television sets, 20 stereo radio/tape players, and five portable refrigerators. The cost of these purchases was divided equally among the 221 students in Old North Dorm. How much did each student contribute?

Cumulative Review Problems

27. Evaluate: 7^3.

28. Perform in the proper order: $3 \times 2^3 + 15 \div 3 - 4 \times 2$.

For Extra Practice Examples and Exercises, turn to page 61.

Solutions to Odd-Numbered Practice Problems

1.
```
   340
   161
    87
+ 477
```
 1065   Total of 1065 tablets from the four bottles

3. (1024 drams in 1 gallon) × (9 gallons) =
```
  1024
×    9
  9216
```
9216 fluid drams in 9 gallons

5. 50 tables at $200 each $= 50 \times 200 = $10,000$ cost of tables
180 chairs at $40 each $= 180 \times 40 = $ 7,200   cost of chairs
6 carts at $65 each 6×65 390   cost of carts
                                        $=$ $17,590   total cost

The total purchase was $17,590.

7.   51,118   odometer at end of trip
    $-\ 50,698$ odometer at start of trip
      420   miles traveled on trip

$$\frac{420 \text{ miles}}{12 \text{ gallons used}} = 12\overline{)420} = 35 \text{ miles per gallon}$$

35 on the trip
36
60
60

Answers to Even-Numbered Practice Problems

2. 853,119   **4.** $36   **6.** $471

EXTRA PRACTICE: EXAMPLES AND EXERCISES

Section 1.1

Write in standard notation.

Example $40,000 + 3000 + 30 + 9$
    Note that there are 0 hundreds.
            43,039

1. $30,000 + 7000 + 40 + 3$
2. $2000 + 600 + 50 + 6$
3. $20,000 + 7000 + 900 + 40$
4. $60,000 + 70 + 9$
5. $800,000 + 30,000 + 400 + 20 + 6$

Write the word name for each number.

Example   **(a)** 2481   **(b)** 505,960
(a) 2481 = two thousand, four hundred eighty-one
(b) 505,960 = five hundred five thousand, nine hundred sixty

6. 789
7. 8921
8. 902,378
9. 32,508,254
10. 570,234,713

Write each number in standard notation.

Example   Two hundred million, sixty-eight thousand, five hundred twenty-three

         200,068,523

11. Two hundred fifty-one
12. Seven hundred thirty
13. Thirty-six thousand, three hundred twenty
14. Twenty-four thousand, three hundred six
15. Five hundred million, twenty-three thousand, six hundred fifty-one

Specify the digit.

Example   A library in a large city may have as many as 1,026,270 books.
(a) What digit tells the number of hundred thousands?
(b) What digit tells the number of hundreds?
(a) There are 0 hundred thousands.
(b) There are two hundreds.

16. The population of a small town is 97,606. What digit tells the number of hundreds?
17. The average home in Orange County, California, costs about $250,000. What digit tells the number of thousands?
18. The sun is approximately 93,450,000 miles from the earth. What digit tells the number of millions?
19. There are 3,484,800 inches in 55 miles. What digit tells the number of ten thousands?
20. If you invest $10,000 in a 30-year retirement account that pays 12% compound interest per year, you will have approximately $299,599 when you retire. What digit tells the number of tens?

Section 1.2

Add.

Example $2 + 7 + 4 + 3$

   2
   7    $2 + 7 = 9$
$+\ 4$ $9 + 4 = 13$
   3              $13 + 3 = 16$
  16

Line the numbers vertically, then mentally add two numbers at a time.
Line up the ones digit of the answer with the ones digit of the numbers that are being added.

1. $6 + 8 + 1 + 0$
2. $3 + 7 + 6 + 1$
3. $21 + 12 + 13$
4. $16 + 31 + 12$
5. $14 + 31 + 11$

Add the following from the top. Then check by adding in the reverse order.

Example $132 + 11 + 121$

   132   *Step 1:* List numbers vertically. Line up the
    11   ones column.
$+\ 121$ *Step 2:* Add the ones column first, then the
  264   tens column, and last the hundreds
                        column. Add from the top down.
   121   *Step 3:* Reverse the order and add again
    11   to check.
$+\ 132$
  264

6. $341 + 111 + 212$

7. $41 + 3 + 214$

8. $411 + 312 + 105$

9.
```
  1231
  2135
+  201
```

10.
```
  6011
    37
+ 3541
```

Example $567 + 34 + 2 + 78$

```
  12
 567
  34
   2
+ 78
 ___
 681
```

11. $4581 + 367 + 953$

12. $32,610 + 72 + 9865 + 230$

13. $25,073 + 702 + 21 + 17,519$

14.
```
  2,287,004
    985,608
+     2,021
```

15.
```
  4,766,501
    159,589
+       548
```

Answer each question.

Example The Walden Furniture Company sold 1231 bedroom sets and 932 living room sets in January. In February, 956 bedroom sets sold and 752 living room sets sold.
(a) What was the total number of bedroom sets sold?
(b) What was the total number of living room sets sold?

(a) The number of bedroom sets:
```
  1231
+  956
 _____
  2187
```

(b) The number of living room sets:
```
   932
 + 752
 _____
  1684
```

16. Joe read 102 pages in his literature book on Monday, 93 pages on Tuesday, and 113 pages on Wednesday. What was the total number of pages read?

17. Wendy paid $3562 in federal income tax in 1985. In 1986 she paid $4102 in taxes. What was the total of the taxes paid in 1985 and 1986?

18. Jerry's Hamburger House sold 470 hamburgers and 342 french fries in June. In July, 657 hamburgers and 541 french fries were sold.
(a) What was the total number of hamburgers sold?
(b) What was the total number of french fries sold?

19. Macormick Department Store sold $2,547,238 in merchandise and paid $1,345,908 in wages in 1980. In 1981 the store sold $3,973,005 in merchandise and paid $2,872,650 in wages.
(a) What was the total in wages paid for the two years?
(b) What was the total dollar amount of merchandise sold for both years?

20. Mr. Sanson made the following transactions at his bank:

	Deposits	Withdrawals
June	$3894	$254, $342
July	2009	$579, $923, $231
August	4198	$2167, $961

(a) What was the total of the deposits for the three-month period?
(b) What was the total of the withdrawals for the three-month period?

Section 1.3

Subtract.

Example $9743 - 422$

```
  9743
-  422
 _____
  9321
```

Note: Because there is no number under the 9 in the thousands column, the number 0 is subtracted.

1.
```
  8753
- 2411
```

2.
```
  97,538
-  4,321
```

3.
```
  4287
- 1157
```

4.
```
  6892
-   71
```

5.
```
  2698
-  256
```

Example $4571 - 2345$

```
      6 11
  4 5 7̶ 1̶
- 2 3 4 5
  _____
  2 2 2 6
```

6.
```
  5631
- 1327
```

7.
```
  8157
- 5029
```

8.
```
  4725
- 1517
```

9.
```
  5477
-  418
```

10.
```
  9742
-   25
```

Example $8002 - 5293$

```
  7 9 9 12
  8̶ 0̶ 0̶ 2̶     When borrowing from a zero, change
- 5 2 9 3     the zero to 9 and borrow from the
  _____     whole number to the left of the zero.
  2 7 0 9
```

11.
```
  7403
-  125
```

12.
```
  3902
-  373
```

13.
```
  90,642
- 48,961
```

14.
```
  60,020
-  4,987
```

15.
```
  14,005
-  9,776
```

Subtract. Check your answer.

Example
```
  7853
- 4320
```

Subtraction	Check
7853	4320
− 4320	+ 3533
_____	_____
3533	7853 ⟵ this checks

16.
```
  6498
-  236
```

17.
```
  2398
-  176
```

18.
```
  587,642
-  23,521
```

19.
```
  75,133
-  6,019
```

20.
```
  91,207
- 67,342
```

Answer each question.

Example Judy, Cindy, and Sara work for Kelly's Answering Service. Last week Judy answered 892 phone calls, Cindy 980, and Sara 786. How many more calls did Judy receive than Sara?

$$\begin{array}{r} \overset{8\ \ 12}{8\ \not{9}\ \not{2}} \\ -\ 7\ 8\ 6 \\ \hline 1\ 0\ 6 \end{array}$$ Judy
Sara
Judy received 106 more calls.

Note: We did not need to consider the number of calls Cindy received.

21. Jack, George, and Lindsey ran for class president. Jack received 466 votes, George 572 votes, and Lindsey 398 votes. How many more votes did George get than Jack?

22. The Pacific Computer Company made $3,087,245 in 1986. In 1987 the company made $3,982,651. How much more money did the company make in 1987?

23. Jack lives 684 miles from the coast. Heather lives 386 miles while Bob lives 592 miles from the coast. How much closer to the coast does Heather live than Bob?

24. Mr. Johnson deposited $1087 in his checking account. He wrote two checks, one for $276 and one for $125. How much money does Mr. Johnson have left in his checking account?

25. Sol-U-Tan Manufactoring Company made $1,684,220 in 1988. Their expenses for that year were $804,599. How much money was left after the company paid their expenses?

Section 1.4

Multiply.

Example
$$\begin{array}{r} 3212 \\ \times\ \ \ \ 3 \\ \hline 9636 \end{array}$$

1. $\begin{array}{r} 34 \\ \times\ 2 \\ \hline \end{array}$

2. $\begin{array}{r} 421 \\ \times\ \ 4 \\ \hline \end{array}$

3. $\begin{array}{r} 3132 \\ \times\ \ \ \ 3 \\ \hline \end{array}$

4. $\begin{array}{r} 42{,}312 \\ \times\ \ \ \ \ \ \ 2 \\ \hline \end{array}$

5. $\begin{array}{r} 431{,}123 \\ \times\ \ \ \ \ \ \ \ \ 3 \\ \hline \end{array}$

Example
$$\begin{array}{r} 268 \\ \times\ \ \ 7 \\ \hline 1876 \end{array}$$

6. $\begin{array}{r} 456 \\ \times\ \ 4 \\ \hline \end{array}$

7. $\begin{array}{r} 623 \\ \times\ \ 5 \\ \hline \end{array}$

8. $\begin{array}{r} 2074 \\ \times\ \ \ \ 7 \\ \hline \end{array}$

9. $\begin{array}{r} 324{,}872 \\ \times\ \ \ \ \ \ \ \ 3 \\ \hline \end{array}$

10. $\begin{array}{r} 641{,}093 \\ \times\ \ \ \ \ \ \ \ 8 \\ \hline \end{array}$

Example
$$\begin{array}{r} 324 \\ \times\ \ 12 \\ \hline 648 \\ 324 \\ \hline 3888 \end{array}$$
648 ⟵ 2×324
324 ⟵ 10×324 We do not need to write the zero.

11. $\begin{array}{r} 421 \\ \times\ \ 22 \\ \hline \end{array}$

12. $\begin{array}{r} 532 \\ \times\ \ 31 \\ \hline \end{array}$

13. $\begin{array}{r} 7089 \\ \times\ \ \ \ 42 \\ \hline \end{array}$

14. $\begin{array}{r} 34{,}061 \\ \times\ \ \ \ \ \ 53 \\ \hline \end{array}$

15. $\begin{array}{r} 511{,}332 \\ \times\ \ \ \ \ \ \ \ 16 \\ \hline \end{array}$

Example 3461×203

$$\begin{array}{r} 3461 \\ \times\ \ 203 \\ \hline 10383 \\ 0000 \\ 6922 \\ \hline 702583 \end{array}$$
We can eliminate the row of zeros; be sure that the last digit in 6922 is lined up with the first digit in 203.

Section 1.5

Divide. Check your answer.

Example $44 \div 6$

$$\begin{array}{r} 7\ R\ 2 \\ 6\overline{)\ 44} \\ -\ 42 \\ \hline 2 \end{array}$$

Check: $6 \times 7 = 42 + 2 = 44$
↑
remainder

1. $5\overline{)\ 26}$

2. $3\overline{)\ 17}$

3. $7\overline{)\ 45}$

4. $4\overline{)\ 59}$

5. $6\overline{)\ 52}$

Example $2819 \div 8$

$$\begin{array}{r} 352\ R\ 3 \\ 8\overline{)\ 2819} \\ -\ 24 \\ \hline 41 \\ -\ 40 \\ \hline 19 \\ -\ 16 \\ \hline 3 \end{array}$$

Check:
$$\begin{array}{r} 352 \\ \times\ \ \ 8 \\ \hline 2816 + 3 = 2819 \end{array}$$

6. $6\overline{)\ 2716}$

7. $4\overline{)\ 1571}$

8. $8\overline{)\ 4972}$

9. $7\overline{)\ 1860}$

10. $3\overline{)\ 2359}$

Example $9338 \div 23$

$$\begin{array}{r} 406\ R\ 1 \\ 23\overline{)\ 9339} \\ -\ 92 \\ \hline 139 \\ -\ 138 \\ \hline 1 \end{array}$$
139 ⟵ 23 cannot be divided into 13, so we place a 0 in the quotient and bring down the 9.

11. $31\overline{)\ 7792}$

12. $19\overline{)\ 1289}$

13. $29\overline{)\ 1945}$

14. $34\overline{)\ 7077}$

15. $16\overline{)\ 9688}$

Example $17{,}976 \div 321$

$$\begin{array}{r} 56\ R\ 2 \\ 321\overline{)\ 17978} \\ -\ 1605 \\ \hline 1928 \\ -\ 1926 \\ \hline 2 \end{array}$$
321 does not divide into 179, so we must divide 321 into 1797 and place the 5 in the quotient above the 7 in 1797.

16. $125\overline{)\ 5255}$

17. $367\overline{)\ 8079}$

18. $432\overline{)\ 9072}$

19. $621\overline{)\ 26{,}703}$

20. $212\overline{)\ 12{,}726}$

Write each number in exponent form.

Example $3 \times 3 \times 3 \times 3 \times 3 \times 3 = 3^6$

1. $5 \times 5 \times 5 \times 5$

2. $6 \times 6 \times 6 \times 6 \times 6 \times 6 \times 6$

3. $13 \times 13 \times 13 \times 13 \times 13 \times 13$

4. 45×45

5. 52

Find the value of each expression.

Example

(a) 4^3 **(b)** 11^2 **(c)** 121^0

(a) $4 \times 4 \times 4 = 64$

(b) $11 \times 11 = 121$

(c) 1; any number raised to a zero power equals 1.

6. 5^3

7. 12^2

8. 1^{12}

9. 22^0

10. 3^4

Round to the place indicated.

Example Round to the nearest thousand: 354,566.

355,000   The round-off place is 4.
The number to the right is 5 or more.
Increase the thousands digit by one.
All digits to the right are replaced by 0.

Round to the nearest thousand.

11. 345,687   **12.** 412,987

Round to the nearest ten thousand.

13. 447,976   **14.** 2,037,659   **15.** 3,918,002

Example Round to the nearest million: 53,247,900.

53,000,000   The round-off place is 3.
The number to the right is less than 5.
The millions digit does not change.
All digits to the right are replaced by 0.

Round to the nearest hundred thousand.

16. 29,830,000   **17.** 948,709,640   **18.** 17,810,000

19. 156,807,779   **20.** 89,022,698

Section 1.7

Example Evaluate $9 - 2 \times 3 + 12$

$$9 - 6 + 12$$
$$3 + 12 = 15$$

1. $9 + 5 \times 4 - 3$ **2.** $4 \times 3 - 1 + 6$

3. $12 \div 6 + 5 \times 4$ **4.** $15 - 4 \times 3 + 5$

5. $9 + 8 \div 2 \times 3$

Example Evaluate $5^2 - 7 + 12 \div 2$

$$25 - 7 + 12 \div 2$$
$$25 - 7 + 6$$
$$18 + 6 = 24$$

6. $2^3 - 3 + 15 \div 3$ **7.** $16 + 2 \times 4^2 \div 2$

8. $13 - 2 + 3^2 \times 4$ **9.** $50 \div 5^2 + 6$

10. $6^2 + 4 \times 2^3 - 1$

Example Evaluate $2 \times 3 + 6 \div 2 + 7$

$$6 + 6 \div 2 + 7$$
$$6 + 3 + 7$$
$$9 + 7 = 16$$

Example Evaluate $2 \times 8^2 + 7 - 16 \div 2^3$

$$2 \times 64 + 7 - 16 \div 8$$
$$128 + 7 - 2$$
$$135 - 2 = 133$$

11. $3 \times 2 + 8 \times 4 - 5 \times 2$ **12.** $5 \times 6 + 4 \times 7 - 2 \times 2$

13. $25 \div 5 \times 2 - 3^2$ **14.** $4^2 \times 3 \div 2 + 6 \times 2$

15. $3 \times 2^3 + 1 - 8 \div 2^2$

Section 1.8

Solve each applied problem. More than one type of operation may be required.

Example Ashley had $353 in her savings account. She put in deposits of $134, $23, and $200. The following month she made two withdrawals, one for $93 and one for $128. How much was left in her saving account after her last withdrawal?

Total Sum of Deposits	*Total Sum of Withdrawals*
$134	$ 93
23	+ 128
+ 200	$221
$357	

Old balance + total deposits − total withdrawal
$353 + $357 − $221 = $710 − $221 = $489

1. Mr. Johnson had $567 in his savings account. He put in deposits of $312, $49, and $278. The following week he made two withdrawals, one for $156 and one for $289. How much was left in his savings account after his last withdrawal?

2. Last month Frank had $502 in his savings account. He made two deposits: one for $623 and one for $154. The bank credited him with $41 interest. Since last month he made three withdrawals. They were $45, $267, and $398. What was the balance in his account?

3. Julie had a balance in her checking account of $678. She then put in deposits of $125, $81, $389, and $23. She also made out three checks. The checks were made out for $59, $219, and $36. When all the deposits are recorded and all the checks clear, what balance will she have in her checking account?

4. Ms. Fray had $230 when she started clothes shopping. She spent $39 on a pair of pants, $23 on a sweater, and $31 on a pair of shoes. How much money did she have left after her purchases?

5. Wendy had $59 and Carme had $62. They put their money together so that they could buy a fish tank and three fish. The fish tank cost $89 and the total cost of the three fish was $21. How much money was left after they purchased the fish and the tank?

Example Crickets chirp faster when the temperature is higher. At 20°C a cricket chirps approximately 103 times in 1 minute. At the same temperature, how many times would it chirp in 2 hours?

60 minutes = 1 hour, so we want to know the number of chirps in 120 minutes: $60 \times 2 = 120$.

Number of chirps per minute	times	total number of minutes	=	total chirps
103	×	120	=	12,360

Example Mr. Sanson bought 985 shares of stock for $44,325. What was his cost per share?

$$44,325 \div 985 \longrightarrow 985 \overline{)\begin{array}{r} 45 \\ 44325 \\ -3940 \\ \hline 4925 \\ -4925 \\ \hline 0 \end{array}}$$

6. Mrs. Mangel bought 125 shares of stock. Each share cost $485. How much did Mrs. Mangel pay for the 125 shares of stock?

7. A 12-ounce can of beans cost 84¢. What is the unit cost of the beans? (How much did they cost per ounce?)

8. Jill can drive 26 miles on 1 gallon of gasoline. How many miles can she drive on 16 gallons of gas?

9. An 8-ounce bottle of peanuts cost 72¢. What is the unit cost of the peanuts? (How much do the peanuts cost per ounce?)

10. Mr. Jackson ordered 8500 sheets of paper. There are 500 sheets in a ream of paper. How many reams did he order?

11. Jamie bought 600 shares of stock for $7200. What was her cost per share?

12. There are 8 ounces in 1 cup. Mr. Jones needs 15 cups of flour. How many ounces does he need?

13. Michelle can type 48 words per minute. Her English essay is 6720 words long. How many minutes will it take her to type the essay?

14. There are 5280 feet in 1 mile. Jack runs 8 miles a day. How many feet does he run daily?

15. A watch has a crystal that vibrates 31,968 times in 1 second. How many times does it vibrate in 3 minutes? (*Hint:* 60 seconds = 1 minutes.)

Example Joyce wants to determine the miles per gallon rating of her car. She filled the tank when the odometer read 30,489 miles. She drove the car for four days. The odometer then read 30,841 and the tank required 16 gallons to be filled. How many miles per gallon did Joyce's car achieve?

Odometer reading at the end of four days	minus	initial reading	=	total miles driven
30,841	−	30,489	=	352

total miles driven	by number of gallons of gasoline used	=	miles per gallon
352	÷ 16	=	22

16. Nathan wants to determine the miles per gallon rating of his truck. He filled the tank when the odometer read 67,249 miles. He drove the truck for a week. The odometer then read 67,513 miles and the tank required 22 gallons to be filled. How many miles per gallon did Nathan's car achieve?

17. Christina owns 210 acres of farmland. She rents it to a farmer for $140 per acre per year. Her property taxes are $38 per acre per year. How much profit does she make on the land each year?

18. Fred wants to determine the miles per gallon rating of his car. He filled the tank when the odometer read 24,610 miles. He then drove the car on a trip. At the end of the trip the odometer read 24,834 miles. It took 14 gallons to fill the tank. How many miles per gallon does his car deliver?

19. Ms. Jansen owns 14 townhouses. She rents each townhouse for $680 per month. Her mortgage payments are $560 per townhouse per month. How much profit does she make on the townhouses each month?

20. Dianne bought five blouses at $28 each, three skirts at $32 each, and three pairs of hose at $2 each. What was the total cost of her shopping spree?

CHAPTER ORGANIZER

Topic	Procedure	Examples
Place value of numbers, p. 3	Each digit has a value depending on location. millions | hundred thousands | ten thousands | thousands | hundreds | tens | ones	In the number 2,896,341, what place value does 9 have? Ten thousands
Writing whole numbers in words, p. 4	Take the number of each group of three digits and indicate if they are (millions)(thousands)(ones) xxx, xxx, xxx	Write in words: 134,718,216. One hundred thirty-four million, seven hundred eighteen thousand, two hundred sixteen
Writing expanded notation, p. 4	Take the number of each digit and multiply it by one, ten, hundred, thousand, if it is in the ones, tens, hundreds, thousands place.	Write in expanded notation: 46,235 $40,000 + 6000 + 200 + 30 + 5$
Adding whole numbers, p. 9	Starting with right column, add each column separately. If a two-digit sum occurs, "carry" the first digit over to the next column.	Add: $\quad$ 2 1 $\quad$ 2 5 8 $\quad$ 3 6 7 $\quad$ 2 9 1 $+$ 4 5 3 1 3 6 9
Subtracting whole numbers, p. 17	Starting with right column, subtract each column separately. If necessary, borrow a unit from column to the left and bring it to right as a "10."	Subtract: $\quad\quad$ 13 $\quad$ 6 $\not{3}$ 12 1 6,$\not{7}$ $\not{4}$ $\not{2}$ $-$ 1 2,3 9 5 $\quad$ 4,3 4 7
Multiplying several factors, p. 28	Keep multiplying from left to right. Take each product and multiply by next factor to right. Continue until all factors are used once. (Since multiplication is commutative and associative, the factors can be multiplied in any order.)	Multiply: $2 \times 9 \times 7 \times 6 \times 3$ $= 18 \times 7 \times 6 \times 3$ $= 126 \times 6 \times 3$ $= 756 \times 3$ $= 2268$
Multiplying several-digit numbers, p. 27	Multiply top factor by ones digit, then by tens digit, then by hundreds digit. Add the partial products together.	Multiply: $\quad\quad$ 5 6 7 $\quad \times$ 2 3 8 $\quad\quad$ 4 5 3 6 $\quad$ 1 7 0 1 1 1 3 4 1 3 4,9 4 6
Dividing by a two- or three-digit number, p. 37	Consider the first digit of each value. Obtain a trial value. Multiply it back to see if it is too large or small. Continue each step of long division until problem is finished.	Divide: $\quad\quad\quad$ 589 238 $\overline{)\,140182}$ $\quad\quad$ 1190 $\quad\quad \overline{\quad\quad}$ $\quad\quad$ 2118 $\quad\quad$ 1904 $\quad\quad \overline{\quad\quad}$ $\quad\quad$ 2142 $\quad\quad$ 2142 $\quad\quad \overline{\quad\quad}$ $\quad\quad\quad$ 0
Exponent form, p. 41	A form to indicate repeated factors.	Evaluate: 6^3. *Base* is 6. *Exponent* is 3. $6 \times 6 \times 6 = 216$

Topic	Procedure	Examples
Rounding, p. 43	1. If the first digit to right of round-off place is less than 5, the digit in round-off place is unchanged. 2. If the first digit to right of round-off place is 5 or more, the digit in round-off place is increased by 1. 3. Digits to right of round-off place are replaced by zeros.	Round to nearest hundred: $5\ 6,7\ \textcircled{4}\ 3$ The digit 4 is less than 5. 56,700 Round to the nearest thousand. $1\ 2\ 8,5\ 1\ 7$ The digit 5 is obviously 5 or greater. We increase the thousands digit by 1. 129,000
Order of operations, p. 47	1. Perform operations inside parentheses. 2. First raise to a power. 3. Then do multiplication and division in order from left to right. 4. Then do addition and subtraction in order from left to right.	Evaluate: $$2^3 + 16 \div 4^2 \times 5 - 3$$ Raise to a power first. $$8 + 16 \div 16 \times 5 = 3$$ Then do multiplication or division from left to right. $$8 + 1 \times 5 - 3$$ $$8 + 5 - 3$$ Then do addition and subtraction. $$13 - 3 = 10$$

Solving applied problems, p. 50	1. Read the problem over carefully. Find out what is asked for. Draw a picture if this helps you to visualize the situation.
	2. Write down which whole numbers are to be used in solving the problem.
	3. Determine which operation needs to be done: addition, subtraction, multiplication, division, or a combination of these.
	4. Estimate your answer to see if it is reasonable.
	5. Perform the necessary computation. State the answer, including the unit of measure.
	6. Compare your answer with the estimate.
		Bob's car gets 29 miles per gallon of gas. How many gallons will he need to drive 406 miles?
		1. How many gallons of gas are needed?
		
		2. 406 miles for trip
		   29 miles per gallon
		3. We need to divide to find out how many 29's are in 406.
		4. Estimate:
		   Round 29 to 30 (nearest 10).
		   Round 406 to 400 (nearest hundred).
		
		    $30 \times ? = 400$
		    $400 \div 30 =$ about 13 (the division
		            has a remainder)
		
		5. $29 \overline{)\ 406}$ with quotient 14
		   $\underline{29}$
		   116
		   $\underline{116}$
		
		14 gallons of gas are needed.
		6. This is close to our answer, 14.

If you have trouble with a particular type of problem, review the sample examples in the section indicated for that group of problems. Answers to all *exercises are located in the answer key.*

1.1 *Write in words.*

1. 376

2. 5082

3. 109,276

4. 423,576,055

Write in expanded notation.

5. 4364

6. 27,986

7. 1,305,128

8. 42,166,037

Write in standard notation.

9. Nine hundred twenty-four

10. Six thousand, ninety-five

11. One million, three hundred twenty-eight thousand, eight hundred twenty-eight

12. Forty-five million, ninety-two thousand, six hundred fifty-one

1.2 *Add.*

13.	**14.**	**15.**	**16.**	**17.**
36	76	127	12	123
+ 94	+ 39	+ 563	28	61
			34	9
			+ 76	84
				+ 123

18.	**19.**	**20.**	**21.**	**22.**
125	937	28,364	1,356	26
364	405	+ 97,059	2,892	503
+ 980	+ 256		561	935
			89	1,257
			+ 9,805	+ 7,861

1.3 *Subtract.*

23.	**24.**	**25.**	**26.**	**27.**
36	54	126	543	1296
− 19	− 48	− 99	− 372	− 1137

28.	**29.**	**30.**	**31.**	**32.**
9821	101,300	201,010	1,986,312	7,216,003
− 4993	− 98,274	− 137,864	− 1,761,555	− 5,985,312

1.4 Multiply.

33. 12
× 3

34. 57
× 2

35. 36
× 0

36. 24
× 1

37. $1 \times 3 \times 6$

38. $2 \times 4 \times 8$

39. $5 \times 7 \times 3$

40. $4 \times 6 \times 5$

41. $8 \times 1 \times 9 \times 2$

42. $7 \times 6 \times 0 \times 4$

43. $3 \cdot 4 \cdot 2 \cdot 2 \cdot 5$

44. $1 \cdot 2 \cdot 7 \cdot 3 \cdot 4$

45. $26{,}121 \times 100$

46. $84{,}312 \times 1000$

47. $832 \times 100{,}000$

48. $563 \times 1{,}000{,}000$

49. 36
× 24

50. 58
× 32

51. 150
× 27

52. 360
× 38

53. 709
× 36

54. 502
× 48

55. 123
× 714

56. 431
× 623

57. 1782
× 305

58. 2057
× 124

59. 300
× 500

60. 400
× 600

61. 1200
× 6000

62. 2500
× 3000

63. 100,000
× 20,000

64. 300,000
× 40,000

1.5 Divide.

65. $20 \div 10$

66. $40 \div 8$

67. $70 \div 5$

68. $36 \div 9$

69. $0 \div 8$

70. $12 \div 1$

71. $7 \div 1$

72. $0 \div 5$

73. $\dfrac{49}{7}$

74. $\dfrac{42}{6}$

75. $\dfrac{5}{0}$

76. $\dfrac{24}{6}$

77. $\dfrac{56}{8}$

78. $\dfrac{48}{8}$

79. $\dfrac{72}{9}$

80. $\dfrac{0}{0}$

Divide. Be sure to indicate the remainder, if one exists.

81. $7 \overline{)875}$

82. $6 \overline{)750}$

83. $5 \overline{)1290}$

84. $4 \overline{)1476}$

85. $3 \overline{)77,622}$

86. $8 \overline{)29,536}$

87. $6 \overline{)221,748}$

88. $5 \overline{)184,605}$

89. $8 \overline{)127,890}$

90. $7 \overline{)250,485}$

91. $67 \overline{)490}$

92. $72 \overline{)325}$

93. $21 \overline{)666}$

94. $22 \overline{)319}$

95. $68 \overline{)2614}$

96. $76 \overline{)4142}$

97. $35 \overline{)9030}$

98. $45 \overline{)4275}$

99. $132 \overline{)7128}$

100. $204 \overline{)3876}$

1.6   *Write in exponent form.*

101. 13×13

102. 24×24

103. $8 \times 8 \times 8 \times 8 \times 8$

104. $9 \times 9 \times 9 \times 9 \times 9$

Evaluate.

105. 2^6

106. 3^4

107. 5^3

108. 2^7

109. 7^2

110. 9^2

111. 6^3

112. 4^4

Round to the nearest ten.

113. 5673

114. 1275

115. 15,305

116. 42,644

Round to the nearest thousand.

117. 12,350

118. 22,986

119. 675,800

120. 202,498

121. Round to the nearest hundred thousand: 5,668,243.

122. Round to the nearest ten thousand: 9,995,312.

1.7 *Perform each operaton in proper order.*

123. $6 \times 2 - 4 + 3$ **124.** $7 + 2 \times 3 - 5$ **125.** $2^3 + 4 - 5 + 3^2$ **126.** $4^3 + 20 \div 2 + 2^3$

127. $3^3 \times 4 - 6 \div 6$ **128.** $20 \div 20 + 5^2 \times 3$ **129.** $2^3 \times 5 \div 8 + 3 \times 4$ **130.** $3^2 \times 6 \div 3 + 5 \times 6$

131. $9 \times 2^2 + 3 \times 4 - 36 \div (4 + 5)$ **132.** $5 \times 3 + 5 \times 4^2 - 14 \div (6 + 1)$

1.8 *Solve each applied problem.*

133. Ward can type 25 words per minute. He types for 7 minutes at that speed. How many words did he type?

134. The soft-drink cans come six to a package. There are 34 packages in the storage room. How many soft-drink cans are there?

135. The Applepickers, Inc., company bought a truck for $26,300, a car for $14,520, and a minivan for $18,650. What was the total purchase price?

136. Alfonso drove 1362 miles last summer, 562 miles at Christmas break, and 473 miles at spring break. How many total miles were driven at these three times?

137. Roberta was billed $11,658 for tuition. She received a $4630 grant. How much did she pay after the grant was deducted?

138. The plane was flying at 14,630 feet. It flew over a mountain 4329 feet high. How many feet was it from the plane to the top of the mountain?

139. Vincent bought 92 shares of stock for $5888. What was the cost per share?

140. The vacation cruise cost a total of $32,544 for 24 paying passengers, who shared the cost equally. What was the cost per passenger?

141. Marcia's checking account balance last month was $436. She made deposits of $16, $98, $125, and $318. She made out checks of $29, $128, $100, and $402. What will be her balance this month?

142. Melissa's savings account balance last month was $810. The bank added $24 interest. Melissa deposited $105, $36, and $177. She made withdrawals of $18, $145, $250. and $461. What will be her balance this month?

143. John began a trip on a full tank of gas with the car odometer at 56,320 miles. He ended the trip at 56,720 miles and added 16 gallons of gas. How many miles per gallon did he get on the trip?

144. Nancy began a trip on a full tank of gas with the car odometer at 24,396 miles. She ended the trip at 24,780 miles and added 16 gallons of gas. How many miles per gallon did she get on the trip?

145. The Rental Supply Center bought three lawn mowers at $279, four power drills at $61, and two riding tractors at $1980. What was the total purchase price for these items?

146. The Book Center bought 24 sets of shelves at $118 each, four desks at $120 each, and six chairs at $24 each. What was the total purchase price for these items?

Write in words.

1. 10,036

2. 42,310,050

Write in expanded notation.

3. 9367

4. 104,760

Write in standard notation.

5. Sixty-four thousand, nine hundred eight-six

6. Twenty million, one hundred fifty-two thousand, eight hundred four

Add.

7.
```
     16
      8
     25
    133
  +   7
```

8.
```
    126
    273
  + 501
```

9.
```
   76,380
   59,020
  +  3,894
```

10.
```
   1,316,204
 + 2,789,300
```

1. _____

2. _____

3. _____

4. _____

5. _____

6. _____

7. _____

8. _____

9. _____

10. _____

11. _____

12. _____

13. _____

14. _____

15. _____

16. _____

17. _____

18. _____

19. _____

20. _____

21. _____

22. _____

Subtract.

11.   976
   − 324

12.   28,306
    − 24,987

13.   1,010,100
    − 986,908

14.   204,368,982
    − 156,278,399

15.   392,835
    − 391,968

Multiply.

16. $3 \times 2 \times 6 \times 0 \times 4$

17.   128
    $\times$ 5

18.   136,254
    $\times$ 7

19.   167
    $\times$ 48

20.   476
    $\times$ 502

21. $9 \times 3 \times 7 \times 2$

22. 584×100

Divide. *If there is a remainder, be sure to include it as part of your answer.*

1. $6\overline{)62{,}460}$

2. $7\overline{)17{,}368}$

3. $12\overline{)9841}$

4. $97\overline{)32{,}592}$

5. Write in exponent form:
$8 \times 8 \times 8 \times 8$.

6. Evaluate: 5^3.

7. Round to the nearest ten: 874.

8. Round to the nearest hundred: 58,647.

9. Round to the nearest ten thousand: 672,100

10. Round to the nearest million: 19,573,564.

1. _____

2. _____

3. _____

4. _____

5. _____

6. _____

7. _____

8. _____

9. _____

10. _____

11. _____

12. _____

13. _____

14. _____

15. _____

16. _____

17. _____

18. _____

19. _____

20. _____

Perform each operation in proper order.

11. $8 - 5 \times 4 \div 10$

12. $3^3 + 3 \times 2^4 - (9 - 3)$

13. $2 \times 7 + 6 \times 4 - 20 \div 2^2$

14. $6 + 4 \times 5 \div 2 - 3 + 7 \div 7$

Solve each applied problem.

15. Bernadette deposited $126, $315, and $82 into her checking account. Her previous balance was $19. What is the new balance?

16. The university had a budget shortfall of $44,244. It will be equally divided among 12 different academic departments. What will the shortage be in each department?

17. Juan's car gets 26 miles per gallon. How far can he travel with 14 gallons of gas?

18. Michelle began a trip when her car odometer read 40,986 miles. At the end of the trip the odometer read 41,237 miles. How many miles did she travel?

19. The total cost of a new car, including finance charges, is $14,318. Alfredo received $4142 in trade for his old car. Alfredo will pay for the rest in 48 equal monthly payments. How much is each monthly payment?

20. The Dickerson Company bought 12 desks at $184 each, 14 chairs at $37 each, and 10 file cabinets at $119 each. What was the total cost of this purchase?

Do the problems without looking at your notes or in the book.

1. Write in words: 26,005,986.   **2.** Write in expanded notation: 38,598.

3. Write in standard notation: two million, three hundred sixty-one thousand, fifty-four.

Add.

4.
```
  126
   83
    5
  298
+  74
```

5.
```
  361
  275
+ 198
```

6.
```
  156,393
    2,481
   64,053
+ 100,311
```

Subtract.

7.
```
  7932
− 4513
```

8.
```
  200,634
− 152,193
```

9.
```
  19,300,200
− 14,163,554
```

Multiply.

10. $1 \times 6 \times 5 \times 8$

11.
```
   82
× 35
```

12.
```
  560
× 129
```

13.
```
  42,163
×      5
```

1.
2.
3.
4.
5.
6.
7.
8.
9.
10.
11.
12.
13.

14. _____

15. _____

16. _____

17. _____

18. _____

19. _____

20. _____

21. _____

22. _____

23. _____

24. _____

25. _____

26. _____

27. _____

28. _____

Divide. If there is a remainder, be sure to state it as part of the answer.

14. $6 \overline{)12,315}$ **15.** $8 \overline{)29,216}$ **16.** $37 \overline{)58,534}$

17. Write in exponent form: $12 \times 12 \times 12$.

18. Evaluate: 3^5.

19. Round to the nearest hundred: 26,453.

20. Round to the nearest ten thousand: 1,673,542.

21. Round to the nearest hundred thousand: 9,784,563.

Perform each operation in proper order.

22. $5 + 6^2 - 2 \times (9 - 6)$

23. $3^3 + 4^4 + 20 \div 4$

24. $6 \times 3 + 2^3 \times 5 + 7 \div 7$

Answer each question.

25. A cruise for 18 people cost $39,402. If each person paid the same amount, how much would it cost one person for the cruise?

26. The river is 548 feet wide at Big Bend Corner. A boat is in the shallow water, 129 feet from the shore. How far is the boat from the other side of the river? (Ignore the width of the boat)

27. At the bookstore, Hector bought three notebooks at $3 each, one textbook for $56, two lamps at $19 each, and two sweatshirts at $16 each. What was his total bill?

28. Patricia is looking at her checkbook. She had a balance last month of $26. She deposited $876 and $403. She made out checks for $965, $116, $35, $28, and $9. What will be her new balance?

Fractions

A stockbroker must continually evaluate changes in the stock market that involve fractions. Only by doing calculations with accurate fractional values can business transactions be completed correctly.

2

PRETEST CHAPTER 2

If you are familiar with the topics in this chapter, take this test now. Check your answers with those in the back of the book. If an answer was wrong or you couldn't do a problem, study the appropriate section of the chapter.

If you are not familiar with the topics in this chapter, don't take this test now. Instead, study the examples, work the practice problems, and then take the test.

This test will help you identify those concepts that you have mastered and those that need more study.

Section 2.1

1. Use a fraction to represent the shaded part of the object.

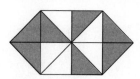

2. Draw a sketch to illustrate by shading in $\frac{3}{7}$ of an object.

3. An inspector inspected 136 books. Of these, 19 were defective. Write a fraction that describes the part of the selection of books that was defective.

Section 2.2   Reduce each fraction.

4. $\frac{12}{18}$ 5. $\frac{28}{35}$ 6. $\frac{18}{108}$ 7. $\frac{30}{48}$ 8. $\frac{21}{98}$

Section 2.3   Change to an improper fraction.

9. $1\frac{3}{8}$ 10. $5\frac{2}{7}$

Change to a mixed number.

11. $\frac{100}{3}$ 12. $\frac{27}{4}$ 13. $\frac{32}{15}$

Section 2.4   Multiply.

14. $\frac{5}{7} \times \frac{2}{3}$ 15. $\frac{5}{8} \times \frac{16}{15}$ 16. $15\frac{1}{3} \times 6\frac{1}{2}$

Section 2.5   Divide.

17. $\frac{3}{7} \div \frac{5}{11}$ 18. $\frac{5}{12} \div \frac{5}{6}$ 19. $7\frac{3}{5} \div 1\frac{7}{12}$ 20. $9 \div \frac{12}{5}$

Section 2.6   Find the least common denominator of each fraction.

21. $\frac{5}{6}, \frac{1}{2}, \frac{5}{3}$ 22. $\frac{3}{7}, \frac{4}{35}$ 23. $\frac{6}{11}, \frac{2}{6}$ 24. $\frac{17}{18}, \frac{3}{24}$

Section 2.7   Add or subtract.

25. $\frac{21}{35} + \frac{5}{14}$ 26. $1\frac{2}{5} + 3\frac{1}{4}$ 27. $6 - 4\frac{2}{3}$

28. $5\frac{3}{14} - 2\frac{9}{28}$ 29. $\frac{11}{24} + \frac{5}{36} + \frac{1}{12}$

30. Miguel and Lee set out to hike $14\frac{1}{2}$ miles from Arlington to Concord. During the first 3 hours they covered $5\frac{3}{8}$ miles going from Arlington to Bedford. How many miles are left to be covered from Bedford to Concord?

31. Robert harvested $12\frac{3}{4}$ tons of wheat. His helper harvested $8\frac{1}{8}$ tons of wheat. How much did they harvest together?

32. The students of Smith Dorm fifth floor contributed money to make a stock purchase of one share of stock each. They paid $678 in all to buy the shares of stock. The cost of buying one share was $\$42\frac{3}{8}$. How many students shared in the purchase?

2.1 UNDERSTANDING FRACTIONS

After studying this section, you will be able to:

1 *Use a fraction to represent a shaded part of an object*

2 *Draw a sketch to illustrate a fraction*

3 *Represent the data in an applied situation in the form of a fraction*

In Chapter 1 we studied whole numbers. However, often we have to represent a fractional part of a whole number. One way to write parts of a whole is with fractions. The word *fraction* (like the word *fracture*) suggests something being broken. In mathematics, fractions are a set of numbers (with a specific appearance) that designate a portion, or a part "broken off" from a whole. That "whole" can be something that "looks like" a unit (a whole pie), or, just as often, the "whole" can be a collection of things that we choose to consider *as a* whole. Here are some examples

Solid object, like a pie

Find $\frac{1}{3}$ of the pie.

A list of items:

Make $\frac{1}{2}$ of the recipe at right.

Groups of people: ACE Company employs 150 men, 200 women.

Women make up what fraction of the total?

Measurements: A board is 8 feet long. We cut shelves from it $3\frac{1}{4}$ feet long

After we cut as many shelves as we can, how many linear feet of shelving will be left over? (See figure at right.)

Money: My stock is selling for $38\frac{1}{8}$, up $\frac{3}{8}$ since yesterday.

By how much did the value of a share of stock improve since yesterday?

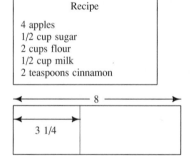

Recipe

4 apples
1/2 cup sugar
2 cups flour
1/2 cup milk
2 teaspoons cinnamon

The most basic concept of fractions is that you can change the way a fraction looks without changing its value. "$\frac{1}{2}$ is the same as $\frac{2}{4}$, which is the same as $\frac{10}{20}$, which is the same as $\frac{50}{100}$." All have the value "one-half." This changing of appearance is crucial to computation, as we will see in this chapter.

1 When you say "$\frac{3}{8}$ of a pizza has been eaten," what you are indicating is that three of eight equal parts of a pizza have been eaten (See figure in margin). When we write the fraction $\frac{3}{8}$, the number on the top, 3, is the **numerator**, and the number on the bottom, 8, is the **denominator**.

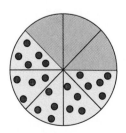

The numerator specifies how many of these parts ⟶ 3
The denominator specifies the total number of parts ⟶ 8

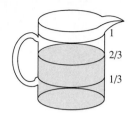

When we say $\frac{2}{3}$ of a cup of flour, we are indicating that

two of these parts are
filled with flour $\longrightarrow \frac{2}{3}$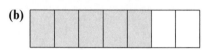
the measuring cup is $\longrightarrow$
divided into 3 parts

Remember that the *numerator* is always the top number and the *denominator* is always the bottom number. The denominator tells *how many equal parts there are in all*; the numerator tells how many of these you are specifically "noticing" in your problem. The parts being noticed may be "shaded" (objects) or perhaps the number of pieces (of pizza) that have been "eaten," or the number that "still remain."

Example 1 Use a fraction to represent the shaded part of the object.

(a)

(b)

(a) Three out of four equal parts are shaded. The fraction is $\frac{3}{4}$.

(b) Five out of seven equal parts are shaded. The fraction is $\frac{5}{7}$. ■

Practice Problem 1 Use a fraction to represent the shaded part of the object.

(a)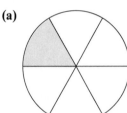

(b) 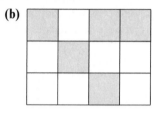 ■

2 Example 2 Draw a sketch to illustrate by shading.

(a) $\frac{7}{11}$ of an object

(b) $\frac{2}{9}$ of an object

(a)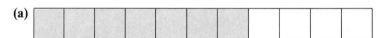

We shade in seven of the eleven equally sized blocks to indicate $\frac{7}{11}$.

(b)

We shade in two of the nine equally sized blocks to indicate $\frac{2}{9}$. ■

Practice Problem 2 Draw a sketch to illustrate by shading.

(a) $\frac{4}{5}$ of an object

(b) $\frac{3}{7}$ of an object ■

3 A number of applied situations can be described with fractions.

Example 3 A baseball player got a hit 5 out of 12 times at bat. What fraction of the times at bat did the player get a hit?

The baseball player got a hit $\frac{5}{12}$ of the times at bat. ■

Practice Problem 3   Nine of the 17 women on the basketball team are on the dean's list for academic excellence. What fraction of the team is this?   ■

Example 4   There are 156 men and 185 women taking psychology this semester. Write a fraction that describes the fractional part of the class that consists only of men.

The total class is $156 + 185 = 341$.

The fractional part that is men is 156 out of 341. Thus we have: $\frac{156}{341}$ of the class is men.   ■

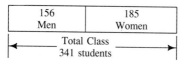

156 Men	185 Women
Total Class 341 students	

Practice Problem 4   The senior class is 382 men and 351 women. Write a fraction that describes what part of the class consists of women.   ■

Example 5   Wanda made 13 calls, out of which she made five sales. Albert made 17 calls, out of which he made six sales. Write a fraction that describes for both people together the number of calls in which a sale was made compared with the total number of calls.

There are $5 + 6 = 11$ calls in which a sale was made. There were $13 + 17 = 30$ total calls. Thus $\frac{11}{30}$ of the calls resulted in a sale.   ■

Practice Problem 5   An inspector found that one out of seven belts was defective. She also found that two out of nine shirts were defective. Write a fraction that describes what part of all the objects examined were defective.   ■

We can think of a fraction as a *division problem*

$$\frac{1}{3} = 1 \div 3 \quad \text{and} \quad 1 \div 3 = \frac{1}{3}$$

The division way of looking asks the question:

What is the result of dividing one whole into three equal parts?   *Answer:* $\frac{1}{3}$

The fraction way of looking asks the question:

How do you represent one shaded part out of three equal parts?   *Answer:* $\frac{1}{3}$

So for most numbers a and b, the fraction $\frac{a}{b}$ means the same as $a \div b$. However, special care must be taken with the number 0.

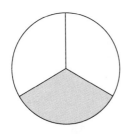

To Think About

How can we use 0 with a fraction? Suppose that we had four equal parts and wanted to take *none* of them. We would want $\frac{0}{4}$ of the parts. Since $\frac{0}{4} = 0 \div 4 = 0$, we see that $\frac{0}{4} = 0$. Any fraction with a 0 numerator equals zero: $\frac{0}{8} = 0, \frac{0}{5} = 0, \frac{0}{13} = 0$, and so on. We cannot have a denominator of zero. A fraction like $\frac{4}{0}$ is not possible. We know that $\frac{4}{0} = 4 \div 0$, which cannot be done. Do you see that it does not make sense to say "I want to pick 4 out of 0 parts"?

$$\frac{3}{0}, \frac{7}{0}, \frac{4}{0} \text{ are not defined.}$$

We cannot divide by 0 and we cannot have a fraction with 0 in the denominator.

   ■

STEPS TOWARD SUCCESS IN MATHEMATICS

Mathematics is a building process, mastered one step at a time. The foundation of this process is built on a few basic requirements. Those who are successful in mathematics realize the absolute necessity for building a study of mathematics on the firm foundation of these six minimum requirements.

1. Attend class every day.
2. Read the textbook.
3. Take notes in class.
4. Do assigned homework every day.
5. Get help immediately when needed.
6. Review regularly.

EXERCISES 2.1

Identify the numerator and the denominator in each fraction.

1. $\dfrac{5}{7}$ **2.** $\dfrac{9}{11}$ **3.** $\dfrac{2}{3}$ **4.** $\dfrac{3}{4}$ **5.** $\dfrac{1}{12}$ **6.** $\dfrac{1}{15}$

In problems 7–26, use a fraction to represent the shaded part of the object or the shaded portion of the set of objects.

7. **8.** **9.** **10.**

11. **12.** **13.** **14.**

15.   **16.**   **17.** **18.**

19. **20.** **21.** ○○○○○○○ **22.**

23. **24.** **25.**   **26.**

Draw a sketch to illustrate each fractional part.

27. $\frac{3}{5}$ of an object

28. $\frac{3}{7}$ of an object

29. $\frac{11}{13}$ of an object

30. $\frac{5}{12}$ of an object

31. $\frac{7}{20}$ of an object

32. $\frac{8}{17}$ of an object

Answer each question.

33. The team consists of nine players, of which two are co-captains. What fractional part of the team are co-captains?

34. The total purchase price was 83¢, of which 5¢ was sales tax. What fractional part of the total purchase price was sales tax?

35. Ali paid his $3651 tuition bill. Part of his payment was $101 that he earned on July 4 weekend. What fractional part of the tuition was paid by his July 4 weekend earnings?

36. Harold flew to Phoenix on business. His total trip cost $329. His taxi fare was $19. What fractional part of his trip cost was the cost of the taxi?

37. The math class has 32 men and 31 women. What fractional part of the class is women?

38. The East dormitory has 111 smokers and 180 non-smokers. What fractional part of the dormitory has nonsmoking residents?

39. In one city the gasoline tax has three parts: 21¢ per gallon is federal tax, 14¢ per gallon is state tax, and 5¢ per gallon is city tax. What fractional part of the total gasoline tax is federal tax?

40. In Roger's paycheck $100 was withheld for federal tax, $24 was withheld for state tax, and $63 was withheld for social security. What fractional part of the withheld money is the federal tax?

41. A jar contains eight black balls, three red balls, five white balls, and seven blue balls. What fractional part of the balls are either blue or white?

42. The university soccer team has 3 freshmen, 6 sophomores, 12 juniors, and 17 seniors. What fractional part of the team are either freshmen or sophomores?

? To Think About

43. Illustrate a real-world example of the fraction $\frac{0}{6}$.

44. What happens when we try to illustrate a real-world example of the fraction $\frac{6}{0}$? Why?

45. The plant manufactured two items: 101 engines and 94 lawn mowers. It was discovered that 19 engines were defective and 3 lawn mowers were defective. Of the engines that were not defective, 40 were excellently constructed but 42 were not of the highest quality. Of the lawn mowers that were not defective, 50 were excellently constructed but 41 were not of the highest quality.
   (a) What fractional part of all items manufactured were of the highest quality?
   (b) What fractional part of all items manufactured were defective?

46. The psychology class had 84 men and 69 women. A total of 33 women were shorter than 62 inches. A total of 7 men were shorter than 62 inches. In this group of 33 women, a subgroup of 16 women were shorter than 59 inches. In this group of 7 men, a subgroup of 2 men were shorter than 59 inches.
   (a) What fractional part of the *class* is shorter than 62 inches?
   (b) What fractional part of the group of *women* is shorter than 59 inches?

47. Add: 18
27
34
16
125
+ 21

48. Subtract: 38,114
− 27,008

49. Multiply: 4136
×   29

50. Divide: $12\overline{)2130}$

For Extra Practice Examples and Exercises, turn to page 133.

Solutions to Odd-Numbered Practice Problems

1. (a) One part out of six is shaded.
The fraction is $\dfrac{1}{6}$.

(b) Five parts out of twelve are shaded.
The fraction is $\dfrac{5}{12}$.

3. $\dfrac{9}{17}$ represents 9 women out of 17.

5. Total number of defective items $1 + 2 = 3$. Total number of items $7 + 9 = 16$.
A fraction that represents the portion of the items that were defective is $\dfrac{3}{16}$.

Answers to Even-Numbered Practice Problems

2. (a) $\dfrac{4}{5}$ of the object is shaded.

(b) $\dfrac{3}{7}$ of the object is shaded.

4. $\dfrac{351}{733}$

☐ After studying this section, you will be able to:

1 *Write a number as a product of prime factors*

2 *Reduce a fraction*

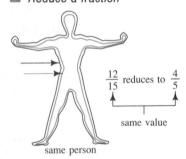

$\dfrac{12}{15}$ reduces to $\dfrac{4}{5}$

same value

same person

2.2 REDUCING FRACTIONS

A fraction is a number with a certain value. That value is the same no matter what the appearance of the fraction. For example, $\dfrac{12}{15}$ equals $\dfrac{4}{5}$. A person can reduce his weight and so change his appearance, but this does not change who he is. At a reduced, less cumbersome weight, however, he may find working easier.

Similarly, when a fraction is *reduced*, we change the way it looks (the numerator and denominator change to smaller numbers), but the worth, or value, of the fraction remains unchanged. And a reduced fraction is often easier, less cumbersome to work with.

This section tells you *how* to reduce a fraction. When will you use reduced fractions? To simplify computations or simplify our results, as we'll see later.

1 A **prime number** is a whole number greater than 1 that can be divided only by itself and 1. If you examine all the whole numbers from 1 to 50, you will find 15 prime numbers.

The First 15 Prime Numbers

2, 3, 5, 7, 11, 13, 17, 19, 23, 29, 31, 37, 41, 43, 47

A **composite number** is a whole number greater than 1 that can be divided by whole numbers other than itself. The number 12 is a composite number since it can be divided by 2, 3, 4, or 6 (as well as 12 and 1). The number 1 is neither a prime nor a composite number. The number 0 is neither a prime nor a composite number.

Recall that factors are numbers that are multiplied together. **Prime factors** are prime numbers. To check to see if a number is prime or composite, simply divide the smaller primes (such as 2, 3, 5, 7, 11, ...) into the given number. If the number can be divided exactly without a remainder by one of the smaller primes, it is a composite and not a prime.

Some students find the following rules helpful when deciding if a number can be divided by 2, 3, or 5.

Divisibility Tests

1. A number is divisible by 2 if the last digit is 0, 2, 4, 6, or 8.

2. A number is divisible by 3 if the sum of the digits is divisible by 3.

3. A number is divisible by 5 if the last digit is 0 or 5.

To illustrate:

1. 478 is divisible by 2 since it ends in 8.
2. 531 is divisible by 3 since when we add the digits of 531 ($5 + 3 + 1 = 9$) we obtain a number divisible by 9.
3. 985 is divisible by 5 since it ends in 5.

Example 1 Express 12 as a product of prime factors.

$12 = \quad 4 \quad \times 3$ Write as the product of any two factors other than 1 and the number itself.

$12 = \quad 4 \quad \times 3$ If either or both of the factors are not prime, factor these as in step 1.
$\quad\quad 2 \times 2 \times 3$

$12 = 2 \times 2 \times 3$ All factors are prime.
$\quad = 2^2 \times 3$ ■

Practice Problem 1 Express 18 as a product of prime factors. ■

Example 2 Express 60 as a product of prime factors.

$60 = \quad 6 \quad \times \quad 10$ Write as the product of any two factors other than 1 and the number itself.
$\quad\quad 3 \times 2 \times 2 \times 5$ If both of these factors are not prime, factor these as in step 1.

$60 = 3 \times 2 \times 2 \times 5$ All factors are prime.

This can also be written as $60 = 3 \times 2^2 \times 5$. ■

Practice Problem 2 Express 72 as a product of prime factors. ■

Example 3 Factor 168 as a product of prime factors.

$168 = \quad 4 \quad \times \quad 42$

$\quad = 2 \times 2 \times 2 \times \quad 21$

$\quad = 2 \times 2 \times 2 \times 3 \times 7$

$168 = 2 \times 2 \times 2 \times 3 \times 7$ All factors are prime.

This can also be written as $2^3 \times 3 \times 7$. ■

Practice Problem 3 Write 400 as a product of prime factors. ■

When you write a composite number as the product of prime factors, is the answer unique? Would we get an equivalent answer if we started in a different way? Let us examine the issue. Suppose we started Example 3 by saying that $168 = 14 \times 12$. Would we obtain the same answer? We place the two approaches side by side.

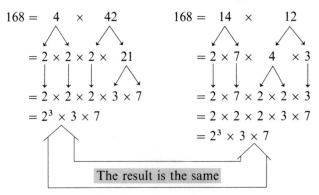

The order of prime factors is not important because multiplication is commutative. No matter how we start to factor a composite number, we get exactly the same prime factors. ∎

> **The Fundamental Theorem of Arithmetic**
>
> Every composite number has a unique product of prime numbers.

You will be able to test this in Exercise Set 2.2, problems 39 and 40.

2 Reducing Fractions

Compare the figures in the margin.

Do you see that the same area of the circle is shaded when $\frac{3}{4}$ of the circle is shaded and when $\frac{6}{8}$ of the circle is shaded? The fractions $\frac{3}{4}$ and $\frac{6}{8}$ are called *equal* fractions. The fraction $\frac{3}{4}$ is a reduced fraction. The fraction $\frac{6}{8}$ is not a reduced fraction.

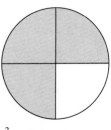

$\frac{3}{4}$ of the circle is shaded.

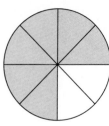

$\frac{6}{8}$ of the circle is shaded.

Any nonzero number divided by itself is 1. Thus $\frac{5}{5} = \frac{17}{17} = \frac{c}{c} = 1.$ For any fraction we can multiply or divide the numerator and the denominator by the same nonzero number. Thus if c is not zero,

$$\frac{a}{b} = \frac{a \times c}{b \times c} \quad \text{and} \quad \frac{a}{b} = \frac{a \div c}{b \div c}$$

A fraction is considered to be a reduced fraction if the numerator and denominator have no common factor other than 1. In the fraction $\frac{3}{4}$, the 3 and the 4 do not have any common factor other than 1. In the fraction $\frac{6}{8}$, the 6 and the 8 both have a common factor of 2. Let's divide the numerator and denominator of $\frac{6}{8}$ by 2.

$$\frac{6}{8} = \frac{6 \div 2}{8 \div 2} = \frac{3}{4}$$

Reducing a Fraction

The numerator and denominator of a fraction can both be divided by a common factor to obtain an equal fraction that has been reduced. When a fraction can no longer be reduced, it is said to be reduced to lowest terms.

Example 4 Reduce $\dfrac{15}{25}$ to lowest terms by dividing numerator and denominator by a common factor.

$$\frac{15}{25} = \frac{15 \div 5}{25 \div 5} = \frac{3}{5} \qquad \text{Dividing numerator and denominator by 5.} \quad \blacksquare$$

Practice Problem 4 Reduce $\dfrac{30}{42}$ to lowest terms by dividing numerator and denominator by a common factor. $\quad \blacksquare$

Example 5 Reduce $\dfrac{42}{56}$ by dividing numerator and denominator by a common factor.

Both numerator and denominator have a common factor of 14.

$$\frac{42}{56} = \frac{42 \div 14}{56 \div 14} = \frac{3}{4}$$

Perhaps you did not think of 14 as a common factor. Note that you can get exactly the same answer in two steps. First observe that the numerator and denominator have a common factor of 2. Then continue by dividing numerator and denominator by 7.

$$\frac{42}{56} = \frac{42 \div 2}{56 \div 2} = \frac{21}{28} = \frac{21 \div 7}{28 \div 7} = \frac{3}{4}$$

Thus we see that the method of reducing a fraction by dividing numerator and denominator by a common factor can be repeated several times to obtain the reduced fraction. $\quad \blacksquare$

Practice Problem 5 Reduce $\dfrac{60}{132}$ to lowest terms by dividing numerator and denominator by a common factor. $\quad \blacksquare$

A second method to reduce fractions is called the method of *prime factors*. We factor the numerator and denominator into prime numbers. We then divide the numerator and denominator by a common prime factor, using a line through each common prime factor to indicate this division.

Example 6 Reduce $\dfrac{35}{42}$ to lowest terms by the method of prime factors.

We factor 35 and 42 into prime factors.

$$\frac{35}{42} = \frac{5 \times 7}{2 \times 3 \times 7}$$

$$\frac{35}{42} = \frac{5 \times \overset{1}{\cancel{7}}}{2 \times 3 \times \underset{1}{\cancel{7}}}$$

We divide the numerator and denominator by 7, the prime factor they have in common. We draw a line through the 7's—to show this division. The 1's replacing the 7's show the result of $7 \div 7$.

Now we multiply the factors in the numerator and denominator:

$$\frac{35}{42} = \frac{5 \times 1}{2 \times 3 \times 1} = \frac{5}{6} \quad \blacksquare$$

Practice Problem 6 Reduce $\dfrac{36}{63}$ to lowest terms by the method of prime factors.
■

Whenever we say "reduce" we mean "reduce to lowest terms."

Example 7 Reduce $\dfrac{70}{110}$ by the method of prime factors.

$$\frac{70}{110} = \frac{2 \times 5 \times 7}{2 \times 5 \times 11} = \frac{\overset{1}{\cancel{2}} \times \overset{1}{\cancel{5}} \times 7}{\underset{1}{\cancel{2}} \times \underset{1}{\cancel{5}} \times 11} = \frac{7}{11}$$

Note that we have a common factor of 5 as well as a common factor of 2. ■

Practice Problem 7 Reduce $\dfrac{120}{135}$ by the method of prime factors. ■

Example 8 Reduce by the method of prime factors: $\dfrac{363}{528}$.

$$\frac{363}{528} = \frac{\overset{1}{\cancel{3}} \times \overset{1}{\cancel{11}} \times 11}{\underset{1}{\cancel{3}} \times \underset{1}{\cancel{11}} \times 2 \times 2 \times 2 \times 2} = \frac{11}{16}$$ ■

Practice Problem 8 Reduce by the method of prime factors: $\dfrac{715}{880}$. ■

How can we check our answers? How can we check that a reduced fraction is *equivalent* to the original fraction? If two fractions are equal, their diagonal products are equal. This is called the *equality test for fractions*. If $\dfrac{3}{4} = \dfrac{6}{8}$, then

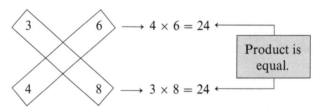

If two fractions are unequal (we use the symbol $\neq$), their diagonal products are unequal. If $\dfrac{5}{6} \neq \dfrac{6}{7}$, then

Since $36 \neq 35$ we know that $\dfrac{5}{6} \neq \dfrac{6}{7}$. The test can be described in this way:

Equality Test for Fractions

For any two fractions where a, b, c, and d are whole numbers and $b \neq 0$, $d \neq 0$,

if $\dfrac{a}{b} = \dfrac{c}{d}$, then $a \times d = b \times c$.

Example 9 Use the equality test for fractions to see if the following fractions are equal.

(a) $\dfrac{2}{11} \overset{?}{=} \dfrac{18}{99}$

(b) $\dfrac{3}{16} \overset{?}{=} \dfrac{12}{62}$

(a)

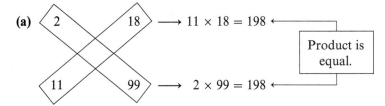

$11 \times 18 = 198$

Product is equal.

$2 \times 99 = 198$

Since $198 = 198$ we know that $\dfrac{2}{11} = \dfrac{18}{99}$.

(b)

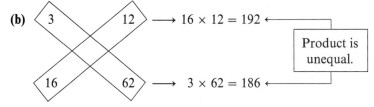

$16 \times 12 = 192$

Product is unequal.

$3 \times 62 = 186$

Since $192 \ne 186$ we know that $\dfrac{3}{16} \ne \dfrac{12}{62}$. ■

Practice Problem 9   Are the following fractions equal?

(a) $\dfrac{84}{108} \overset{?}{=} \dfrac{7}{9}$

(b) $\dfrac{3}{7} \overset{?}{=} \dfrac{79}{182}$ ■

To Think About

What do we mean by equal fractions? For example, is $\dfrac{3}{4} = \dfrac{6}{8}$? The answer is yes!

$$\frac{3}{4} \times 1 = \frac{3}{4} \times \frac{2}{2} = \frac{6}{8}$$

These fractions are equal.

What if they are not equal? For example, is $\dfrac{5}{6} = \dfrac{6}{7}$? No, because

$$\frac{5}{6} \times 1 = \frac{5}{6} \times \frac{7}{7} = \frac{35}{42} \qquad \text{and} \qquad \frac{6}{7} \times 1 = \frac{6}{7} \times \frac{6}{6} = \frac{36}{42}$$

Since $35 < 36$ it is true that $\dfrac{35}{42} < \dfrac{36}{42}$. Do you see why? We will discuss "building up" fractions in more detail later in this chapter. ■

1. Identify the prime numbers in this list of 12 whole numbers.

   4, 12, 11, 15, 6, 19, 1, 41, 38, 24, 5, 46

2. Identify the prime numbers in this list of 12 whole numbers.

   2, 6, 8, 20, 1, 23, 28, 15, 14, 37, 38, 47

Express each number as a product of prime factors.

3. 15 **4.** 35 **5.** 6 **6.** 8 **7.** 49 **8.** 25 **9.** 64

10. 81 **11.** 20 **12.** 28 **13.** 45 **14.** 36 **15.** 75 **16.** 125

17. 54 **18.** 90 **19.** 84 **20.** 76 **21.** 98 **22.** 65

State which of the following numbers is prime. If it is composite, express it as the product of prime factors.

23. 31 **24.** 37 **25.** 57 **26.** 51

27. 67 **28.** 71 **29.** 77 **30.** 91

31. 89 **32.** 97 **33.** 127 **34.** 137

35. 112 **36.** 134 **37.** 161 **38.** 169

 To Think About

Answer each question.

39. Obtain the product of prime factors for 304 by starting with $304 = 19 \times 16$. Do it again starting with $304 = 8 \times 38$. Is your result the same? Why?

40. Obtain the product of prime factors for 338 by starting with $338 = 13 \times 26$. Do it again starting with $338 = 2 \times 169$. Is your result the same? Why?

Reduce each fraction by dividing the numerator and denominator by a common factor.

41. $\dfrac{14}{21}$ **42.** $\dfrac{16}{24}$ **43.** $\dfrac{54}{90}$ **44.** $\dfrac{30}{42}$

45. $\dfrac{16}{24}$ **46.** $\dfrac{48}{64}$ **47.** $\dfrac{42}{48}$ **48.** $\dfrac{35}{80}$

Reduce each fraction by the method of prime factors.

49. $\dfrac{9}{15}$ **50.** $\dfrac{15}{21}$ **51.** $\dfrac{39}{52}$ **52.** $\dfrac{63}{84}$

53. $\dfrac{12}{42}$ **54.** $\dfrac{65}{91}$ **55.** $\dfrac{27}{45}$ **56.** $\dfrac{28}{42}$

Reduce each fraction by any method.

57. $\dfrac{33}{36}$ **58.** $\dfrac{56}{96}$ **59.** $\dfrac{65}{169}$ **60.** $\dfrac{21}{98}$ **61.** $\dfrac{120}{192}$

62. $\dfrac{165}{180}$ **63.** $\dfrac{112}{140}$ **64.** $\dfrac{96}{108}$ **65.** $\dfrac{119}{210}$ **66.** $\dfrac{99}{189}$

Is each set of fractions equal?

67. $\dfrac{3}{17} \overset{?}{=} \dfrac{15}{85}$ **68.** $\dfrac{12}{13} \overset{?}{=} \dfrac{72}{78}$ **69.** $\dfrac{12}{40} \overset{?}{=} \dfrac{9}{30}$ **70.** $\dfrac{24}{72} \overset{?}{=} \dfrac{15}{45}$

71. $\dfrac{23}{27} \overset{?}{=} \dfrac{92}{107}$ **72.** $\dfrac{70}{120} \overset{?}{=} \dfrac{41}{73}$ **73.** $\dfrac{35}{55} \overset{?}{=} \dfrac{207}{330}$ **74.** $\dfrac{18}{24} \overset{?}{=} \dfrac{23}{28}$

75. $\dfrac{65}{91} \overset{?}{=} \dfrac{195}{273}$ **76.** $\dfrac{56}{96} \overset{?}{=} \dfrac{168}{288}$

Reduce each fraction, if possible. If it cannot be reduced further, state "irreducible."

77. $\dfrac{892}{936}$ **78.** $\dfrac{752}{835}$

Cumulative Review Problems

79. Multiply: 386×425.   **80.** Divide: $15{,}552 \div 12$.

For Extra Practice Examples and Exercises, turn to page 134.

1. $18 = \quad 9 \quad \times 2$

$= 3 \times 3 \times 2$

so $\quad 18 = 3 \times 3 \times 2$
$\qquad 18 = 2 \times 3 \times 3 \quad$ or $\quad 2 \times 3^2$

3. $400 = \quad 10 \quad \times \quad 40$

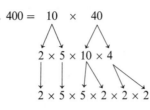

$2 \times 5 \times 10 \times 4$

$2 \times 5 \times 5 \times 2 \times 2 \times 2 \qquad$ so $\quad 400 = 2 \times 2 \times 2 \times 2 \times 5 \times 5 \quad$ or $\quad 2^4 \times 5^2$

5. $\dfrac{60}{132} = \dfrac{60 \div 12}{132 \div 12} = \dfrac{5}{11} \quad$ or by repeated division $\quad \dfrac{60 \div 2}{132 \div 2} = \dfrac{30 \div 6}{66 \div 6} = \dfrac{5}{11}$

7. $\dfrac{120}{135} = \dfrac{2 \times 2 \times 2 \times 3 \times 5}{3 \times 3 \times 3 \times 5} = \dfrac{2 \times 2 \times 2 \times \cancel{3} \times \cancel{5}}{3 \times 3 \times \cancel{3} \times \cancel{5}} = \dfrac{8}{9}$

9. (a) $\dfrac{84}{108} \overset{?}{=} \dfrac{7}{9}$

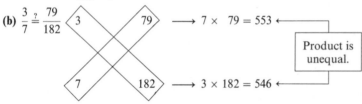

$108 \times 7 = 756 \leftarrow$

$84 \times 9 = 756 \leftarrow$

Product is equal.

Therefore, $\dfrac{84}{108} = \dfrac{7}{9}$.

(b) $\dfrac{3}{7} \overset{?}{=} \dfrac{79}{182}$

$7 \times 79 = 553 \leftarrow$

$3 \times 182 = 546 \leftarrow$

Product is unequal.

Therefore, $\dfrac{3}{7} \neq \dfrac{79}{182}$.

Answers to Even-Numbered Practice Problems

2. $2 \times 2 \times 2 \times 3 \times 3$ or $2^3 \times 3^2$ **4.** $\dfrac{5}{7}$ **6.** $\dfrac{4}{7}$ **8.** $\dfrac{13}{16}$

☐ After studying this section, you will be able to:

1 *Change a mixed number to an improper fraction*

2 *Change an improper fraction to a mixed number*

3 *Reduce a mixed number or an improper fraction*

2.3 IMPROPER FRACTIONS AND MIXED NUMBERS

Let's look at the *value* of a fraction. The value of any fraction can be

- Less than 1
- Equal to 1
- Greater than 1

We have different names for each of these kinds of fractions. If the value is less than 1 $\left(\text{for example, } \dfrac{3}{4}\right)$, we call the fraction "proper". If the value is equal to 1 $\left(\text{for example, } \dfrac{4}{4}\right)$, we call the fraction the "unit fraction". If the value is greater than 1, the fraction can have different names. We call the $\dfrac{5}{4}$ an "improper" fraction. We call its equivalent form $1\dfrac{1}{4}$ a "mixed number" because it is a "mix" of a whole number and a proper fraction.

Compare the values, the names, and the diagrams in the chart. See if you can see the pattern.

Value less than 1	Value equal to 1	Value greater than 1		
Proper Fraction	Unit Fraction	Improper Fraction	or	Mixed Number
$\dfrac{3}{4}$	$\dfrac{4}{4}$	$\dfrac{5}{4}$	or	$1\dfrac{1}{4}$
$\dfrac{7}{8}$	$\dfrac{8}{8}$	$\dfrac{17}{8}$	or	$2\dfrac{1}{8}$
$\dfrac{3}{100}$	$\dfrac{100}{100}$	$\dfrac{109}{100}$	or	$1\dfrac{9}{100}$

1 Often, we deal with a quantity that has an integer part and a fraction part. Suppose that we have $1\dfrac{1}{6}$ of a pizza. It consists of 1 whole pizza plus $\dfrac{1}{6}$ of a pizza.

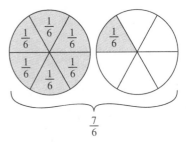

Another way to write $1\dfrac{1}{6}$ is to write $\dfrac{7}{6}$ of a pizza. The quantity $\dfrac{7}{6}$ is called an improper fraction. If the numerator is greater than or equal to the denominator, the fraction is an **improper fraction**. The fractions $\dfrac{3}{5}, \dfrac{5}{7}, \dfrac{1}{8}$ are called proper fractions. If the numerator is smaller than the denominator, the fraction is called a **proper fraction**. The quantity $1\dfrac{1}{2}$ is called a mixed number. A **mixed number** is a sum of a whole number greater than zero and a proper fraction. The notation $1\dfrac{1}{6}$ actually means $1 + \dfrac{1}{6}$. The plus sign is omitted.

Improper fractions are more convenient to add, subtract, multiply, and divide than mixed numbers. We need to be able to change a mixed number to an improper fraction.

Changing a Mixed Number to an Improper Fraction

1. Multiply the whole number by the denominator of the fraction.
2. Add the numerator of the fraction to the product found in step 1.
3. Write the sum found in step 2 over the denominator of the fraction.

Example 1 Change the mixed number to an improper fraction: $3\frac{2}{5}$.

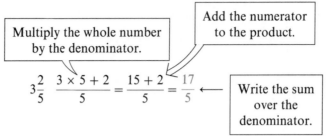

Multiply the whole number by the denominator.

Add the numerator to the product.

$$3\frac{2}{5} \quad \frac{3 \times 5 + 2}{5} = \frac{15 + 2}{5} = \frac{17}{5} \longleftarrow$$

Write the sum over the denominator. ∎

Practice Problem 1 Change the mixed number to an improper fraction: $4\frac{3}{7}$. ∎

Example 2 Change each mixed number to an improper fraction.

(a) $5\frac{4}{9}$ **(b)** $18\frac{3}{5}$

(a) $5\frac{4}{9} = \frac{9 \times 5 + 4}{9} = \frac{45 + 4}{9} = \frac{49}{9}$ **(b)** $18\frac{3}{5} = \frac{5 \times 18 + 3}{5} = \frac{90 + 3}{5} = \frac{93}{5}$ ∎

Practice Problem 2 Change the mixed number to an improper fraction.

(a) $6\frac{2}{3}$ **(b)** $19\frac{4}{7}$ ∎

2 There are several occasions where we need to change an improper fraction to a mixed number.

> **Changing an Improper Fraction to a Mixed Number**
>
> **1.** Divide the numerator by the denominator.
> **2.** Write the quotient followed by the fraction with the remainder over the denominator.
>
> $$\text{quotient } \frac{\text{remainder}}{\text{denominator}}$$

Example 3 Write $\frac{13}{5}$ as a mixed number.

We divide the denominator 5 into 13.

$$\begin{array}{r} 2 \\ 5\overline{)13} \\ 10 \\ \hline 3 \end{array}$$

2 ← quotient

3 ← remainder

The answer is in the form $\quad \text{quotient } \dfrac{\text{remainder}}{\text{denominator}}$

$$\frac{13}{5} \qquad \begin{array}{r} 2 \\ 5\overline{)13} \\ 10 \\ \hline 3 \end{array} \qquad 2\frac{3}{5}$$

quotient

remainder

denominator

Thus $\frac{13}{5} = 2\frac{3}{5}$. ∎

Practice Problem 3 Write $\frac{17}{4}$ as a mixed number. ■

Example 4 Write as a mixed number or a whole number.

(a) $\frac{29}{7}$

(b) $\frac{105}{31}$

(c) $\frac{85}{17}$

(a) $7\overline{)29}$ → $\begin{array}{r} 4 \\ 7\overline{)29} \\ 28 \\ \hline 1 \end{array}$ $\frac{29}{7} = 4\frac{1}{7}$

(b) $\begin{array}{r} 3 \\ 31\overline{)105} \\ 93 \\ \hline 12 \end{array}$ $\frac{105}{31} = 3\frac{12}{31}$

(c) $\begin{array}{r} 5 \\ 17\overline{)85} \\ 85 \\ \hline 0 \end{array}$ The remainder is 0, thus $\frac{85}{17} = 5$, a whole number. ■

Practice Problem 4 Write as a mixed number or a whole number.

(a) $\frac{36}{5}$

(b) $\frac{116}{27}$

(c) $\frac{91}{13}$ ■

3 Either mixed numbers or improper fractions may need to be reduced.

Example 5 Reduce the improper fraction $\frac{22}{8}$.

$$\frac{22}{8} = \frac{\overset{1}{\cancel{2}} \times 11}{\underset{1}{\cancel{2}} \times 2 \times 2} = \frac{11}{4} \quad ■$$

Practice Problem 5 Reduce the improper fraction: $\frac{51}{15}$. ■

Example 6 Reduce the mixed number: $4\frac{21}{28}$.

We do not need to reduce the whole number 4, only the fraction $\frac{21}{28}$.

$$\frac{21}{28} = \frac{3 \times \overset{1}{\cancel{7}}}{4 \times \underset{1}{\cancel{7}}} = \frac{3}{4}$$

Therefore, $4\frac{21}{28} = 4\frac{3}{4}$. ■

Practice Problem 6 Reduce the mixed number: $3\frac{16}{80}$. ■

If an improper fraction contains a very large numerator and denominator, it is best to change the fraction to a mixed number before reducing.

Example 7 Reduce $\dfrac{945}{567}$ by first changing to a mixed number.

$$567\overline{)945} \qquad \text{so} \qquad \dfrac{945}{567} = 1\dfrac{378}{567}$$
$$\underline{567}$$
$$378$$

To reduce the fraction we write

$$\dfrac{378}{567} = \dfrac{3 \times 3 \times 3 \times 2 \times 7}{3 \times 3 \times 3 \times 3 \times 7} = \dfrac{\overset{1}{3} \times \overset{1}{3} \times \overset{1}{3} \times 2 \times \overset{1}{7}}{\underset{1}{3} \times \underset{1}{3} \times \underset{1}{3} \times 3 \times \underset{1}{7}} = \dfrac{2}{3}$$

So $\dfrac{945}{567} = 1\dfrac{378}{567} = 1\dfrac{2}{3}$. ■

Practice Problem 7 Reduce $\dfrac{1001}{572}$ by first changing to a mixed number. ■

To Think About

A student concluded that just by looking at the denominator he could tell that the fraction $\dfrac{1655}{97}$ cannot be reduced unless $1655 \div 97$ is a whole number. How did he make that conclusion? Note that 97 is a prime number. The only factors of 97 are 97 and 1. Therefore, *any* fraction with 97 in the denominator can be reduced only if 97 is a factor of the numerator. Since $1655 \div 97$ is not a whole number (see the division below), it is therefore impossible to reduce $\dfrac{1655}{97}$.

$$97\overline{)1655}$$
$$\underline{97}$$
$$685$$
$$\underline{679}$$
$$6$$

You may explore this idea in problems 76 and 77. ■

Change each mixed number to an improper fraction.

1. $3\frac{1}{2}$

2. $2\frac{1}{4}$

3. $4\frac{2}{3}$

4. $3\frac{5}{6}$

5. $2\frac{3}{7}$

6. $3\frac{5}{8}$

7. $5\frac{3}{10}$

8. $4\frac{7}{10}$

9. $6\frac{7}{8}$

10. $8\frac{1}{3}$

11. $21\frac{2}{3}$

12. $13\frac{1}{3}$

13. $47\frac{1}{2}$

14. $56\frac{1}{2}$

15. $28\frac{1}{6}$

16. $4\frac{6}{7}$

17. $10\frac{11}{12}$

18. $13\frac{5}{7}$

19. $7\frac{9}{10}$

20. $4\frac{1}{50}$

21. $8\frac{1}{25}$

22. $106\frac{3}{4}$

23. $203\frac{1}{2}$

24. $118\frac{2}{3}$

25. $164\frac{2}{3}$

26. $5\frac{13}{15}$

27. $7\frac{14}{15}$

28. $4\frac{26}{27}$

29. $5\frac{13}{25}$

30. $6\frac{18}{19}$

Change each improper fraction to a mixed number or a whole number.

31. $\frac{7}{5}$

32. $\frac{8}{3}$

33. $\frac{9}{4}$

34. $\frac{11}{6}$

35. $\frac{15}{6}$

36. $\frac{21}{6}$

37. $\frac{27}{8}$

38. $\frac{42}{6}$

39. $\frac{65}{13}$

40. $\frac{56}{15}$

41. $\frac{86}{9}$

42. $\frac{47}{2}$

43. $\frac{28}{13}$

44. $\frac{54}{17}$

45. $\frac{51}{16}$

46. $\frac{19}{3}$

47. $\frac{28}{3}$

48. $\frac{33}{4}$

49. $\frac{35}{2}$

50. $\frac{132}{11}$

51. $\frac{91}{7}$

52. $\frac{183}{7}$

53. $\frac{305}{9}$

54. $\frac{195}{13}$

55. $\frac{102}{17}$

Reduce each mixed number.

56. $1\frac{9}{12}$

57. $2\frac{10}{15}$

58. $4\frac{17}{34}$

59. $3\frac{15}{90}$

60. $10\frac{33}{36}$

61. $12\frac{17}{85}$

Reduce each improper fraction.

62. $\dfrac{24}{6}$ **63.** $\dfrac{32}{10}$ **64.** $\dfrac{36}{15}$ **65.** $\dfrac{32}{18}$ **66.** $\dfrac{78}{9}$ **67.** $\dfrac{130}{22}$

Change to a mixed number and reduce.

68. $\dfrac{340}{126}$ **69.** $\dfrac{386}{226}$ **70.** $\dfrac{986}{424}$ **71.** $\dfrac{764}{328}$ **72.** $\dfrac{455}{364}$ **73.** $\dfrac{1365}{1155}$

To Think About

74. Reduce: $\dfrac{3952}{3705}$.

75. Change to a mixed number: $\dfrac{8913}{31}$.

76. Can $\dfrac{5687}{101}$ be reduced? Why?

77. Can $\dfrac{9810}{157}$ be reduced? Why?

78. The designer's fall wardrobe calls for $126\frac{2}{3}$ yards of wool tweed cloth. Change this number to an improper fraction.

79. The barn holds $136\frac{3}{4}$ bales of hay. Change this number to an improper fraction.

Cumulative Review Problems

80. Add: $16,385 + 4126 + 8056$.

81. Subtract: $1,398,210 - 1,137,963$.

For Extra Practice Examples and Exercises, turn to page 134.

Solutions to Odd-Numbered Practice Problems

1. $4\dfrac{3}{7} = \dfrac{7 \times 4 + 3}{7} = \dfrac{28 + 3}{7} = \dfrac{31}{7}$

3. $4\overline{)17}$ so $\dfrac{17}{4} = 4\dfrac{1}{4}$
$\quad\quad \underline{16}$
$\quad\quad\;\; 1$

5. $\dfrac{51}{15} = \dfrac{3 \times 17}{3 \times 5} = \dfrac{\overset{1}{\cancel{3}} \times 17}{\cancel{3} \times 5} = \dfrac{17}{5}$

7. $\dfrac{1001}{572} = 1\dfrac{429}{572}$ Now the fraction $\dfrac{429}{572} = \dfrac{3 \times \overset{1}{\cancel{11}} \times \overset{1}{\cancel{13}}}{2 \times 2 \times \underset{1}{\cancel{11}} \times \underset{1}{\cancel{13}}} = \dfrac{3}{4}$. Thus $\dfrac{1001}{572} = 1\dfrac{429}{572} = 1\dfrac{3}{4}$.

Answers to Even-Numbered Practice Problems

2. (a) $\dfrac{20}{3}$ **(b)** $\dfrac{137}{7}$ **4. (a)** $7\dfrac{1}{5}$ **(b)** $4\dfrac{8}{27}$ **(c)** 7 **6.** $3\dfrac{1}{5}$

2.4 MULTIPLICATION OF FRACTIONS

After studying this section, you will be able to:

1 Multiply two fractions that are proper or improper

2 Multiply an integer by a fraction

3 Multiply mixed numbers

We need to be able to compute with fractions. We start with multiplication. When would you multiply with fractions? Suppose that you are laying a carpet in a hallway: The area—the measure of how much carpeting will fit the hallway—is found by *multiplying* $1\frac{1}{5}$ by $4\frac{5}{6}$ (see Practice Problem 6 and the figure in the margin). Or suppose you want to cook an amount equal to half of what the recipe shown calls for.

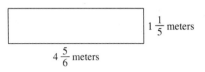

$1\frac{1}{5}$ meters

$4\frac{5}{6}$ meters

You would be multiplying with fractions:

$\frac{1}{2}$ of 2 cups sugar $\frac{1}{2}$ of $\frac{1}{4}$ teaspoon salt

$\frac{1}{2}$ of 4 oz chocolate $\frac{1}{2}$ of 1 teaspoon vanilla

$\frac{1}{2}$ of $\frac{1}{2}$ cup butter $\frac{1}{2}$ of 1 cup all-purpose flour

$\frac{1}{2}$ of 4 eggs $\frac{1}{2}$ of 1 cup nutmeats

Fudge Squares	
Ingredients:	
2 cups sugar	1/4 teaspoon salt
4 oz. chocolate	1 teaspoon vanilla
1/2 cup butter	1 cup all-
4 eggs	purpose flour
	1 cup nut meats

When you multiply two whole numbers together you get a greater whole number. But what are one-half of one-half $\left(\frac{1}{2} \times \frac{1}{2}\right)$ and one-half of one-fourth $\left(\frac{1}{2} \times \frac{1}{4}\right)$?

1 To multiply two fractions, we multiply the numerators and multiply the denominators.

$$\frac{2}{3} \times \frac{5}{7} = \frac{10}{21} \quad \begin{matrix} \longleftarrow 2 \times 5 = 10 \\ \longleftarrow 3 \times 7 = 21 \end{matrix}$$

In general, for all positive whole numbers a, b, c, and d,

$$\frac{a}{b} \times \frac{c}{d} = \frac{a \times c}{b \times d}$$

In words: multiply the numbers in the numerators to get the numerator of the result. Multiply the numbers in the denominators to get the denominator of the result.

We often use multiplication of fractions to describe taking a fractional part of something. To find $\frac{1}{2}$ of $\frac{3}{7}$, we multiply

$$\frac{1}{2} \times \frac{3}{7} = \frac{3}{14}$$

Here $a = 1$, $b = 2$, $c = 3$, and $d = 7$. We begin with a bar that is $\frac{3}{7}$ shaded.

To find $\frac{1}{2}$ of $\frac{3}{7}$ we divide the bar in half and take $\frac{1}{2}$ of the shaded section.

$\frac{1}{2}$ of $\frac{3}{7}$ yields 3 out of 14 squares. Thus

$$\frac{1}{2} \times \frac{3}{7} = \frac{3}{14}$$

Example 1 Multiply.

(a) $\frac{3}{8} \times \frac{5}{7}$

(b) $\frac{1}{11} \times \frac{2}{13}$

(a) $\frac{3}{8} \times \frac{5}{7} = \frac{15}{56}$

(b) $\frac{1}{11} \times \frac{2}{13} = \frac{2}{143}$ ∎

Practice Problem 1 Multiply.

(a) $\dfrac{6}{7} \times \dfrac{3}{13}$

(b) $\dfrac{1}{5} \times \dfrac{11}{12}$ ∎

Some products may be reduced.

$$\frac{12}{35} \times \frac{25}{18} = \frac{300}{630} = \frac{10}{21}$$

By canceling before multiplication, the reducing can be done more easily. In a multiplication problem, a factor in the numerator can be canceled with a common factor in the denominator of the same or a different fraction.

Example 2 Simplify by cancellation first and then multiply.

$$\frac{12}{35} \times \frac{25}{18}$$

$\dfrac{\overset{2}{\cancel{12}}}{35} \times \dfrac{25}{\underset{3}{\cancel{18}}}$ First dividing 12 by 6 and 18 by 6.

$\dfrac{\overset{2}{\cancel{12}}}{\underset{7}{\cancel{35}}} \times \dfrac{\overset{5}{\cancel{25}}}{\underset{3}{\cancel{18}}}$ Next dividing 25 by 5 and 35 by 5.

$\dfrac{2}{7} \times \dfrac{5}{3} = \dfrac{10}{21}$ Then multiplying.

The answer is $\dfrac{10}{21}$ ∎

Practice Problem 2 Simplify by cancellation first and then multiply.

$$\frac{55}{72} \times \frac{16}{33}$$ ∎

2 When multiplying a fraction by a whole number it is more convenient to express the whole number as a fraction with a denominator of 1. We know that $5 = \dfrac{5}{1}$, $7 = \dfrac{7}{1}$, and so on.

Example 3 Multiply.

(a) $5 \times \dfrac{3}{8}$

(b) $\dfrac{22}{7} \times 14$

(a) $5 \times \dfrac{3}{8} = \dfrac{5}{1} \times \dfrac{3}{8} = \dfrac{15}{8}$ or $1\dfrac{7}{8}$

(b) $\dfrac{22}{7} \times 14 = \dfrac{22}{\underset{1}{\cancel{7}}} \times \dfrac{\overset{2}{\cancel{14}}}{1} = \dfrac{44}{1} = 44$ ∎

Practice Problem 3 Multiply.

(a) $7 \times \dfrac{5}{13}$

(b) $\dfrac{13}{16} \times 8$ ∎

3 To multiply a fraction by a mixed number or to multiply two mixed numbers, it is necessary to change each mixed number to an improper fraction.

Example 4 Multiply: $4\frac{1}{3} \times 2\frac{1}{4}$.

Change mixed numbers to improper fractions.

$$4\frac{1}{3} = \frac{13}{3} \qquad 2\frac{1}{4} = \frac{9}{4}$$

Thus $4\frac{1}{3} \times 2\frac{1}{4} = \frac{13}{\overset{1}{\cancel{3}}} \times \frac{\overset{3}{\cancel{9}}}{4} = \frac{39}{4}$ or $9\frac{3}{4}$ ■

Practice Problem 4 Multiply: $3\frac{1}{5} \times 2\frac{1}{2}$. ■

Example 5 Multiply.

(a) $\frac{5}{7} \times 3\frac{1}{4}$
(b) $20\frac{2}{5} \times 6\frac{2}{3}$
(c) $\frac{3}{4} \times 1\frac{1}{2} \times \frac{4}{7}$

(a) $\frac{5}{7} \times 3\frac{1}{4} = \frac{5}{7} \times \frac{13}{4} = \frac{65}{28}$ or $2\frac{9}{28}$

(b) $20\frac{2}{5} \times 6\frac{2}{3} = \frac{\overset{34}{\cancel{102}}}{\underset{1}{\cancel{5}}} \times \frac{\overset{4}{\cancel{20}}}{\underset{1}{\cancel{3}}} = \frac{136}{1} = 136$

(c) $\frac{3}{4} \times 1\frac{1}{2} \times \frac{4}{7} = \frac{3}{\underset{1}{\cancel{4}}} \times \frac{3}{2} \times \frac{\overset{1}{\cancel{4}}}{7} = \frac{9}{14}$ ■

Practice Problem 5 Multiply.

(a) $2\frac{1}{6} \times \frac{4}{7}$
(b) $10\frac{2}{3} \times 13\frac{1}{2}$
(c) $\frac{3}{5} \times 1\frac{1}{3} \times \frac{5}{8}$ ■

Example 6 Find the area in square miles of a rectangle with width $1\frac{1}{3}$ miles and length $12\frac{1}{4}$ miles.

Length $= 12\frac{1}{4}$ miles

Width $= 1\frac{1}{3}$ miles

We find the area of a rectangle by multiplying the width times the length.

$$1\frac{1}{3} \times 12\frac{1}{4} = \frac{4}{3} \times \frac{49}{\underset{1}{\cancel{4}}} = \frac{49}{3} \text{ or } 16\frac{1}{3}$$

The area is $16\frac{1}{3}$ square miles. ■

Practice Problem 6 Find the area in square meters of a rectangle with width $1\frac{1}{5}$ meters and length $4\frac{5}{6}$ meters. ■

To Think About

Is it possible to multiply two mixed numbers without changing to improper fractions? It is if you are very careful. Suppose that we investigate the two mixed numbers in Example 4. First we note that $4\frac{1}{3} = 4 + \frac{1}{3}$ and $2\frac{1}{4} = 2 + \frac{1}{4}$. So the problem is the same as $\left(4 + \frac{1}{3}\right) \times \left(2 + \frac{1}{4}\right)$:

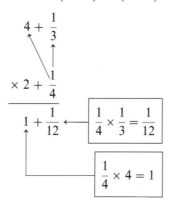

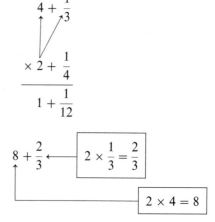

Now we add

$$8 + \frac{2}{3} + 1 + \frac{1}{12} = 9 + \frac{2}{3} + \frac{1}{12}$$

Now $\frac{8}{12} = \frac{2}{3}$. We can see this since $\frac{8 \div 4}{12 \div 4} = \frac{2}{3}$. We will explain this concept is detail in Section 2.6. For now we see that $\frac{2}{3}$ and $\frac{8}{12}$ are equivalent, so we can replace $\frac{2}{3}$ by $\frac{8}{12}$.

$$9 + \frac{2}{3} + \frac{1}{12} = 9 + \frac{8}{12} + \frac{1}{12}$$

$$= 9 + \frac{9}{12} = 9 + \frac{3}{4}$$

The answer is $9\frac{3}{4}$. This agrees with our result in Example 4. You can try this method in problems 49 and 50.   ■

DEVELOPING YOUR STUDY SKILLS

HOW TO SOLVE WORD PROBLEMS

There are three steps to solving word problems. You will be delighted with the success you will achieve if you follow these steps.

Step 1 Understand the Problem

- Read the problem for understanding. What does it say? Pick out all the important information and write it down.
- Draw any diagrams, charts, or pictures that will help you visualize the problem.
- Write down the question the problem is asking.

The point of Step 1 is to understand the problem. Do not be concerned about *how* to solve the problem at this time. Unfortunately, many students misunderstand or ignore this step. Instead of reading for understanding, they read the problem with the thought, "How am I going to solve this problem?" Watch out for this, as it will block your reasoning process.

Step 2 Set up the Problem

- Identify the variables to be used for the unknown quantity. Decide what you will let "*x*" be in your solution. Be specific, and include the units of measure, if applicable.
- Identify other unknown quantities if called for. Sometimes two or more quantities are to be found. Let "*x*" be one unknown and show the other in terms of "*x*."

Be careful in Step 2. Carelessness or incompleteness here leads to confusion. You need to know exactly what "*x*" *means* to understand your solutions.

Step 3 Solve the Problem

- Write the equation for the relationship that exists in the problem, using the unknowns you have identified in Step 2. Be sure that the equation makes sense and that it shows the relationship accurately.
- Solve the equation. Be careful with decimal points and negative signs—little errors here can cause *big* problems.
- Read the problem again to be sure you have found all the solutions the problem asks for.
- Check your answer. Does it work? Does it make sense? Watch for negative or fractional solutions. Can they exist?

Remember that the key to success with word problems is to practice solving them. KEEP TRYING, AND DON'T GIVE UP!

EXERCISES 2.4

Multiply.

1. $\dfrac{1}{7} \times \dfrac{3}{5}$

2. $\dfrac{1}{8} \times \dfrac{5}{11}$

3. $\dfrac{3}{4} \times \dfrac{5}{13}$

4. $\dfrac{4}{7} \times \dfrac{3}{5}$

5. $\dfrac{6}{5} \times \dfrac{10}{12}$

6. $\dfrac{7}{8} \times \dfrac{16}{21}$

7. $\dfrac{15}{7} \times \dfrac{8}{25}$

8. $\dfrac{7}{9} \times \dfrac{12}{21}$

9. $\dfrac{15}{28} \times \dfrac{7}{9}$

10. $\dfrac{5}{24} \times \dfrac{18}{15}$

11. $\dfrac{9}{10} \times \dfrac{35}{12}$

12. $\dfrac{12}{17} \times \dfrac{3}{24}$

13. $7 \times \dfrac{3}{5}$

14. $\dfrac{2}{11} \times 4$

15. $\dfrac{5}{16} \times 8$

16. $10 \times \dfrac{7}{25}$

17. $\dfrac{3}{7} \times \dfrac{2}{5} \times \dfrac{14}{9}$

18. $\dfrac{1}{5} \times \dfrac{3}{13} \times \dfrac{10}{27}$

19. $\dfrac{4}{5} \times \dfrac{1}{8} \times \dfrac{35}{7}$

20. $\dfrac{10}{13} \times \dfrac{26}{15} \times \dfrac{2}{3}$

Multiply. Change any mixed number to an improper fraction before multiplying.

21. $3\dfrac{1}{5} \times \dfrac{7}{8}$

22. $\dfrac{8}{11} \times 4\dfrac{3}{4}$

23. $1\dfrac{1}{4} \times 3\dfrac{2}{3}$

24. $2\dfrac{3}{5} \times 1\dfrac{4}{7}$

25. $2\dfrac{1}{2} \times 6$

26. $4\dfrac{1}{3} \times 9$

27. $2\dfrac{3}{10} \times 1\dfrac{2}{5}$

28. $4\dfrac{3}{5} \times 2\dfrac{1}{10}$

29. $1\dfrac{3}{16} \times 0$

30. $3\dfrac{7}{8} \times 1$

31. $4\dfrac{1}{5} \times 12\dfrac{2}{9}$

32. $5\dfrac{1}{4} \times 10\dfrac{3}{7}$

33. $8\dfrac{5}{6} \times \dfrac{2}{5}$

34. $\dfrac{3}{4} \times 9\dfrac{5}{7}$

35. $1 \times 12\dfrac{5}{13}$

36. $0 \times 7\dfrac{1}{4}$

Solve each applied problem.

37. Find the area of a rectangular park that is $6\dfrac{1}{3}$ miles long and $2\dfrac{1}{4}$ miles wide. (The area of a rectangle is the product of the length times the width.)

38. Find the area of a wheat-harvest complex that is $5\dfrac{1}{2}$ miles long and $3\dfrac{1}{6}$ miles wide. (The area of a rectangle is the product of the length times the width.)

39. A car has $12\dfrac{1}{5}$ gallons of gas. The car averages 25 miles per gallon. How far can the car go?

40. A car has $14\dfrac{3}{4}$ gallons of gas. The car averages 18 miles per gallon. How far can the car go?

41. A dress requires $3\dfrac{3}{8}$ yards of material. How many yards of material are required for 14 of the same dresses?

42. A capsule contains $4\dfrac{1}{3}$ grams of a certain chemical. How many grams of the chemical would be needed to fill 24 capsules?

43. A recipe requires $15\frac{3}{4}$ ounces of sugar. Wally wants to make $\frac{3}{4}$ of that recipe. How many ounces of sugar will he need?

44. The wheel on a car is turning $29\frac{1}{4}$ revolutions per minute. How fast would it turn at $\frac{3}{4}$ of that speed?

45. If one share of Mobil stock costs $\$48\frac{3}{8}$, how much money will be needed to buy 72 shares?

46. If one share of Polaroid stock costs $\$45\frac{7}{8}$, how much money will be needed to buy 64 shares?

 To Think About

Using the method discussed in the text, multiply the mixed number without changing to improper fractions.

47. $2\frac{3}{4} \times 3\frac{1}{6}$

48. $4\frac{2}{3} \times 5\frac{1}{4}$

49. Of the eligible voters in Springfield, $\frac{5}{6}$ were registered. Of those registered, $\frac{3}{4}$ actually voted.
  (a) What fractional portion of the voters of Springfield actually voted?
  (b) If Springfield has 12,000 eligible voters, how many people actually voted?

50. Roberto has $\frac{3}{8}$ of his income withheld for taxes, retirement, dues, and medical coverage. He found that $\frac{3}{5}$ of the withheld amount goes for federal taxes.
  (a) What fractional portion of his salary goes to federal taxes?
  (b) What amount is withheld each month for feddderal taxes if his rate of pay is $1440 per month?

Multiply.

51. $\frac{5}{8} \times \frac{3}{4} \times 2\frac{5}{7} \times \frac{16}{17} \times \frac{3}{25} \times \frac{2}{19}$

52. $\frac{5}{7} \times \frac{2}{5} \times \frac{3}{11} \times 4\frac{1}{3} \times \frac{7}{9} \times \frac{22}{25}$

Cumulative Review Problems

53. A total of 16,399 cars used a toll bridge in January (31 days). What is the average number of cars using the bridge in one day? (Divide the number of cars by the number of days.)

54. The Office of Investors services has 15,456 calls made per month from the sales personnel. There are 42 sales personnel in the office. What is the average number of calls made per month by one salesperson? (Divide the number of calls by the number of sales personnel.)

55. A computer printer can print 146 lines per minute. How many lines can it print in 12 minutes?

56. At cruising speed a new commercial jetplane uses 12,360 gallons of fuel per hour. How many gallons will be used in 14 hours of flying time?

For Extra Practice Examples and Exercises, turn to page 134.

1. (a) $\dfrac{6}{7} \times \dfrac{3}{13} = \dfrac{18}{91}$ **(b)** $\dfrac{1}{5} \times \dfrac{11}{12} = \dfrac{11}{60}$ **3. (a)** $7 \times \dfrac{5}{13} = \dfrac{7}{1} \times \dfrac{5}{13} = \dfrac{35}{13}$ or $2\dfrac{9}{13}$ **(b)** $\dfrac{13}{16} \times 8 = \dfrac{13}{\cancel{16}_{2}} \times \dfrac{\cancel{8}^{1}}{1} = \dfrac{13}{2}$ or $6\dfrac{1}{2}$

5. (a) $2\dfrac{1}{6} \times \dfrac{4}{7} = \dfrac{13}{\cancel{6}_{3}} \times \dfrac{\cancel{4}^{2}}{7} = \dfrac{26}{21}$ or $1\dfrac{5}{21}$ **(b)** $10\dfrac{2}{3} \times 13\dfrac{1}{2} = \dfrac{\cancel{32}^{16}}{\cancel{3}_{1}} \times \dfrac{\cancel{27}^{9}}{\cancel{2}_{1}} = \dfrac{144}{1} = 144$ **(c)** $\dfrac{3}{5} \times 1\dfrac{1}{3} \times \dfrac{5}{8} = \dfrac{\cancel{3}^{1}}{\cancel{5}_{1}} \times \dfrac{\cancel{4}^{1}}{\cancel{3}_{1}} \times \dfrac{\cancel{5}^{1}}{\cancel{8}_{2}} = \dfrac{1}{2}$

2. $\dfrac{10}{27}$ **4.** 8   **6.** $5\dfrac{4}{5}$ square meters

---

1 *Divide two proper or improper fractions*

2 *Divide a whole number and a fraction*

3 *Divide two mixed numbers or one mixed number and a fraction*

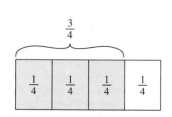

2.5 DIVISION OF FRACTIONS

Why would you divide fractions? Consider these problems.

- A copper pipe that is $9\dfrac{1}{4}$ feet long is to be cut into 14 equal pieces. How long will each piece be?

- Wanda traveled 140 miles for $2\dfrac{1}{3}$ hours. What was her average speed (in miles per hour)?

- A butcher takes $21\dfrac{1}{4}$ pounds of steak that she wishes to cut up into packages that average $1\dfrac{1}{4}$ pounds each. How many packages can she make?

In each of these situations, you recognize that you must *divide* certain numbers to arrive at the result. To do the actual computations, we need a technique for correctly manipulating the fractions. The method used depends on knowing that *a division problem can be transformed into an equivalent multiplication problem.* "Stay tuned" to learn this method and use it to solve the three problems above.

1 Division of Two Proper or Improper Fractions

How do we divide two fractions? We invert the second fraction and multiply.

$$\frac{3}{4} \div \frac{1}{4} = \frac{3}{\cancel{4}} \times \frac{\cancel{4}^{1}}{1} = \frac{3}{1} = 3$$

Suppose that you want to know how many $\dfrac{1}{4}$'s are in $\dfrac{3}{4}$. In this sketch $\dfrac{3}{4}$ of the rectangle is shaded.

The answer is 3. We notice that three squares are shaded.

To find out how many $\dfrac{1}{4}$'s are in $\dfrac{3}{4}$, we divide $\dfrac{3}{4} \div \dfrac{1}{4}$. We invert the second fraction and multiply:

$$\frac{3}{4} \div \frac{1}{4} = \frac{3}{\cancel{4}} \times \frac{\cancel{4}^{1}}{1} = 3.$$

In the following sketch, there are three $\dfrac{1}{4}$'s in $\dfrac{3}{4}$.

To **invert** a fraction is to interchange the numerator and the denominator. If we invert $\frac{5}{9}$, we obtain $\frac{9}{5}$. If we invert $\frac{6}{1}$, we obtain $\frac{1}{6}$. Numbers such as $\frac{5}{9}$ and $\frac{9}{5}$ are called **reciprocals** of each other.

Rule for Division of Fractions

To divide two fractions, we invert the second fraction and multiply.

$$\frac{a}{b} \div \frac{c}{d} = \frac{a}{b} \times \frac{d}{c}$$

(when a, b, c, and d are not zero).

Example 1 Divide.

(a) $\frac{3}{11} \div \frac{2}{5}$

(b) $\frac{5}{8} \div \frac{25}{16}$

(a) $\frac{3}{11} \div \frac{2}{5} = \frac{3}{11} \times \frac{5}{2} = \frac{15}{22}$

(b) $\frac{5}{8} \div \frac{25}{16} = \frac{\overset{1}{5}}{\underset{1}{8}} \times \frac{\overset{2}{16}}{\underset{5}{25}} = \frac{2}{5}$ ∎

Practice Problem 1 Divide.

(a) $\frac{7}{13} \div \frac{3}{4}$

(b) $\frac{16}{35} \div \frac{24}{25}$ ∎

② Division Involving a Whole Number and a Fraction

When dividing with whole numbers it is helpful to remember that for any whole number a, $a = \frac{a}{1}$.

Example 2 Divide.

(a) $\frac{3}{7} \div 2$

(b) $5 \div \frac{10}{13}$

(a) $\frac{3}{7} \div 2 = \frac{3}{7} \div \frac{2}{1} = \frac{3}{7} \times \frac{1}{2} = \frac{3}{14}$

(b) $5 \div \frac{10}{13} = \frac{5}{1} \div \frac{10}{13} = \frac{\overset{1}{5}}{1} \times \frac{13}{\underset{2}{10}} = \frac{13}{2}$ or $6\frac{1}{2}$ ∎

Practice Problem 2 Divide.

(a) $\frac{3}{17} \div 6$

(b) $14 \div \frac{7}{15}$ ∎

Example 3 Divide, if possible.

(a) $\frac{23}{25} \div 1$ (b) $1 \div \frac{7}{5}$ (c) $0 \div \frac{4}{9}$ (d) $\frac{3}{17} \div 0$

(a) $\frac{23}{25} \div 1 = \frac{23}{25} \times \frac{1}{1} = \frac{23}{25}$ ← Any fraction can be multiplied by 1 without changing the value of the fraction.

(b) $1 \div \frac{7}{5} = \frac{1}{1} \times \frac{5}{7} = \frac{5}{7}$ ←

(c) $0 \div \frac{4}{9} = \frac{0}{1} \times \frac{9}{4} = \frac{0}{4} = 0$
Zero divided by any nonzero number is zero.

(d) $\frac{3}{17} \div 0$
You can never divide by zero.
This problem cannot be done. ∎

$\times$ 0 Low

Practice Problem 3 Divide, if possible.

(a) $1 \div \dfrac{11}{13}$ (b) $\dfrac{14}{17} \div 1$ (c) $\dfrac{3}{11} \div 0$ (d) $0 \div \dfrac{9}{16}$ ∎

To Think About

Why is it that we divide by inverting the second fraction and multiply? What is really going on when we divide fractions in that manner? We are actually multiplying by 1. Let us see why. Consider

$$\frac{3}{7} \div \frac{2}{3} = \frac{\frac{3}{7}}{\frac{2}{3}}$$

We write the division by using a fraction

$$= \frac{\frac{3}{7}}{\frac{2}{3}} \times 1$$

Any fraction can be multiplied by 1 without changing the value of the fraction. This is the fundamental rule of fractions.

$$= \frac{\frac{3}{7}}{\frac{2}{3}} \times \frac{\frac{3}{2}}{\frac{3}{2}}$$

Any nonzero number divided by itself equals 1.

$$= \frac{\frac{3}{7} \times \frac{3}{2}}{\frac{2}{3} \times \frac{3}{2}}$$

Definition of multiplication of fractions.

$$= \frac{\frac{3}{7} \times \frac{3}{2}}{1} = \frac{3}{7} \times \frac{3}{2}$$

Any number can be written as a fraction with a denominator of 1 without changing its value.

Thus

$$\frac{3}{7} \div \frac{2}{3} = \frac{3}{7} \times \frac{3}{2} = \frac{9}{14} \quad ∎$$

◧ Division of Mixed Numbers

If one or more mixed numbers is involved in the division, they should be converted to improper fractions first.

Example 4 Divide.

(a) $3\dfrac{7}{15} \div 1\dfrac{1}{25}$ (b) $\dfrac{3}{5} \div 2\dfrac{1}{7}$

(a) $3\dfrac{7}{15} \div 1\dfrac{1}{25} = \dfrac{52}{15} \div \dfrac{26}{25} = \dfrac{\overset{2}{\cancel{52}}}{\underset{3}{\cancel{15}}} \times \dfrac{\overset{5}{\cancel{25}}}{\underset{1}{\cancel{26}}} = \dfrac{10}{3}$ or $3\dfrac{1}{3}$

(b) $\dfrac{3}{5} \div 2\dfrac{1}{7} = \dfrac{3}{5} \div \dfrac{15}{7} = \dfrac{\overset{1}{\cancel{3}}}{5} \times \dfrac{7}{\underset{5}{\cancel{15}}} = \dfrac{7}{25}$ ∎

Practice Problem 4 Divide.

(a) $1\dfrac{1}{5} \div \dfrac{7}{10}$
(b) $2\dfrac{1}{4} \div 1\dfrac{7}{8}$ ■

The division of two fractions may be indicated by a fraction bar.

Example 5 Divide.

(a) $\dfrac{10\dfrac{2}{9}}{2\dfrac{1}{3}}$
(b) $\dfrac{1\dfrac{1}{15}}{3\dfrac{1}{3}}$

(a) $\dfrac{10\dfrac{2}{9}}{2\dfrac{1}{3}} = 10\dfrac{2}{9} \div 2\dfrac{1}{3} = \dfrac{92}{9} \div \dfrac{7}{3} = \dfrac{92}{\overset{3}{\cancel{9}}} \times \dfrac{\overset{1}{\cancel{3}}}{7} = \dfrac{92}{21}$ or $4\dfrac{8}{21}$

(b) $\dfrac{1\dfrac{1}{15}}{3\dfrac{1}{3}} = 1\dfrac{1}{15} \div 3\dfrac{1}{3} = \dfrac{16}{15} \div \dfrac{10}{3} = \dfrac{\overset{8}{\cancel{16}}}{\underset{5}{\cancel{15}}} \times \dfrac{\overset{1}{\cancel{3}}}{\underset{5}{\cancel{10}}} = \dfrac{8}{25}$ ■

Practice Problem 5 Divide.

(a) $\dfrac{5\dfrac{2}{3}}{7}$
(b) $\dfrac{1\dfrac{2}{5}}{2\dfrac{1}{3}}$ ■

Example 6 There are 117 milligrams of cholesterol in $4\dfrac{1}{3}$ cups of milk. How much cholesterol would be contained in 1 cup of milk?

We want to divide the 117 by $4\dfrac{1}{3}$ to find out how much is in 1 cup.

$$117 \div 4\dfrac{1}{3} = 117 \div \dfrac{13}{3} = \dfrac{\overset{9}{\cancel{117}}}{1} \times \dfrac{3}{\underset{1}{\cancel{13}}} = \dfrac{27}{1} = 27$$

Thus there are 27 milligrams of cholesterol in 1 cup of milk. ■

Practice Problem 6 A copper pipe that is $19\dfrac{1}{4}$ feet long is going to be cut into 14 equal pieces. How long will each piece be? ■

Divide, if possible.

1. $\dfrac{2}{5} \div \dfrac{7}{2}$

2. $\dfrac{3}{4} \div \dfrac{5}{6}$

3. $\dfrac{3}{13} \div \dfrac{5}{26}$

4. $\dfrac{7}{8} \div \dfrac{2}{3}$

5. $\dfrac{5}{6} \div \dfrac{25}{27}$

6. $\dfrac{3}{14} \div \dfrac{6}{7}$

7. $\dfrac{7}{10} \div \dfrac{14}{25}$

8. $\dfrac{8}{15} \div \dfrac{35}{24}$

9. $\dfrac{4}{5} \div 1$

10. $1 \div \dfrac{3}{7}$

11. $2 \div \dfrac{7}{8}$

12. $\dfrac{3}{11} \div 4$

13. $\dfrac{2}{3} \div \dfrac{5}{6}$

14. $\dfrac{3}{4} \div \dfrac{2}{3}$

15. $\dfrac{3}{18} \div \dfrac{3}{18}$

16. $\dfrac{4}{11} \div \dfrac{4}{11}$

17. $\dfrac{3}{7} \div \dfrac{2}{6}$

18. $\dfrac{11}{12} \div \dfrac{1}{5}$

19. $\dfrac{4}{3} \div \dfrac{7}{27}$

20. $\dfrac{8}{9} \div \dfrac{5}{81}$

21. $0 \div \dfrac{3}{17}$

22. $0 \div \dfrac{5}{16}$

23. $\dfrac{18}{19} \div 0$

24. $\dfrac{24}{29} \div 0$

25. $\dfrac{9}{16} \div \dfrac{3}{4}$

26. $\dfrac{3}{4} \div \dfrac{9}{16}$

27. $\dfrac{3}{7} \div \dfrac{15}{28}$

28. $\dfrac{5}{6} \div \dfrac{15}{18}$

29. $\dfrac{12}{35} \div \dfrac{35}{21}$

30. $\dfrac{2}{45} \div \dfrac{7}{90}$

31. $12 \div \dfrac{3}{4}$

32. $18 \div \dfrac{3}{7}$

33. $\dfrac{7}{8} \div 4$

34. $\dfrac{5}{6} \div 12$

35. $5400 \div \dfrac{6}{5}$

36. $2300 \div \dfrac{4}{7}$

37. $\dfrac{\frac{2}{5}}{3}$

38. $\dfrac{\frac{3}{7}}{4}$

39. $\dfrac{\frac{5}{8}}{\frac{25}{7}}$

40. $\dfrac{\frac{3}{16}}{\frac{5}{8}}$

? To Think About

41. Division of fractions is not commutative. For example,

$$\dfrac{2}{3} \div \dfrac{5}{7} \neq \dfrac{5}{7} \div \dfrac{2}{3}$$

In general

$$\dfrac{a}{b} \div \dfrac{c}{d} \neq \dfrac{c}{d} \div \dfrac{a}{b} \qquad (b \neq 0, d \neq 0)$$

But sometimes there are exceptions. Can you think of any numbers a, b, c, d so that for fractions $\dfrac{a}{b}$ and $\dfrac{c}{d}$ it would be true that $\dfrac{a}{b} \div \dfrac{c}{d} = \dfrac{c}{d} \div \dfrac{a}{b}$?

42. Can you think of a way to divide $3\frac{7}{12} \div \frac{1}{4}$ *without* changing $3\frac{7}{12}$ to an improper fraction? Try your method to see if it works.

Divide.

43. $3\frac{1}{4} \div \frac{1}{2}$

44. $2\frac{1}{8} \div \frac{1}{4}$

45. $2\frac{1}{3} \div 4$

46. $6\frac{1}{2} \div 3$

47. $1\frac{7}{9} \div 2\frac{2}{3}$

48. $1\frac{1}{15} \div 3\frac{1}{3}$

49. $5 \div 1\frac{1}{4}$

50. $7 \div 1\frac{2}{5}$

51. $12\frac{1}{2} \div 5\frac{5}{6}$

52. $14\frac{2}{3} \div 3\frac{1}{2}$

53. $8\frac{1}{4} \div 2\frac{3}{4}$

54. $2\frac{3}{8} \div 5\frac{3}{7}$

55. $3\frac{1}{2} \div 1\frac{7}{9}$

56. $1\frac{1}{8} \div 2\frac{2}{7}$

57. $4\frac{3}{5} \div 10$

58. $2\frac{6}{7} \div 5$

59. $\dfrac{4\frac{1}{2}}{\frac{3}{8}}$

60. $\dfrac{6\frac{1}{2}}{\frac{3}{4}}$

61. $\dfrac{1\frac{1}{4}}{1\frac{7}{8}}$

62. $\dfrac{2\frac{3}{5}}{1\frac{7}{10}}$

63. $\dfrac{\frac{5}{9}}{2\frac{1}{3}}$

64. $\dfrac{\frac{14}{15}}{3\frac{1}{5}}$

Answer each question.

65. A television cable $3\frac{1}{8}$ miles long is divided into 10 equal pieces. How long is each piece?

66. A tank that holds $24\frac{3}{5}$ gallons is used to fill eight containers of equal size. How much does each container hold?

67. Wanda traveled the interstate highway for 140 miles in $2\frac{1}{3}$ hours. What was her average speed (in miles per hour)?

68. Roberto drove his truck to Cedarville, a distance of 150 miles, in $3\frac{1}{3}$ hours. What was his average speed (in miles per hour)?

69. A shirt requires $1\frac{3}{8}$ yards of fabric. The manufacturer has $31\frac{5}{8}$ yards of fabric available for shirts. How many shirts can he make?

70. A butcher has $21\frac{1}{4}$ pounds, of steak that she wishes to cut up into packages that average $1\frac{1}{4}$ pounds each. How many packages can she make?

Divide. Simplify your answer.

71. $126\frac{3}{4} \div 27\frac{1}{8}$

72. $89\frac{5}{9} \div 34\frac{3}{5}$

Cumulative Review Problem

73. Write in words: 39,576,304.

74. Write in expanded form: 459,273.

75. Add: $126 + 34 + 9 + 891 + 12 + 27$.

76. Write in symbols: eighty-seven million, five hundred ninety-five thousand, six hundred thirty-one.

For Extra Practice Examples and Exercises, turn to page 135.

Solutions to Odd-Numbered Practice Problems

1. (a) $\dfrac{7}{13} \div \dfrac{3}{4} = \dfrac{7}{13} \times \dfrac{4}{3} = \dfrac{28}{39}$ **(b)** $\dfrac{16}{35} \div \dfrac{24}{25} = \dfrac{\overset{2}{\cancel{16}}}{\underset{7}{\cancel{35}}} \times \dfrac{\overset{5}{\cancel{25}}}{\underset{3}{\cancel{24}}} = \dfrac{10}{21}$

3. (a) $1 \div \dfrac{11}{13} = \dfrac{1}{1} \times \dfrac{13}{11} = \dfrac{13}{11}$ or $1\dfrac{2}{11}$ **(b)** $\dfrac{14}{17} \div 1 = \dfrac{14}{17} \times \dfrac{1}{1} = \dfrac{14}{17}$

(c) $\dfrac{3}{11} \div 0$ Division by zero cannot be done. This problem cannot be done. **(d)** $0 \div \dfrac{9}{16} = \dfrac{0}{1} \times \dfrac{16}{9} = \dfrac{0}{9} = 0$

5. (a) $\dfrac{5\frac{2}{3}}{7} = 5\dfrac{2}{3} \div 7 = \dfrac{17}{3} \times \dfrac{1}{7} = \dfrac{17}{21}$ **(b)** $\dfrac{1\frac{2}{5}}{2\frac{1}{3}} = 1\dfrac{2}{5} \div 2\dfrac{1}{3} = \dfrac{7}{5} \div \dfrac{7}{3} = \dfrac{\cancel{7}}{5} \times \dfrac{3}{\underset{1}{\cancel{7}}} = \dfrac{3}{5}$

Answers to Even-Numbered Practice Problems

2. (a) $\dfrac{1}{34}$ **(b)** 30   **4. (a)** $\dfrac{12}{7}$ or $1\dfrac{5}{7}$ **(b)** $\dfrac{6}{5}$ or $1\dfrac{1}{5}$ **6.** Each piece will be $1\dfrac{3}{8}$ inches long.

□ After studying this section, you will be able to:

1 *Find the least common denominator given two or three fractions*

2 *Build up a fraction so that it has as a denominator a given least common denominator*

2.6 THE LEAST COMMON DENOMINATOR AND BUILDING UP FRACTIONS

When we reduced a fraction earlier, we made the numerator and denominator *smaller* numbers, but the value of the fraction remained the same. So

$$\frac{12}{16} \quad \text{was} \quad \frac{3}{4}, \quad \text{when } reduced$$

Now, in **building up** a fraction, we proceed in the reverse direction. We make the numerator and denominator larger numbers, but the value of the fraction remains the same. So

$$\frac{3}{4} \text{ can be built up to } \frac{6}{8}$$

$$\frac{3}{4} \text{ can be built up to } \frac{9}{12}$$

$$\frac{3}{4} \text{ can be built up to } \frac{15}{20} \quad \text{and so on}$$

You can see the pattern: the given number $\left(\text{here it is } \frac{3}{4} \right)$ in each case is multiplied by $\frac{2}{2}$, or $\frac{3}{3}$, or $\frac{5}{5}$ to get a built-up number. The reason this does not change the value of the original fraction is that $\frac{2}{2}$, $\frac{3}{3}$, $\frac{5}{5}$, and so on, each equal 1. When you multiply the original fraction by 1, the result is equal to the number you started with.

Why build up a fraction? Actually, you want to build up not one fraction by itself but two (or more) fractions simultaneously. You want the built-up numbers to have the same denominator, preferably the smallest one they can each be built up to. This *least common denominator* will allow you to add and subtract fractions.

In this section we master building up fractions and the least common denominator. In the next section we use these skills in actual computations of addition and subtraction of fractions.

■ Find the Least Common Denominator

We need some way to determine which of two fractions is larger and to add and subtract them. Suppose that Marcia and Melissa each have some pizza left.

 (a)
Marcia's Pizza
$\frac{1}{3}$ of a pizza left

Melissa's Pizza
$\frac{1}{4}$ of a pizza left
(b)

Who has more pizza left? How much more? Comparing the amounts of pizza left would be easy if each pizza had been cut into equal-sized pieces. If the original pizzas had each been cut into 12 pieces

 (a)
Marcia's Pizza
$\left(\begin{array}{c} \text{We know these fractions are} \\ \text{equal by reducing } \frac{4}{12}. \end{array} \right)$

Melissa's Pizza
$\left(\begin{array}{c} \text{We know that} \\ \frac{3}{12} = \frac{1}{4} \text{ by reducing} \end{array} \right)$
(b)

The denominator 12 is common to the fractions $\frac{4}{12}$ and $\frac{3}{12}$. We call a denominator that allows us to compare and to add and subtract fractions the least common denominator, abbreviated LCD. The number 12 is the least common denominator for the fractions $\frac{1}{3}$ and $\frac{1}{4}$.

Definition

The least common denominator (LCD) of two or more fractions is the smallest number that can be divided without remainder by each of the original denominators.

Let's check. $12 \div 3 = 4$. $12 \div 4 = 3$. We have no remainder, so 12 is the answer. Suppose we had thought the number 6 might be the LCD. $6 \div 3 = 2$ with no remainder. We see that $6 \div 4 = 1$ with a remainder of 2, however, so 6 cannot be the LCD. In this case 12 is the LCD of $\frac{1}{3}$ and $\frac{1}{4}$. The number 12 is the smallest number that can be divided by 3 and by 4 without remainder.

In some problems you may be able to guess the LCD quite quickly. By experience you can often determine the LCD mentally. For example, you now know that if the denominators of two fractions are 3 and 4, the LCD is 12. For the fractions $\frac{1}{2}$ and $\frac{1}{4}$, the LCD is 4; for the fractions $\frac{1}{3}$ and $\frac{1}{6}$, the LCD is 6. We can see that if the denominator of one fraction divides without remainder into the denominator of another, the LCD of the two fractions is the larger of the denominators.

Example 1 Determine the LCD of the fraction.

(a) $\frac{7}{15}$ and $\frac{4}{5}$

(b) $\frac{2}{3}$ and $\frac{5}{27}$

(a) Since 5 can be divided into 15, the LCD of $\frac{7}{15}$ and $\frac{4}{5}$ is 15.

(b) Since 3 can be divided into 27, the LCD of $\frac{2}{3}$ and $\frac{5}{27}$ is 27. ■

Practice Problem 1 Determine the LCD of the fraction.

(a) $\frac{3}{4}$ and $\frac{11}{12}$

(b) $\frac{1}{7}$ and $\frac{8}{35}$ ■

In a few cases (but not most) the LCD is the product of the two denominators.

Example 2 Find the LCD for $\frac{1}{4}$ and $\frac{3}{5}$.

We see that $4 \times 5 = 20$. Also, 20 is the *smallest* number that can be divided without remainder by 4 and by 5. So the LCD $= 20$. ■

Practice Problem 2 Find the LCD for $\frac{3}{7}$ and $\frac{5}{6}$. ■

There will be cases when we need to follow an exact procedure to determine the LCD for fractions.

Three-Step Procedure for Finding the Least Common Denominator

1. Write each denominator as the product of prime factors.
2. List all the prime factors that appear in either product.
3. Form a product of those prime factors, using each factor the greatest number of times it appears in any one denominator.

Example 3 Find the LCD of $\frac{5}{6}$ and $\frac{4}{15}$ by the three-step procedure.

Step 1: Write each denominator as a product of prime factors.

$$6 = 2 \times 3 \qquad 15 = 5 \times 3$$

Step 2: LCD will contain the factors 2, 3, and 5.

$$6 = 2 \times 3 \qquad 15 = 5 \times 3$$
$$LCD = 2 \times 3 \times 5$$
$$= 30$$

Step 3 is not needed. No factor was repeated in any one denominator. ■

Practice Problem 3   Find the LCD of $\dfrac{3}{14}$ and $\dfrac{1}{10}$. ■

Example 4   Find the LCD of $\dfrac{7}{18}$ and $\dfrac{7}{30}$.

Step 1: Write each denominator as a product of prime factors.

$$18 = 2 \times 9 = 2 \times 3 \times 3$$
$$30 = 3 \times 10 = 2 \times 3 \times 5$$

Step 2: The LCD will be a product containing 2, 3, and 5.
Step 3: The LCD will contain the factor 3 twice since it occurs twice in the denominator 18.

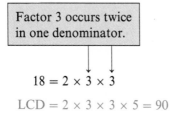

Factor 3 occurs twice in one denominator.

$$18 = 2 \times 3 \times 3$$
$$LCD = 2 \times 3 \times 3 \times 5 = 90 \quad ■$$

Practice Problem 4   Find the LCD of $\dfrac{1}{15}$ and $\dfrac{7}{50}$. ■

Example 5   Find the LCD of $\dfrac{10}{27}$ and $\dfrac{5}{18}$.

$$27 = 3 \times 3 \times 3 \qquad 18 = 3 \times 3 \times 2$$

Factor 3 occurs three times

The LCD will contain the factor 2 once but the factor 3 three times.

$$LCD = 2 \times 3 \times 3 \times 3 = 54 \quad ■$$

Practice Problem 5   Find the LCD of $\dfrac{3}{16}$ and $\dfrac{5}{12}$. ■

A similar procedure can be used if three fractions are involved.

Example 6   Find the LCD of $\dfrac{7}{12}, \dfrac{1}{15}, \dfrac{11}{30}$.

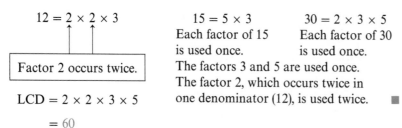

$$12 = 2 \times 2 \times 3$$

Factor 2 occurs twice.

$$LCD = 2 \times 2 \times 3 \times 5$$
$$= 60$$

$15 = 5 \times 3$
Each factor of 15 is used once.

$30 = 2 \times 3 \times 5$
Each factor of 30 is used once.
The factors 3 and 5 are used once.
The factor 2, which occurs twice in one denominator (12), is used twice. ■

Practice Problem 6 Find the LCD of $\dfrac{3}{49}$, $\dfrac{5}{21}$, and $\dfrac{6}{7}$. ■

❷ Building Fractions

We cannot add fractions with unlike denominators. We must follow a two-step process. We need (1) to determine the LCD and (2) to build up the addends—the fractions being added—into equivalent fractions that have the LCD as the denominator. We know now how to determine the LCD. Let's look at how we build up fractions. We know, for example, that

$$\frac{1}{2} = \frac{2}{4} = \frac{50}{100}, \qquad \frac{1}{4} = \frac{25}{100}, \qquad \text{and} \qquad \frac{3}{4} = \frac{75}{100}$$

In these cases we have mentally multiplied the given fraction by 1, in the form of a certain number, c, in the numerator and that same number, c, in the denominator.

$$\frac{1}{2} \times \boxed{\frac{c}{c}} = \frac{2}{4} \qquad \text{Here } c = 2, \quad \frac{2}{2} = 1$$

$$\frac{1}{2} \times \boxed{\frac{c}{c}} = \frac{50}{100} \qquad \text{Here } c = 50, \quad \frac{50}{50} = 1$$

This property is called the building fraction property. We can summarize the property this way:

Building Fraction Property

For whole numbers a, b, and c where $b \neq 0$, $c \neq 0$,

$$\frac{a}{b} = \frac{a}{b} \times 1 = \frac{a}{b} \times \boxed{\frac{c}{c}} = \frac{a \times c}{b \times c}$$

Example 7 Build $\dfrac{3}{4}$ to an equivalent fraction with a denominator of 28.

$$\frac{3}{4} \times \boxed{\frac{c}{c}} = \frac{?}{28}. \qquad \text{We know that } 4 \times 7 = 28, \text{ so the value } c \text{ that we multiply numerator and denominator by is 7.}$$

$$\frac{3}{4} \times \frac{7}{7} = \frac{21}{28} \qquad ■$$

Practice Problem 7 Build $\dfrac{3}{5}$ to an equivalent fraction with a denominator of 40. ■

Example 8 Assume that the LCD $= 45$. Build $\dfrac{4}{5}$ to an equivalent fraction with the LCD as the denominator.

$$\frac{4}{5} \times \boxed{\frac{c}{c}} = \frac{?}{45}$$

We know that $5 \times 9 = 45$, so $c = 9$.

$$\frac{4}{5} \times \frac{9}{9} = \frac{36}{45} \qquad ■$$

Practice Problem 8 Assume that LCD $= 44$. Build $\dfrac{7}{11}$ to an equivalent fraction with the LCD as a denominator. ■

Example 9 The LCD of $\frac{1}{3}$ and $\frac{4}{5}$ is 15. Build $\frac{1}{3}$ and $\frac{4}{5}$ to equivalent fractions that have the LCD as the denominator.

$$\frac{1}{3} = \frac{?}{15}$$ We know that $3 \times 5 = 15$, so we multiply numerator and denominator by 5.

$$\frac{1}{3} \times \boxed{\frac{5}{5}} = \frac{5}{15}$$

$$\frac{4}{5} = \frac{?}{15}$$ We know that $5 \times 3 = 15$, so we multiply numerator and denominator by 3.

$$\frac{4}{5} \times \frac{3}{3} = \frac{12}{20}$$ ■

Practice Problem 9 The LCD of $\frac{2}{7}$ and $\frac{3}{4}$ is 28. Build $\frac{2}{7}$ and $\frac{3}{4}$ to equivalent fractions that have the LCD as the denominators. ■

Example 10

(a) Find the LCD of $\frac{1}{32}$ and $\frac{7}{48}$.

(b) Build the fractions to equivalent fractions that have the LCD as their denominators.

(a) First we find the prime factors of 32 and 48.

$$32 = 2 \times 2 \times 2 \times 2 \times 2$$

$$48 = 2 \times 2 \times 2 \times 2 \times 3$$

Thus the LCD will require a factor of 2 five times and a factor of 3 one time.

$$LCD = 2 \times 2 \times 2 \times 2 \times 2 \times 3 = 96$$

(b) $\frac{1}{32} = \frac{?}{96}$ Since $32 \times 3 = 96$ we multiply by 3.

$$\frac{1}{32} = \frac{1}{32} \times \boxed{\frac{3}{3}} = \frac{3}{96}$$

$\frac{7}{48} = \frac{?}{96}$ Since $48 \times 2 = 96$, we multiply by 2.

$$\frac{7}{48} = \frac{7}{48} \times \boxed{\frac{2}{2}} = \frac{14}{96}$$ ■

Practice Problem 10

(a) Find the LCD of $\frac{3}{20}$ and $\frac{11}{15}$.

(b) Build the fractions to equivalent fractions that have the LCD as their denominators. ■

To Think About   ?

Is there another method that can be used to find the least common denominator for two or more fractions? Yes. You can find the least common multiple (LCM) of the two denominators. Suppose that we want to find the least common multiple of 10 and 12. List the multiples of 10 (multiply by 1, 2, 3, 4, 5, and so on).

$$10, 20, 30, 40, 50, \boxed{60,} 70, 80, \dots .$$

List the multiples of 12.

$$12, 24, 36, 48, \boxed{60,} 72, 84, 96, \ldots$$

The first multiple that appears on each list is the least common multiple. It is also the LCD for two fractions that have 10 and 12 for denominators. Problems 65 and 66 will give you a chance to use this method. ■

EXERCISES 2.6

Find the LCD for each pair of fractions.

1. $\frac{1}{5}$ and $\frac{3}{10}$ **2.** $\frac{3}{8}$ and $\frac{5}{16}$ **3.** $\frac{3}{7}$ and $\frac{1}{4}$ **4.** $\frac{5}{6}$ and $\frac{3}{5}$ **5.** $\frac{2}{5}$ and $\frac{3}{8}$

6. $\frac{1}{16}$ and $\frac{2}{3}$ **7.** $\frac{4}{9}$ and $\frac{5}{6}$ **8.** $\frac{1}{4}$ and $\frac{3}{14}$ **9.** $\frac{5}{12}$ and $\frac{11}{30}$ **10.** $\frac{9}{14}$ and $\frac{2}{49}$

11. $\frac{1}{16}$ and $\frac{3}{4}$ **12.** $\frac{2}{39}$ and $\frac{1}{13}$ **13.** $\frac{5}{10}$ and $\frac{11}{45}$ **14.** $\frac{13}{20}$ and $\frac{17}{30}$ **15.** $\frac{7}{12}$ and $\frac{7}{30}$

16. $\frac{5}{6}$ and $\frac{7}{15}$ **17.** $\frac{5}{21}$ and $\frac{8}{35}$ **18.** $\frac{5}{18}$ and $\frac{11}{12}$ **19.** $\frac{1}{20}$ and $\frac{7}{8}$ **20.** $\frac{1}{24}$ and $\frac{7}{40}$

Find the LCD for each set of three fractions.

21. $\frac{2}{3}, \frac{3}{4}, \frac{5}{6}$ **22.** $\frac{1}{5}, \frac{2}{3}, \frac{7}{10}$ **23.** $\frac{1}{12}, \frac{7}{18}, \frac{13}{20}$ **24.** $\frac{7}{15}, \frac{1}{24}, \frac{11}{30}$ **25.** $\frac{5}{16}, \frac{11}{18}, \frac{1}{24}$

26. $\frac{5}{6}, \frac{1}{2}, \frac{5}{22}$ **27.** $\frac{7}{12}, \frac{1}{21}, \frac{3}{14}$ **28.** $\frac{1}{30}, \frac{3}{40}, \frac{7}{8}$ **29.** $\frac{7}{15}, \frac{11}{12}, \frac{7}{8}$ **30.** $\frac{5}{36}, \frac{2}{48}, \frac{1}{24}$

Build each fraction to an equivalent fraction with the denominator specified.

31. $\frac{1}{4} = \frac{?}{20}$ **32.** $\frac{1}{5} = \frac{?}{35}$ **33.** $\frac{5}{6} = \frac{?}{54}$ **34.** $\frac{4}{7} = \frac{?}{28}$ **35.** $\frac{4}{11} = \frac{?}{55}$ **36.** $\frac{2}{13} = \frac{?}{39}$

37. $\frac{7}{24} = \frac{?}{48}$ **38.** $\frac{3}{50} = \frac{?}{150}$ **39.** $\frac{8}{9} = \frac{?}{108}$ **40.** $\frac{6}{7} = \frac{?}{147}$ **41.** $\frac{13}{42} = \frac{?}{126}$ **42.** $\frac{14}{15} = \frac{?}{180}$

The LCD of each pair of fractions is listed. Build each fraction to an equivalent fraction that has the LCD as the denominator.

43. LCD = 36, $\dfrac{7}{12}$ and $\dfrac{5}{9}$

44. LCD = 40, $\dfrac{9}{10}$ and $\dfrac{3}{4}$

45. LCD = 200, $\dfrac{3}{25}$ and $\dfrac{7}{40}$

46. LCD = 108, $\dfrac{5}{54}$ and $\dfrac{7}{36}$

47. LCD = 432, $\dfrac{9}{108}$ and $\dfrac{19}{144}$

48. LCD = 480, $\dfrac{43}{120}$ and $\dfrac{7}{160}$

For each set of fractions:

(a) Find the LCD.
(b) Build up the fractions to equivalent fractions having the LCD as the denominator.

49. $\dfrac{5}{7}$ and $\dfrac{7}{42}$

50. $\dfrac{3}{8}$ and $\dfrac{13}{40}$

51. $\dfrac{5}{12}$ and $\dfrac{1}{16}$

52. $\dfrac{7}{15}$ and $\dfrac{4}{25}$

53. $\dfrac{9}{10}$ and $\dfrac{13}{18}$

54. $\dfrac{13}{20}$ and $\dfrac{11}{16}$

55. $\dfrac{7}{12}$ and $\dfrac{23}{30}$

56. $\dfrac{3}{16}$ and $\dfrac{23}{24}$

57. $\dfrac{5}{6}$, $\dfrac{11}{12}$, $\dfrac{3}{4}$

58. $\dfrac{1}{30}$, $\dfrac{7}{15}$, $\dfrac{1}{45}$

59. $\dfrac{3}{56}$, $\dfrac{7}{8}$, $\dfrac{5}{7}$

60. $\dfrac{5}{9}$, $\dfrac{4}{63}$, $\dfrac{6}{7}$

61. $\dfrac{5}{63}$, $\dfrac{4}{21}$, $\dfrac{8}{9}$

62. $\dfrac{3}{8}$, $\dfrac{5}{14}$, $\dfrac{13}{16}$

63. Suppose that you wish to compare the lengths of the three portions of a stainless steel pin that came out of the door.
   (a) What is the LCD for each of the three fractions?
   (b) Build up each fraction to an equivalent fraction that has the LCD as a denominator.

64. Suppose that you want to prepare a report on a plant that grew the following lengths during each week of a 3-week experiment.
   (a) What is the LCD for the three fractions?
   (b) Build up each fraction to an equivalent fraction that has the LCD for a denominator.

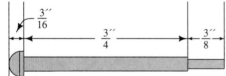

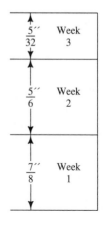

Using the discussion in the text on least common multiples, find the least common multiple of each pair of numbers.

65. 24 and 42

66. 30 and 36

67. Find the LCD of $\frac{4}{51}$ and $\frac{11}{119}$.

68. Find the LCD of $\frac{3}{95}, \frac{2}{171}$.

Cumulative Review Problems

69. Divide: $32\overline{)5699}$.

70. Divide: $182\overline{)659568}$.

For Extra Practice Examples and Exercises, turn to page 136.

Solutions to Odd-Numbered Practice Problems

1. (a) The LCD of $\frac{3}{4}$ and $\frac{11}{12}$ is 12.
12 can be divided by 4 and 12.
(b) The LCD of $\frac{1}{7}$ and $\frac{8}{35}$ is 35.
35 can be divided by 7 and 35.
3. $14 = 2 \times 7$ Each factor 2, 5, 7 is used once.
$10 = 2 \times 5$
The LCD $= 2 \times 5 \times 7 = 70$.

5. $16 = 2 \times 2 \times 2 \times 2$ The factor 2 is used four times in LCD.
$12 = 2 \times 2 \times 3$ The factor 3 is used once.
The LCD $= 2 \times 2 \times 2 \times 2 \times 3 = 48$.
7. $\frac{3}{5} = \frac{?}{40}$ We know that $5 \times 8 = 40$, so we multiply by 8.
$\frac{3}{5} = \frac{3}{5} \times \frac{8}{8} = \frac{24}{40}$

9. $\frac{2}{7} = \frac{?}{28}$ We know that $7 \times 4 = 28$, so we multiply by 4.
$\frac{2}{7} = \frac{2 \times 4}{7 \times 4} = \frac{8}{28}$ $\frac{3}{4} = \frac{?}{28}$ $\frac{3}{4} = \frac{3 \times 7}{4 \times 7} = \frac{21}{28}$ Now we multiply by 7.

Answers to Even-Numbered Practice Problems

2. LCD $= 42$ **4.** LCD $= 150$ **6.** LCD $= 147$ **8.** $\frac{7}{11} = \frac{28}{44}$ **10. (a)** LCD $= 60$ **(b)** $\frac{3}{20} = \frac{9}{60}, \frac{11}{15} = \frac{44}{60}$

After studying this section, you will be able to:

1 *Add or subtract fractions with a common denominator*

2 *Add or subtract fractions without a common denominator*

3 *Add or subtract mixed numbers*

2.7 ADDITION AND SUBTRACTION OF FRACTIONS

You must have common denominators (denominators that are alike) to add or subtract fractions.

If your problem has fractions without a common denominator or if it has mixed numbers, you must use what you already know about changing each fraction's form (how the fractions looks) but not changing its value. Only after all the fractions have a common denominator can you add, as on the signpost.

An important distinction: You must have common denominators to add or subtract fractions, but you need not have common denominators to multiply or divide them.

$$\frac{1}{3} \times \frac{1}{4} = \frac{1}{12}$$

$$\frac{1}{3} + \frac{1}{4} =$$

Notice the
built-up
fractions
using the LCD.

$$\frac{4}{12} + \frac{3}{12} = \frac{7}{12}$$

1 Adding and Subtracting Fractions with a Common Denominator

To add two fractions that have the same denominator, add the numerators and write the sum over the common denominator. To illustrate we use $\frac{1}{5} + \frac{2}{5} = \frac{3}{5}$. The sketches show evidence that $\frac{1}{5} + \frac{2}{5} = \frac{3}{5}$.

Example 1 Add: $\frac{5}{13} + \frac{7}{13}$.

$$\frac{5}{13} + \frac{7}{13} = \frac{12}{13}\qquad\blacksquare$$

Practice Problem 1 Add: $\frac{3}{17} + \frac{12}{17}$. $\blacksquare$

The answer may need to be reduced. Sometimes the answer may be written as a mixed number.

Example 2 Add.

(a) $\frac{4}{9} + \frac{2}{9}$

(b) $\frac{5}{7} + \frac{6}{7}$

(a) $\frac{4}{9} + \frac{2}{9} = \frac{6}{9} = \frac{2}{3}$

(b) $\frac{5}{7} + \frac{6}{7} = \frac{11}{7}$ or $1\frac{4}{7}$ $\blacksquare$

Practice Problem 2 Add.

(a) $\frac{1}{12} + \frac{5}{12}$

(b) $\frac{13}{15} + \frac{7}{15}$ $\blacksquare$

A similar rule is followed for subtraction except that the numerators are subtracted and the result placed over a common denominator. Be sure to reduce all answers when possible.

Example 3 Subtract.

(a) $\frac{5}{13} - \frac{4}{13}$

(b) $\frac{17}{20} - \frac{3}{20}$

(a) $\frac{5}{13} - \frac{4}{13} = \frac{1}{13}$

(b) $\frac{17}{20} - \frac{3}{20} = \frac{14}{20} = \frac{7}{10}$ $\blacksquare$

Practice Problem 3 Subtract.

(a) $\frac{5}{19} - \frac{2}{19}$

(b) $\frac{21}{25} - \frac{6}{25}$ $\blacksquare$

2 If the two fractions do not have a common denominator, we follow the procedure in Section 2.6 and find the LCD and then build up each fraction so that it has the LCD as the denominator.

Example 4   Add: $\dfrac{7}{12} + \dfrac{1}{4}$.

The LCD is 12. $\dfrac{7}{12}$ already has the correct denominator.

$$\begin{array}{rcl} \dfrac{7}{12} & = & \dfrac{7}{12} \\[2ex] + \ \dfrac{1}{4} \times \dfrac{3}{3} & = & + \ \dfrac{3}{12} \\[2ex] \hline & & \dfrac{10}{12} = \dfrac{5}{6} \quad \blacksquare \end{array}$$

Practice Problem 4   Add: $\dfrac{2}{15} + \dfrac{2}{5}$. $\blacksquare$

Example 5   Add: $\dfrac{7}{20} + \dfrac{4}{15}$.

LCD $= 60$.

$$\dfrac{7}{20} \times \dfrac{3}{3} = \dfrac{21}{60} \qquad \dfrac{4}{15} \times \dfrac{4}{4} = \dfrac{16}{60}$$

Thus

$$\dfrac{7}{20} + \dfrac{4}{15} = \dfrac{21}{60} + \dfrac{16}{60} = \dfrac{37}{60} \quad \blacksquare$$

Practice Problem 5   Add: $\dfrac{5}{12} + \dfrac{5}{16}$. $\blacksquare$

A similar procedure holds for the addition of three or more fractions.

Example 6   Add: $\dfrac{3}{8} + \dfrac{5}{6} + \dfrac{1}{4}$.

LCD $= 24$.

$$\dfrac{3}{8} \times \dfrac{3}{3} = \dfrac{9}{24} \qquad \dfrac{5}{6} \times \dfrac{4}{4} = \dfrac{20}{24} \qquad \dfrac{1}{4} \times \dfrac{6}{6} = \dfrac{6}{24}$$

Thus

$$\dfrac{3}{8} + \dfrac{5}{6} + \dfrac{1}{4} = \dfrac{9}{24} + \dfrac{20}{24} + \dfrac{6}{24} = \dfrac{35}{24} \quad \text{or} \quad 1\dfrac{11}{24} \quad \blacksquare$$

Practice Problem 6   Add: $\dfrac{3}{16} + \dfrac{1}{8} + \dfrac{1}{12}$. $\blacksquare$

Example 7   Subtract: $\dfrac{17}{25} - \dfrac{3}{35}$.

LCD $= 175$.

$$\dfrac{17}{25} \times \dfrac{7}{7} = \dfrac{119}{175} \qquad \dfrac{3}{35} \times \dfrac{5}{5} = \dfrac{15}{175}$$

Thus

$$\dfrac{17}{25} - \dfrac{3}{35} = \dfrac{119}{175} - \dfrac{15}{175} = \dfrac{104}{175} \quad \blacksquare$$

Practice Problem 7 Subtract: $\dfrac{9}{48} - \dfrac{5}{32}$. ■

3 When adding or subtracting mixed numbers, it is best to add or subtract fractions separately and whole numbers separately.

Example 8 Add: $3\dfrac{1}{8} + 2\dfrac{5}{8}$.

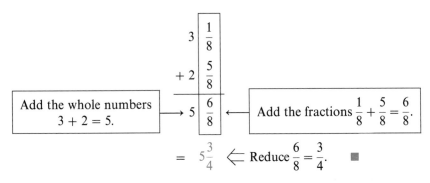

Add the whole numbers
$3 + 2 = 5$.

Add the fractions $\dfrac{1}{8} + \dfrac{5}{8} = \dfrac{6}{8}$.

$= 5\dfrac{3}{4} \ \Leftarrow$ Reduce $\dfrac{6}{8} = \dfrac{3}{4}$. ■

Practice Problem 8 Add: $5\dfrac{1}{12} + 9\dfrac{5}{12}$. ■

If the fraction portions of the mixed numbers do not have a common denominator, three steps must be taken to build up the fraction part to obtain a common denominator for each fraction before adding. If the sum of the fractions is an improper fraction, we convert it to a mixed number and add the whole numbers together.

Example 9 Add: $6\dfrac{5}{6} + 4\dfrac{3}{8}$.

The LCD of $\dfrac{5}{6}$ and $\dfrac{3}{8}$ is 24.

$$6 \boxed{\dfrac{5}{6} \times \dfrac{4}{4}} = 6 \boxed{\dfrac{20}{24}}$$

$$+ 4 \boxed{\dfrac{3}{8} \times \dfrac{3}{3}} = + 4 \boxed{\dfrac{9}{24}}$$

Add the fractions
$\dfrac{20}{24} + \dfrac{9}{24} = \dfrac{29}{24}$.

$\boxed{6 + 4 = 10} \longrightarrow 10\,\dfrac{29}{24}$

$= 10 + \boxed{1\dfrac{5}{24}}$ Write $\dfrac{29}{24} = 1\dfrac{5}{24}$.

$= 11\dfrac{5}{24}$ Add $10 + 1 = 11$. ■

Practice Problem 9 Add: $7\dfrac{1}{4} + 3\dfrac{5}{6}$. ■

Similar procedures are followed for subtraction.

Example 10   Subtract: $8\frac{5}{7} - 5\frac{5}{14}$.

The LCD of $\frac{5}{7}$ and $\frac{5}{14}$ is 14.

$$8\;\boxed{\frac{5}{7} \times \frac{2}{2}} = 8\frac{10}{14}$$

$$5\frac{5}{14} = 5\frac{5}{14}$$

Subtract $8 - 5 = 3 \longrightarrow 3\frac{5}{14} \longleftarrow$ Subtract $\dfrac{10}{14} - \dfrac{5}{14} = \dfrac{5}{14}$ ∎

Practice Problem 10   Subtract: $12\frac{5}{6} - 7\frac{5}{12}$.   ∎

Sometimes borrowing is necessary in the subtraction process.

Example 11   Subtract: $9\frac{1}{4} - 6\frac{5}{14}$.

The LCD of $\frac{1}{4}$ and $\frac{5}{14}$ is 28.

$$9\;\boxed{\frac{1}{4} \times \frac{7}{7}} = 9\frac{7}{28} \longleftarrow$$

$$-6\;\boxed{\frac{5}{14} \times \frac{2}{2}} = -6\frac{10}{28} \longleftarrow$$

We cannot subtract $\dfrac{7}{28} - \dfrac{10}{28}$, so we will need to borrow.

$$9\frac{7}{28} = 8\frac{35}{28} \longleftarrow$$

We borrow 1 from 9 to obtain
$$9\frac{7}{28} = 8 + 1\frac{7}{28} = 8 + \frac{35}{28} = 8\frac{35}{28}$$

$$-6\frac{10}{28} \quad -6\frac{10}{28}$$

$$8 - 6 = 2 \Longrightarrow 2\frac{25}{28} \longleftarrow \boxed{\dfrac{35}{28} - \dfrac{10}{28} = \dfrac{25}{28}}$$ ∎

Practice Problem 11   Subtract: $9\frac{1}{8} - 3\frac{2}{3}$.   ∎

Example 12   Subtract: $15 - 9\frac{3}{16}$.

The LCD $= 16$

$$15 \quad = \quad 14\frac{16}{16} \longleftarrow$$

We borrow 1 from 15 to obtain
$$15 = 14 + 1 = 14 + \frac{16}{16} = 14\frac{16}{16}$$

$$- 9\frac{3}{16} \quad - 9\frac{3}{16}$$

$$14 = 9 - 5 \Longrightarrow 5\frac{13}{16} \longleftarrow \dfrac{16}{16} - \dfrac{3}{16} = \dfrac{13}{16}$$ ∎

Practice Problem 12   Subtract: $18 - 6\frac{7}{18}$.   ∎

Can mixed numbers be added and subtracted as improper fractions? Yes. Consider the results of Example 11.

$$9\frac{1}{4} - 6\frac{5}{14} = 2\frac{25}{28}$$

If we write $9\frac{1}{4} - 6\frac{5}{14}$ using improper fractions, we have $\frac{37}{4} - \frac{89}{14}$. Now we build up each of these improper fractions so that they both have the LCD for their denominators.

$$\frac{37}{4} \boxed{\times \frac{7}{7}} = \frac{259}{28}$$

$$-\frac{89}{14} \boxed{\times \frac{2}{2}} = \frac{178}{28}$$

$$= \frac{81}{28} = 2\frac{25}{28}$$

The same result is obtained as in Example 11. For more practice, see problems 59 and 60. ■

EXERCISES 2.7

Add or subtract. Simplify all answers.

1. $\frac{3}{8} + \frac{2}{8}$

2. $\frac{4}{7} + \frac{2}{7}$

3. $\frac{11}{30} + \frac{11}{30}$

4. $\frac{3}{28} + \frac{7}{28}$

5. $\frac{21}{23} - \frac{3}{23}$

6. $\frac{5}{24} - \frac{3}{24}$

7. $\frac{53}{88} - \frac{19}{88}$

8. $\frac{103}{110} - \frac{3}{110}$

Add or subtract. Simplify all answers.

9. $\frac{3}{5} + \frac{1}{2}$

10. $\frac{4}{7} + \frac{1}{14}$

11. $\frac{3}{10} + \frac{3}{20}$

12. $\frac{4}{9} + \frac{1}{2}$

13. $\frac{3}{10} + \frac{7}{100}$

14. $\frac{13}{100} + \frac{7}{10}$

15. $\frac{3}{25} + \frac{1}{35}$

16. $\frac{3}{15} + \frac{1}{25}$

17. $\frac{7}{8} + \frac{5}{12}$

18. $\frac{5}{6} + \frac{7}{8}$

19. $\frac{1}{12} + \frac{4}{15}$

20. $\frac{1}{10} + \frac{7}{16}$

21. $\dfrac{5}{12} - \dfrac{1}{6}$

22. $\dfrac{37}{20} - \dfrac{1}{4}$

23. $\dfrac{3}{7} - \dfrac{1}{5}$

24. $\dfrac{7}{8} - \dfrac{5}{6}$

25. $\dfrac{7}{8} - \dfrac{5}{16}$

26. $\dfrac{9}{50} - \dfrac{2}{25}$

27. $\dfrac{5}{12} - \dfrac{7}{30}$

28. $\dfrac{9}{24} - \dfrac{3}{8}$

29. $\dfrac{20}{35} - \dfrac{4}{7}$

30. $\dfrac{11}{20} - \dfrac{3}{8}$

31. $\dfrac{4}{5} + \dfrac{1}{20} + \dfrac{3}{4}$

32. $\dfrac{1}{18} + \dfrac{2}{9} + \dfrac{1}{2}$

33. $\dfrac{5}{30} + \dfrac{3}{40} + \dfrac{1}{8}$

34. $\dfrac{1}{12} + \dfrac{3}{14} + \dfrac{4}{21}$

Add or subtract. Express the answer as a mixed number.

35. $15\dfrac{4}{15}$
$+\ 26\dfrac{8}{15}$

36. $22\dfrac{1}{8}$
$+\ 14\dfrac{3}{8}$

37. $4\dfrac{1}{3}$
$+\ 2\dfrac{1}{4}$

38. $6\dfrac{1}{8}$
$+\ 7\dfrac{3}{4}$

39. $2\dfrac{1}{15}$
$+\ 14\dfrac{3}{5}$

40. $8\dfrac{1}{7}$
$+\ 3\dfrac{11}{14}$

41. $47\dfrac{3}{10}$
$+\ 26\dfrac{5}{8}$

42. $34\dfrac{1}{20}$
$+\ 45\dfrac{8}{15}$

43. $19\dfrac{5}{6}$
$-\ 14\dfrac{1}{3}$

44. $27\dfrac{11}{12}$
$-\ 21\dfrac{1}{3}$

45. $4\dfrac{1}{15}$
$-\ 2\dfrac{3}{5}$

46. $7\dfrac{1}{14}$
$-\ 4\dfrac{3}{7}$

47. $12\dfrac{3}{20}$
$-\ 7\dfrac{7}{15}$

48. $8\dfrac{5}{12}$
$-\ 5\dfrac{9}{10}$

49. 12
$-\ 3\dfrac{7}{15}$

50. 19
$-\ 6\dfrac{3}{7}$

51. 120
$-\ 17\dfrac{3}{8}$

52. 98
$-\ 89\dfrac{15}{17}$

Solve each problem.

53. Mr. Appleby repaired two violins. The first took $3\dfrac{1}{4}$ hours. The second took $4\dfrac{5}{6}$ hours. What was the total time taken to repair the instruments?

54. Manuel jogged $3\dfrac{1}{8}$ miles yesterday and $2\dfrac{7}{10}$ miles today. What was his total jogging distance?

55. Juanita purchased stock at $\$57\frac{3}{8}$ per share and sold it at $\$71\frac{1}{8}$ per share. How much money did she make per share?

56. John bought $5\frac{5}{16}$ pounds of finish nails. He used $3\frac{5}{8}$ pounds on his carpentry projects. How many pounds of nails were left over?

57. Heather set a goal of $46\frac{1}{8}$ kilometers to be covered on three days. She hiked $16\frac{1}{4}$ kilometers on the first day and $14\frac{3}{8}$ kilometers on the second day.

(a) How far did she hike on the first two days?
(b) How far must she hike on the third day to reach her goal?

58. A young man has been under a doctor's care to lose weight. His doctor wanted him to lose 46 pounds in the first three months. He lost $17\frac{5}{8}$ pounds the first month and $13\frac{1}{2}$ pounds the second month.

(a) How much did he lose during the first two months?
(b) How much would he need to lose in the third month to reach the goal?

? **To Think About**

Using the discussion in the text describing adding and subtracting improper fractions as a basis, perform the following calculations without changing to mixed numbers.

59. Add: $\dfrac{379}{8} + \dfrac{89}{5}$.

60. Subtract: $\dfrac{151}{6} - \dfrac{130}{7}$.

Add.

61.
$$3\frac{5}{18}$$
$$6\frac{7}{15}$$
$$+\,2\frac{7}{30}$$

62.
$$4\frac{7}{18}$$
$$8\frac{3}{20}$$
$$+\,2\frac{6}{45}$$

Cumulative Review Problems

Multiply.

63.
$$\begin{array}{r} 12{,}367 \\ \times\quad\ 9 \\ \hline \end{array}$$

64.
$$\begin{array}{r} 304 \\ \times\ 128 \\ \hline \end{array}$$

65.
$$\begin{array}{r} 6737 \\ \times\ \ 76 \\ \hline \end{array}$$

66.
$$\begin{array}{r} 4050 \\ \times\ 2106 \\ \hline \end{array}$$

For Extra Practice Examples and Exercises, turn to page 136.

1. $\frac{3}{17} + \frac{12}{17} = \frac{15}{17}$ **3. (a)** $\frac{5}{19} - \frac{2}{19} = \frac{3}{19}$ **(b)** $\frac{21}{25} - \frac{6}{25} = \frac{15}{25} = \frac{3}{5}$

5. LCD = 48 $\frac{5}{12} \times \frac{4}{4} = \frac{20}{48}$ $\frac{5}{16} \times \frac{3}{3} = \frac{15}{48}$ $\frac{5}{12} + \frac{5}{16} = \frac{20}{48} + \frac{15}{48} = \frac{35}{48}$

7. LCD = 96 $\frac{9}{48} \times \frac{2}{2} = \frac{18}{96}$ $\frac{5}{32} \times \frac{3}{3} = \frac{15}{96}$ $\frac{9}{48} - \frac{5}{32} = \frac{18}{96} - \frac{15}{96} = \frac{3}{96} = \frac{1}{32}$

9. LCD = 12 $7\boxed{\frac{1}{4} \times \frac{3}{3}} = 7\frac{3}{12}$

$\quad\quad\quad\quad + 3\boxed{\frac{5}{6} \times \frac{2}{2}} \quad + 3\frac{10}{12}$

$\quad\quad\quad\quad\quad\quad\quad\quad\quad\quad 10\frac{13}{12} = 10 + 1\frac{1}{12} = 11\frac{1}{12}$

11. LCD = 24 $9\boxed{\frac{1}{8} \times \frac{3}{3}} = 9\frac{3}{24} = 8\frac{27}{24}$ ← Borrow 1 from 9:

$\quad\quad\quad\quad\quad\quad\quad\quad\quad\quad\quad\quad\quad\quad 9\frac{3}{24} = 8 + 1 + \frac{3}{24} = 8\frac{27}{24}$

$\quad\quad\quad\quad - 3\boxed{\frac{2}{3} \times \frac{8}{8}} \quad - 3\frac{16}{24} \quad\quad 3\frac{16}{24}$

$\quad\quad\quad\quad\quad\quad\quad\quad\quad\quad\quad\quad\quad\quad 5\frac{11}{24}$

2. (a) $\frac{1}{2}$ **(b)** $\frac{4}{3}$ or $1\frac{1}{3}$ **4.** $\frac{8}{15}$ **6.** $\frac{19}{48}$ **8.** $14\frac{1}{2}$ **10.** $5\frac{5}{12}$ **12.** $11\frac{11}{18}$

After studying this section, you will be able to:

1 *Solve applied problems with fractions*

2.8 APPLIED PROBLEMS INVOLVING FRACTIONS

Problem solving requires the same thought processes no matter what the applied situation. In this section we have a chance to combine problem-solving skills with our new computational skills with fractions. Sometimes the challenge of solving the problem is in figuring out what must be done, sometimes it is in actually doing the fractional work. Remember that *estimating* is a key way to stay on track with problem solving. We start with a step-by-step approach.

1 The six-step outline discussed in Section 1.8 is once again helpful.

Steps to Solving an Applied Problem

1. Read over the problem carefully. Find out what is asked for. Draw a picture if this helps you to visualize the situation.
2. Write down which numbers (both whole numbers and fractions) are to be used in solving the problem.
3. Estimate the answer.
4. Determine what operation needs to be done: addition, subtraction, multiplication division, or a combination of these.
5. Perform the necessary computations. State the answer, including the unit of measure.
6. Compare your answer with the estimate. See if the answer is reasonable.

Example 1 Maria bought Bank of New England stock at $26\frac{1}{8}$. The stock's value went up $2\frac{5}{8}$ the first week. The second week it went up $3\frac{3}{4}$. What was its value after these two increases?

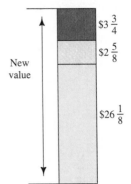

New value

$3\frac{3}{4}$

$2\frac{5}{8}$

$26\frac{1}{8}$

1. We are trying to find the stock value after two increases.

2. We will use the values $26\frac{1}{8}$, $2\frac{5}{8}$, and $3\frac{3}{4}$.

3. Estimate the sum by rounding to the nearest dollar and adding, $26 + 3 + 4 = \$33$.

4. To solve for an exact answer, we *add* up the three values to find the *total* value.

5. $$\text{LCD} = 8 \qquad 26\frac{1}{8} \qquad = \qquad 26\frac{1}{8}$$

$$2\frac{5}{8} \qquad\qquad 2\frac{5}{8}$$

$$+ \; 3\;\boxed{\frac{3}{4} \times \frac{2}{2}} \qquad + \; 3\frac{6}{8}$$

$$31\frac{12}{8} = 31\frac{3}{2} = \$32\frac{1}{2}$$

The total value of the stock is $32\frac{1}{2}$.

6. This is close to our estimate of $33. The answer is reasonable. ∎

Practice Problem 1 Nichole purchased the following amounts of gas for her car in last three fill-ups: $18\frac{7}{10}$ gallons, $15\frac{2}{5}$ gallons, and $14\frac{1}{2}$ gallons. How many gallons did she buy altogether? ∎

We have two common meanings of the word *diameter*. First, it refers to a line segment that passes through the center and intersects the circle twice. Second, it refers to the *length* of this segment. Usually in measurement problems when we ask for the diameter, we want to know how long it is.

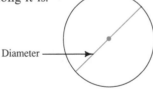

Diameter

Example 2 What is the inside diameter (distance across) of a cement storm drain pipe that has an outside diameter of $4\frac{1}{8}$ feet and is $\frac{3}{8}$ feet thick?

1. We try to visualize a sketch of a cross section of the pipe.

2. We need to use the outside diameter of $4\frac{1}{8}$ feet and the measure of the pipe thickness (the same on each side) of $\frac{3}{8}$ foot.

3. We estimate. The outside diameter is about 4 feet. The two thickness values *added together* are about 1 foot.

$$4 \text{ feet} - 1 \text{ foot} = 3 \text{ feet}$$

4. We need to *add* the two thickness measurements and subtract from the outside diameter.

5. Adding $\frac{3}{8} + \frac{3}{8} = \frac{6}{8}$ foot total of thickness of pipe on each side. We will not reduce $\frac{6}{8}$ since the LCD in our subtraction is 8.

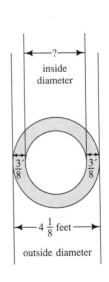

—?—
inside diameter

$\frac{3}{8}$ $\frac{3}{8}$

← $4\frac{1}{8}$ feet →

outside diameter

$$4\frac{1}{8} = \quad 3\frac{9}{8}$$ ←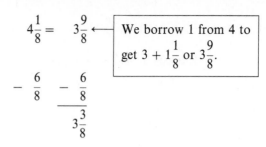

We borrow 1 from 4 to get $3 + 1\frac{1}{8}$ or $3\frac{9}{8}$.

$$-\frac{6}{8} \quad -\frac{6}{8}$$

$$\overline{\qquad 3\frac{3}{8}}$$

The inside diameter is $3\frac{3}{8}$ feet.

6. This is close to our estimate of 3 feet. The answer is reasonable. ■

Practice Problem 2   The outside dimension of a poster can have a total overall length of $12\frac{1}{4}$ inches. We want a $1\frac{3}{8}$-inch border on the top and a 2-inch border on the bottom. What is the length of the inside portion of the poster where writing will be placed? ■

Example 3   Michael earns $\$8\frac{1}{4}$ per hour, working on Tuesday for 8 hours. Then he earns overtime pay, which is $1\frac{1}{2}$ times his regular rate of $\$8\frac{1}{4}$. He earns overtime pay for 4 hours on Tuesday. How much pay did he earn on Tuesday?

1. Michael's earnings on Tuesday are the sum of two parts:

Pay at regular pay rate		pay at overtime pay rate		total pay for the day
	+		=	

2. The amount of money at the regular rate is 8 hours at $\$8\frac{1}{4}$ per hour. The amount of money at the overtime rate is 4 hours at $1\frac{1}{2} \times \$8\frac{1}{4}$.

3. Estimate. Round values to nearest whole number.

$$8 \text{ hours} \times \$8 \text{ per hour} = \$64 \text{ regular pay}$$

$$\$8 \text{ per hour} \times 1\frac{1}{2} = \$12 \text{ per hour.}$$

$$\$12 \times 4 = \$48 \text{ overtime pay.}$$

Estimated sum: $\$64 + \$48 = \$112$.

4. We need to multiply to get the actual pay at each rate. Then we add the two pay rates together.

5.

Pay at regular pay rate $8 \times \$8\frac{1}{4}$		pay at overtime pay rate $4 \times 1\frac{1}{2} \times \$8\frac{1}{4}$		total pay
	+		=	

His pay at regular pay rate: $8 \times 8\frac{1}{4} = \overset{2}{\cancel{8}} \times \frac{33}{\underset{1}{\cancel{4}}} = \$66$.

His overtime pay rate: $1\frac{1}{2} \times \$8\frac{1}{4} = \frac{3}{2} \times \frac{33}{4} = \frac{\$99}{8}$ per hour.

For 4 hours at overtime pay he earned $\overset{1}{\cancel{4}} \times \frac{99}{\underset{2}{\cancel{8}}} = \frac{99}{2} = \$49\frac{1}{2}$

Step 4: We multiply $6 \times 8\frac{1}{4}$ for regular tents and $16 \times 1\frac{1}{2} \times 8\frac{1}{4}$ for large tents. Then add total yardage.

Step 5:
$$6 \times 8\frac{1}{4} \text{ (regular tents)} = \overset{3}{6} \times \frac{33}{\underset{2}{4}} = \frac{99}{2} = 49\frac{1}{2} \text{ yards.}$$

$$\text{(Large tents)} \quad 16 \times 1\frac{1}{2} \times 8\frac{1}{4} = \overset{8}{16} \times \frac{3}{\underset{1}{2}} \times \frac{33}{\underset{1}{4}} = \frac{198}{1} = 198 \text{ yards.}$$

$$\text{Total yards for all tents } 198 + 49\frac{1}{2} = 247\frac{1}{2} \text{ yards.}$$

Step 6: This is close to our estimate, 240 yards. The answer is reasonable.

Answers to Even-Numbered Practice Problems

2. $8\frac{7}{8}$ inches
4. (a) He will need to buy two 12-foot lengths of pine shelving. **(b)** He actually needs 22 linear feet of shelving.
(c) He will have 2 feet of shelving left over.

EXTRA PRACTICE: EXAMPLES AND EXERCISES

Section 2.1

Use a fraction to represent the shaded part of the object.

Example
2 out of 3 parts. The fraction is $\frac{2}{3}$.

1.

3.

2.

4.

5.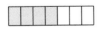

Draw a sketch to illustrate each statement.

Example $\frac{4}{7}$ of an object ▯▯▯▯▯▯▯

6. $\frac{3}{8}$ of an object

7. $\frac{6}{11}$ of an object

8. $\frac{5}{6}$ of an object

9. $\frac{9}{20}$ of an object

10. $\frac{7}{13}$ of an object

Express each answer as a fraction.

Example A soccer team consists of 11 players, of which 3 are forwards. What fractional part of the team are forwards?

3 of the 11 players—$\frac{3}{11}$

11. A team consists of 13 players, of which 5 were nominated for the All-Stars. What fractional part of the team was nominated?

12. The total purchase price was 95¢, of which 6¢ was sales tax. What fractional part of the total purchase price was sales tax?

13. A teacher gives five exams in a history course, three of which are multiple choice. What fractional part of the total exams are multiple choice?

14. John paid $897 for his bills last month, of which $500 was rent. What fractional part of his bills was rent?

15. A basketball player made 6 out of the 11 baskets shot. What fractional part of the baskets shot were made?

Example A company employs 412 women and 321 men. What fractional part of the company's employees are men?

321 men + 412 women
= 733 employees, of which 321 are men

$$\frac{321}{733}$$

16. There are 103 men and 69 women business majors at a college. What fractional part of the college's business majors are women?

17. There are 15 nickels, 19 dimes, and 29 quarters in a coin collection. What fractional part of the collection are dimes?

18. Heather went to Palm Springs for spring break. Her total trip cost $977. Her plane fare was $192. What fractional part of her trip cost was the cost of the plane fare?

19. There were 15 A's, 12 B's, 21 C's, 7 D's, and 2 F's on an exam in a history class. What fractional part of the grades were either A's or B's?

20. A track team has 12 freshmen, 9 sophomores, 18 juniors, and 19 seniors. What fractional part of the team are either juniors or seniors?

Express each number as a product of prime factors.

Example 210

$$
\begin{array}{c}
210 \\
\diagup \quad \diagdown \\
10 \quad \times \quad 21 \\
\diagup \diagdown \quad \diagup \diagdown \\
2 \times 5 \times 3 \times 7
\end{array}
$$

1. 48   **2.** 72   **3.** 140

4. 126   **5.** 105

Reduce by dividing the numerator and denominator by a common factor.

Example $\dfrac{20}{35}$

$$\dfrac{20 \div 5}{35 \div 5} = \dfrac{4}{7}$$

6. $\dfrac{30}{55}$ **7.** $\dfrac{12}{28}$ **8.** $\dfrac{46}{26}$

9. $\dfrac{24}{27}$ **10.** $\dfrac{42}{50}$

Reduce by methods of prime factor.

Example $\dfrac{150}{195}$

$$\dfrac{150}{195} = \dfrac{2 \times 3 \times 5 \times 5}{3 \times 5 \times 13} = \dfrac{10}{13}$$

11. $\dfrac{24}{42}$ **12.** $\dfrac{28}{105}$ **13.** $\dfrac{54}{126}$

14. $\dfrac{66}{110}$ **15.** $\dfrac{52}{169}$

Use the equality test of fractions to see if the fraction is equal.

Example $\dfrac{9}{32} \overset{?}{=} \dfrac{26}{128}$

$$\dfrac{9}{32} = \dfrac{26}{128} \qquad \begin{array}{l} 32 \times 26 = 832 \\ 9 \times 128 = 1152 \end{array}$$

832 does not equal 1152, so the fractions are not equal.

16. $\dfrac{3}{4} \overset{?}{=} \dfrac{27}{40}$ **17.** $\dfrac{12}{13} \overset{?}{=} \dfrac{144}{156}$ **18.** $\dfrac{21}{25} \overset{?}{=} \dfrac{84}{100}$

19. $\dfrac{32}{41} \overset{?}{=} \dfrac{96}{82}$ **20.** $\dfrac{65}{72} \overset{?}{=} \dfrac{130}{216}$

Change the following to improper fractions.

Example $14\dfrac{3}{8}$

$$14\dfrac{3}{8} = \dfrac{14 \times 8 + 3}{8} = \dfrac{112 + 3}{8} = \dfrac{115}{8}$$

1. $4\dfrac{3}{7}$ **2.** $12\dfrac{2}{3}$ **3.** $22\dfrac{3}{17}$

4. $17\dfrac{2}{9}$ **5.** $31\dfrac{2}{3}$

Write as a mixed number or a whole number.

Example $\dfrac{23}{7}$

$$\dfrac{23}{7} \longrightarrow 7\overline{)23} = 3\dfrac{2}{7}$$
$$\phantom{\dfrac{23}{7} \longrightarrow 7)} \underline{-21} $$
$$\phantom{\dfrac{23}{7} \longrightarrow 7)23=3} 2$$

6. $\dfrac{47}{6}$ **7.** $\dfrac{35}{7}$ **8.** $\dfrac{57}{5}$

9. $\dfrac{106}{9}$ **10.** $\dfrac{201}{10}$

Reduce each mixed number.

Example $3\dfrac{4}{12}$

$$3\dfrac{4}{12} \qquad 3\dfrac{4}{12} = \dfrac{\overset{1}{\cancel{2}} \times \cancel{2}}{\cancel{2} \times \cancel{2} \times 3} = 3\dfrac{1}{3}$$

11. $5\dfrac{10}{30}$ **12.** $6\dfrac{18}{24}$ **13.** $2\dfrac{21}{42}$

14. $12\dfrac{22}{33}$ **15.** $9\dfrac{20}{35}$

Reduce each improper fraction.

Example $\dfrac{36}{12}$

$$\dfrac{36}{12} = \dfrac{\overset{1}{\cancel{2}} \times \overset{1}{\cancel{3}} \times \overset{1}{\cancel{2}} \times 3}{\underset{1}{\cancel{2}} \times \underset{1}{\cancel{3}} \times \underset{1}{\cancel{2}}} = \dfrac{3}{1} = 3$$

16. $\dfrac{35}{25}$ **17.** $\dfrac{42}{14}$ **18.** $\dfrac{66}{11}$

19. $\dfrac{28}{4}$ **20.** $\dfrac{91}{26}$

Multiply each fraction. Reduce your answer if necessary.

Example $\dfrac{15}{36} \times \dfrac{9}{25}$

$$\dfrac{\overset{3}{\cancel{15}}}{\underset{4}{\cancel{36}}} \times \dfrac{\overset{1}{\cancel{9}}}{\underset{5}{\cancel{25}}} = \dfrac{3 \times 1}{4 \times 5} = \dfrac{3}{20}$$

1. $\dfrac{2}{9} \times \dfrac{1}{5}$ **2.** $\dfrac{3}{7} \times \dfrac{4}{11}$ **3.** $\dfrac{35}{42} \times \dfrac{6}{7}$

4. $\dfrac{12}{3} \times \dfrac{9}{36}$ **5.** $\dfrac{18}{25} \times \dfrac{5}{2}$

Example $5 \times \dfrac{1}{7} \times \dfrac{2}{15}$

$$\dfrac{5}{1} \times \dfrac{1}{7} \times \dfrac{2}{15} = \dfrac{\overset{1}{\cancel{5}} \times 1 \times 2}{1 \times 7 \times \underset{3}{\cancel{15}}} = \dfrac{2}{21}$$

6. $3 \times \dfrac{2}{5}$

7. $4 \times \dfrac{1}{12}$

8. $7 \times \dfrac{2}{21} \times \dfrac{5}{8}$

9. $11 \times \dfrac{15}{9} \times \dfrac{3}{5}$

10. $9 \times \dfrac{14}{81} \times \dfrac{1}{2}$

Multiply each fraction. Change any mixed number to an improper fraction before multiplying.

Example $4\dfrac{2}{3} \times \dfrac{5}{2}$

$$4\dfrac{2}{3} = \dfrac{14}{3} \qquad \dfrac{\overset{7}{\cancel{14}}}{3} \times \dfrac{5}{\underset{1}{\cancel{2}}} = \dfrac{7 \times 5}{3 \times 1} = \dfrac{35}{3} \quad \text{or} \quad 11\dfrac{2}{3}$$

11. $4\dfrac{1}{5} \times \dfrac{10}{7}$

12. $\dfrac{2}{35} \times 3\dfrac{3}{4}$

13. $2\dfrac{1}{3} \times 2\dfrac{8}{14}$

14. $5\dfrac{2}{15} \times \dfrac{11}{7}$

15. $1 \times 13\dfrac{1}{2}$

Solve each problem.

Example Find the area of a parking lot that is $2\dfrac{1}{2}$ miles long and $1\dfrac{1}{3}$ miles wide.

Length $= 2\dfrac{1}{2}$, width $= 1\dfrac{1}{3}$. Multiply the length times the width to find area.

$$2\dfrac{1}{2} \times 1\dfrac{1}{3} = \dfrac{5}{\underset{1}{\cancel{2}}} \times \dfrac{\overset{2}{\cancel{4}}}{3} = \dfrac{10}{3} \quad \text{or} \quad 3\dfrac{1}{3} \text{ square miles}$$

16. Find the area of a playground that is $3\dfrac{3}{5}$ miles long and $2\dfrac{1}{3}$ miles wide.

17. Find the area of a wheat field that is $15\dfrac{3}{4}$ miles long and $10\dfrac{2}{9}$ miles wide.

18. A car has $13\dfrac{1}{2}$ gallons of gas. The car averages 28 miles per gallon. How far can the car go?

19. A bus has $20\dfrac{2}{3}$ gallons of gas. The bus averages 10 miles per gallon. How far can the bus go?

20. If $3\dfrac{1}{3}$ cups of flour are required to bake one loaf of bread, how much flour would be needed to bake 15 loaves of bread?

Section 2.5

Divide, if possible.

Example

(a) $\dfrac{3}{5} \div \dfrac{15}{22}$ **(b)** $\dfrac{14}{17} \div 0$

(a) Invert the fraction on the right and simplify.

$$\dfrac{\overset{1}{\cancel{3}}}{5} \times \dfrac{22}{\underset{5}{\cancel{15}}} = \dfrac{1 \times 22}{5 \times 5} = \dfrac{22}{25}$$

(b) $\dfrac{14}{17} \div 0$ You can never divide by zero. This problem cannot be done.

1. $\dfrac{2}{7} \div \dfrac{3}{5}$

2. $\dfrac{8}{21} \div \dfrac{2}{7}$

3. $\dfrac{16}{29} \div 0$

4. $0 \div \dfrac{2}{3}$

5. $\dfrac{7}{12} \div \dfrac{35}{24}$

Example $\dfrac{4}{7} \div 2$

$$\dfrac{4}{7} \div \dfrac{2}{1} \longrightarrow \dfrac{\overset{2}{\cancel{4}}}{7} \times \dfrac{1}{\underset{1}{\cancel{2}}} = \dfrac{2 \times 1}{7 \times 1} = \dfrac{2}{7}$$

6. $\dfrac{3}{8} \div 6$

7. $5 \div \dfrac{25}{27}$

8. $13 \div \dfrac{26}{28}$

9. $7 \div \dfrac{14}{19}$

10. $250 \div \dfrac{5}{6}$

Example $4\dfrac{2}{3} \div 3\dfrac{3}{7}$

$$4\dfrac{2}{3} = \dfrac{14}{3} \qquad 3\dfrac{3}{7} = \dfrac{24}{7} \qquad \dfrac{14}{3} \div \dfrac{24}{7} = \dfrac{\overset{7}{\cancel{14}}}{3} \times \dfrac{7}{\underset{12}{\cancel{24}}} = \dfrac{7 \times 7}{3 \times 12} = \dfrac{49}{36}$$

11. $2\dfrac{1}{2} \div 3\dfrac{1}{7}$

12. $1\dfrac{2}{9} \div 4\dfrac{1}{3}$

13. $5\dfrac{7}{9} \div 2\dfrac{2}{5}$

14. $8\dfrac{1}{3} \div 10\dfrac{5}{7}$

15. $1\dfrac{5}{7} \div 1\dfrac{11}{21}$

Example $\dfrac{2\dfrac{3}{8}}{1\dfrac{15}{16}}$

$$2\dfrac{3}{8} \div 1\dfrac{15}{16} \qquad \dfrac{19}{8} \div \dfrac{31}{16} \qquad \dfrac{19}{\underset{1}{\cancel{8}}} \times \dfrac{\overset{2}{\cancel{16}}}{31} = \dfrac{19 \times 2}{1 \times 31} = \dfrac{38}{31} \quad \text{or} \quad 1\dfrac{7}{31}$$

16. $\dfrac{1\dfrac{7}{9}}{2\dfrac{2}{3}}$

17. $\dfrac{\dfrac{3}{4}}{3\dfrac{3}{5}}$

18. $\dfrac{3\dfrac{1}{2}}{2\dfrac{7}{8}}$

19. $\dfrac{6\dfrac{2}{3}}{\dfrac{2}{7}}$

20. $\dfrac{5\dfrac{1}{3}}{4}$

Example Find the LCD for

(a) $\dfrac{2}{7}, \dfrac{5}{14}$ (b) $\dfrac{1}{9}, \dfrac{3}{5}$

In these problems you can guess the LCD quickly

(a) 14   7 can be divided into 14.
(b) 45 $9 \times 5 = 45$ and 45 is the smallest number that can be divided without remainder by 9 or 5.

1. $\dfrac{1}{2}, \dfrac{3}{4}$ 2. $\dfrac{4}{5}, \dfrac{7}{20}$ 3. $\dfrac{1}{3}, \dfrac{2}{7}$

4. $\dfrac{3}{4}, \dfrac{9}{11}$ 5. $\dfrac{13}{28}, \dfrac{6}{7}$

Example Find the LCD of $\dfrac{3}{30}$ and $\dfrac{7}{20}$.

$30 = 2 \times 3 \times 5$ The LCD must contain 2's, 3's, and
$20 = 2 \times 2 \times 5$ 5's. The LCD must contain the
                             factor 2 twice since it appears twice
LCD $= 2 \times 2 \times 3 \times 5$ as a factor of 20.
    $= 60$

6. $\dfrac{3}{18}, \dfrac{21}{75}$ 7. $\dfrac{13}{30}, \dfrac{23}{50}$ 8. $\dfrac{13}{40}, \dfrac{7}{12}$

9. $\dfrac{9}{25}, \dfrac{7}{10}$ 10. $\dfrac{19}{60}, \dfrac{1}{18}$

Find the LCD for each set of fractions.

Example $\dfrac{1}{8}, \dfrac{17}{20}, \dfrac{3}{4}$

$8 = 2 \times 2 \times 2$ 2 appears three times as a factor.
$20 = 2 \times 2 \times 5$ LCD $= 2 \times 2 \times 2 \times 5 = 40$.
$4 = 2 \times 2$

11. $\dfrac{4}{5}, \dfrac{7}{30}, \dfrac{1}{6}$ 12. $\dfrac{11}{18}, \dfrac{2}{3}, \dfrac{1}{12}$ 13. $\dfrac{6}{15}, \dfrac{9}{20}, \dfrac{2}{9}$

14. $\dfrac{5}{21}, \dfrac{7}{8}, \dfrac{13}{14}$ 15. $\dfrac{17}{36}, \dfrac{2}{45}, \dfrac{9}{10}$

For each set of fractions:
(a) Find the LCD.
(b) Build up the fractions to equivalent fractions having the same LCD as the denominator.

Example $\dfrac{7}{12}$ and $\dfrac{9}{20}$

(a) $12 = 3 \times 2 \times 2$ 2 is a factor of 12 and 20 twice.
$20 = 5 \times 2 \times 2$ LCD $= 2 \times 2 \times 5 \times 3 = 60$.

(b) $\dfrac{7}{12} \times \dfrac{5}{5} = \dfrac{35}{60}$ Because $12 \times 5 = 60$, multiply by 5.

$\dfrac{9}{20} \times \dfrac{3}{3} = \dfrac{27}{60}$ Because $20 \times 3 = 60$, multiply by 3.

16. $\dfrac{2}{3}, \dfrac{7}{15}$ 17. $\dfrac{3}{7}, \dfrac{5}{21}$

18. $\dfrac{7}{40}, \dfrac{11}{56}$ 19. $\dfrac{9}{16}, \dfrac{3}{20}$

20. $\dfrac{1}{9}, \dfrac{3}{45}, \dfrac{13}{24}$

Add or subtract each fraction.

Example $\dfrac{5}{16} + \dfrac{7}{16}$

$\dfrac{5}{16} + \dfrac{7}{16} = \dfrac{\overset{3}{\cancel{12}}}{\underset{4}{\cancel{16}}} = \dfrac{3}{4}$

1. $\dfrac{3}{7} + \dfrac{2}{7}$ 2. $\dfrac{4}{19} + \dfrac{2}{19}$ 3. $\dfrac{21}{29} - \dfrac{7}{29}$

4. $\dfrac{15}{31} - \dfrac{3}{31}$ 5. $\dfrac{11}{41} - \dfrac{3}{41}$ 6. $\dfrac{13}{80} - \dfrac{3}{80}$

Example $\dfrac{19}{20} - \dfrac{1}{15}$

LCD $= 60$ $\dfrac{19}{20} \times \dfrac{3}{3} = \dfrac{57}{60}$ $\dfrac{1}{15} \times \dfrac{4}{4} = \dfrac{4}{60}$

$\dfrac{57}{60} - \dfrac{4}{60} = \dfrac{53}{60}$

7. $\dfrac{3}{14} + \dfrac{1}{2}$ 8. $\dfrac{7}{18} + \dfrac{1}{12}$ 9. $\dfrac{7}{12} - \dfrac{3}{8}$

10. $\dfrac{9}{24} - \dfrac{3}{16}$ 11. $\dfrac{32}{35} - \dfrac{2}{5}$

Example $16\dfrac{3}{8} + 3\dfrac{3}{4}$

LCD $= 8$ $16\dfrac{3}{8}$

$+ \ 3\dfrac{6}{8}$

$19\dfrac{9}{8}$ or $19 + 1\dfrac{1}{8} = 20\dfrac{1}{8}$

12. $13\dfrac{2}{7} + 15\dfrac{9}{35}$ 13. $14\dfrac{1}{3} + 32\dfrac{1}{4}$ 14. $5\dfrac{1}{15} + 4\dfrac{3}{25}$

15. $32\dfrac{3}{10} + 24\dfrac{5}{20}$ 16. $13\dfrac{1}{12} + 21\dfrac{5}{14}$

Example $9\dfrac{1}{8} - 5\dfrac{15}{24}$

LCD $= 24$ 15 cannot be subtracted from 3, so we must borrow 1 from the 9.

$9\dfrac{3}{24}$ $8 + 1\dfrac{3}{24} = 8\dfrac{27}{24}$

$- 5\dfrac{15}{24}$ $- 5\dfrac{15}{24}$

$3\dfrac{12}{24} = 3\dfrac{1}{2}$

17. $7\dfrac{1}{4} - 1\dfrac{1}{3}$ 18. $14\dfrac{1}{15} - 2\dfrac{3}{20}$ 19. $21\dfrac{3}{5} - 17\dfrac{7}{15}$

Example $16\dfrac{3}{8} + 3\dfrac{3}{4}$

20. $18\dfrac{5}{12} - 5\dfrac{9}{14}$ 21. $15 - 3\dfrac{7}{11}$

Example Danielle needs $200 to buy a stereo. She has $79 in her savings account and received $47 for her birthday. How much money does she need to buy the stereo?

$$
\begin{array}{rl}
\$\ 79 & \\
+\ \ 47 & \\
\hline
\$126 & \text{total amount she has}
\end{array}
\qquad
\begin{array}{rl}
\$200 & \text{cost of stereo} \\
-\ 126 & \text{money has} \\
\hline
\$\ 74 & \text{amount of} \\
 & \text{money needed}
\end{array}
$$

1. The truck hauled $6\frac{1}{8}$ tons of gravel on Wednesday, $5\frac{3}{4}$ tons on Thursday, and $7\frac{1}{8}$ tons on Friday. How many tons were hauled over the three days?

2. The baker used $7\frac{1}{3}$ pounds of sugar Tuesday, $9\frac{1}{4}$ pounds on Wednesday, and $4\frac{7}{10}$ pounds on Thursday. How many pounds were used over the three days?

3. Joan wants to run 17 miles. On Friday she ran $5\frac{1}{6}$ miles. On Saturday she ran $7\frac{1}{4}$ miles. How many more miles must she run to make her goal?

4. Janice wants a bush in her front yard to be 7 feet tall. When the bush was planted it was $3\frac{1}{4}$ feet above the ground in height. It grew $2\frac{1}{3}$ feet the first year after being planted. How many more feet does it need to grow to reach Janice's 7-foot goal?

5. Jerry bought a can of green beans that weighed in total $18\frac{2}{3}$ ounces. It contained $2\frac{1}{4}$ ounces of water. The empty metal can weighed $1\frac{1}{3}$ ounces.

   (a) How many ounces of green beans did he buy?

   (b) The can stated net weight $15\frac{1}{2}$ ounces (which means he should have found $15\frac{1}{2}$ ounces of green beans in the can). How much in error was this measurement?

Example Bob bought $5\frac{2}{3}$ pounds of ham at $\$4\frac{1}{2}$ per pound. How much did the purchase cost?

$$
5\frac{2}{3} \times 4\frac{1}{2} = \frac{17}{3} \times \frac{\overset{3}{\cancel{9}}}{2} = \frac{51}{2} \quad \text{or} \quad \$25\frac{1}{2}
$$

6. Julie purchased $6\frac{1}{4}$ pounds of potato salad for a party at $\$1\frac{1}{2}$ per pound. How much did the purchase cost?

7. Karen bought $15\frac{1}{4}$ yards of fabric selling for $\$2\frac{1}{2}$ per yard. How much did she pay?

8. A car is traveling 50 miles per hour. It travels for $3\frac{1}{4}$ hours at that speed. How far does it travel?

9. How many gallons can a tank hold that has a volume of $29\frac{2}{3}$ cubic feet? (Assume that 1 cubic foot holds about $7\frac{1}{2}$ gallons.)

10. How much does a gallon of water containing $6\frac{1}{9}$ cubic feet weigh? (Assume that 1 cubic foot of water weighs $62\frac{1}{2}$ pounds.)

Example Jan's car gets $22\frac{1}{2}$ miles per gallon. How many gallons of gasoline would Jan need to drive 260 miles?

$$
260 \div 22\frac{1}{2}
$$

$$
260 \div \frac{45}{2} \longrightarrow \frac{\overset{52}{\cancel{260}}}{1} \times \frac{2}{\underset{9}{45}} = \frac{104}{9} \quad \text{or} \quad 11\frac{5}{9} \text{ gallons}
$$

11. A piece of wood that is $134\frac{4}{9}$ feet long is to be cut into pieces that are $6\frac{1}{9}$ feet long. How many pieces will be made?

12. Ellen bought 22 shares of stock for $\$445\frac{1}{2}$. How much did one share of stock cost?

13. Frank's car gets $27\frac{1}{4}$ miles per gallon. How many gallons of gasoline would Frank need to drive 327 miles?

14. Wendy purchased $16\frac{4}{7}$ yards of fabric to make scarves. It takes $\frac{4}{7}$ yard of fabric to make scarves. How many scarves can she make?

15. The Gourmet Food Shop purchased $21\frac{7}{8}$ ounces of tea. The shop places the tea in $\frac{5}{8}$-ounce bags. How many $\frac{5}{8}$-ounce bags can they make?

Example Matt earns $\$7\frac{1}{2}$ an hour and earns overtime after working 8 hours. Overtime pay is $1\frac{1}{2}$ times his regular rate of $\$7\frac{1}{2}$ an hour. How much does Matt earn for working 10 hours?

$$
10 \text{ hours} - 8 \text{ hours} = 2 \text{ hours overtime}
$$

$$
\text{Overtime pay rate is } 1\frac{1}{2} \times 7\frac{1}{2} = \frac{3}{2} \times \frac{15}{2} = \frac{45}{4} = \$11\frac{1}{4}
$$

$$
2 \text{ hours overtime} \times \text{overtime pay rate} = \text{overtime pay}
$$

$$
2 \times 11\frac{1}{4} = \frac{2}{1} \times \frac{45}{\underset{2}{\cancel{4}}} = \frac{45}{2} \text{ or } \$22\frac{1}{2}
$$

8 hours regular pay × regular pay rate = regular pay

$$8 \times 7\frac{1}{2} = \frac{\overset{4}{\cancel{8}}}{1} \times \frac{15}{\underset{1}{\cancel{2}}} = \$60$$

Pay for 8 hours + pay for overtime = total pay

$$\$60 \quad + \quad \$22\frac{1}{2} \quad = \quad \$82\frac{1}{2}$$

16. Les bought four tires for his car. Each tire cost $45\frac{1}{2}$. The auto shop charged Les $15\frac{1}{2}$ to mount all four tires on his car. How much did Les pay for the tires and mounting?

17. Joe earns $575 per week. He has $\frac{3}{10}$ of his income withheld for federal taxes, $\frac{1}{25}$ of his income withheld for state taxes, and $\frac{1}{23}$ of his income withheld for medical coverage. How much per week is left for Joe after those three deductions?

18. Bob earns $7\frac{1}{2}$ an hour. He worked 8 hours Monday, 6 hours Tuesday, and $5\frac{1}{2}$ hours Wednesday. How much money did Bob earn for the three days?

19. Sally earns $6\frac{3}{4}$ an hour and earns overtime pay after 8 working hours. Overtime pay is $1\frac{1}{2}$ times her regular rate of $6\frac{3}{4}$. How much does Sally earn for working 10 hours?

20. Mary is making a recipe that requires $15\frac{3}{4}$ tablespoons of sugar. She wants to triple the recipe but has only 35 tablespoons of sugar. How much sugar will Mary need to buy to triple the recipe?

CHAPTER ORGANIZER

Topic	Procedure	Examples
Concept of a fractional part, p. 78	The numerator is the number of parts selected. The denominator is the number of total parts.	What part of this sketch is shaded? $\frac{7}{10}$
Prime factorization, p. 82	Prime factorization is the writing of a number as the product of prime numbers.	Write the prime factorization of 36. $36 = 4 \times 9$ $2 \times 2 \quad 3 \times 3$ $= 2 \times 2 \times 3 \times 3$
Reducing fractions, p. 84	1. Factor numerator and denominator into prime factors. 2. Divide out factors common to numerator and denominator.	Reduce: $\frac{54}{90}$. $\frac{54}{90} = \frac{\overset{1}{\cancel{3}} \times \overset{1}{\cancel{3}} \times 3 \times \overset{1}{\cancel{2}}}{\underset{1}{\cancel{3}} \times \underset{1}{\cancel{3}} \times \underset{1}{\cancel{2}} \times 5} = \frac{3}{5}$
Changing an improper fraction to a mixed number, p. 92	1. Divide numerator into denominator. 2. The quotient is the whole number. 3. The fraction is the remainder over the divisor.	Change to a mixed number: $\frac{32}{5}$. $5\overline{)32} = 6\frac{2}{5}$ $\underline{30}$ 2
Changing mixed numbers to improper fractions, p. 91	1. Multiply whole number by denominator. 2. Add product to numerator. 3. Place sum over denominator.	Write as an improper fraction: $7\frac{3}{4} = \frac{4 \times 7 + 3}{4} \quad \frac{28 + 3}{4} = \frac{31}{4}$
Multiplying fractions, p. 97	1. Divide out a factor in the numerator and the same factor in the denominator whenever possible. 2. Multiply numerators. 3. Multiply denominators.	Multiply: $\frac{3}{7} \times \frac{5}{13} = \frac{15}{91}$ Multiply: $\frac{\overset{1}{\cancel{5}}}{\cancel{8}} \times \frac{\overset{2}{\cancel{16}}}{\underset{3}{\cancel{15}}} = \frac{2}{3}$

Topic	Procedure	Examples
Multiplying mixed and/or whole numbers, p. 98	1. Change any whole numbers to a fraction with a denominator of 1. 2. Change any mixed numbers to improper fractions. 3. Use multiplication rule for fractions.	Multiply: $7 \times 3\frac{1}{4}$ $$\frac{7}{1} \times \frac{13}{4} = \frac{91}{4}$$ $$= 22\frac{3}{4}$$
Dividing fractions, p. 105	To divide two fractions, we invert the second fraction and multiply.	Divide: $\frac{3}{7} \div \frac{2}{9} = \frac{3}{7} \times \frac{9}{2} = \frac{27}{14}$ or $1\frac{13}{14}$
Dividing mixed numbers and/or whole numbers, p. 105	1. Change any whole numbers to a fraction with a denominator of 1. 2. Change any mixed numbers to improper fractions. 3. Use rule for division of fractions.	Divide: $8\frac{1}{3} \div 5\frac{5}{9} = \frac{25}{3} \div \frac{50}{9}$ $$= \frac{\overset{1}{\cancel{25}}}{\cancel{3}} \times \frac{\overset{3}{\cancel{9}}}{\cancel{50}} = \frac{3}{2} \text{ or } 1\frac{1}{2}$$
Finding the least common denominator, p. 112	1. Write each denominator as the product of prime factors. 2. List all the prime factors that appear in either product. 3. Form a product of those factors, using each factor the greatest number of times it appears in any one denominator.	Find LCD of $\frac{1}{10}, \frac{3}{8}$, and $\frac{7}{25}$. $10 = 2 \times 5$ $8 = 2 \times 2 \times 2$ $25 = 5 \times 5$ LCD $= 2 \times 2 \times 2 \times 5 \times 5$ $= 200$
Building fractions, p. 114	1. Find how many times the original denominator can be divided into the new denominator. 2. Multiply that value by numerator and denominator of original fracton.	Build $\frac{5}{7}$ to an equivalent fraction with a denominator of 42. First we find $7\overline{)42}$ (quotient 6). Then we multiply by 6 for numerator and denominator. $$\frac{5}{7} \times \frac{6}{6} = \frac{30}{42}$$
Adding or subtracting fractions with a common denominator, p. 119	1. Add or subtract the numerators. 2. Keep the common denominator.	Add: $\frac{3}{13} + \frac{5}{13} = \frac{8}{13}$. Subtract: $\frac{15}{17} - \frac{12}{17} = \frac{3}{17}$.
Adding or subtracting fractions without a common denominator, p. 120	1. Find the LCD of the fractions. 2. Build up each fraction, if necessary, to obtain the LCD in the denominator. 3. Follow the steps for adding and subtracting fractions with the same denominator.	Add: $\frac{1}{4} + \frac{3}{7} + \frac{5}{8}$. LCD $= 56$ $\frac{1 \times 14}{4 \times 14} + \frac{3 \times 8}{7 \times 8} + \frac{5 \times 7}{8 \times 7}$ $$= \frac{14}{56} + \frac{24}{56} + \frac{35}{56} = \frac{73}{56} = 1\frac{17}{56}$$
Adding or subtracting mixed numbers, p. 121	1. Change fractional parts to equivalent fractions with LCD as a denominator, if necessary. 2. Follow the steps for adding and subtracting fractions with the same denominator.	Subtract: $37\frac{1}{7} = 36\frac{8}{7}$ $\underline{-23\frac{4}{7}} \quad \underline{23\frac{4}{7}}$ $\qquad\qquad 13\frac{4}{7}$

Topic	Procedure	Examples
Solving applied problems, p. 126	1. Read the problem over carefully. Find out what is asked for. Draw a picture if this helps you to visualize the situation. 2. Write down which fractions and whole numbers are to be used in solving the problem. 3. Estimate. Round off the values and do a simple calculation. 4. Determine which operation needs to be done: addition, subtraction, multiplication, division, or a combination of these. 5. Perform the necessary computation. State the answer, including the unit of measure. 6. Compare answer with estimate to see if the answer is reasonable.	A wire is $95\frac{1}{3}$ feet long. It is cut up into smaller equal-sized pieces each $4\frac{1}{3}$ feet long. How many pieces will there be? 1. How many small wires of $4\frac{1}{3}$ feet are in $95\frac{1}{3}$ feet? 2. Total wire $95\frac{1}{3}$ feet. Each piece $4\frac{1}{3}$ feet. 3. Estimate. Round $95\frac{1}{3}$ to nearest ten $= 100$. Round $4\frac{1}{3}$ to nearest integer $= 4$. $100 \div 4 = 25$. 4. We need to divide $95\frac{1}{3} \div 4\frac{1}{3}$ 5. $\dfrac{286}{3} \div \dfrac{13}{3} = \dfrac{\overset{22}{\cancel{286}}}{\underset{1}{\cancel{3}}} \times \dfrac{\overset{1}{\cancel{3}}}{\underset{1}{\cancel{13}}} = \dfrac{22}{1} = 22$ There will be 22 pieces of wire. 6. This is close to our estimate; answer is reasonable.

REVIEW PROBLEMS CHAPTER 2

2.1 *In each problem, use a fraction to represent the shaded part of the object.*

1.

2.

Draw a sketch to illustrate the fraction.

3. $\frac{4}{7}$ of an object

4. $\frac{7}{9}$ of an object

5. An inspector looked at 31 parts and found 6 of them to be defective. What fractional part of these items was defective?

6. The dean asked 100 of the freshman if they would be staying in the dorm over the holidays. A total of 87 said they would not. What fractional part of the freshmen said they would not?

2.2 *Express each number as a product of prime factors.*

7. 42

8. 54

9. 168

Identify which of the following numbers is prime. If it is composite, express it as the product of prime factors.

10. 59

11. 78

12. 167

Reduce the fraction.

13. $\frac{13}{52}$

14. $\frac{12}{42}$

15. $\frac{21}{36}$

16. $\frac{26}{34}$

17. $\frac{168}{192}$

18. $\frac{51}{105}$

2.3 *Change each mixed number to an improper fraction.*

19. $4\dfrac{3}{8}$

20. $2\dfrac{19}{23}$

Change each improper fraction to a mixed number.

21. $\dfrac{19}{7}$

22. $\dfrac{63}{13}$

23. Reduce and leave your answer as a mixed number: $3\dfrac{15}{55}$.

24. Reduce and leave your answer as an improper fraction: $\dfrac{234}{16}$.

25. Change to a mixed number and then reduce: $\dfrac{385}{240}$.

2.4 *Multiply each fraction.*

26. $\dfrac{4}{7} \times \dfrac{5}{11}$

27. $\dfrac{7}{9} \times \dfrac{21}{35}$

28. $12 \times \dfrac{3}{7} \times 0$

29. $\dfrac{3}{5} \times \dfrac{2}{7} \times \dfrac{10}{27}$

30. $12 \times 8\dfrac{1}{5}$

31. $5\dfrac{3}{8} \times 3\dfrac{4}{5}$

32. $5\dfrac{1}{4} \times 4\dfrac{6}{7}$

33. Find the area of a rectangle that is $18\dfrac{1}{5}$ inches wide and $26\dfrac{3}{4}$ inches long.

34. One share of stock costs $\$37\dfrac{5}{8}$. How much money will 18 shares cost?

2.5 *Divide, if possible.*

35. $\dfrac{3}{7} \div \dfrac{2}{5}$

36. $\dfrac{9}{17} \div \dfrac{18}{5}$

37. $1200 \div \dfrac{5}{8}$

38. $\dfrac{2\dfrac{1}{4}}{3\dfrac{1}{3}}$

39. $\dfrac{20}{4\dfrac{4}{5}}$

40. $2\dfrac{1}{8} \div 20\dfrac{1}{2}$

41. $0 \div 3\dfrac{7}{5}$

42. $4\dfrac{2}{11} \div 3$

43. There are 420 calories in $2\dfrac{1}{4}$ cans of grape soda. How many calories are contained in 1 can of soda?

44. A dress requires $3\dfrac{1}{8}$ yards of fabric. Amy has $21\dfrac{7}{8}$ yards available. How many dresses can she make?

2.6 *Find the LCD for each set of fractions.*

45. $\dfrac{7}{14}$ and $\dfrac{3}{49}$

46. $\dfrac{7}{40}$ and $\dfrac{11}{30}$

47. $\dfrac{5}{18}, \dfrac{1}{6}, \dfrac{7}{45}$

Build each fraction to an equivalent fraction with the specified denominator.

48. $\dfrac{3}{7} = \dfrac{?}{56}$

49. $\dfrac{11}{24} = \dfrac{?}{72}$

50. $\dfrac{9}{43} = \dfrac{?}{172}$

51. $\dfrac{17}{18} = \dfrac{?}{198}$

2.7 *Add or subtract.*

52. $\dfrac{3}{7} - \dfrac{5}{14}$

53. $\dfrac{1}{2} + \dfrac{1}{3} + \dfrac{1}{4}$

54. $\dfrac{4}{7} + \dfrac{7}{9}$

55. $\dfrac{7}{8} - \dfrac{3}{5}$

56. $3\dfrac{3}{12} + 1\dfrac{5}{8}$

57. $7\dfrac{3}{16} - 2\dfrac{5}{6}$

58. $120 - 16\dfrac{2}{3}$

59. $22\dfrac{2}{3} + 48\dfrac{3}{4}$

60. Bob jogged $1\dfrac{7}{8}$ miles on Monday, $2\dfrac{3}{4}$ miles on Tuesday, and $4\dfrac{1}{10}$ miles on Wednesday. How many miles did he jog on these three days?

2.8 *Answer each question.*

61. The stock was purchased at $\$88\dfrac{3}{8}$. However, it was sold at $\$79\dfrac{5}{8}$. How much did the value of the stock decrease during that time period?

62. Rafael traveled in a car that gets $24\dfrac{1}{4}$ miles per gallon. He had $8\dfrac{1}{2}$ gallons of gas in the gas tank. Approximately how far could he drive?

63. How many lengths of pipe $3\dfrac{1}{5}$ inches long can be cut from a pipe 48 inches long?

64. A car radiator holds $15\dfrac{3}{4}$ liters. If it contains $6\dfrac{1}{8}$ liters of antifreeze and the rest water, how much is water?

65. Delbert typed 366 words in $12\dfrac{1}{5}$ minutes. How many words per minute did he type?

66. Alicia earns $\$4\dfrac{1}{2}$ dollars per hour for regular pay and $1\dfrac{1}{2}$ times that rate of pay for overtime. On Monday she worked 8 hours at regular pay and 3 hours at overtime. How much did she earn on Monday?

67. A 3-inch bolt passes through $1\dfrac{1}{2}$ inches of pine board, a $\dfrac{1}{16}$-inch washer, and a $\dfrac{1}{8}$-inch nut. How many inches extend beyond the board, washer, and nut if the head of the bolt is $\dfrac{1}{4}$ inch long?

68. A recipe calls for $3\dfrac{1}{3}$ cups of sugar and $4\dfrac{1}{4}$ cups of flour. How much flour and how much sugar would be needed for $\dfrac{1}{2}$ of that recipe?

69. Francine has a take-home pay of $880 per month. She gives $\dfrac{1}{10}$ of it to her church, spends $\dfrac{1}{2}$ of it for rent and food, and spends $\dfrac{1}{8}$ of it on electricity, heat, and telephone. How many dollars per month does she have left for other things?

70. Manuel's new car used $18\dfrac{2}{5}$ gallons of gas on a 460-mile trip.
   (a) How many miles can his car travel on 1 gallon of gas?
   (b) How much did his trip cost him in gasoline expense if the average cost of gasoline was $\$1\dfrac{1}{5}$ per gallon?

Write a fraction that indicates the part of the whole that is shaded.

1.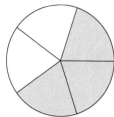

2.

Draw a sketch to illustrate each fraction.

3. $\dfrac{7}{11}$

4. $\dfrac{5}{6}$

5. Six out of 17 men were over 6 feet tall. Write a fraction that describes the part of the group that is over 6 feet tall.

Write each number as a product of prime factors.

6. 40

7. 225

Reduce each fraction.

8. $\dfrac{14}{21}$

9. $\dfrac{75}{100}$

10. $\dfrac{18}{48}$

Transform each improper fraction to a mixed number.

11. $\dfrac{47}{4}$

12. $\dfrac{38}{7}$

1. _____

2. _____

3. _____

4. _____

5. _____

6. _____

7. _____

8. _____

9. _____

10. _____

11. _____

12. _____

Transform each mixed number to an improper fraction.

13. $4\frac{2}{5}$

14. $23\frac{3}{4}$

Multiply.

15. $\frac{3}{7} \times \frac{5}{11}$

16. $\frac{7}{36} \times \frac{45}{21}$

17. $2\frac{1}{3} \times 5$

18. $3\frac{2}{5} \times 1\frac{1}{4}$

Divide.

19. $\dfrac{\frac{3}{8}}{\frac{2}{7}}$

20. $\dfrac{\frac{1}{4}}{\frac{3}{16}}$

21. $4\frac{1}{2} \div \frac{3}{7}$

22. $2\frac{5}{8} \div 1\frac{2}{3}$

23. A man purchased $5\frac{1}{3}$ sacks of flour at $\$8\frac{1}{4}$ per sack. How much did he pay?

24. How many dresses can be made from $52\frac{1}{2}$ yards of material if each dress requires $4\frac{3}{8}$ yards of material?

Find the least common denominator (LCD) for each group of fractions.

1. $\dfrac{2}{5}, \dfrac{5}{6}$

2. $\dfrac{1}{4}, \dfrac{3}{20}$

3. $\dfrac{1}{36}, \dfrac{5}{9}, \dfrac{1}{2}$

4. $\dfrac{2}{3}, \dfrac{3}{4}, \dfrac{4}{5}$

5. $\dfrac{7}{40}, \dfrac{5}{32}$

6. $\dfrac{7}{12}, \dfrac{19}{21}$

7. $\dfrac{3}{8}, \dfrac{7}{12}, \dfrac{1}{16}$

8. $\dfrac{1}{9}, \dfrac{5}{27}, \dfrac{5}{18}$

Add or subtract.

9. $\dfrac{3}{8} + \dfrac{1}{5}$

10. $\dfrac{7}{12} + \dfrac{2}{3}$

11. $\dfrac{4}{7} + \dfrac{3}{35}$

12. $\dfrac{7}{19} + \dfrac{2}{3}$

13. $5\dfrac{4}{5} + 2\dfrac{1}{4}$

14. $7 + 3 + 4\dfrac{2}{3}$

1. _____

2. _____

3. _____

4. _____

5. _____

6. _____

7. _____

8. _____

9. _____

10. _____

11. _____

12. _____

13. _____

14. _____

15. _____

16. _____

17. _____

18. _____

19. _____

20. _____

21. _____

22. _____

23. _____

24. _____

25. _____

26. _____

15. $\dfrac{4}{15} + \dfrac{1}{20}$

16. $\dfrac{11}{40} + \dfrac{6}{25}$

17. $\dfrac{1}{5} + \dfrac{1}{6} + \dfrac{3}{10}$

18. $3\dfrac{11}{16} + 2\dfrac{5}{24}$

Solve each problem.

19. A triangle has three sides. The lengths are $2\dfrac{1}{4}$ inches, $3\dfrac{1}{3}$ inches, and $4\dfrac{1}{2}$ inches. Find the perimeter (total distance around) of the triangle.

20. The recipe calls for $\dfrac{1}{8}$ teaspoon ginger, $1\dfrac{2}{3}$ teaspoons sugar, $\dfrac{3}{4}$ teaspoon salt, and $1\dfrac{1}{3}$ teaspoons cornstarch. In total, how many teaspoons of ingredients are called for?

21. Fred drove for 8 hours from Boston to Trenton. He drove $1\dfrac{3}{4}$ hours in Massachusetts and $4\dfrac{5}{6}$ hours in Connecticut and New York. The rest of the trip was in New Jersey. How long did Fred drive in New Jersey?

22. Hillary has a new car that gets $31\dfrac{1}{4}$ miles per gallon in city driving. Her old car got $24\dfrac{9}{10}$ miles per gallon. How much farther can her new car go on a gallon of gas?

23. An office has seven file cabinets placed tightly together on one wall. Each one is $3\dfrac{1}{4}$ feet wide. The entire wall is 24 feet long. How many feet of wall space is not covered by these file cabinets?

24. The stock that Robert purchased opened at $\$46\dfrac{5}{8}$ and closed at $\$41\dfrac{3}{4}$. How much did the stock go down that day?

25. A car can travel $24\dfrac{3}{10}$ miles on a gallon of gas. Approximately how far can it be driven on $2\dfrac{2}{3}$ gallons?

26. A 15-story building is $168\dfrac{3}{4}$ feet high. What is the height of each story?

1. Use a fraction to present the shaded portion of the object.

2. A basketball star shot at the hoop 364 times. The ball went in 307 times. Write a fraction that describes the part of the time that his shot went in.

Reduce each fraction.

3. $\dfrac{15}{80}$

4. $\dfrac{21}{39}$

5. $\dfrac{209}{38}$

6. Change to an improper fraction: $7\dfrac{4}{5}$.

7. Change to a mixed number: $\dfrac{113}{13}$.

Multiply.

8. $18 \times \dfrac{5}{6}$

9. $\dfrac{8}{9} \times \dfrac{2}{3}$

10. $3\dfrac{1}{5} \times 6\dfrac{1}{4}$

Divide.

11. $\dfrac{8}{9} \div \dfrac{3}{5}$

12. $4\dfrac{1}{6} \div 3$

13. $3\dfrac{3}{4} \div 1\dfrac{1}{24}$

14. $\dfrac{30}{77} \div \dfrac{45}{91}$

Find the least common denominator of each set of fractions.

15. $\dfrac{5}{12}$ and $\dfrac{7}{18}$

16. $\dfrac{3}{14}$ and $\dfrac{16}{21}$

1. _____

2. _____

3. _____

4. _____

5. _____

6. _____

7. _____

8. _____

9. _____

10. _____

11. _____

12. _____

13. _____

14. _____

15. _____

16. _____

17. _____

18. _____

19. _____

20. _____

21. _____

22. _____

23. _____

24. _____

25. _____

26. _____

27. _____

28. _____

17. $\dfrac{1}{6}, \dfrac{3}{8}$, and $\dfrac{4}{9}$

18. Build the fraction to an equivalent fraction with the specified denominator.

$$\frac{7}{9} = \frac{?}{72}$$

Add or subtract.

19. $\dfrac{5}{9} - \dfrac{5}{12}$

20. $\dfrac{1}{15} + \dfrac{7}{12}$

21. $\dfrac{2}{5} + \dfrac{3}{4} + \dfrac{3}{10}$

22. $9\dfrac{3}{4} + 7\dfrac{5}{6}$

23. $24\dfrac{7}{8} - 16\dfrac{11}{12}$

Answers each question.

24. A hallway measures $1\dfrac{1}{3}$ yards by $7\dfrac{1}{2}$ yards. How many square yards is the area of the hallway?

25. A butcher has $16\dfrac{1}{3}$ pounds of steak that he wishes to place into packages that average $2\dfrac{1}{3}$ pounds each. How many packages can he make?

26. From central parking it is $\dfrac{7}{10}$ of a mile to the science building. Bob started at central parking and walked $\dfrac{2}{5}$ of a mile toward the science building. He stopped for coffee. When he finishes his coffee, how far did he have to walk?

27. Robin jogged $4\dfrac{1}{8}$ miles on Monday, $3\dfrac{1}{2}$ miles on Tuesday, and $6\dfrac{3}{4}$ miles on Wednesday. How far did she jog on those three days?

28. Mr. and Mrs. Samuel visited Florida and purchased 120 oranges. They gave $\dfrac{1}{2}$ to relatives, ate $\dfrac{1}{12}$ of them in the hotel, and gave $\dfrac{1}{10}$ of them to friends. They shipped the rest home to Illinois.

(a) How many oranges did they ship?

(b) If it costs 24¢ for each orange to be shipped to Illinois, what was the total of the shipping bill?

One-half of this test is based on Chapter 1 material. The remainder is based on material covered in Chapter 2.

1. Write in words: 84, 361, 208.

2. Add: 128
452
178
34
+ 77

3. Add: 156,200
364,700
+ 198,320

4. Subtract: 5718
− 3643

5. Subtract: 1,000,361
− 983,145

6. Multiply: 126
× 38

7. Multiply: 16,908
× 12

8. Divide: $7\overline{)30150}$.

9. Divide: $18\overline{)6642}$.

10. Evaluate: 7^2.

11. Round to the nearest thousand: 6,037,452.

12. Perform the operations in their proper order:

$$4 \times 3^2 + 12 \div 6$$

13. Roone bought three shirts at $26 and two pairs of pants at $48. What was his total bill?

14. Leslie had a balance of $64 in her checking account. She deposited $1160. She made checks out for $516, $199, and $203. What will be her new balance?

15. Eighty-four students enrolled in psychology. Fifty-five were women. Write a fraction that describes the part of the class that was made up of women.

1. _____

2. _____

3. _____

4. _____

5. _____

6. _____

7. _____

8. _____

9. _____

10. _____

11. _____

12. _____

13. _____

14. _____

15. _____

16. _____	**16.** Reduce: $\dfrac{28}{52}$.
17. _____	
18. _____	**18.** Write as a mixed number: $\dfrac{100}{7}$.
19. _____	**20.** Divide: $\dfrac{44}{49} \div 2\dfrac{13}{21}$
20. _____	
21. _____	*Add or subtract.*
22. _____	**22.** $\dfrac{7}{18} + \dfrac{5}{27}$
23. _____	
24. _____	**24.** $12\dfrac{1}{5} - 4\dfrac{2}{3}$
25. _____	
26. _____	
27. _____	
28. _____	

17. Write as an improper fraction: $18\dfrac{3}{4}$.

19. Multiply: $3\dfrac{7}{8} \times 2\dfrac{5}{6}$.

21. Find the least common denominator of $\dfrac{6}{13}$ and $\dfrac{5}{39}$.

23. $2\dfrac{1}{8} + 6\dfrac{3}{4}$

25. $\dfrac{11}{14} - \dfrac{9}{28}$

26. A truck hauled $9\dfrac{1}{2}$ tons of gravel on Monday. On Tuesday it hauled $6\dfrac{3}{8}$ tons and on Wednesday, $7\dfrac{1}{4}$ tons. How many tons were hauled on the three days?

27. Melinda traveled $221\dfrac{2}{5}$ miles on 9 gallons of gas. How many miles per gallon did her car achieve?

28. A recipe requires $3\dfrac{1}{4}$ cups of sugar and $2\dfrac{1}{3}$ cups of flour. Marcia wants to make $2\dfrac{1}{2}$ times that recipe. How many cups of each will she need?

Decimals

Environmentalist—those of us who are involved with protecting our environment must be able to understand decimal measurements and calculations. Serious danger may exist if certain harmful pollutants are present even in amounts as small as of one part per ten million.

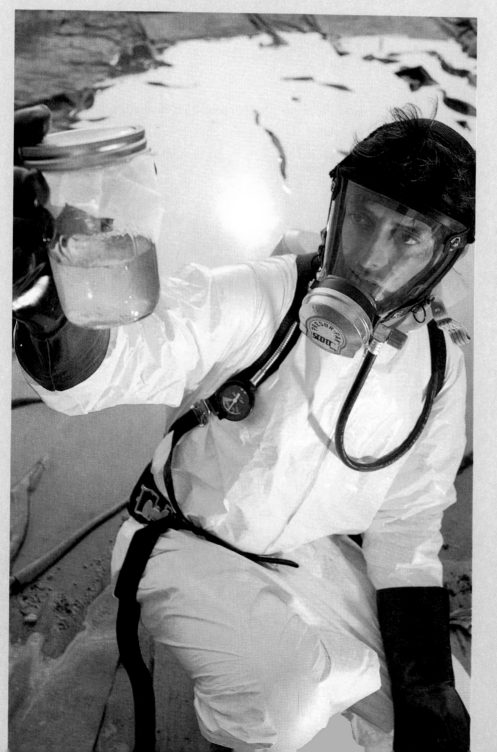

3

PRETEST CHAPTER 3

If you are familiar with the topics in this chapter, take this test now. Check your answers with those in the back of the book. If an answer was wrong or you couldn't do a problem, study the appropriate section of the chapter.

If you are not familiar with the topics in this chapter, don't take this test now. Instead, study the examples, work the practice problems, and then take the test.

This test will help you identify those concepts that you have mastered and those that need more study.

Section 3.1

1. Write a word name for the decimal: 36.524.

2. Express this fraction as a decimal: $\dfrac{1234}{10,000}$.

Write in fractional notation. Reduce whenever possible.

3. 1.39

4. 0.465

Section 3.2

5. Place the set of numbers in the proper order from smallest to largest: 2.7, 2.69, 2.71, 2.701.

6. Round to the nearest tenth: 158.253.

7. Round to the nearest thousandth: 0.381476.

Section 3.3   Add.

8.
```
  1.82
  3.6
  5.71
+ 2.3
```

9.
```
 24.613
  0.518
+ 2.305
```

Subtract.

10.
```
  98.35
- 64.98
```

11.
```
 12
- 3.814
```

Section 3.4   Multiply.

12.
```
  12.34
× 0.06
```

13. 2.8643×1000

14. 0.0007918×10^4

Section 3.5   Divide.

15. $0.07\,\overline{)\,0.00903}$

16. $2.6\,\overline{)\,33.28}$

Section 3.6

17. Write as a decimal: $\dfrac{5}{16}$.

18. Write as a repeating decimal: $\dfrac{3}{22}$.

19. Perform the operations in the proper order:

$$(0.3)^2 + 5.2 \times 0.6 - 2.82$$

Section 3.7

20. Marcia began her trip with an odometer reading of 43,298.6 and ended with an odometer reading of 43,538.6. She used 10.5 gallons of gas. How many miles per gallon did her car achieve? Round your answer to the nearest tenth.

21. A rectangular room is 10.5 yards long and 3.6 yards wide. How much will it cost to install carpeting for the entire room at $12.30 per square yard?

22. Tony worked 37 hours last week. He was paid $177.60. How much was he paid per hour?

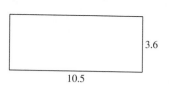

3.1 DECIMAL NOTATION

☐ After studying this section, you will be able to:

1 Write a word name for a decimal fraction

2 Change from fractional notation to decimal notation

3 Change from decimal notation to fractional notation

In Chapter 2 we discussed *fractions*—the set of numbers such as $\frac{1}{2}, \frac{2}{3}, \frac{1}{10}, \frac{6}{7}, \frac{18}{100}$, and so on. Now we will take a closer look at **decimal fractions**—that is, fractions with 10, 100, 1000, and so on, in the denominator such as $\frac{1}{10}, \frac{18}{100}$, and $\frac{43}{1000}$.

Why, of all fractions, do we take special notice of these? Our hands have *ten* digits; our U.S. money system is based on the unit of "dollar," which has 100 equal parts, "cents." And the international system of measurement called the metric system is based on 10 and powers of 10.

As with other numbers, these decimal fractions can be written in different ways (forms). For example, the shaded part of the whole can be written

In words (one-tenth)

In fraction form $\left(\frac{1}{10}\right)$

In a special notation using a decimal point (0.1)

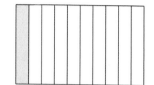

All mean the same quantity, namely 1 out of 10 equal parts of the whole. We'll see that when we use decimal notation, computations can be easily done based on the old rules for whole numbers and a few new rules about "where to place the decimal point." In a world where calculators and computers are commonplace, many of the fractions we encounter are decimal fractions. A decimal fraction is a fraction whose denominator is a power of 10.

$\frac{7}{10}$ is a decimal fraction. $\qquad \frac{89}{10^2} = \frac{89}{100}$ is a decimal fraction.

Decimal fractions can be written with numerals in two ways: fractional form or decimal form.

Fractional Form		Decimal Form
$\frac{3}{10}$	=	0.3
$\frac{59}{100}$	=	0.59
$\frac{171}{1000}$	=	0.171

The zero in front of the decimal point is not actually required. We place it there simply to make sure that we don't miss the decimal point. When a number is written in decimal form, the first digit to the right of the decimal point represents tenths, the next digit hundredths, the next digit thousands, and so on.

1 Using Words

0.9 is "nine-tenths" and is equivalent to $\frac{9}{10}$. 0.51 is "fifty-one hundredths" and is equivalent to $\frac{51}{100}$. Some decimals are larger than 1. 1.683 is "one and six hundred eighty-three thousandths." It is equivalent to $1\frac{683}{1000}$. Note that word "and" is used to indicate the decimal point. A place-value chart is helpful.

Decimal Place Values

Hundreds	Tens	Ones	Decimal Point	Tenths	Hundredths	Thousandths	Ten Thousandths
1	5	6	.	2	8	7	4
100	10	1	"and"	$\frac{1}{10}$	$\frac{1}{100}$	$\frac{1}{1,000}$	$\frac{1}{10,000}$

So, for 156.2878 we can write the word name "one hundred fifty-six and two thousand eight hundred seventy-eight ten-thousandths."

Example 1 Write a word name for each decimal.

(a) 0.79   (b) 0.5308   (c) 1.6   (d) 23.765

(a) 0.79 =   Seventy-nine hundredths
(b) 0.5308 = Five thousand three hundred eight ten-thousandths
(c) 1.6 =   One and six-tenths
(d) 23.765 = Twenty-three and seven hundred sixty-five thousandths   ∎

Practice Problem 1 Write a word name for each decimal.
(a) 0.073   (b) 4.68   (c) 0.0017   (d) 561.78   ∎

Decimal notation is commonly used with money. When writing a check we often write the amount that is less than 1 dollar, such as 23¢, as $\frac{23}{100}$ dollar.

```
Alice J. Jennington                                        3680
208 Barton Springs                                      37-86
Austin, TX   78704                                      110
                                  18 March  19 91
Pay To
the Order of   Rosetta Ramirez                    $  59.23
       fifty-nine and  23/100                        DOLLARS
Austin Central Bank
Austin, Texas
For    Textbooks
                                        Void
```

Example 2 Write a word name for the amount on a check of $672.89.

Six hundred seventy-two and $\frac{89}{100}$ dollars   ∎

Practice Problem 2 Write a word name for the amount of a check of $7863.04.   ∎

2 It is helpful to be able to write decimals in both decimal notation and fractional notation.

Example 3 Write each fraction as a decimal.

(a) $\frac{8}{10}$ (b) $\frac{74}{100}$ (c) $1\frac{3}{10}$ (d) $2\frac{56}{1000}$

(a) $\frac{8}{10} = 0.8$ (b) $\frac{74}{100} = 0.74$

(c) $1\frac{3}{10} = 1.3$ (d) $2\frac{56}{1000} = 2.056$ ∎

Practice Problem 3 Write each fraction as a decimal.

(a) $\dfrac{9}{10}$ 　　　　(b) $\dfrac{136}{1000}$ 　　　　(c) $2\dfrac{56}{100}$ 　　　　(d) $34\dfrac{86}{1000}$ ∎

3 Example 4 Write in fractional notation.

(a) 0.51 　　　　(b) 18.1 　　　　(c) 0.7611 　　　　(d) 1.363

(a) $0.51 = \dfrac{51}{100}$ 　　　　　　　　(b) $18.1 = 18\dfrac{1}{10}$

(c) $0.7611 = \dfrac{7611}{10,000}$ 　　　　(d) $1.363 = 1\dfrac{363}{1000}$ ∎

Practice Problem 4 Write in fractional notation.

(a) 0.37 　　　　(b) 182.3 　　　　(c) 0.7131 　　　　(d) 42.019 ∎

When we convert from decimal form to fractional form it is assumed that we will reduce whenever possible.

Example 5 Write in fractional form. Reduce whenever possible.

(a) 2.6 　　　　(b) 0.38 　　　　(c) 0.525 　　　　(d) 361.007

(a) $2.6 = 2\dfrac{6}{10} = 2\dfrac{3}{5}$ 　　　　(b) $0.38 = \dfrac{38}{100} = \dfrac{19}{50}$

(c) $0.525 = \dfrac{525}{1000} = \dfrac{105}{200} = \dfrac{21}{40}$

(d) $361.007 = 361\dfrac{7}{1000}$ (cannot be reduced) ∎

Practice Problem 5 Write in fractional form. Reduce whenever possible.

(a) 8.5 　　　　(b) 0.58 　　　　(c) 36.25 　　　　(d) 106.013 ∎

Example 6 A chemist measured the concentration of lead in a water sample to be 5 parts per million. What fraction would represent the concentration of lead?

Five parts per million, means 5 parts out of 1,000,000. As a fraction, this is $\dfrac{5}{1,000,000}$. We can reduce this by dividing numerator and denominator by five. Thus $\dfrac{5}{1,000,000} = \dfrac{1}{200,000}$.

The water sample is $\dfrac{1}{200,000}$ in concentration of lead. ∎

Practice Problem 6 A chemist measures the concentration of PCB in a water sample to be 2 parts per billion. What fraction would represent the concentration of PCB? ∎

Write a word name for each decimal.

1. 0.57   **2.** 0.78   **3.** 1.6   **4.** 9.8

5. 0.124   **6.** 0.367   **7.** 28.0007   **8.** 54.0003

Write a word name as you would on a check.

9. $87.36   **10.** $54.98   **11.** $1236.08

12. $7652.02   **13.** $10,000.76   **14.** $20,000.67

Express each fraction as a decimal.

15. $\dfrac{5}{10}$ **16.** $\dfrac{9}{10}$ **17.** $\dfrac{76}{100}$ **18.** $\dfrac{84}{100}$ **19.** $\dfrac{771}{1000}$ **20.** $\dfrac{652}{1000}$

21. $\dfrac{9}{10,000}$ **22.** $\dfrac{7}{10,000}$ **23.** $8\dfrac{7}{10}$ **24.** $6\dfrac{3}{10}$ **25.** $84\dfrac{13}{100}$ **26.** $52\dfrac{77}{100}$

27. $1\dfrac{19}{1000}$ **28.** $2\dfrac{23}{1000}$ **29.** $126\dfrac{571}{10,000}$ **30.** $198\dfrac{333}{10,000}$

Write in fractional notation. Reduce whenever possible.

31. 0.18   **32.** 0.56   **33.** 3.6   **34.** 7.4

35. 0.121   **36.** 0.143   **37.** 12.625   **38.** 29.875

39. 7.0015   **40.** 4.0016   **41.** 307.1206   **42.** 405.6432

43. 0.0187   **44.** 0.0209   **45.** 8.0108   **46.** 7.0605

47. 289.376   **48.** 423.814

Write each decimal as a fraction in lowest terms.

49. 0.000150

50. 0.0000125

51. A chemist measures 15 parts per one hundred million of water in a lake sample to be a harmful pollutant. Write as a fraction in lowest terms.

52. A chemist measures 35 parts per billion of a sample of milk to contain a harmful chemical. Write as a fraction in lowest terms.

Cumulative Review Problems

53. Add:
156
84
39
463
+ 76

54. Subtract:
12,843
− 11,905

55. Round to the nearest *hundred*: 56,758.

56. Round to the nearest *thousand*: 8,069,482.

For Extra Practice Examples and Exercises, turn to page 193.

Solutions to Odd-Numbered Practice Problems

1. (a) 0.073   seventy-three thousandths
(b) 4.68   four and sixty-eight hundredths
(c) 0.0017   seventeen ten-thousandths
(d) 561.78   five hundred sixty-one and seventy-eight hundredths

3. (a) $\frac{9}{10} = 0.9$ **(b)** $\frac{136}{1000} = 0.136$ **(c)** $2\frac{56}{100} = 2.56$ **(d)** $34\frac{86}{1000} = 34.086$

5. (a) $8.5 = 8\frac{5}{10} = 8\frac{1}{2}$ **(b)** $0.58 = \frac{58}{100} = \frac{29}{50}$ **(c)** $36.25 = 36\frac{25}{100} = 36\frac{1}{4}$ **(d)** $106.013 = 106\frac{13}{1000}$

Answers to Even-Numbered Practice Problems

2. Seven thousand eight hundred sixty-three and $\frac{4}{100}$ dollars   **4. (a)** $\frac{37}{100}$ **(b)** $182\frac{3}{10}$ **(c)** $\frac{7131}{10,000}$ **(d)** $42\frac{19}{1000}$

6. $\frac{1}{500,000,000}$

3.2 ORDERING AND ROUNDING DECIMALS

As in rounding whole numbers, when we round a *decimal* number we want to replace a given number, say

128.37448

with another number, one that is *near* the original but that does not have as many digits. For 128.37448, we might want to round to the nearest *thousandth*, which means that we replace 128.37448 (by rounding) with 128.374.

In this section we discuss a method of telling the relative size of two decimal numbers, and the commonly accepted rules for rounding. Money calculations in our society are usually rounded to the nearest hundredth (nearest cent). In scientific work, however, you may have to round to some other specified place.

After studying this section, you will be able to:

1 Place two or more decimal fractions in order from smallest to largest

2 Round off decimal fractions to a specified decimal place

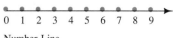
Number Line

1 All of the numbers we studied have a specific order. If we placed all of the whole numbers on a number line, the arrow points in the direction of the largest number. If one number is to the right of a second number, it is larger than the second number. Thus 5 is greater than 2 (written $5 > 2$) since 5 lies to the right of 2. In the opposite sense 4 is less than 6 (written $4 < 6$) since 4 lies to the left of 6 on the number line. The symbols "$<$" and "$>$" are called **inequality symbols**.

$a < b$ is read "a is less than b."

$a > b$ is read "a is greater than b."

We can assign a unique point on the number line to each decimal number. When two decimal numbers are placed on a number line, the one farther to the right is the larger. Thus we could say that $0.5 < 1.0$, $1.8 < 2.2$, $3.4 > 2.7$, and $4.3 > 4.0$. Do you see why?

Comparing Two Numbers in Decimal Notation

1. Start at the left and compare corresponding digits. If the digits are the same, move one place to the right.
2. When two digits are different, the larger number is the one with the larger digit.

Example 1 Write an inequality statement with 0.167 and 0.166.

The number 1 in the tenths place is the same.

$$0.\boxed{1}\,6\,7 \qquad\qquad 0.\boxed{1}\,6\,6$$

The number 6 in the hundredths place is the same.

$$0.1\,\boxed{6}\,7 \qquad\qquad 0.1\,\boxed{6}\,6$$

The numbers in the thousandths place differ.

$$0.1\,6\,\boxed{7} \qquad\qquad 0.1\,6\,\boxed{6}$$

Since $7 > 6$ we know that $0.167 > 0.166$ ■

Practice Problem 1 Write an inequality statement with 5.74 and 5.75. ■

Whenever necessary, extra zeros can be written to the right of the decimal point or the last nonzero digit to the right of the decimal without changing the value of the decimal fraction. Thus

$$0.56 = 0.56000 \qquad \text{and} \qquad 0.7768 = 0.77680$$

The zero to the left of the decimal point is optional. Thus $0.56 = .56$, and both notations are used. You are encouraged to place a zero to the left of the decimal point so that you don't "miss" the decimal point when doing calculations involving decimals.

Example 2 Fill in the blank with one of the symbols $<$, $=$, or $>$.

$$0.77 \underline{\hspace{1cm}} 0.777$$

If we add a zero to the first decimal

$$0.77\underline{0} \qquad\qquad 0.77\underline{7}$$

we see that the tenths and hundredths digits are equal. But the thousandths digits differ. Since $0 < 7$ we have $0.770 < 0.777$ ■

Practice Problem 2 Fill in the blank with one of the symbols $<$, $=$, or $>$. ■

$$0.894 \underline{\hspace{1cm}} 0.89$$

Example 3 Place the following five decimal numbers in the proper order from smallest to largest.

1.834, 1.83, 1.381, 1.38, 1.8

First we add zeros to make the comparison more simple.

1.834, 1.830, 1.381, 1.380, 1.800

Now we rearrange with smallest first.

1.380, 1.381, 1.800, 1.830, 1.834 ∎

Practice Problem 3 Place the following five decimal numbers in the proper order from smallest to largest.

2.45, 2.543, 2.46, 2.54, 2.5 ∎

❷ Rounding Off Decimals

Sometimes in calculations involving money we see numbers like $386.432 and $29.5986. To make these useful, we usually round them to the nearest cent. $386.432 is rounded to $386.43. $29.5986 is rounded to $29.60. A general rule for rounding decimals follows:

To Round a Decimal:

1. Locate the position (units, tenths, hundredths, and so on) for which rounding off is required.
2. If the first digit to the right of this place is less than 5, drop it and all digits to the right of it.
3. If the first digit to the right of this place is 5 or greater, increase the digit by one. Drop all digits to the right of this place.

Example 4 Round 156.37 to the nearest tenth.

156. 3 7

└── We locate the tenths position.

Note that 7 is greater than 5. We round up to 156. 4 and drop the digits to the right. The answer is 156.4 ∎

Practice Problem 4 Round 723.88 to the nearest tenth. ∎

Example 5 Round to the nearest thousandth.

(a) 0.06358   **(b)** 128.37448

(a) 0.06 3 58

↑
└── We locate the thousandths position.

We round up to 0.064 and drop all the digits to the right. Note that the digit to the right of the thousandths place is 5.

(b) 128.37 4 48

── We locate the thousandths place.

We round to 128.374 and drop all the digits to the right. Note that the digit to the right of thousandths place is less than 5. ∎

Practice Problem 5 Round to the nearest thousandth.
(a) 12.92647   **(b)** 0.007892 ∎

It is helpful to remember that rounding up to the next digit in a position may result in several digits being changed.

Example 6   Round to the nearest hundredth: 203.9964.

We locate the hundredth position ⌐

203.9 9 64

Since the digit to the right of hundredths is greater than 5, we round up. This affects the next two positions. Do you see why? The result is 204.00   ■

Practice Problem 6   Round to the nearest tenth: 15,699.953.   ■

Sometimes, we round decimals to the nearest whole numbers. For example, when completing figures for income tax forms, a taxpayer may round all figures to the nearest dollar.

Example 7   To complete her income tax return, Marge needs to round these figures to the nearest whole dollar.

Rent $1684.50   Medical bills $779.86
Taxes $563.49   Retirement contributions $674.38
Contributions to charity $534.77
Complete the necessary rounding.

	Original Figure	*Rounded to Nearest Dollar*
Rent	$1684.50	$1685
Medical bills	779.86	780
Taxes	563.49	563
Retirement	674.38	674
Charity	534.77	535 ■

Practice Problem 7   Round to the nearest whole dollar the following figures.

Rent $2351.47   Medical bills $375.50
Taxes $981.39   Retirement $980.49
Charity $817.65   ■

To Think About

Why is it so important to consider only *one* digit to the right of the desired round-off position? What is wrong with rounding in steps? Suppose that Mark wanted to round 1.349 to the nearest tenth in steps. First he rounded 1.349 to 1.35 (nearest hundredth). Then he rounded 1.35 to 1.4 (nearest tenth). What is wrong with this reasoning?
  Rounding is creating an approximate answer. When you create one approximate answer and do further work based on it, the amount of error is increased.

To round 1.349 to the nearest tenth, we ask if 1.349 is closer to 1.3 or to 1.4. It is closer to 1.3. If we "round in steps" by first moving to 1.35, we increase the error and move in the wrong direction. To control rounding errors we consider *only* the first digit to the right of the decimal place to which we are rounding.   ■

PROBLEMS WITH ACCURACY

Strive for accuracy. Mistakes are often made as a result of human error rather than by lack of understanding. Such mistakes are frustrating. A simple arithmetic or sign error can lead to an incorrect answer.

These five steps will help you cut down on errors.

1. Work carefully, and take your time. Do not rush through a problem just to get it done.
2. Concentrate on one problem at a time. Sometimes problems become mechanical, and your mind begins to wander. You become careless and make a mistake.
3. Check your problem. Be sure that you copied it correctly from the book.
4. Check your computations from step to step. Check the solution in the problem. Does it work? Does it make sense?
5. Keep practicing new skills. Remember the old saying "Practice makes perfect." An increase in practice results in an increase in accuracy. Many errors are due simply to lack of practice.

There is no magic formula for eliminating all errors, but these five steps will be a tremendous help in reducing them.

EXERCISES 3.2

Fill in the blank with one of the symbols $<$, $=$, or $>$.

1. 1.3 _____ 1.29

2. 2.6 _____ 2.58

3. 0.25 _____ 0.250

4. 23.62 _____ 23.64

5. 18.92 _____ 18.93

6. 0.460 _____ 0.46

7. 0.0006 _____ 0.0005

8. 0.0037 _____ 0.003

9. 1.0024 _____ 1.003

10. 1.003 _____ 1.004

11. 126.34 _____ 125.35

12. 406.78 _____ 407.75

13. 16.0572 _____ 16.0574

14. 18.00039 _____ 18.00038

15. $\dfrac{8}{10}$ _____ 0.08

16. $\dfrac{5}{100}$ _____ 0.005

In each case place the set of numbers in the proper order from smallest to largest.

17. 12.6, 12.8, 12.65

18. 18.32, 18.4, 18.3

19. 0.0053, 0.003, 0.005

20. 0.0096, 0.0069, 0.009

21. 1.8, 1.1, 1.81, 1.79

22. 2.7, 2.5, 2.53, 2.48

23. 26.034, 26.003, 26.04, 26.033

24. 33.082, 33.02, 33.088, 33.079

25. 18.006, 18.060, 18.066, 18.606, 18.065

26. 15.020, 15.002, 15.001, 15.018, 15.022

Round each term to the nearest tenth.

27. 5.67 **28.** 8.35 **29.** 29.49 **30.** 38.48 **31.** 197.053 **32.** 281.076 **33.** 2176.83 **34.** 4082.74

Round each term to the nearest hundredth.

35. 26.032 **36.** 47.071 **37.** 5.76582 **38.** 2.98613

39. 156.1197 **40.** 283.2168 **41.** 2786.716 **42.** 4609.285

Round to the given place value.

43. 1.06132 thousandths **44.** 8.10263 thousandths

45. 0.091263 ten-thousandths **46.** 0.028946 ten-thousandths

47. 5.00761238 hundred-thousandths **48.** 4.01062378 hundred-thousandths

49. 0.00753682 millionths **50.** 0.00964983 millionths

51. 129.08939 nearest whole number **52.** 208.4372 nearest whole number

Round each amount to the nearest dollar.

53. $7812.78 **54.** $5319.62 **55.** $10,098.47 **56.** $20,159.48

Round each amount to the nearest cent.

57. $56.9832 **58.** $28.7619 **59.** $5783.716 **60.** $3928.649

 To Think About

61. A person wants to round 86.23498 to the nearest hundredth. He first rounds 86.23498 to 86.2350. He then rounds to 86.235. Finally, he rounds to 86.24. What is wrong with his reasoning?

62. Fred is checking the calculations on his bank statement. An interest charge of $16.3724 was rounded to $16.38. An interest charge of $43.7214 was rounded to $43.73. What rule does the bank use for rounding off to the nearest cent?

63. Arrange in order from smallest to largest.

$$0.61, \quad 0.062, \quad \frac{6}{10}, \quad 0.006, \quad 0.0059,$$
$$\frac{6}{100}, \quad 0.0601, \quad 0.0519, \quad 0.0612.$$

64. Arrange in order from smallest to largest.

$$1.05, \quad 1.512, \quad \frac{15}{10}, \quad 1.0513, \quad 0.049,$$
$$\frac{151}{100}, \quad 0.0515, \quad 0.052, \quad 1.051.$$

Practice Problem 3 A car odometer read 93,521.8 miles before a trip of 1634.8 miles. What was the final odometer reading? ∎

Example 4 Kelvey deposited checks into his checking account in the amounts of $98.64, $157.32, $204.81, $36.07, and $229.89. What was the sum of his five checks?

$$
\begin{array}{r}
232\ 2 \\
\$\ 98.64 \\
157.32 \\
204.81 \\
36.07 \\
+\ 229.89 \\
\hline
\$726.73
\end{array}
$$ ∎

Practice Problem 4 Will deposited the following checks into his account: $80.95, $133.91, $256.47, $53.08, and $381.32. What was the sum of his five checks? ∎

2 When we subtract mixed numbers with common denominators, sometimes it is necessary to borrow from the whole number.

$$
\begin{array}{rcr}
5\frac{4}{10} & = & 4\frac{14}{10} \\
-2\frac{7}{10} & = & -2\frac{7}{10} \\
\hline
& & 2\frac{7}{10}
\end{array}
$$

We could write the same problem in decimal form as

$$
\begin{array}{r}
{\scriptstyle 4\ 14} \\
\cancel{5}.4 \\
-\ 2.7 \\
\hline
2.7
\end{array}
$$

Subtraction of decimals is thus similar to subtraction of fractions (we get the same result), but it's usually considered easier to subtract with decimals than to subtract with fractions.

Subtracting Decimals

1. Write the numbers to be subtracted vertically and line up the decimal points. Additional zeros may be placed to the right if each number does not have the same number of decimal places.
2. Subtract all digits with the same place value, starting with the right column, moving to the left. Borrow when necessary.
3. Place the decimal point of the difference in line with the decimal point of the two numbers being subtracted.

Example 5 Subtract.

(a)
$$
\begin{array}{r}
84.8 \\
-27.3 \\
\end{array}
$$

(b)
$$
\begin{array}{r}
1076.320 \\
-983.518 \\
\end{array}
$$

(a)
$$
\begin{array}{r}
{\scriptstyle 7\ 14} \\
8\cancel{4}.8 \\
-2\ 7.3 \\
\hline
5\ 7.5
\end{array}
$$

(b)
$$
\begin{array}{r}
{\scriptstyle 9} \\
{\scriptstyle 10\ 17\ 5\ 13\ 1\ 10} \\
\cancel{1}\ \cancel{0}\ \cancel{7}\ \cancel{6}.\cancel{3}\ \cancel{2}\ \cancel{0} \\
-\ 9\ 8\ 3.5\ 1\ 8 \\
\hline
9\ 2.8\ 0\ 2
\end{array}
$$ ∎

Practice Problem 5 Subtract.

(a)
$$
\begin{array}{r}
38.8 \\
-26.9 \\
\end{array}
$$

(b)
$$
\begin{array}{r}
2034.908 \\
-1986.325 \\
\end{array}
$$ ∎

---

65. Add: $3\frac{1}{4} + 2\frac{1}{2} + 6\frac{3}{8}$

66. Subtract: $27\frac{1}{5} - 16\frac{3}{4}$

67. Mary drove her car on a trip. At the start of the trip the odometer (which measures distance) read 46,381. At the end of the trip it read 47,073. How many miles long was the trip?

68. The cost of four items was $1736, $2714, $892, and $4316. What was their total cost?

For Extra Practice Examples and Exercises, turn to page 194.

Solutions to Odd-Numbered Practice Problems

1. Since $4 < 5$

$$5.7\boxed{4} \qquad\qquad 5.7\boxed{5} \qquad \text{therefore, } 5.74 < 5.75$$

3. 2.45, 2.543, 2.46, 2.54, 2.5
It is helpful to add extra zeros and to place the decimals that begin with 2.4 in a group and the decimals that begin with 2.5 in the other.

$$2.450, \quad 2.460 \qquad 2.543, \quad 2.540, \quad 2.500$$

In order, we have from smallest to largest

$$2.450, \quad 2.460, \quad 2.500, \quad 2.540, \quad 2.543$$

It is OK to leave the extra terminal zeros in our answer.

5. (a) $12.92\,\underset{\uparrow}{6}\,47$ — Since digit to right of thousandth is less than 5, we drop the digits 4 and 7.   **12.926**

(b) $0.00\,\underset{\uparrow}{7}\,892$ — Since digit to right of thousandths is greater than 5, we round up.   **0.008**

		Rounded to Nearest Dollar
7. Rent	$2351.47	$2351
Medical bills	375.50	376
Taxes	981.39	981
Retirement	980.49	980
Charity	817.65	818

Answers to Even-Numbered Practice Problems

2. $0.894 > 0.890$, so $0.894 > 0.89$ **4.** 723.9   **6.** 15,700.0

3.3 ADDITION AND SUBTRACTION OF DECIMALS

☐ After studying this section, you will be able to:

Our money system is written with decimals. The decimal point separates the dollars and the cents. *Buying* something and *receiving change* in our society uses addition and subtraction of decimals, whether or not you are always aware of it.

If you buy desserts for a total of $7.14, you've used addition of decimals. If you receive $2.86 as change from a 10-dollar bill, you've used subtraction of decimals.

Whether you buy dessert for yourself or food/shelter/clothing for your family or a pollution-control device for your multinational corporation, you use addition and subtraction of decimals.

Most people find it easy to apply the rules of adding and subtracting whole numbers to those computations with decimals. You can, if necessary, double-check your work with a calculator.

1 *Add two or more decimal fractions*

2 *Subtract two decimal fractions*

1 We have learned how to add fractions with

$$\frac{3}{10} + \frac{6}{10} = \frac{9}{10} \quad \text{and}$$

These same problems can be written more effi

$$
\begin{array}{r}
0.3 \\
+ 0.6 \\
\hline
0.9
\end{array}
\quad +
$$

Some steps to follow are listed below.

Addition of Decimals

1. Write the numbers to be added vertically a
   ditional zeros may be placed to right of th
2. Add all the digits with the same place valu
   moving to the left.
3. Place the decimal point of the sum in line
   numbers added.

Example 1 Add.

(a) $2.8 + 5.6 + 3.2$ (b)

$$
\begin{array}{r}
\overset{1}{} \\
\text{(a)} \quad 2.8 \\
5.6 \\
+ 3.2 \\
\hline
11.6
\end{array}
\quad \text{(b)}
$$

Practice Problem 1 Add.

$$
\begin{array}{r}
\text{(a)} \quad 9.8 \\
3.6 \\
+ 5.4 \\
\hline
\end{array}
\quad \text{(b)}
$$

It is helpful to add extra zeros to the right c
number to be added has the same number of dec

Example 2 Add: $5.3 + 26.182 + 0.0007 + 6.24.$

$$
\begin{array}{r}
\overset{1}{}\ \overset{1}{} \\
5.3000 \\
26.1820 \\
0.0007 \\
+ 6.2400 \\
\hline
37.7227
\end{array}
\quad
\begin{array}{l}
\text{Extra zeros hav} \\
\text{make the probl}
\end{array}
$$

Practice Problem 2 Add: $8.9 + 37.056 + 0.002$

An odometer has a final digit that is measure
shown reads 38,516.2 miles.

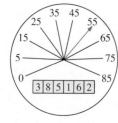

Example 3 Barbara checked her odometer be
49,645.8 miles. She traveled 3852.6 miles that su
odometer reading at the end of the summer?

$$
\begin{array}{r}
\overset{11}{}\ \overset{1}{} \\
49,645.8 \\
+ \ 3,852.6 \\
\hline
53,498.4
\end{array}
$$

It read 53,498.4 miles. ■

In cases where the two numbers being subtracted do not have the same number of decimal places, add the necessary number of zeros.

Example 6 Subtract.

(a) $12 - 8.362$ (b) $156.381 - 99.82$

$$
\text{(a)} \quad
\begin{array}{r}
\overset{11}{}\ \overset{9}{}\ \overset{9}{}\ \overset{10}{} \\
\cancel{1}\ \cancel{2}\ .\cancel{0}\ \cancel{0}\ \cancel{0} \\
- \quad 8.3\ 6\ 2 \\
\hline
3.6\ 3\ 8
\end{array}
\qquad
\text{(b)} \quad
\begin{array}{r}
\overset{14}{}\ \overset{15}{} \\
\cancel{1}\ \cancel{5}\ \cancel{6}.\cancel{3}\ 8\ 1 \\
- \quad 9\ 9\ .\ 8\ 2\ 0 \\
\hline
5\ 6\ .\ 5\ 6\ 1
\end{array}
\quad ■
$$

Practice Problem 6 Subtract.

(a) $19 - 12.579$ (b) $283.076 - 96.38$ ■

Example 7 On Tuesday Don Ling filled the gas tank in his car. The odometer read 56,098.5. He drove for four days and the next time he filled the tank the odometer read 56,420.2. How many miles had he driven?

$$
\begin{array}{r}
\overset{11}{} \\
\cancel{5}\ \cancel{4}\ \overset{9}{}\ \overset{12}{} \\
5\ 6,\cancel{4}\ \cancel{2}\ \cancel{0}.\cancel{2} \\
- 5\ 6,0\ 9\ 8\ .\ 5 \\
\hline
3\ 2\ 1\ .\ 7
\end{array}
$$

He had driven 321.7 miles. ■

Practice Problem 7 Abdul had his car oil changed when his car odometer read 82,370.9 miles. When he changed oil again the odometer read 87,160.1 miles. How many miles did he drive between oil changes? ■

To Think About

When we add decimals, like $3.1 + 2.16 + 4.007$, we may add zeros to obtain

$$
\begin{array}{r}
3.100 \\
2.160 \\
+ 4.007 \\
\hline
9.267
\end{array}
$$

What are we really doing? What right do we have to add these extra zeros?
    "Decimals" means "decimal fractions." If we look at the numbers as fractions, we are actually using the property of multiplying a fraction by 1 in order to obtain common denominators. Look at the problem this way:

$$3.1 \ = 3\frac{1}{10}$$

$$2.16 \ = 2\frac{16}{100}$$

$$4.007 = 4\frac{7}{1000}$$

The least common denominator is 1000.
To obtain the common denominator for the first two fractions, we multiply.

$$3\ \frac{1}{10} \times \frac{100}{100} = 3\frac{100}{1000}$$

$$2\ \frac{16}{100} \times \frac{10}{10} = 2\frac{160}{1000}$$

Once we obtain a common denominator, we can add the three fractions.

$$+ 4\frac{7}{1000} \ = 4\frac{7}{1000}$$

$$9\frac{267}{1000} = 9.267$$

The adding of extra zeros to a decimal fraction is really a very simple way of transforming fractions to equivalent fractions with a common denominator. Working with decimal fractions is easier than working with other fractions. ■

Add.

1.	44.6 + 28.2	**2.**	18.6 + 23.2	**3.**	718.98 + 496.57	**4.**	813.47 + 629.86

5.	1.806 + 3.725	**6.**	9.360 + 5.786	**7.**	79.061 + 57.783	**8.**	26.905 + 87.453

9.	6.5 12.6 + 304.8	**10.**	18.2 7.6 + 199.8	**11.**	5.6 9.23 + 8.17	**12.**	2.65 3.2 + 7.76

13.	4.9637 28.12 + 3.645	**14.**	7.0276 3.451 + 16.98	**15.**	12 3.62 + 51.8	**16.**	17 6.89 + 48.3

17. $753.61 + 28.75 + 162.3 + 100.5 + 67.05$

18. $432.51 + 16.08 + 892.1 + 301.2 + 84.07$

Calculate the perimeter of each triangle.

19.

6.1M   5.62M
8.14M

20.

5.09M   6.7M
9.28M

21. On a three-day trip Jan purchased 12.6, 14.8, 8.3, and 11.9 gallons of gas. How many gallons of gas did he buy?

22. Alberta drove from Philadelphia to Cleveland. On the way she purchased 10.2, 11.6, 9.3, and 12.8 gallons of gas. How many gallons of gas did she buy?

23. Mia purchased items at a grocery store that cost $1.89, $0.43, $3.69, $1.12, and $5.18. What was her grocery bill?

24. Brent purchased items at a grocery store that cost $7.18, $5.29, $0.61, $1.15, and $1.49. What was his grocery bill?

A portion of a bank checking account deposit slip is shown below. Add the numbers to determine the total deposit in each case. The line drawn between the dollars and the cents column serves as the decimal point.

25.

		Dollars	Cents
SUNSHINE BANK	Cash	52	89
PHOENIX ARIZONA	Checks	105	37
DEPOSIT SLIP		76	04
		25	00
		167	82
	Total		

26.

		Dollars	Cents
FOOTHILLS BANK	Cash	18	42
AUSTIN TEXAS	Checks	706	15
DEPOSIT SLIP		21	03
		45	00
		621	37
	Total		

Subtract.

27. 6.8
− 2.9

28. 3.6
− 2.8

29. 123.51
− 96.34

30. 161.78
− 89.29

31. 76.8
− 12.62

32. 82.5
− 43.93

33. 186.418
− 78.3

34. 243.967
− 84.2

35. 1.00782
− 0.98631

36. 7.00278
− 6.34125

37. 24.0079
− 19.3614

38. 52.0708
− 41.9312

39. 8
− 1.263

40. 12
− 7.981

41. 7362.14
− 6173.07

42. 4986.71
− 3615.93

Solve each problem.

43. A new car has a sticker price of $8964.32. The dealer will reduce the price by $658.32. How much will the car cost?

44. A builder will put an addition on your house for $17,540.00. If you pay all cash in advance, he will reduce the price by $2,648.15. How much would the addition then cost?

45. Normal rainfall for this month is 14.895 centimeters. The amount recorded for this month is 12.739 centimeters. How far short of normal is the amount recorded this month?

46. A conducting wire should be 3.468 meters long. By mistake it was made 2.97 meters long. By how much is the conducting wire too short?

47. Luis began a trip with a car odometer reading of 18,439.6 miles. At the end of the trip the odometer read 19,295.8 miles. How long was the trip?

48. Tamika drove on a summer trip. When she began the odometer read 26,052.3 miles. At the end of the trip the odometer read 28,715.1 miles. How long was the trip?

49. Vanessa paid her hotel bill of $62.86 with a $100 traveler's check. How much change did she receive?

50. Michael bought $76.49 worth of supplies at the store and gave the clerk $80.00. How much change should he receive?

51. The outside radius of a pipe is 9.39 centimeters. The inside radius is 7.93 centimeters. What is the thickness of the pipe?

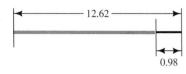

52. An insulated wire measures 12.62 centimeters. The last 0.98 centimeter of the wire is exposed. How long is the part of the wire that is not exposed?

53. The state budget called for an income of $12,563,784.56. Unfortunately, the income for the year was only $11,962,375.49. What was the revenue shortage for the state?

54. For this year, the state income was $11,962,375.49. The yearly expenses of the state were $12,427,396.84. How much in state expenses could not be covered by the income?

 To Think About

Calculate.

55. $1.0234 + 5.072 - 1.02 - 1.0056 + 1.0101$

56. $2.078 + 27.808 + 2.00087 + 207.8 + 7.082$

57. $186.0032 + 257.3716 + 982.46 + 1873.7618 + 57.764 + 9840.3219$

58. $25.671 + 198.323 + 200.14 - 71.36 - 24.861 - 14.9$

Cumulative Review Problems

Multiply.

59.
$$\begin{array}{r} 2536 \\ \times \quad 8 \\ \hline \end{array}$$

60.
$$\begin{array}{r} 467 \\ \times \quad 39 \\ \hline \end{array}$$

 Calculator Problems

Add.

61.
$$\begin{array}{r} 12.5062 \\ 65.201 \\ 0.9976 \\ 122.2 \\ + \quad 75.822 \\ \hline \end{array}$$

62.
$$\begin{array}{r} 33.05 \\ 0.6722 \\ 121.2 \\ 46.801 \\ + \quad 462.9 \\ \hline \end{array}$$

Subtract.

63.
$$\begin{array}{r} 76.00211 \\ - \quad 19.87246 \\ \hline \end{array}$$

64.
$$\begin{array}{r} 200.0110 \\ - \quad 99.8896 \\ \hline \end{array}$$

For Extra Practice Examples and Exercises, turn to page 194.

Solutions to Odd-Numbered Practice Problems

1. (a)
$$\begin{array}{r} 1 \\ 9.8 \\ 3.6 \\ + \ 5.4 \\ \hline 18.8 \end{array}$$
(b)
$$\begin{array}{r} 1\ 1\ 1 \\ 300.72 \\ 163.75 \\ + \ 291.09 \\ \hline 755.55 \end{array}$$

3.
$$\begin{array}{r} 1 \quad 1 \\ 93,521.8 \\ + \ 1,634.8 \\ \hline 95,156.6 \end{array}$$
95,156.6 miles

5. (a)
$$\begin{array}{r} 7\ 18 \\ 3\ 8.8 \\ - 2\ 6.9 \\ \hline 1\ 1.9 \end{array}$$
(b)
$$\begin{array}{r} 1\ 9\ 12\ 14\ 8\ 10 \\ 2\ 0\ 3\ 4.9\ 0\ 8 \\ - 1\ 9\ 8\ 6.3\ 2\ 5 \\ \hline 4\ 8.5\ 8\ 3 \end{array}$$

7.
$$\begin{array}{r} 10\ 15 \\ 6\ 0\ 8\ 9\ 11 \\ 8\ 7,1\ 6\ 0.1 \\ - 8\ 2,3\ 7\ 0.9 \\ \hline 4,7\ 8\ 9.2 \end{array}$$
4789.2 miles between oil changes

Answers to Even-Numbered Practice Problems

2. 55.4083 **4.** $905.73 **6. (a)** 6.421 **(b)** 186.696

3.4 MULTIPLICATION OF DECIMALS

Examples of multiplication when one or both factors is a decimal are so common they are easy to overlook. Here are two typical situations.

- You are paid by the hour at $5.50 per hour. You work 18 hours this week. How much do you earn? *Solution:* Multiply 18×5.50.
- You are carpeting a room 12.5 feet long by 10.4 feet wide. What is the area of the room (how many square feet of carpeting are required)? *Solution:* Area, in square feet, is 12.5×10.4.

In this section we learn how to do these types of problems, where one or both factors in multiplication is a decimal. The other factor may be a whole number. If the whole number is 10 or a power of 10, special shortcut rules can apply.

1 We have previously learned that the product of two fractions is the product of the numerators over the product of the denominators. For example,

$$\frac{3}{10} \times \frac{7}{100} = \frac{21}{1000}$$

In decimal form this product would be written

$$0.3 \quad \times \quad 0.07 \quad = \quad 0.021$$

| one decimal place | two decimal places | three decimal places |

Multiplication of Decimal Fractions

1. Multiply the numbers just as you would multiply whole numbers.
2. Find the sum of the decimal places in the two factors.
3. Place the decimal point in the product so that the product has the same number of decimal places as the sum in step 2. It is sometimes necessary to insert zeros to the left of the number found in step 1.

Let us try this procedure with the problem we just multiplied.

Example 1 Multiply 0.07×0.3.

$$
\begin{array}{r}
0.07 \\
\times \ 0.3 \\
\hline
0.021
\end{array}
$$

2 decimal places
1 decimal place
3 decimal places in product $(2 + 1 = 3)$ ∎

Practice Problem 1 Multiply 0.09×0.6. ∎

When performing the calculation, it is usually easier to place the factor with the fewest number of nonzero digits underneath the other factor.

Example 2 Multiply. **(a)** 0.8×2.6 **(b)** 0.3×1.672

(a)
$$
\begin{array}{r}
2.6 \\
\times \ 0.8 \\
\hline
2.08
\end{array}
$$
1 decimal place
1 decimal place
2 decimal places
$(1 + 1 = 2)$

(b)
$$
\begin{array}{r}
1.672 \\
\times \ \ 0.3 \\
\hline
0.5016
\end{array}
$$
3 decimal places
1 decimal place
4 decimal places
$(3 + 1 = 4)$ ∎

Practice Problem 2 Multiply. **(a)** 0.9×5.3 **(b)** 2.831×0.7 ∎

Example 3 Multiply. **(a)** 0.38×0.26 **(b)** 12.64×0.572

(a)

$$
\begin{array}{r}
0.38 \\
\times\ 0.26 \\
\hline
228 \\
76\ \ \\
\hline
0988
\end{array}
$$

2 decimal places
2 decimal places

4 decimal places $(2 + 2 = 4)$
Note that we need to insert a
zero before the 988.

(b)

$$
\begin{array}{r}
12.64 \\
\times\ 0.572 \\
\hline
2528 \\
8848\ \ \\
6\ 320\ \ \ \\
\hline
7.23008
\end{array}
$$

2 decimal places
3 decimal places

5 decimal places
$(2 + 3 = 5)$ ■

Practice Problem 3 Multiply.

(a)

$$
\begin{array}{r}
0.47 \\
\times\ 0.28 \\
\hline
\end{array}
$$

(b)

$$
\begin{array}{r}
18.39 \\
\times\ 0.436 \\
\hline
\end{array}
$$ ■

When multiplying decimal fractions by a whole number, you need to remember that a whole number has zero decimal places.

Example 4 Multiply: 5.261×45.

$$
\begin{array}{r}
5.261 \\
\times\ \ \ \ 45 \\
\hline
26\ 305 \\
210\ 44\ \ \\
\hline
236.745
\end{array}
$$

3 decimal places
0 decimal places

3 decimal places $(3 + 0 = 3)$ ■

Practice Problem 4 Multiply: 0.4264×38 ■

② Multiplication by a Power of 10

Observe the following pattern:

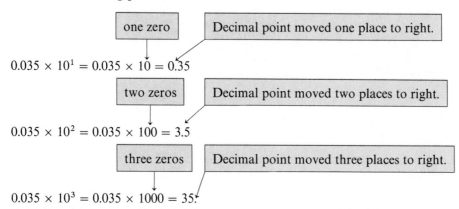

one zero — Decimal point moved one place to right.

$0.035 \times 10^1 = 0.035 \times 10 = 0.35$

two zeros — Decimal point moved two places to right.

$0.035 \times 10^2 = 0.035 \times 100 = 3.5$

three zeros — Decimal point moved three places to right.

$0.035 \times 10^3 = 0.035 \times 1000 = 35.$

> **Multiplication of a Decimal by a Power of 10**
>
> To multiply a decimal fraction by a power of 10, move the decimal point to the right the same number of places as there are zeros in the power of 10.

Example 5 Multiply. **(a)** 2.671×10 **(b)** 37.85×100

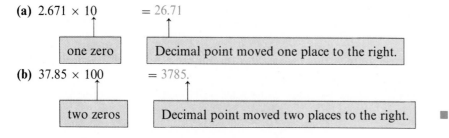

(a) 2.671×10 $= 26.71$

one zero   Decimal point moved one place to the right.

(b) 37.85×100 $= 3785.$

two zeros   Decimal point moved two places to the right.   ■

Practice Problem 5   Multiply.

(a) 0.0561×10 **(b)** 1462.37×100 ■

Sometimes it is necessary to add extra zeros before placing the decimal point in the answer.

Example 6   Multiply.   **(a)** 4.8×1000 **(b)** $0.076 \times 10{,}000$

(a) $4.8 \times 1000 = 4800.$

(three zeros) | Decimal point moved three places to the right. Two extra zeros were needed.

(b) $0.076 \times 10{,}000 = 760.$

(four zeros) | Decimal point moved four places to the right. One extra zero was needed.   ■

Practice Problem 6   Multiply.

(a) 0.26×1000 **(b)** $5862.89 \times 10{,}000$ ■

If the number that is a power of 10 is in exponent form, move the decimal point to the right the same number of places as the number that is the exponent.

Example 7   Multiply.   **(a)** 3.68×10^3 **(b)** 0.00027×10^4

exponent of 3 | Decimal point moved three places to right.

(a) $3.68 \times 10^3 = 3680.$

exponent of 4 | Decimal point moved four places to right.

(b) $0.00027 \times 10^4 = 2.7$ ■

Practice Problem 7   Multiply.

(a) 7.684×10^4 **(b)** 0.00073×10^2 ■

Some problems in conversion can be handled by multiplying by 10, 100, 1000, and so on.

Example 8   To convert from kilometers to meters, multiply by 1000. How many meters in 2.96 kilometers?

$$2.96 \times 1000 = 2960$$

There are 2960 meters in 2.96 kilometers.   ■

Practice Problem 8   One kilometer is 1000 meters. How many meters are there in 156.2 kilometers?   ■

?

To Think About

Could you devise a quick rule to use when multiplying a decimal fraction $\frac{1}{10}$, $\frac{1}{100}$, $\frac{1}{1000}$, and so on? How is it similar to the rules developed in this section? Consider a few examples.

Original Problem	Change Fraction to Decimal	Decimal Multiplication	Observation
$86 \times \dfrac{1}{10}$	86×0.1	$\begin{array}{r} 86 \\ \times\ 0.1 \\ \hline 8.6 \end{array}$	Decimal point moved one place to left
$86 \times \dfrac{1}{100}$	86×0.01	$\begin{array}{r} 86 \\ \times\ 0.01 \\ \hline 0.86 \end{array}$	Decimal point moved two places to left
$86 \times \dfrac{1}{1000}$	86×0.001	$\begin{array}{r} 86 \\ \times\ 0.001 \\ \hline 0.086 \end{array}$	Decimal point moved three places to left

Can you think of a way to describe a rule that you could use in solving this type of problem without going through all the foregoing steps? See problems 55 and 56. ■

EXERCISES 3.4

Multiply.

1. $\begin{array}{r} 0.6 \\ \times\ 0.2 \end{array}$ 2. $\begin{array}{r} 0.9 \\ \times\ 0.3 \end{array}$ 3. $\begin{array}{r} 0.12 \\ \times\ 0.5 \end{array}$ 4. $\begin{array}{r} 0.17 \\ \times\ 0.4 \end{array}$ 5. $\begin{array}{r} 0.0028 \\ \times\ 0.7 \end{array}$ 6. $\begin{array}{r} 0.056 \\ \times\ 0.08 \end{array}$ 7. $\begin{array}{r} 0.079 \\ \times\ 0.09 \end{array}$

8. $\begin{array}{r} 0.0034 \\ \times\ 0.5 \end{array}$ 9. $\begin{array}{r} 0.025 \\ \times\ 0.012 \end{array}$ 10. $\begin{array}{r} 0.071 \\ \times\ 0.031 \end{array}$ 11. $\begin{array}{r} 12.36 \\ \times\ 0.06 \end{array}$ 12. $\begin{array}{r} 18.07 \\ \times\ 0.05 \end{array}$ 13. $\begin{array}{r} 7986 \\ \times\ 0.32 \end{array}$ 14. $\begin{array}{r} 5167 \\ \times\ 0.19 \end{array}$

15. $\begin{array}{r} 1.892 \\ \times\ 0.007 \end{array}$ 16. $\begin{array}{r} 2.163 \\ \times\ 0.008 \end{array}$ 17. $\begin{array}{r} 0.7613 \\ \times\ 2003 \end{array}$ 18. $\begin{array}{r} 0.6178 \\ \times\ 5004 \end{array}$ 19. $\begin{array}{r} 9630 \\ \times\ 0.51 \end{array}$ 20. $\begin{array}{r} 7980 \\ \times\ 0.46 \end{array}$

21. Melissa makes $5.85 per hour for a 40-hour week. How much does she make in one week?

22. Marcia makes $6.35 per hour for a 40-hour week. How much does she make in one week?

23. Elva works for a company that manufactures textile products. She earns $7.20 per hour for a 40-hour week. How much does she earn in 1 week? (The average wage, for 1987, for U.S. textile mill industries was $7.18 per hour for an average of 41.9 hours, for a total of $300.84 weekly earnings.)

24. Mee Lee works for a company that manufactures electric and electronic equipment and earns $9.90 per hour for a 40-hour week. How much does she earn in one week? (The average wage, for 1987, for U.S. electric/electronic equipment manufacturers was $9.90 per hour for an average of 40.9 hours, for a total of $404.91 per week.)

25. A sheet of paper has dimensions of 8.2 inches by 10.7 inches. What is the area of the paper in square inches? (The area of a rectangle is the length measurement times the width measurement.)

26. The living room has dimensions of 20.3 feet by 11.6 feet. What is the area of the living room in square feet? (The area of a rectangle is the length measurement times the width measurement.)

27. Dwight is paying off a student loan with payments of $36.20 per month for the next 18 months. How much will he pay off during the next 18 months?

28. Sandy is making car payments of $230.50 per month for 16 more months. How much will she pay for car payments in the next 16 months?

29. Maria bought 1.4 pounds of bananas at $0.55 per pound. How much did she pay for the bananas?

30. Angela purchased 2.6 pounds of tomatoes at $0.65 per pound. How much did she pay for the tomatoes?

31. Steve's car gets approximately 26.4 miles per gallon. His gas tank holds 19.5 gallons. Approximately how many miles can he travel on a full tank of gas?

32. Jim's 4 × 4 truck gets approximately 18.6 miles per gallon. His gas tank holds 20.5 gallons. Approximately how many miles can he travel on a full tank of gas?

Multiply.

33. 2.86×10

34. 1.98×10

35. 0.236×100

36. 0.798×100

37. 128.65×1000

38. 204.37×1000

39. $1.27986 \times 10{,}000$

40. $5.60982 \times 10{,}000$

41. $280{,}560.2 \times 10^2$

42. 7163.241×10^2

43. 763.49×10^4

44. 816.32×10^3

45. 0.6718×10^3

46. 0.7153×10^4

47. 0.0007163×10^5

48. 0.00981376×10^5

49. A store purchased 100 items at $19.64 each. How much did the order cost?

50. To convert from meters to centimeters, multiply by 100. How many centimeters are in 5.932 meters?

51. To convert from kilometers to meters, multiply by 1000. How many meters are in 2.98 kilometers?

52. To convert from kilograms to grams, multiply by 1000. How many grams are in 9.64 kilograms?

53. The city school system pays approximately $3640.50 per student for schooling. What would the city spend for 10,000 students?

54. An automobile manufacturer makes a profit of $2984.60 per car. What would be the profit for 10,000 cars?

 To Think About

55. Explain in words a rule to determine the answer when multiplying a decimal fraction by $\frac{1}{10}$, $\frac{1}{100}$, $\frac{1}{1000}$, $\frac{1}{10,000}$, and so on. Use your rule to evaluate.

(a) $1.3684 \times \frac{1}{100}$

(b) $258.7193 \times \frac{1}{10,000}$

(c) $0.061834 \times \frac{1}{1,000,000}$

56. Explain in words a rule to determine the answer when multiplying a decimal fraction by 0.1, 0.01, 0.001, 0.0001, and so on. Use your rule to evaluate.
(a) 7.6134×0.01
(b) 716.0521×0.0001
(c) $0.0032145 \times 0.0000001$

57. The college is purchasing new carpeting for the learning center. What is the price of a carpet that is 19.6 yards wide and 254.2 yards long if the cost is $12.50 per square yard?

58. The new college athletic field has artificial turf. The portion that has artificial turf is 21.6 yards wide and 128.2 yards long. The artificial turf costs $18.50 per square yard. What is the cost of the artificial turf?

Cumulative Review Problems

Divide. Be sure to include any remainder as part of your answer.

59. $12 \overline{)1176}$

60. $14 \overline{)1204}$

61. $37 \overline{)4629}$

62. $29 \overline{)3745}$

 Calculator Problems

Multiply. The number of decimal places in your answer will depend on the calculator used.

63. $\begin{array}{r} 32.76041 \\ \times\ 13.12663 \\ \hline \end{array}$

64. $\begin{array}{r} 6.02799 \\ \times\ 0.76443 \\ \hline \end{array}$

65. $\begin{array}{r} 33120.12 \\ \times\ \ \ \ \ 72.01 \\ \hline \end{array}$

For Extra Practice Examples and Exercises, turn to page 194.

1. 0.09   2 decimal places
   × 0.6   1 decimal place
   ─────
   0.054   3 decimal places in product (2 + 1 = 3)

3. (a)   0.47   2 decimal places   **(b)**   18.39   2 decimal places
    × 0.28   2 decimal places   × 0.436   3 decimal places
    ─────   ──────
    376   11034
    94   5517
    ─────   7 356
    0.1316   4 decimal places in product (2 + 2 = 4)   ──────
                                            8.01804   5 decimal places in product (2 + 3 = 5)

5. (a) $0.0561 \times 10 = 0.561$ (move decimal point one place to right)
  (b) $1462.37 \times 100 = 146,237$ (move decimal point two places to right)

7. (a) $7.684 \times 10^4 = 76,840$ (four places to right)   **(b)** $0.00073 \times 10^2 = 0.073$ (two places to right)

Answers to Even-Numbered Practice Problems

2. (a) 4.77   **(b)** 1.9817   **4.** 16.2032   **6. (a)** 260   **(b)** 58,628,900   **8.** 156,200

After studying this section, you will be able to:

1 *Divide a decimal number by a whole number*

2 *Divide a decimal number by a decimal number*

3.5 DIVISION OF DECIMALS

In real-life situations you may know a product and one of the factors. To find the other factor, you *divide*. Here are some examples in which you *divide* with decimals.

- You know you've traveled 250 miles and you've used 12.5 gallons of gas. How many miles per gallon have you averaged?   *Solution:* Number of miles per gallon equals 250 divided by 12.5.

- You know your total cost of an item that sells by the pound, such as $9.98 for 10 pounds of oranges. How much does 1 pound cost?   *Solution:* Cost of 1 pound of oranges equals the cost of 10 pounds, $9.98, divided by 10.

To divide when one number is a decimal ($250 \div 12.5$ or $9.98 \div 10$), or even when both are decimals is much like dividing whole numbers. But you must use extra knowledge in order to place the decimal point properly in your result. And very often the answer to such a division problem should be "rounded off" so that the answer makes sense. These are the skills we'll learn in this section.

1 When you divide a decimal by a whole number, place the decimal point for the quotient directly above the decimal point in the dividend. Then divide as if the numbers were whole numbers.

  To divide $26.8 \div 4$, we place the decimal point of our answer (the quotient) directly *above* the decimal point in the dividend.

$$4\,\overline{)\,26.8}$$

The decimal points are directly above each other.

Then we divide as if we were dividing whole numbers.

```
        6.7
    4 ) 26.8        The answer is 6.7.
        24
        ──
         28
         28
         ──
          0
```

The quotient to a problem may have all digits to the right of the decimal point. In some cases you will have to put a zero in the quotient as a "place holder." Thus if you divide

$$
\begin{array}{r}
.067 \\
4\overline{\smash{\big)}\ 0.268} \\
\underline{24} \\
28 \\
\underline{28} \\
0
\end{array}
$$

You have a zero after the decimal point in .067.

Example 1   Divide.

(a) $9\overline{\smash{\big)}\ 0.3204}$

(b) $14\overline{\smash{\big)}\ 36.12}$

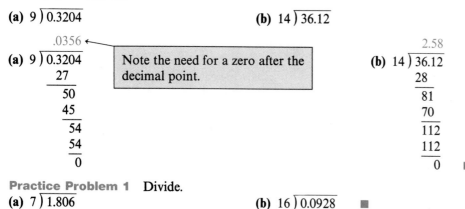

(a)
$$
\begin{array}{r}
.0356 \\
9\overline{\smash{\big)}\ 0.3204} \\
\underline{27} \\
50 \\
\underline{45} \\
54 \\
\underline{54} \\
0
\end{array}
$$

Note the need for a zero after the decimal point.

(b)
$$
\begin{array}{r}
2.58 \\
14\overline{\smash{\big)}\ 36.12} \\
\underline{28} \\
81 \\
\underline{70} \\
112 \\
\underline{112} \\
0
\end{array}
$$

Practice Problem 1   Divide.

(a) $7\overline{\smash{\big)}\ 1.806}$

(b) $16\overline{\smash{\big)}\ 0.0928}$ ■

Some problems do not yield a remainder of zero. In such cases we may be asked to round off the answer to a specified place value. To round off, we carry out the division until our answer contains a digit that is one place value to the right of that to which we intend to round off. Then we round our answer to the specified place value. For example, to round to the nearest thousandth, we carry out division to the ten-thousandths place. In some division problems, you will need to add zeros to the dividend so that this division can be carried out.

Example 2   Divide and round the answer to the nearest thousandth.

$$12.67 \div 39$$

We will carry out our division to the ten-thousandths place. Then we round our answer to the nearest thousandth.

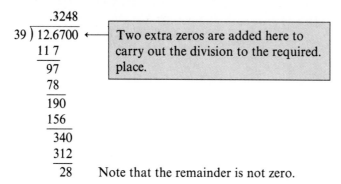

$$
\begin{array}{r}
.3248 \\
39\overline{\smash{\big)}\ 12.6700} \\
\underline{11\ 7} \\
97 \\
\underline{78} \\
190 \\
\underline{156} \\
340 \\
\underline{312} \\
28
\end{array}
$$

Two extra zeros are added here to carry out the division to the required place.

Note that the remainder is not zero.

Now we round 0.3248 to 0.325. The answer is rounded to the nearest thousandth.

■

Practice Problem 2   Divide and round the answer to the nearest hundredth: 23.82 ÷ 46   ■

2 When the divisor is not a whole number, we can convert the division problems to an equivalent problem that has a whole number as a divisor. Think about the reasons this procedure will work. We will ask you about it after you study Examples 3 and 4.

> **Dividing by a Decimal**
>
> 1. Make the divisor a whole number by moving the decimal point to the right. Mark that position with a caret ($\wedge$).
> 2. Move the decimal point in the dividend to the right the same number of places. Mark that position with a caret.
> 3. Place the decimal point of your answer directly above the caret marking the decimal point of the dividend.
> 4. Divide as with whole numbers.

Example 3 Divide: $0.08\overline{)1.632}$.

$$0.08._{\curvearrowright}\overline{)1.63_{\curvearrowright}2}$$

Moving each decimal point two places to the right.

Place the decimal point of the answer directly above the caret.

$$0.08_{\wedge}\overline{)1.63_{\wedge}2}$$

Marking the new position by a caret ($\wedge$).

$$
\begin{array}{r}
20.4 \\
0.08_{\wedge}\overline{)1.63_{\wedge}2} \\
16 \\
\hline
3\ 2 \\
3\ 2 \\
\hline
0
\end{array}
$$

Performing the division.

The answer is 20.4 ■

Practice Problem 3 Divide: $0.09\overline{)0.1008}$. ■

Example 4 Divide: $1.352 \div 0.026$.

$$
\begin{array}{r}
52. \\
0.026_{\wedge}\overline{)1.352_{\wedge}} \\
1\ 30 \\
\hline
52 \\
52 \\
\hline
0
\end{array}
$$

Moving the decimal point three places to the right and marking the new position by a caret ($\wedge$).

The answer is 52 ■

Practice Problem 4 Divide: $1.702 \div 0.037$. ■

?

To Think About

Why do we move the decimal point so many places to the right in divisor and dividend? What rule allows us to do this? How do we know the answer will be valid? We are actually using the property that multiplication of a fraction by 1 leaves the fraction unchanged. This is called the multiplicative identity. Let us examine Example 4 again. We will write $1.352 \div 0.026$ as a fraction.

$$\frac{1.352}{0.026} \times 1 \qquad \text{Multiplication of a fraction by 1 does not change the value of the fraction.}$$

$$= \frac{1.352}{0.026} \times \frac{1000}{1000} \qquad \text{We know that } \frac{1000}{1000} = 1.$$

$$= \frac{1352}{26} \qquad \text{Multiplication by 1000 can be done by moving the decimal point three places to the right.}$$

$$= 52 \qquad \text{Division of whole numbers.}$$

Thus in Example 4 when we moved the decimal point three places to the right in the divisor and the dividend, we were actually creating an equivalent fraction where numerator and denominator were multiplied by 1000. Do you see why? ■

Example 5 Divide.

(a) $1.7 \overline{)0.0323}$

(b) $0.0032 \overline{)7.68}$

(a) $1.7_{\wedge} \overline{)0.0_{\wedge}323}$

$$\begin{array}{r} 019 \\ \underline{17} \\ 153 \\ \underline{153} \\ 0 \end{array}$$

Moving the decimal point in the divisor and dividend one place to the right and marking that position with a caret

(b) $0.0032_{\wedge} \overline{)7.6800_{\wedge}}$

$$\begin{array}{r} 2400 \\ \underline{6\,4} \\ 1\,28 \\ \underline{1\,28} \\ 000 \end{array}$$

Note that two extra zeros are needed in the dividend as we move the decimal point four places to the right. ■

Practice Problem 5 Divide.

(a) $1.8 \overline{)0.0414}$

(b) $0.0036 \overline{)8.316}$ ■

Example 6

(a) Find $2.9 \overline{)431.2}$ rounded to the nearest tenth.
(b) Find $2.17 \overline{)0.08}$ rounded to the nearest thousandth.

(a) $2.9_{\wedge} \overline{)431.2_{\wedge}00}$

$$\begin{array}{r} 148\,.68 \\ \underline{29} \\ 141 \\ \underline{116} \\ 252 \\ \underline{232} \\ 200 \\ \underline{174} \\ 260 \\ \underline{232} \\ 28 \end{array}$$

Calculating to the hundred place and rounding our answer to the nearest tenth.

The answer rounded to the nearest tenth is 148.7

(b) $2.17_{\wedge} \overline{)0.08_{\wedge}0000}$

$$\begin{array}{r} .0368 \\ \underline{6\,51} \\ 1\,490 \\ \underline{1\,302} \\ 1880 \\ \underline{1736} \\ 144 \end{array}$$

Calculating to the ten-thousandths place and then rounding the answer.

Rounding 0.0368 to the nearest thousandth we obtain 0.037 ■

Practice Problem 6

(a) Find $3.8 \overline{)521.6}$ rounded to the nearest tenth.
(b) Find $8.05 \overline{)0.17}$ rounded to the nearest thousandth. ■

Divide until there is a remainder of zero.

1. $6\overline{)13.08}$ **2.** $8\overline{)17.28}$ **3.** $4\overline{)0.1476}$ **4.** $5\overline{)0.0129}$ **5.** $7\overline{)123.34}$

6. $9\overline{)221.58}$ **7.** $64\overline{)3.616}$ **8.** $56\overline{)1.624}$ **9.** $21\overline{)0.0609}$ **10.** $23\overline{)0.0805}$

11. $0.8\overline{)9.76}$ **12.** $0.6\overline{)8.16}$ **13.** $0.5\overline{)32.15}$ **14.** $0.4\overline{)47.28}$ **15.** $0.09\overline{)0.7209}$

16. $0.07\overline{).8113}$ **17.** $3.6\overline{)75.6}$ **18.** $91.2 \div 4.8$ **19.** $95.20 \div 0.28$ **20.** $40.30 \div 0.31$

Divide and round your answer to the nearest tenth.

21. $7\overline{)36.92}$ **22.** $9\overline{)47.31}$ **23.** $1.8\overline{)4.16}$

24. $1.9\overline{)2.36}$ **25.** $0.95\overline{)32.067}$ **26.** $0.85\overline{)41.903}$

Divide and round your answer to the nearest hundredth.

27. $4\overline{)123.82}$ **28.** $5\overline{)471.03}$ **29.** $1.9\overline{)21.9}$

30. $1.8\overline{)47.9}$ **31.** $0.27\overline{)6.729}$ **32.** $0.41\overline{)8.378}$

Divide and round your answer to the nearest thousandth.

33. $8\overline{)238.162}$ **34.** $6\overline{)409.387}$ **35.** $0.69\overline{)8.45}$ **36.** $0.87\overline{)79.40}$

Divide and round your answer to the nearest whole number.

37. $0.075\overline{)3.729}$ **38.** $0.065\overline{)4.398}$ **39.** $0.55\overline{)7.00}$ **40.** $0.39\overline{)5.00}$

41. Joanne is paying $116.52 per year in life insurance premiums. If she makes 12 equal monthly payments, how much does she pay per month?

42. John is paying $123.84 per year in life insurance premiums. If he makes 12 equal monthly payments, how much does he pay per month?

43. If a car travels 248 miles on 14.2 gallons of gas, how many miles per gallon does it achieve? (Round your answer to the nearest tenth.)

44. If a car travels 316 miles on 15.3 gallons of gas, how many miles per gallon does it achieve? (Round your answer to the nearest tenth.)

45. Willy paid $8.98 for 12 pounds of oranges. How much did he pay per pound? (Round your answer to the nearest cent.)

46. Cathy paid $9.12 for 14 pounds of apples. How much did she pay per pound? (Round your answer to the nearest cent.)

47. Sanchez has to pay off $3570.25 in car payments at $142.81 per month. How many more car payments must he make?

48. Juanita has to pay off $4533.13 in car payments that amount to $146.23 per month. How many more car payments must she make?

49. Aaron has a job in quality control. He inspects boxes of ballpoint pens. Each pen weighs 6.8 grams. The contents of a box of pens weigh 326.4 grams. How many pens are contained in the box? If the box is labeled "CONTENTS: 50 PENS," how great an error was made packing the box?

50. Rahim has a job in quality control inspecting boxes of textbooks. A certain textbook weighs 5.4 kilograms. How many books are contained in a box in which the contents weigh 226.8 kilograms? If the box is labeled "CONTENTS: 45 TEXTBOOKS," how great an error was made packing the box?

 To Think About

Review the method explaining how to divide two decimals in fraction form in the To Think About section immediately following Example 4. Use this method to divide.

51. $\dfrac{3.8702}{0.0523}$

52. $\dfrac{2.9356}{0.0716}$

Divide. Round to the nearest millionth.

53. $21.8 \overline{)0.38612527}$

54. $32.6 \overline{)0.41289369}$

Cumulative Review Problems

55. Add: $\dfrac{3}{8} + 1\dfrac{2}{5}$.

56. Subtract: $2\dfrac{13}{16} - 1\dfrac{7}{8}$.

57. Multiply: $3\dfrac{1}{2} \times 2\dfrac{1}{6}$.

58. Divide: $4\dfrac{1}{3} \div 2\dfrac{3}{5}$.

 Calculator Problems

Divide. The number of decimal places in your answer will depend on the calculator used.

59. $2.0976 \overline{)588.1445}$

60. $0.014127 \div 0.00203$

61. $112.776 \div 0.0015$

62. The Garcias are retiring this year. They have $30,250.60 in their savings account and $62,650.00 in their retirement account from their employers. They sold property for $26,750.80.
(a) How much money total do the Garcias have for retirement?
(b) If they want this money to last 20 years, how much of this money should they use each year?

For Extra Practice Examples and Exercises, turn to page 195.

1. (a)
$$\begin{array}{r} 0.258 \\ 7\overline{)1.806} \\ \underline{14} \\ 40 \\ \underline{35} \\ 56 \\ \underline{56} \\ 0 \end{array}$$

(b)
$$\begin{array}{r} 0.0058 \\ 16\overline{)0.0928} \\ \underline{80} \\ 128 \\ \underline{128} \\ 0 \end{array}$$

3.
$$\begin{array}{r} 1.12 \\ 0.09_\wedge\overline{)0.10_\wedge08} \\ \underline{9} \\ 10 \\ \underline{9} \\ 18 \\ \underline{18} \\ 0 \end{array}$$

5. (a)
$$\begin{array}{r} 0.023 \\ 1.8_\wedge\overline{)0.0_\wedge414} \\ \underline{36} \\ 54 \\ \underline{54} \\ 0 \end{array}$$

(b)
$$\begin{array}{r} 2310. \\ 0.0036_\wedge\overline{)8.3160_\wedge} \\ \underline{72} \\ 111 \\ \underline{108} \\ 36 \\ \underline{36} \\ 0 \end{array}$$

2. 0.52   **4.** 46   **6. (a)** 137.3   **(b)** 0.021

After studying this section, you will be able to:

1 *Convert any fraction to decimal form*

2 *Use the correct order of operations when several decimals are combined with various operations*

3.6 CONVERTING FRACTIONS TO DECIMALS

For a coin, heads and tails are two sides of the same object, two ways of identifying a single quantity. For numbers, fractions and decimals are two equivalent forms for the same quantity.

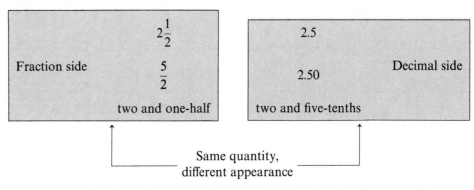

Same quantity, different appearance

Every decimal can be expressed as an equivalent fraction. For example,

$$0.75 = \frac{75}{100} \quad \text{or} \quad \frac{3}{4}$$

$$0.5 = \frac{5}{10} \quad \text{or} \quad \frac{1}{2}$$

$$2.5 = 2\frac{5}{10} = 2\frac{1}{2} \quad \text{or} \quad \frac{5}{2}$$

Decimal form $\Longrightarrow$ fraction form

And every fraction can be expressed as an equivalent decimal, as we will learn in this section. For example,

$$\frac{1}{5} = 0.20 \quad \text{or} \quad 0.2$$

$$\frac{3}{8} = 0.375$$

$$\frac{5}{11} = 0.4545\ldots \quad \text{(the "45" keeps repeating)}$$

Fraction form $\Longrightarrow$ decimal form

Certain of these decimal equivalencies are so common that people find it helpful to memorize those for $\frac{1}{2}, \frac{1}{3}, \frac{1}{4}, \frac{1}{5}, \frac{1}{6}, \frac{1}{7}, \frac{1}{8}, \frac{1}{9}, \frac{1}{10}$. How many of these do you know already?

1 We have previously studied how to transfer some fractions with a denominator of 10, 100, 1000, and so on, to decimal form. For example, $\frac{3}{10} = 0.3$ and $\frac{7}{100} = 0.07$. We need to develop a procedure to write other fractions, such as $\frac{3}{8}$ and $\frac{5}{16}$, in decimal form.

> **Converting a Fraction to an Equivalent Decimal**
>
> Divide the denominator into the numerator until
> **a.** the remainder becomes zero, or
> **b.** the remainder repeats itself, or
> **c.** the desired number of decimal places is achieved.

Example 1 Write $\frac{3}{8}$ as an equivalent decimal.

$$
\begin{array}{r}
.375 \\
8\,)\overline{3.000} \\
\underline{2\,4} \\
60 \\
\underline{56} \\
40 \\
\underline{40} \\
0
\end{array}
$$

Therefore, $\frac{3}{8} = 0.375$ ∎

Practice Problem 1 Write $\frac{5}{16}$ as an equivalent decimal. ∎

Example 2 Write $\frac{31}{40}$ as an equivalent decimal.

$$
\begin{array}{r}
.775 \\
40\,)\overline{31.000} \\
\underline{280} \\
300 \\
\underline{280} \\
200 \\
\underline{200} \\
0
\end{array}
$$

Therefore, $\frac{31}{40} = 0.775$ ∎

Practice Problem 2 Write $\frac{11}{80}$ as an equivalent decimal. ∎

Decimals such as 0.375 and 0.775 are called **terminating decimals**. When converting $\frac{3}{8}$ to 0.375 or $\frac{31}{40}$ to 0.775 the division operation eventually yields a remainder of zero. Other fractions yield a repeating pattern. For example, $\frac{1}{3} = 0.3333\ldots$ and $\frac{2}{3} = 0.6666\ldots$ have a pattern of repeating digits. Decimals that have a digit or a group of digits that repeats are called **repeating decimals**. We often indicate the repeating pattern with a bar over the repeating group of digits:

$$0.3333\ldots = 0.\overline{3} \qquad 0.\,\boxed{74}\,\boxed{74}\,\boxed{74}\ldots = 0.\overline{74}$$

$$0.\,\boxed{218}\,\boxed{218}\,\boxed{218}\ldots = 0.\overline{218} \qquad 0.\,\boxed{8942}\,\boxed{8942}\ldots = 0.\overline{8942}$$

When converting fractions to decimal form and the remainder repeats itself, we know that we have a repeating decimal.

Example 3   Write $\dfrac{5}{11}$ as an equivalent decimal.

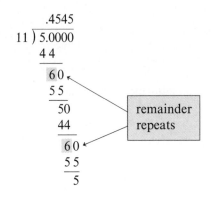

Thus $\dfrac{5}{11} = 0.4545\ldots = 0.\overline{45}$ ■

Practice Problem 3   Write $\dfrac{7}{11}$ as an equivalent decimal.   ■

Example 4   Write as an equivalent decimal.   **(a)** $\dfrac{13}{22}$ **(b)** $\dfrac{5}{37}$

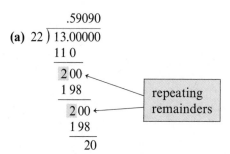

Thus $\dfrac{11}{22} = 0.5909090\ldots = 0.5\overline{90}$. Notice that the bar is over the digits 9 and 0 but *not* over the digit 5.

```
      .1351
(b) 37 ) 5.000
       3 7
       1 3 0
       1 1 1
         1 9 0
         1 8 5
             5 0
             3 7
             1 3
```

repeating remainders

Thus $\dfrac{5}{37} = 0.135135135\ldots = 0.\overline{135}$ ■

Practice Problem 4 Write as an equivalent decimal. **(a)** $\frac{8}{15}$ **(b)** $\frac{13}{44}$ ■

Example 5 Write as an equivalent decimal. **(a)** $3\frac{7}{15}$ **(b)** $\frac{20}{11}$

(a) $3\frac{7}{15}$ means $3 + \frac{7}{15}$.

$$\begin{array}{r} 0.466 \\ 15 \overline{)\ 7.000} \\ \underline{60} \\ 100 \\ \underline{90} \\ 100 \\ \underline{90} \\ 10 \end{array}$$

(b)
$$\begin{array}{r} 1.818 \\ 11 \overline{)\ 20.000} \\ \underline{11} \\ 90 \\ \underline{88} \\ 20 \\ \underline{11} \\ 90 \\ \underline{88} \\ 2 \end{array}$$

Thus $\frac{7}{15} = 0.4\overline{6}$ and $3\frac{7}{15} = 3.4\overline{6}$

Thus $\frac{20}{11} = 1.818181\ldots = 1.\overline{81}$ ■

Practice Problem 5 Write as an equivalent decimal. **(a)** $2\frac{11}{18}$ **(b)** $\frac{28}{27}$ ■

In some cases, the pattern of repeating is quite long. For example,

$$\frac{1}{7} = 0.142857142857\ldots = 0.\overline{142857}$$

Such problems often are rounded to a certain value.

Example 6 Express $\frac{5}{7}$ as a decimal rounded to the nearest thousandth.

$$\begin{array}{r} .7142 \\ 7 \overline{)\ 5.0000} \\ \underline{49} \\ 10 \\ \underline{7} \\ 30 \\ \underline{28} \\ 20 \\ \underline{14} \\ 6 \end{array}$$

Rounding to the nearest thousandth, we round 0.7142 to 0.714. (In repeating form, $\frac{5}{7} = 0.714285714285\ldots = 0.\overline{714285}$.) ■

Practice Problem 6 Express $\frac{19}{24}$ as a decimal rounded to the nearest thousandth. ■

② Order of Operations

The rules we discussed in Section 1.7 are appropriate for operations with decimals.

Order of Operations
Do first   **1.** Perform operations inside parentheses.
↑   **2.** Simplify any expressions with exponents.
↓   **3.** Multiply or divide from left to right.
Do last   **4.** Add or subtract from left to right.

Example 7 Evaluate: $0.5 + 0.7 \times 0.21 - 0.63$.

$$= 0.5 + 0.147 - 0.63 \qquad \text{Multiply} \quad \begin{array}{r} 0.21 \\ \times\ 0.7 \\ \hline 0.147 \end{array}$$

$$= 0.647 - 0.63 \qquad \text{Add} \quad \begin{array}{r} 0.5 \\ +\ 0.147 \\ \hline 0.647 \end{array}$$

$$= 0.017 \qquad \text{Subtract} \quad \begin{array}{r} 0.647 \\ -\ 0.630 \\ \hline 0.017 \end{array} \quad \blacksquare$$

Practice Problem 7 Evaluate: $0.9 - 0.3 + 0.8 \times 0.3$. $\quad\blacksquare$

Sometimes exponents are used with decimals. In such cases, we merely evaluate using repeated multiplication.

$$(0.2)^2 = 0.2 \times 0.2 = 0.04$$

$$(0.2)^3 = 0.2 \times 0.2 \times 0.2 = 0.008$$

$$(0.2)^4 = 0.2 \times 0.2 \times 0.2 \times 0.2 = 0.0016$$

Example 8 Evaluate.

$$(0.3)^3 + 0.6 \times 0.2 + 0.013$$

First we need to evaluate $(0.3)^3 = 0.3 \times 0.3 \times 0.3 = 0.027$. Can you see that this is true? Thus

$$(0.3)^3 + 0.6 \times 0.2 + 0.013$$

$$= 0.027 + 0.6 \times 0.2 + 0.013$$

$$= 0.027 + 0.12 + 0.013 \longleftarrow \text{ When addends have a different number} \qquad \begin{array}{r} 0.027 \\ 0.120 \\ +\ 0.013 \\ \hline 0.160 \end{array}$$
of decimal places, writing the problem
in column form makes finding the
solution easier.

The result is 0.16. $\quad\blacksquare$

Practice Problem 8 Evaluate: $0.3 \times 0.5 + (0.4)^3 - 0.036$. $\quad\blacksquare$

To Think About

Can you imagine how to predict if you will get terminating or repeating decimals when adding or subtracting repeating decimals? When we subtract $0.\overline{6} - 0.\overline{3} = 0.\overline{3}$ we obtain a repeating decimal. You can see this by lining up the decimals:

$$\begin{array}{r} 0.6666\ldots \\ -\ 0.3333\ldots \\ \hline 0.3333\ldots \end{array}$$

However, $\dfrac{2}{3} + \dfrac{1}{3} = \dfrac{3}{3} = 1$. Therefore, $\dfrac{2}{3} = 0.\overline{6}$ and $\dfrac{1}{3} = 0.\overline{3}$. Thus $0.\overline{6} + 0.\overline{3} = 1.0$, a terminating decimal. Notice when we try to add it up by lining up decimals:

$$\begin{array}{r} 0.6666\ldots \\ +\ 0.3333\ldots \\ \hline 0.9999\ldots \end{array}$$

we get a repeating decimal. In fact, although it seems hard to believe it is true that $0.9999\ldots = 1.0$! These are two ways to write the same number. For more problems with repeating decimals, see problems 59 and 60. $\quad\blacksquare$

Write as an equivalent decimal. If an infinite repeating decimal is obtained, use notation such as $0.\overline{7}$, $0.\overline{16}$, or $0.\overline{245}$.

1. $\dfrac{1}{4}$ 2. $\dfrac{3}{4}$ 3. $\dfrac{5}{8}$ 4. $\dfrac{7}{8}$ 5. $\dfrac{7}{16}$ 6. $\dfrac{1}{16}$ 7. $\dfrac{7}{20}$ 8. $\dfrac{3}{40}$

9. $\dfrac{31}{50}$ 10. $\dfrac{23}{25}$ 11. $\dfrac{7}{4}$ 12. $\dfrac{5}{2}$ 13. $2\dfrac{1}{8}$ 14. $3\dfrac{3}{16}$ 15. $1\dfrac{1}{40}$ 16. $1\dfrac{7}{16}$

17. $\dfrac{2}{3}$ 18. $\dfrac{1}{3}$ 19. $\dfrac{5}{9}$ 20. $\dfrac{5}{6}$ 21. $\dfrac{13}{18}$ 22. $\dfrac{7}{11}$ 23. $\dfrac{8}{11}$ 24. $\dfrac{7}{15}$

25. $\dfrac{14}{15}$ 26. $\dfrac{8}{9}$ 27. $\dfrac{5}{33}$ 28. $\dfrac{7}{22}$ 29. $\dfrac{41}{36}$ 30. $\dfrac{25}{24}$ 31. $2\dfrac{5}{18}$ 32. $6\dfrac{1}{6}$

Write as an equivalent decimal or a decimal approximation. Round your answer to the nearest thousandth if necessary.

33. $\dfrac{4}{7}$ 34. $\dfrac{6}{7}$ 35. $\dfrac{20}{21}$ 36. $\dfrac{19}{21}$ 37. $\dfrac{7}{48}$ 38. $\dfrac{5}{48}$ 39. $\dfrac{47}{33}$ 40. $\dfrac{29}{32}$

41. $\dfrac{13}{81}$ 42. $\dfrac{7}{82}$ 43. $\dfrac{17}{18}$ 44. $\dfrac{5}{13}$ 45. $\dfrac{17}{14}$ 46. $\dfrac{23}{14}$ 47. $\dfrac{9}{19}$ 48. $\dfrac{11}{17}$

Evaluate.

49. $(0.3)^2 + 0.6 \times 0.2 + 5.8$

50. $(0.2)^3 + 5.9 \times 1.3 - 2.6$

51. $18.6 \times 13.2 - 5.8 + 3.2 \div 0.2$

52. $1.9 \times 14.5 - 7.7 + 28.4 \div 1.6$

53. $12 \div 0.03 - 15 \times (0.9 + 0.7)$

54. $61.95 \div 1.05 + 3.6 \times (2.1 + 0.31)$

55. $116.32 + (0.12)^2 + 18.06 \times 2.2$

56. $105.08 + (0.21)^2 - 0.05 \times 123.4$

57. $(1.1)^3 + 2.6 \div 0.13 + 0.083$

58. $(1.1)^3 + 8.6 \div 2.15 + 0.086$

To Think About

59. Subtract: $0.\overline{16} - 0.00\overline{16}$.
 (a) What do you obtain?
 (b) Now subtract $0.\overline{16} - 0.01\overline{6}$.
   What do you obtain?
 (c) What is different about these results?

60. Subtract: $1.\overline{89} - 0.01\overline{89}$.
 (a) What do you obtain?
 (b) Now subtract $1.\overline{89} - 0.18\overline{9}$.
   What do you obtain?
 (c) What is different about these results?

61. Do you think that $1.\overline{9} = 2$? Yes
 Why or why not?

62. Can you show an example of why it is logical to assume that $2.\overline{9} = 3$?

Evaluate.

63. $(1.6)^3 + (2.4)^2 + 18.666 \div 3.05 + 4.86$

64. $5.9 \times 3.6 \times 2.4 - 0.1 \times 0.2 \times 0.3 \times 0.4$

Cumulative Review Problems

65. What is the area of a rectangle that measures 12 feet by 26 feet?

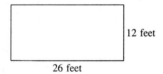

66. What is the difference in sales between selling 186,352 units and selling 180,579 units?

67. If a person makes deposits of $56, $81, $42 and $198, what is the total amount deposited in the account?

68. If a profit of $14,352 is shared equally among 12 sales people, how much profit will each person get?

 Calculator Problems

Write each fraction as a decimal. Round your answer to six decimal places.

69. $\dfrac{5236}{8921}$

70. $\dfrac{17,359}{19,826}$

For Extra Practice Examples and Exercises, turn to page 195.

Solutions to Odd-Numbered Practice Problems

1.
```
      .3125
16 ) 5.0000
     48
     ──
      20
      16
      ──
       40
       32
       ──
        80
        80
        ──
         0
```
$\dfrac{5}{16} = 0.3125$

3.
```
      .6363
11 ) 7.0000
     66
     ──
      40
      33
      ──
       70
       66
       ──
        40
        33
        ──
         7
```
$\dfrac{7}{11} = 0.\overline{63}$

5. (a) $2\dfrac{11}{18} = 2 + \dfrac{11}{18}$

$$\begin{array}{r} .611 \\ 18\overline{)11.000} \\ \underline{108} \\ 20 \\ \underline{18} \\ 20 \\ \underline{18} \\ 2 \end{array}$$

Since $\dfrac{11}{18} = 0.6\overline{1}$. Therefore, $2\dfrac{11}{18} = 2.6\overline{1}$.

(b) $\dfrac{28}{27}$

$$\begin{array}{r} 1.03703 \\ 27\overline{)28.00000} \\ \underline{27} \\ 100 \\ \underline{81} \\ 190 \\ \underline{189} \\ 100 \\ \underline{81} \\ 19 \end{array}$$

Thus $\dfrac{28}{27} = 1.\overline{037}$.

7. $0.9 - 0.3 + 0.8 \times 0.3 = 0.9 - 0.3 + 0.24 = 0.6 + 0.24 = 0.84$

Answers to Even-Numbered Practice Problems

2. 0.1375   **4. (a)** $0.5\overline{3}$ **(b)** $0.29\overline{54}$ **6.** 0.792   **8.** 0.178

3.7 APPLICATIONS

After studying this section, you will be able to:

1 *Solve applied problems involving one operation with decimal numbers*

2 *Solve applied problems involving more than one operation with decimal numbers*

If you are a worker, you may need or want to compute your earnings, and this usually means problem solving with decimals. The six-step method is the same as before except that now the numbers are decimals. Let's take the case of an electrician who is paid $14.30 per hour for a 40-hour week. She is paid time and a half for overtime (1.5 times the hourly wage) for every hour above 40 hours worked in that same week. If she works 48 hours in one week, what will she earn for that week?

Be sure you understand the problem and the information given. Now *estimate* the result!

Approximately $14 per hour for 40 hours.

Approximately $20 per hour for 8 hours.

(20 is an estimated wage for 1.5 times 14)

So multiplying gives $560 for 40 hours of work and $160 for 8 hours of work.

The estimated earnings are $720 total.

You'll do the exact computation in Exercise Set 3.7, problem 17. Be sure to check whether your answer is close to this estimate to see if your computed answer is reasonable.

1 When solving applied problems with decimals the six-step procedure discussed in Section 1.8 is helpful. We repeat that procedure here for reference.

Steps to Solving an Applied Problem

1. Read over the problem carefully. Find out what is asked for. Draw a picture if this helps you to visualize the situation.
2. Write down which numbers are to be used in solving the problem.
3. Estimate your answer to see if it is reasonable.
4. Determine what operation needs to be done: addition, subtraction, multiplication, division, or a combination of these.
5. Perform the necessary computations. State the answer, including the unit of measure.
6. See if your answer is reasonable compared to your estimate.

---

2 **Example 1** Brendon bought a camera for $89.59, and the sales tax was $4.48. He also bought a camera case for $16.99, and the sales tax was $0.85. He brought $120 for the purchase. How much money did he have left over?

1. We want a total cost for two items plus tax. Then we want to see what is left over from the $120.00.

2. The expenses Brendon paid were $89.59, $4.48, $16.99, and $0.85. He has available $120.00.

3. Estimate: The purchases were about $90 and $15. The tax amounts were about $5 and $1.

$$
\begin{array}{ll}
\text{Estimate sum:} & \begin{array}{r} \$\ 90 \\ 15 \\ 5 \\ +\quad 1 \\ \hline \$111 \end{array}
\end{array}
\qquad
\begin{array}{ll}
\text{Estimate difference:} & \begin{array}{r} \$120 \\ -\ 111 \\ \hline \$\ 9 \end{array}
\end{array}
$$

We estimate an answer of $9.

4. We *add* the four expenses exactly and subtract the result from $120.00.

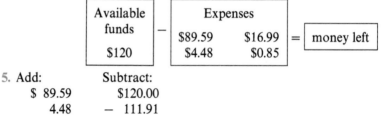

5. Add: Subtract:

$$
\begin{array}{r} \$\ 89.59 \\ 4.48 \\ 16.99 \\ +\quad 0.85 \\ \hline \$111.91 \end{array}
\qquad
\begin{array}{r} \$120.00 \\ -\ 111.91 \\ \hline 8.09 \end{array}
$$

He had $8.09 left over.

6. This is close to our estimate of $9. Our answer is reasonable. ■

Practice Problem 1   Renée purchased a coat for $198.95 which had a sales tax of $9.95. She also purchased boots for $59.95, which had a sales tax of $3.00. She brought $280 to make the purchase. How much money will she have left? ■

Example 2   A laborer is paid $7.38 per hour for a 40-hour week and 1.5 times that wage for all hours worked above 40 hours in one week. If he works 47 hours in a week, what will he earn?

1. We want to compute his regular pay and his overtime pay and add the results together.

$$\boxed{\text{Regular pay}} + \boxed{\text{overtime pay}} = \boxed{\text{total pay}}$$

2. He works 40 hours at $7.38. He works 7 hours at $1.5 \times \$7.38$.

3. We estimate $40 \times 7 = \$280$ and $7 \times 10 = \$70$ for overtime. $280 + 70 = \$350$ is an estimate answer.

4. Exactly, we multiply 40×7.38, we multiply $7 \times 1.5 \times 7.38$, then we add the results.

5.
$$
\begin{array}{r} 7.38 \\ \times\quad 40 \\ \hline 295.20 \end{array}
\qquad
\begin{array}{r} 7.38 \\ \times\quad 1.5 \\ \hline 3\ 690 \\ 7\ 38 \\ \hline 11.070 \end{array}
$$

He earns $295.20 at $7.38 per hour.

He earns $11.07 per hour in overtime.

$$
\begin{array}{r} 11.07 \\ \times\qquad 7 \\ \hline 77.49 \end{array}
$$

For 7 hours overtime he earns $77.49.

$$
\begin{array}{r} \$295.20 \\ +\quad 77.49 \\ \hline \$372.69 \end{array}
$$

regular 40-hour week earnings
overtime earnings
total earnings

He earns $372.69 for the week.

6. This is close to our estimate of the answer of $350. ■

Practice Problem 2 Melinda works for the phone company as a line repair technician. She earns $9.36 per hour. She worked 51 hours last week. If she gets time and a half for all hours above 40 hours per week, how much did she earn last week? ■

EXERCISES 3.7

1. Renée made deposits into her checking account of $156.42, $87.35, $101.89, and $64.03. What was the total of these four deposits?

2. Lou made deposits into her savings account of $202.97, $43.67, $121.46, and $88.25. What was the total of the four deposits?

3. When Todd began his summer trip, the odometer read 36,382.5 miles. At the end of the trip it read 39,107.3 miles. How many miles did he travel?

4. Rashad drove south to visit his aunt. At the beginning of the trip, the odometer read 51,671.8 miles. At the end of the trip, it read 53,265.9 miles. How many miles did he travel?

5. A rectangular field measures 121.82 meters long by 80.05 meters wide. What is its area, in square meters?

6. An engineer works with a circuit board that is 116.34 centimeters long by 70.05 centimeters wide. What is the area of the board, in square centimeters?

7. A butcher has 17.25 pounds of ground beef. He divides the meat into packages of equal weight each containing 0.75 pound. How many packages can he make?

8. A chemist has 22.96 liters of a liquid she is testing. If she wishes to place it into several containers that each hold 0.82 liters, how many containers will she need?

9. Alfredo went to buy a tire that cost $65.35. The federal excise tax was $2.12, the state sales tax was $3.54 and the local sales tax was $1.25. Alfredo took $80 with him. How much money will he have left after he pays for the tire and the taxes?

10. Alicia bought a blouse for $28.69 and paid $1.43 tax. She bought a pair of slacks for $49.89 and paid $2.49 tax. She brought $90 to pay for these purchases (including tax). How much money will she have left after paying for the clothes and the tax?

11. Donald took a trip. His odometer was 42,380.6 at the beginning and 42,505.6 at the end. He used 6.2 gallons of gas. How many miles per gallon did his car get on the trip? (Round your answer to the nearest tenth.)

12. Sean took a trip. His odometer was 67,571.4 at the beginning and 67,711.4 at the end. He used 7.4 gallons of gas. How many miles per gallon did his car get on the trip? (Round your answer to the nearest tenth.)

13. A 55-gallon drum of cleaning solvent is being used at the rate of 2.5 gallons per day. How many days will it last?

14. A gasoline-powered generator used 1.4 gallons of gas per hour. The tank holds 42 gallons of gas. How many hours can the generator run on a full tank of gas?

15. The Williams family is planning to put carpeting in their house. They need 23.8 square yards in the living room, 10.5 square yards in the hall, and 16.7 square yards in the dining room. The carpeting will cost $14.35 per square yard. What will the total bill be?

16. The Williams family is installing insulation in their house. They are installing 98.6 square yards of insulation in the attic, 24.3 square yards of insulation in the basement, and 36.1 square yards of insulation in the family room. The insulation costs $6.40 per square yard. How much will all the insulation cost?

17. An electrician is paid $14.30 per hour for a 40-hour week. She is paid time and a half for overtime (1.5 times the hourly wage) for every hour above 40 hours worked in that same week. If she works 48 hours in one week, what will she earn for that week?

18. A factory worker is paid $12.60 per hour for a 40-hour week. He is paid time and a half for overtime (1.5 times the hourly wage) for every hour above 40 hours worked in that same week. If he works 52 hours in one week, what will he earn for that week? (Assume that no taxes have yet been deducted.)

19. Barbara had $420.13 in her savings account last month. Since then she made deposits of $116.32 and $318.57. The bank credited her with interest of $1.86 for the month. She wrote three checks, withdrawing $16.50, $36.89, and $376.94. What is her current balance? (Assume that no bank fees have been charged.)

20. Harvey had $113.08 in his checking account last month. He made two deposits: $612.83 and $86.54. The bank deducted a monthly fee this month of $4.60. He made out three checks: $100.00, $216.34, and $398.92. What will his new balance be?

 To Think About

21. The Hiking Club wants to rent a motor home for a week. The cost is $410 per week plus $0.36 per mile. They plan a round trip to the mountains, a trip of 378 miles. In this area, the average cost of gasoline is $1.20 per gallon and the motor home gets 12 miles to the gallon. How much will the Hiking Club spend to rent the motor home for the week and take the trip? (Include the cost of gasoline.)

22. The Outing Club wants to rent a 22-foot inboard boat for the week. The cost is $265 per week plus $12.40 per hour of engine running time. (The engine is equipped with a timer that activates when the engine is turned on.) The club plans the boat trip to take a week and estimates it will take about 26.5 hours of engine running time. The boat uses an average of 9 gallons of gas per hour of running time (at cruising speed), and marine gas costs $1.40 per gallon in the club's area. How much will it cost to rent the boat and pay for the fuel for one week? (Assume 26.5 hours of running time at cruising speed.)

Cumulative Review Problems

Add.

23. $\dfrac{1}{5} + \dfrac{3}{7} + \dfrac{1}{2}$

24. $\dfrac{10}{17} + \dfrac{1}{34} - \dfrac{5}{17}$

Multiply.

25. $\dfrac{5}{12} \times \dfrac{36}{27}$

26. $2\dfrac{2}{3} \times 8\dfrac{1}{4}$

27. If General Motors sells 6,340 new Chevrolet Camaros next month and makes a profit of $2,318.20 on each car what profit will they make on Camaros next month?

28. A jet fuel tank containing 17316.8 gallons is being emptied at the rate of 126.4 gallons per minute. How many minutes will it take to empty the tank?

For Extra Practice Examples and Exercises, turn to page 196.

Solution to Odd-Numbered Practice Problem

1. *Step 1:* We want to find the total cost of the coat, the boots, and both tax amounts. Then we find how much is left over from the $280.00.

Step 2: Purchases: Coat   198.95   Boots   59.95
                              tax   9.95   tax   3.00

Money available: $280.00
Step 3: Estimate costs: 200 + 10 + 60 + 5 = $275.00; estimate difference: 280 − 275 = $5.
Step 4: We need to *add* expenses and *subtract* result from $280.00.
Step 5: Add to obtain total expenses

$198.95	Then subtract: $280.00 available funds
9.95	− 271.85
59.95	8.15
3.00	
$271.85	

He has $8.15 left over.
Step 6: Our answer seems reasonable, since we estimated that $5.00 would be left over.

Answer to Even-Numbered Practice Problem

2. $528.84

EXTRA PRACTICE: EXAMPLES AND EXERCISES

Section 3.1

Write a word name for each decimal.

Example   21.098

          Twenty-one and ninety-eight thousandths

1. 0.96

2. 0.267

3. 19.807

4. 1.2791

5. 29.0003

Express each fraction as a decimal.

Example $16\frac{24}{1000}$

                    16.024

6. $\frac{6}{10}$

7. $\frac{4}{100}$

8. $3\frac{161}{1000}$

9. $98\frac{17}{100}$

10. $171\frac{376}{10,000}$

Write in fractional notation. Reduce whenever possible.

Example   3.145

$3\frac{145}{1000}$, which reduces to $3\frac{29}{200}$

11. 0.65

12. 2.7

13. 0.156

14. 13.125

15. 5.0017

Section 3.2

Fill in the blank with one of the symbols <, =, >.

Example 3.0672 _____ 3.06

               3.0672 _____ 3.0600   Add zeros

672 > 600, so 3.0672 > 3.06.

1. 2.4 _____ 1.93   **2.** 0.52 _____ 0.520

3. 21.002 _____ 21.003   **4.** 5.011 _____ 5.0121

5. 18.0754 _____ 18.0751

Round to the nearest hundredth.

Example 20.061

      20.061   1 is less than 5 so 20.061 rounds to 20.06.

6. 21.053   **7.** 3.9772

8. 352.111   **9.** 4703.275

10. 3910.0662

Round to the place value given.

Example 7.072997   hundred-thousandths

7.072997 7 > 5, so round the two 9's up 1 to 0
    ↑   (really 10, but the 1 is carried to
              the thousandths place)

7.072997 rounds to 7.07300
       ↑↑ be sure to include the zeros
          because the rounding is to the
          hundred-thousandths place

11. 2.07122   thousandths

12. 0.081373   ten-thousandths

13. 15.00841338   hundred-thousandths

14. 6.0499   thousandths

15. 132.8877   whole number

Round to the nearest dollar or cent as indicated.

Example $8621.392   cent

    $8621.392   2 < 5 so $8621.392 rounds to $8621.39

16. $6212.7813   cent

17. $8109.72   dollar

18. $11,098.43   dollar

19. $30,249.513   cent

20. $75.9863   cent

Section 3.3

Add.

Example 3.708 + 5.275 + 2.81

     3.708
     5.275   Line up the decimal points and add zeros
 + 2.810   where necessary.

 11.793

1. 2.589 + 8.4 + 5.178   **2.** 3.9 + 2.08 + 8.125

3. 14 + 0.45 + 2.67   **4.** 18.984 + 0.22 + 45.78

5. 59.077 + 23 + 21.9

Example Susan purchased items at the market that cost $2.78, $0.39, $1.23, $4, and $6.12. What was her total bill?

             1 2
          $2.78
          0.39
          1.23
          4.00
        + 6.12

        $14.52

6. Nancy purchased items at the drugstore that cost $3.77, $0.89, $5.22, $1.75, and $0.92. What was her total bill?

7. Brian made the following three deposits to his savings account: $45.78, $98, and $59.44. What was the total of his three deposits?

8. Jamie drove to San Diego for a vacation. On the way she purchased 11.4, 10.8, 7.5, and 11.9 gallons of gas. How many gallons of gas did she buy?

Example 23.0089 − 1.822

          2 9 10
        23.Ø Ø89
       − 1.8220

       21.1869

9. 3.9 − 2.1   **10.** 78.92 − 68.22

11. 26.981 − 13.39   **12.** 67.234 − 35.889

13. 14.0665 − 5.733

Use your calculator to subtract.

Example 987.98243 − 587.9077

                400.07473

14. 278.00982 − 34.9879   **17.** 45 − 0.984221

15. 700.80701 − 677.965   **18.** 89.01 − 44.6788

16. 29.003103 − 0.7896

Section 3.4

Multiply.

Example 45.82 × 2.03

        45.82   2 decimals
     ×   2.03   2 decimals
     ----
     1 3746
     0 000   this row of zeros can be deleted
     91 64

     93.0146   2 + 2 = 4 decimals

1. 22.98 × 3.07   **2.** 38.77 × 4.06

3. 13.98 × 2.7   **4.** 3.722 × 8.2

5. 67.924 × 39

Example 27.6792 × 1000

                       Decimal point moves
                       three places.

     27.6792 × 1000 = 27,679.2
                   └ three zeros

3.7 *Solve each problem.*

53. An office manager paid $18.50 for a desk pad set. She paid $29.95 for a lamp. She also paid $2.42 in sales tax for the two items. She took $60 from petty cash to make the purchase at a local supply store. How much will she have left after the purchase?

54. Terri had a balance of $62.36 in her checking account. She made deposits of $108.19 and $73.56. She made out checks for $102.14, $37.56, and $19.95. What will her new balance be?

55. Jill earns $6.88 per hour in her new job. She earns 1.5 times that amount for every hour over 40 hours per week. She worked 49 hours last week. How much did she earn?

56. Dan drove to the mountains. His odometer read 26,005.8 miles at the start, 26,325.8 miles at the end of the trip. He used 12.9 gallons of gas on the trip. How many miles per gallon did his car achieve? (Round your answer to the nearest tenth.)

57. Robert is considering buying a car and making installment payments of $189.60 for 48 months. The cash price of the car is $6,930.50. How much extra does he pay if he uses the installment plan as compared with buying the car in one payment?

58. Mr. Zeno has a choice of working as an assistant manager at ABC company at $315.00 per week or receiving an hourly salary of $8.26 per hour at a different company. He learned from several previous assistant managers that they usually worked 38 hours per week. At which company will he probably earn more money?

PUTTING YOUR SKILLS TO WORK

Usually, people with a personal checking account keep a record of their banking with a check register. When the bank sends a copy of their records (showing the amount in the account), you should look over what the bank's total is, compare it to your register, and be sure the two are the same. This is called *balancing* the checkbook. It requires the use of decimals because the amount of money is given in dollars and cents.

Suppose that Regina had a checking account containing $23.69. Suppose she then made deposits of $116.84 on March 3 and $89.46 on March 15. Then suppose she made out check 156 to Joe's Market for $16.80 on March 8 and check 157 to Marshall Field for $129.95 on March 18. She would record her transactions as follows:

Check Number	Date	Transaction	Amount of Withdrawal (−)		Amount of Deposit (+)		New Balance	
							23	69
	3/3	Deposit			116	84	140	53
156	3/8	Joe's Market	16	80			123	73
	3/15	Deposit			89	46	213	19
157	3/18	Marshall Field	129	95			83	24

Her ending balance is $83.24 as of March 18. Do you see why?

A Challenge for You

Now suppose that Regina has a balance on March 18 of $83.24. See if you can find her balance after the following eight transactions.

1. A monthly service fee of $3.50 is charged to her account on March 20.
2. A deposit of $301.64 is made on March 22.
3. Check 158 is made out to Sears for $159.95 on March 30.
4. A deposit of $450.36 is made on March 31.
5. Check 159 is made out to Gulf Oil for $46.52 on April 2.
6. Check 160 is made out to Western Bell for $39.64 on April 3.
7. Check 161 is made out to IRS for $561.18 on April 14.
8. A bank charge for $6.95 is billed to Regina for printing new checks on April 18.

Check Number	Date	Transaction	Amount of Withdrawal (−)		Amount of Deposit (+)		New Balance	
							83	24

Write in words.

1. 0.62

2. 136.413

Write in symbols as decimals.

3. Thirty-seven ten-thousandths

4. Twelve and forty-nine thousands

Round as indicated.

5. 1.248 to the nearest tenth

6. 18.621 to the nearest hundredth

7. 7.0698 to the nearest thousandth

8. 28.07639 to the nearest ten-thousandth

9. Write in decimal form: $\dfrac{6}{100}$

10. Write as a common fraction: 0.591

1. _____

2. _____

3. _____

4. _____

5. _____

6. _____

7. _____

8. _____

9. _____

10. _____

11. _____	
12. _____	
13. _____	
14. _____	
15. _____	
16. _____	
17. _____	
18. _____	
19. _____	
20. _____	
21. _____	
22. _____	

Add.

11.
```
    2.63
    1.984
   62.1
 +  4.37
```

12.
```
   18.3
   27.6
   41.2
 + 16.8
```

Subtract.

13. $12 - 0.7639$

14.
```
   15.08
 - 14.67
```

15.
```
   20013.629
 - 18037.098
```

Multiply.

16. 182.36×10

17. $2.0075 \times 10,000$

18.
```
    1.36
 ×  0.4
```

19.
```
    2.9
 × 3.7
```

20.
```
   142.03
 ×   0.52
```

Write in order from smallest to largest.

21. 0.741, 0.714, 0.7, 0.711, 0.74

22. 1.629, 1.63, 1.063, 1.0063, 1.163, 1.613

Divide until there is a remainder of zero.

1. $5\overline{)62.81}$ **2.** $0.008\overline{)0.029232}$

3. $0.28\overline{)1.008}$ **4.** $1.2\overline{)36.84}$

Divide and round your answer to the nearest tenth.

5. $1.7\overline{)62.58}$

Divide and round your answer to the nearest thousandth.

6. $0.9\overline{)0.023}$

Write each fraction as a decimal.

7. $\dfrac{9}{200}$ **8.** $\dfrac{15}{32}$ **9.** $2\dfrac{7}{8}$

1. _____

2. _____

3. _____

4. _____

5. _____

6. _____

7. _____

8. _____

9. _____

10. _____

Write each fraction as a decimal rounded to the nearest thousandth.

10. $\dfrac{12}{7}$

11. $\dfrac{13}{17}$

11. _____

Write each fraction as a repeating decimal.

12. $\dfrac{5}{18}$

13. $\dfrac{13}{22}$

12. _____

13. _____

14. What is the area of a rectangle that measures 1.56 centimeters long and 1.04 centimeters wide?

15. Juanita works for 37 hours at $6.15 per hour. How much does she get paid?

14. _____

16. On a 3-day trip Won Lin purchased 13.6, 12.5, and 17.2 gallons of gas. He traveled 1055 miles. How many miles per gallon did his car get? Round your answer to the nearest tenth of a mile per gallon.

17. Bob can buy a stereo system for $379.95 cash, or he can make installment payments of $42.50 for 12 months. How much more does it cost to buy the stereo using installment payments?

15. _____

16. _____

18. Thomson paid $4.20 for 12 pounds of apples. Mary paid $5.04 for 14 pounds of apples. How much more _per pound_ did Mary pay than Thomson?

17. _____

18. _____

1. Write a word name for the decimal: 0.263.

2. Express this fraction as a decimal: $\frac{7899}{10,000}$.

Write in fractional notation. Reduce whenever possible.

3. 5.62

4. 0.385

5. Place the set of numbers in the proper order from smallest to largest.

1.09, 1.91, 1.9, 1.903

6. Round to the nearest hundredth: 69.9793.

7. Round to the nearest ten thousandth: 0.0791846.

Add.

8.
```
   36.152
    0.178
 +  1.32
```

9.
```
    18
     9.6
    15.2
     0.03
 +   1.46
```

1. _____

2. _____

3. _____

4. _____

5. _____

6. _____

7. _____

8. _____

9. _____

10. _____

Subtract.

10.   1.0068
    − 0.9079

11.   26.5
    − 3.218

11. _____

Multiply.

12.   7.92
    × 0.04

13. 5.836 × 100

12. _____

13. _____

Write as a decimal.

14. $\dfrac{17}{15}$

15. $\dfrac{5}{32}$

14. _____

16. Perform the operations in the proper order to evaluate.

$$(0.5)^3 + 2.18 \div 0.05 - 1.08$$

15. _____

16. _____

Solve each problem.

17. A beef roast weighing 5.6 pounds costs $4.25 per pound. How much will the roast cost?

18. Frank traveled from the city to the shore. His odometer was 36,180.5 miles at the start and 36,390.5 miles at the end of the trip. He used 8.5 gallons of gas. How many miles per gallon did his car achieve? Round to the nearest tenth.

17. _____

18. _____

19. The rainfall for this March in Central City was 7.16 centimeters; for April, 4.08 centimeters and for May, 7.49 centimeters. The "normal rainfall" for these three months is 21 centimeters. How much less rain fell during these three months than usual; that is, how does this year's figure compare with the figure for "normal rainfall"?

19. _____

Approximately one-half of this test is based on Chapter 3 material. The remainder is based on material covered in Chapters 1 and 2.

1. Write in words: 38,056,954.

2. Add: 156,028
301,579
+ 21,980

3. Subtract: 1,091,000
− 1,036,520

4. Multiply: 589
×   67

5. Divide: $17 \overline{)\, 4386}$.

6. Evaluate: $20 \div 4 + 2^5 - 7 \times 3$.

7. Reduce the fraction: $\dfrac{33}{88}$.

8. Add: $4\dfrac{1}{3} + 3\dfrac{1}{6}$.

9. Subtract: $\dfrac{23}{35} - \dfrac{2}{5}$.

10. Multiply: $\dfrac{7}{10} \times \dfrac{5}{3}$.

11. Divide: $52 \div 3\dfrac{1}{4}$.

12. Divide: $1\dfrac{3}{8} \div \dfrac{5}{12}$.

1. _____

2. _____

3. _____

4. _____

5. _____

6. _____

7. _____

8. _____

9. _____

10. _____

11. _____

12. _____

13. _____

14. _____

15. _____

16. _____

17. _____

18. _____

19. _____

20. _____

21. _____

22. _____

23. _____

24. _____

13. Write as a decimal: $\dfrac{571}{1000}$.

14. Place in order from smallest to largest: 2.1, 20.1, 2.01, 2.12, 2.11.

15. Round to the nearest thousandth: 26.07984.

16.
1.9
2.36
15.2
+ 0.08

17.
28.007
− 19.368

18.
56.8
× 0.02

19. 365.123 × 100

20. $0.06 \overline{\smash{)}0.06348}$

21. Write as a decimal: $\dfrac{13}{16}$.

22. Perform the operations in the proper order.

$$1.44 \div 0.12 + (0.3)^3 + 1.57$$

23. A car gets 28.5 miles per gallon. The gas tank holds 16.0 gallons of gas. Approximately how many miles can the car be driven on one tank of gas?

24. Sue has a savings account balance of $199.36. This month she received interest of $1.03. She deposited $166.35 and $93.50. She withdrew money three times, once each for $90.00, $37.49, and $137.18. What will her balance be at the start of the next month?

Ratio and Proportion

A cook sometimes has to use proportions in order to determine how to modify a recipe for a larger or a smaller group.

4

PRETEST CHAPTER 4

If you are familiar with the topics in this chapter, take this test now. Check your answers with those in the back of the book. If an answer was wrong or you couldn't do a problem, study the appropriate section of the chapter.

If you are not familiar with the topics in this chapter, don't take this test now. Instead, study the examples, work the practice problems, and then take the test.

This test will help you identify those concepts that you have mastered and those that need more study.

Section 4.1   Write each *ratio* in simplest form.

1. 15 to 17   **2.** 35 to 165

3. $68 to $14   **4.** 124 kilograms to 136 kilograms

5. Sam earns $210 per week take-home pay, from which $65 per week is withheld for federal taxes and $18 per week is withheld for state taxes.
  (a) Find the ratio of federal withholding to take-home pay.
  (b) Find the ratio of state withholding to take-home pay.

Write each *rate* in simplest form.

6. $156 for 48 cabinets

7. 580 gallons of water for every 720 square feet of lawn.

Write as a *unit rate*. Round to the nearest tenth if necessary.

8. Traveled 126 miles in 4 hours. What is the rate in miles per hour?

9. Purchased 14 radios for $546. What is the cost per radio?

Section 4.2   Write a proportion for each of the following.

10. 23 is to 50 as 69 is to 150.   **11.** 54 is to 66 as 27 is to 33.

Determine if each proportion is true or false.

12. $\dfrac{17}{38} = \dfrac{51}{114}$ **13.** $\dfrac{12}{42} = \dfrac{16}{54}$

Section 4.3   Solve for *n* in each equation.

14. $7 \times n = 112$ **15.** $195 = 13 \times n$

Solve for *n* in each proportion.

16. $\dfrac{32}{20} = \dfrac{8}{n}$ **17.** $\dfrac{3}{165} = \dfrac{n}{275}$ **18.** $\dfrac{n}{600} = \dfrac{10}{12.5}$

Section 4.4   Solve each problem using a proportion.

19. A recipe for four people needs 1.5 cups of flour. How many cups of flour are needed for a recipe for 18 people?

20. Maria's car can travel 63 miles on 2 gallons of gas. How many miles can she travel on 7 gallons of gas?

21. Two cities are 7 inches apart on a map, but the actual distance between them is 385 miles. What is the actual distance between two other cities that are 3 inches apart on the map?

22. A shipment of 144 light bulbs had seven defective ones. How many defective ones would we expect, if the same rate holds, in a shipment of 1296 light bulbs?

4.1 RATIOS AND RATES

After studying this section, you will be able to:
1 *Use a ratio to compare two quantities with the same unit*
2 *Use a rate to compare two quantities with different units*

We have learned how to add, subtract, multiply, and divide with whole numbers, fractions, and decimals. We learn now how to compare quantities. The notion of comparison leads us to two important ideas. A *ratio* compares two quantities that have the same units. A *rate* compares two quantities that have different units.

Assume that you earn 13 dollars an hour and your friend earns 10 dollars per hour. The *ratio* 13:10 compares what you and your friend make. This ratio means simply that for every 13 dollars you earn, your friend earns 10. The *rate* you are paid is 13 dollars per hour, which measures one numerical quantity—13 dollars—in terms of another quantity—1 hour. In this section we see how to use both rates and ratios. Later we use them to solve many everyday problems.

1 Using a Ratio to Compare Two Quantities with the Same Unit

Suppose that we want to compare an object weighing 20 pounds to an object weighing 23 pounds. The ratio of their weight would be 20 to 23. A **ratio** is the comparison of two quantities that have the same units.

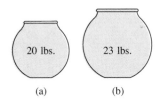

20 lbs.   23 lbs.

(a)   (b)

We can express the ratio three ways:

We can write "the ratio of 20 to 23."

We can write "20:23" using a colon.

We can write $\dfrac{\text{“20”}}{23}$ using a fraction.

All are valid ways to *compare* 20 to 23. Each is read as "20 to 23."

> We always want to write a ratio in simplest form. A ratio is in *simplest form* when the two numbers do not have a common factor.

Example 1 Write each ratio in simplest form. Express your answer as a fraction.

(a) The ratio of 15 hours to 20 hours
(b) The ratio of 36 hours to 30 hours
(c) 125:150

(a) $\dfrac{15}{20} = \dfrac{3}{4}$ **(b)** $\dfrac{36}{30} = \dfrac{6}{5}$ **(c)** $\dfrac{125}{150} = \dfrac{5}{6}$

Notice that in each case the two numbers *do* have a common factor. When we form the fraction, that is, the ratio, we take the extra step to *reduce* the fraction. ■

Practice Problem 1 Write each ratio in simplest form. Express your answer as a fraction.
(a) The ratio of 36 feet to 40 feet
(b) The ratio of 18 feet to 15 feet
(c) 220:270 ■

Example 2 Marlin earns $350 weekly. However, he takes home only $250 per week from his paycheck.

$350.00   gross pay   (what Marlin earns)

45.00	withheld for federal tax ⎫	⎛ what is taken out ⎞
20.00	withheld for state tax ⎬	⎝ of Marlin's earnings ⎠
35.00	withheld for retirement ⎭	

$250.00   take-home pay   (what Marlin has left)

(a) What is the ratio of the amount withheld for federal tax to gross pay?

(b) What is the ratio of the amount withheld for state tax to the amount withheld for federal tax?

(a) The ratio of the amount withheld for federal tax compared to gross pay is

$$\frac{45}{350} = \frac{9}{70}$$

(b) The ratio of the amount withheld for state tax compared to the amount withheld for federal tax is

$$\frac{20}{45} = \frac{4}{9}$$ ∎

Practice Problem 2 Professor Fowler has 90 students in his course on introductory psychology. The class has 18 freshman students, 25 sophomore students, 36 junior students, and 11 senior students.

(a) What is the ratio of freshman students to junior students?

(b) What is the ratio of sophomore students to the entire number of students in the class? ∎

2 A **rate** is a comparison of two quantities with different units. Usually, to avoid misunderstanding, we express a rate as a fraction with the units included.

Example 3 The college has 2112 students with 128 faculty. What is the rate of students to faculty?

The rate is

$$\frac{2112 \text{ students}}{128 \text{ faculty}} = \frac{33 \text{ students}}{2 \text{ faculty}}$$ ∎

Practice Problem 3 A farmer is charged $44 storage for every 900 tons of grain. What is the rate of storage dollars to tons of grain? ∎

Often, we want the rate for a single unit, which is the unit rate. A *unit rate* is a rate in which the denominator is the number 1.

Example 4 The car traveled 301 miles in 7 hours. Find the unit rate.

$\dfrac{301}{7}$ can be simplified. We divide $301 \div 7 = 43$.

Thus

$$\frac{301 \text{ miles}}{7 \text{ hours}} = \frac{43 \text{ miles}}{1 \text{ hour}}$$

The denominator is 1. We write our answer as 43 miles/hour. The fraction line is read as the word "per," so our answer here is read "43 miles per hour." "Per" means "for every," so a rate of 43 miles per hour means 43 miles traveled for every hour traveled. ∎

Practice Problem 4 The car traveled 212 miles in 4 hours. Find the unit rate. ∎

Example 5   A grocer purchased 200 pounds of apples for $34. He sold the 200 pounds of apples for $86. How much profit did he make per pound of apples?

$$
\begin{array}{rl}
\$86 & \text{selling price} \\
-\ 34 & \text{cost} \\
\hline
\$52 & \text{profit}
\end{array}
$$

The rate that compares profit to pounds of apples sold is $\dfrac{\$52 \text{ dollars}}{200 \text{ pounds}}$. We will divide $52 \div 200$:

$$
\begin{array}{r}
.26 \\
200\overline{)\,52.00} \\
\underline{400} \\
1200 \\
\underline{1200} \\
0
\end{array}
$$

The unit rate of profit is $0.26 per pound.   ■

Practice Problem 5   A retailer purchased 120 nickel-cadmium batteries for flash-lights for $129.60. She sold them for $170.40. What was her profit per battery?   ■

Example 6   Hamburger at a local butchery is packaged in large and extra-large packages. A large package costs $7.86 for 6 pounds and an extra-large package is $10.08 for 8 pounds.

(a) What is the unit rate in dollars per pound for each size package?
(b) How much per pound does a person save by buying the extra-large package?

(a) $\dfrac{7.86 \text{ dollars}}{6 \text{ pounds}} = \$1.31/\text{pound}$ for the large package

$\dfrac{10.08 \text{ dollars}}{8 \text{ pounds}} = \$1.26/\text{pound}$ for the extra-large package

(b)
$$
\begin{array}{rl}
\$1.31 & \\
-\ 1.26 & \\
\hline
\$0.05 & \text{A person saves \$0.05/pound by buying the extra-large package.}
\end{array}
$$

   ■

Practice Problem 6   A 12-ounce package of Fred's favorite cereal costs $2.04. A 20-ounce package of the same cereal costs $2.80.
(a) What is the cost per ounce of each size of cereal?
(b) How much per ounce does a person save by buying the larger size?   ■

To Think About

Perhaps you have heard statements like "a certain jet plane travels at Mach 2.2." Let's investigate what that means. A Mach number is the ratio of the velocity of an object to the velocity of sound, and the resulting fraction is expressed with a denominator of 1.
   What is the Mach number of a jet traveling at 690 meters per second? Sound travels at about 330 meters per second.

$$\text{Mach number of jet} = \frac{690 \text{ meters per second}}{330 \text{ meters per second}}$$

$$= 2.09090909 \ldots \quad \text{or} \quad 2.\overline{09}$$

Rounded to the nearest tenth, the Mach number of the jet is 2.1.

What is the Mach number of a rocket traveling at 2145 meters per second?

$$\text{Mach number of rocket} = \frac{2145 \text{ meters per second}}{330 \text{ meters per second}}$$

$$= 6.5$$

The Mach number of the rocket is 6.5.
Problems 63 and 64 also deal with this concept. ■

EXERCISES 4.1

Write each ratio in simplest form. Express your answer as a fraction.

1. 16:24   **2.** 36:30   **3.** 21:18   **4.** 42:50   **5.** 55:121   **6.** 78:104

7. 150:225   **8.** 360:480   **9.** 60 to 64   **10.** 33 to 54   **11.** 28 to 42   **12.** 21 to 98

13. 32 to 20   **14.** 90 to 63   **15.** 8 ounces to 12 ounces

16. 24 minutes to 50 minutes   **17.** 12 yards to 40 yards   **18.** 15 miles to 35 miles

19. 50 years to 85 years   **20.** 39 kilograms to 26 kilograms   **21.** $86 to $120

22. $45 to $63   **23.** 153 inches to 17 inches   **24.** 143 feet to 13 feet

25. 91 tons to 133 tons   **26.** 99 pounds to 126 pounds

Use the following table to answer problems 27–30.

ROBIN'S WEEKLY PAYCHECK

Total (Gross Pay)	Federal Withholding	State Withholding	Retirement	Insurance	Take-Home Pay
$285	$35	$20	$28	$16	$165

27. What is the ratio of take-home pay to total (gross) pay?

28. What is the ratio of retirement to insurance?

29. What is the ratio of federal withholding to take-home pay?

30. What is the ratio of retirement to total (gross pay)?

An automobile insurance company prepared the following analysis for its clients. Use this table for problems 31–34.

ANALYSIS OF NUMBER OF YEARS THAT FOUR-DOOR SEDANS ARE DRIVEN

Sedans That Lasted 2 Years or Less	Sedans that Lasted 4 Years or Less but More Than 2 Years	Sedans that Lasted 6 Years or Less but More Than 4 Years	Sedans That Lasted More Than 6 Years	Total Number of Sedans
205	255	450	315	1225

31. What is the ratio of sedans that lasted two years or less to the total number of sedans?

32. What is the ratio of sedans that lasted more than six years to the total number of sedans?

33. What is the ratio of the number of sedans that lasted six years or less but more than four years to the number of sedans that lasted two years or less?

34. What is the ratio of the number of sedans that lasted more than six years to the number of sedans that lasted four years or less but more than two years?

35. A builder constructs a new home for $128,000. His labor costs were $72,000. What is the ratio of labor cost to total cost?

36. A company produces a new model television set for a total cost of $245.00. The research and development cost for each set is $70. What is the ratio of research and development cost to the total cost?

Write as a rate in simplest form.

37. $160 for 12 chairs

38. $135 for 20 cabinets

39. 82 pounds for 18 people

40. 98 pounds for 22 people

41. 310 gallons of water for every 625 square feet of lawn

42. 460 gallons of water for every 800 square feet of lawn

43. 6150 revolutions for every 15 miles

44. 9540 revolutions for every 18 miles

45. Snowplow 18 miles of road every 8 hours

46. Snowplow 26 miles of road every 12 hours

Write as a unit rate. Round to the nearest tenth when necessary.

47. Earn $520 in 40 hours

48. Earn $266 in 38 hours

49. Travel 192 miles on 12 gallons of gas

50. Travel 322 miles on 14 gallons of gas

51. Pump 3675 gallons in 15 hours

52. Pump 3520 gallons in 16 hours

53. 2760 words on 12 pages

54. 1935 words on 9 pages

55. Travel 3619 kilometers in 7 hours

56. Travel 2992 kilometers in 11 hours

57. $4320 was spent for 720 shares of stock. Find the cost per share.

58. $5120 was spent for 640 shares of stock. Find the cost per share.

59. A retailer purchased 136 portable radios for $1496. He sold them for $2040. How much profit did he make per radio?

60. A retailer purchased 26 compact disc players for $3380. She sold them for $4030. How much profit did she make per compact disc player?

61. A 12-ounce jar of jelly costs $1.32. An 18-ounce jar of the same jelly costs $1.62.
   (a) What is the cost of each jar of jelly per ounce?
   (b) How much does the consumer save per ounce with the larger jar?

62. A 16-ounce can of beef stew costs $2.88. A 26-ounce can of the same beef stew costs $4.16.
   (a) What is the cost of each can of stew per ounce?
   (b) How much does the consumer save per ounce with the larger can?

 To Think About

For Problems 63 and 64, recall that the speed of sound is 330 meters per second. Round your answers to the nearest tenth.

63. A jet plane was originally designed to fly at 750 meters per second. It was modified to fly at 810 meters per second. By how much was its Mach number *increased*?

64. A rocket was first flown at 1960 meters per second. It proved unstable and unreliable at that speed. It is now flown at a maximum of 1920 meters per second. By how much was its Mach number decreased?

65. A room 11 yards × 4.8 yards was installed with a carpet. The bill was $554.40. What was the cost of the installed carpet per yard?

66. A room 12 yards × 5.2 yards was installed with a carpet. The bill was $764.40. What was the cost of the installed carpet per yard?

Cumulative Review Problems

Calculate.

67. $2\frac{1}{4} + \frac{3}{8}$

68. $\frac{5}{7} \div \frac{3}{21}$

69. $\frac{8}{23} \times \frac{5}{16}$

70. $3\frac{1}{16} - 2\frac{3}{4}$

 Calculator Problems

71. Mr Jenson bought 525 shares of stock for $12,876.50. How much did he pay per share? Round to the nearest cent.

72. A retailer purchased 275 television sets for $97,650. How much did the retailer pay per set?

For Extra Practice Examples and Exercises, turn to page 237.

Solutions to Odd-Numbered Practice Problems

1. (a) $\frac{36}{40} = \frac{9}{10}$ **(b)** $\frac{18}{15} = \frac{6}{5}$ **(c)** $\frac{220}{270} = \frac{22}{27}$ **3.** $\frac{44\ \text{dollars}}{900\ \text{tons}} = \frac{11\ \text{dollars}}{225\ \text{tons}}$

5.
Selling cost	$170.40
− purchase price	− 129.60
= profit	$ 40.80

She made a profit of $40.80 on 120 batteries.

$$120\overline{)40.80}$$
$$\underline{360}$$
$$480$$
$$\underline{480}$$
$$0$$

Her profit was $0.34 per battery.

Answers to Even-Numbered Practice Problems

2. (a) $\frac{1}{2}$ **(b)** $\frac{5}{18}$ **4.** 53 miles/hour

6. (a) 12-ounce package costs $0.17 per ounce; 20-ounce package costs $0.14 per ounce.
(b) A person saves $0.03 per ounce by buying the larger size.

4.2 THE CONCEPT OF PROPORTIONS

☐ After studying this section, you will be able to:

"A car traveled 550 miles in 15 hours. A bus traveled 220 miles in 6 hours. Did they travel at the same rate?" Here is a question in which we want to know if two rates are equal. In this section, we'll learn two methods to set up a *proportion*. When you do problem 43 in this section, you'll see that these rates are not equal.

1 *Write a proportion*

2 *Test to see if a proportion is true or false*

1 A **proportion** is a true or false statement that two ratios or two rates are equal. For example, $\frac{5}{8} = \frac{15}{24}$ is a proportion and $\frac{7\ \text{feet}}{8\ \text{dollars}} = \frac{35\ \text{feet}}{40\ \text{dollars}}$ is also a proportion.

A proportion can be read two ways. The proportion $\frac{5}{8} = \frac{15}{24}$ can be read "five eighths equals fifteen twenty-fourths," or it can be read "five is to eight as fifteen is to twenty-four."

Example 1 Write a proportion for the following: 5 is to 7 as 15 is to 21.

$$\frac{5}{7} = \frac{15}{21}$$ ■

Practice Problem 1 Write a proportion for the following: 6 is to 8 as 9 is to 12.

2 Some proportions are true and some are false. A **proportion is *true*** if each side can be reduced to the same fraction. Is $\frac{4}{14} = \frac{6}{21}$? We can reduce $\frac{4}{14} = \frac{2}{7}$ and also reduce $\frac{6}{21} = \frac{2}{7}$. Therefore, $\frac{4}{14} = \frac{6}{21}$ is a *true* proportion. You recall that in Section 2.2 we developed the equality test for fractions. ■

> **Equality Test for Fractions**
>
> For any two fractions where a, b, c, and d are whole numbers and $b \neq 0$, $d \neq 0$:
>
> $$\text{If } \frac{a}{b} = \frac{c}{d}, \text{ then } a \times d = b \times c$$

Thus, to see if $\frac{4}{14} = \frac{6}{21}$, we can multiply:

$$\frac{4}{14} \bowtie \frac{6}{21}$$

$14 \times 6 = 84 \longleftarrow$ Products are
$4 \times 21 = 84 \longleftarrow$ equal.

Thus the equality test for fractions can be used to see if we have true or false proportions.

Example 2 Is each proportion true or false? **(a)** $\frac{14}{18} = \frac{35}{45}$ **(b)** $\frac{16}{21} = \frac{174}{231}$

(a) $\frac{14}{18} \stackrel{?}{=} \frac{35}{45}$

$$\frac{14}{18} \bowtie \frac{35}{45}$$

$18 \times 35 = 630$
↑
products equal
↓
$14 \times 45 = 630$

Thus $\frac{14}{18} = \frac{35}{45}$ is true.

(b) $\frac{16}{21} \stackrel{?}{=} \frac{174}{231}$

$$\frac{16}{21} \bowtie \frac{174}{231}$$

$21 \times 174 = 3654$
↑
products not equal
↓
$16 \times 231 = 3696$

Thus $\frac{16}{21} = \frac{174}{231}$ is false. ■

Practice Problem 2 Is each proportion true or false?

(a) $\frac{10}{18} = \frac{25}{45}$ **(b)** $\frac{42}{100} = \frac{22}{55}$ ■

The proportions may involve fractions or decimals.

Example 3 Is each proportion true or false? **(a)** $\dfrac{5.5}{7} = \dfrac{33}{42}$ **(b)** $\dfrac{5}{8\frac{3}{4}} = \dfrac{40}{72}$

(a) $\dfrac{5.5}{7} \stackrel{?}{=} \dfrac{33}{42}$

$$7 \times 33 = 231$$

$\dfrac{5.5}{7} \diagdown\!\!\!\!\diagup \dfrac{33}{42}$ products equal   Thus $\dfrac{5.5}{7} = \dfrac{33}{42}$ is true.

$$5.5 \times 42 = 231$$

(b) $\dfrac{5}{8\frac{3}{4}} \stackrel{?}{=} \dfrac{40}{72}$. To multiply $8\dfrac{3}{4} \times 40$ we use

$$\dfrac{35}{\cancel{4}} \times \cancel{40}^{10} = 35 \times 10 = 350$$
$$\;_1$$

Thus

$\dfrac{5}{8\frac{3}{4}} \diagdown\!\!\!\!\diagup \dfrac{40}{72}$ $8\dfrac{3}{4} \times 40 = 350$ products not equal   Thus $\dfrac{5}{8\frac{3}{4}} = \dfrac{40}{72}$ is false. ■

$$5 \times 72 = 360$$

Practice Problem 3   Is each following proportion true or false?

(a) $\dfrac{2.4}{3} = \dfrac{12}{15}$ **(b)** $\dfrac{2\frac{1}{3}}{6} = \dfrac{14}{38}$ ■

Example 4   Is the rate $\dfrac{\$86}{13 \text{ tons}}$ equal to the rate $\dfrac{\$79}{12 \text{ tons}}$?

We want to know whether $\dfrac{86}{13} \stackrel{?}{=} \dfrac{79}{12}$.

$$13 \times 79 = 1027$$

$\dfrac{86}{13} \diagdown\!\!\!\!\diagup \dfrac{79}{12}$ products not equal

$$86 \times 12 = 1032$$

Thus the two rates are not equal. ■

Practice Problem 4   Is the rate $\dfrac{1260 \text{ words}}{7 \text{ pages}}$ equal to the rate $\dfrac{3530 \text{ words}}{20 \text{ pages}}$? ■

To Think About

?

Why don't we reduce each fraction instead of using the equality test for fractions to see if a proportion is true or false? Certainly, we can use either procedure. However, reducing fractions often takes longer. Suppose that you want to know if $\dfrac{78}{91} \stackrel{?}{=} \dfrac{102}{119}$. It takes most people a lot longer to reduce $\dfrac{78}{91}$ and $\dfrac{102}{119}$ than it does to multiply 78×119 and 91×102. Try it each way, and time yourself to the nearest second. What do you conclude? Problems 45 and 46 provide more practice problems of this type. ■

EXERCISES 4.2

Write a proportion for each.

1. 48 is to 32 as 3 is to 2.

2. 40 is to 36 as 10 is to 9.

3. 9 is to 26 as 18 is to 52.

4. 27 is to 48 as 18 is to 32.

5. 20 is to 36 as 5 is to 9.

6. 60 is to 64 as 15 is to 16.

7. 27 is to 15 as 9 is to 5.

8. 42 is to 28 as 6 is to 4.

9. 44 is to 60 as 22 is to 30.

10. 28 is to 44 as 14 is to 22.

11. 45 is to 135 as 9 is to 27.

12. 54 is to 72 as 12 is to 16.

13. 5.5 is to 10 as 11 is to 20.

14. 6.5 is to 14 as 13 is to 28.

Determine if each proportion is true or false.

15. $\dfrac{21}{35} = \dfrac{15}{25}$

16. $\dfrac{12}{42} = \dfrac{10}{35}$

17. $\dfrac{18}{12} = \dfrac{42}{28}$

18. $\dfrac{14}{10} = \dfrac{56}{40}$

19. $\dfrac{27}{45} = \dfrac{22}{40}$

20. $\dfrac{48}{56} = \dfrac{40}{45}$

21. $\dfrac{12}{16} = \dfrac{15}{20}$

22. $\dfrac{8}{9} = \dfrac{48}{54}$

23. $\dfrac{92}{100} = \dfrac{47}{50}$

24. $\dfrac{17}{75} = \dfrac{22}{100}$

25. $\dfrac{102}{120} = \dfrac{85}{100}$

26. $\dfrac{315}{2100} = \dfrac{15}{100}$

27. $\dfrac{6}{14} = \dfrac{4.5}{10.5}$

28. $\dfrac{8}{9} = \dfrac{6}{4.5}$

29. $\dfrac{11}{12} = \dfrac{9.5}{10}$

30. $\dfrac{3}{17} = \dfrac{4.5}{24.5}$

31. $\dfrac{2\frac{1}{3}}{3} = \dfrac{7}{15}$

32. $\dfrac{2}{4\frac{1}{3}} = \dfrac{6}{13}$

33. $\dfrac{9}{22} = \dfrac{3}{7\frac{1}{3}}$

34. $\dfrac{8}{19} = \dfrac{2}{4\frac{3}{4}}$

35. $\dfrac{75 \text{ feet}}{4 \text{ rolls}} = \dfrac{150 \text{ feet}}{8 \text{ rolls}}$

36. $\dfrac{68 \text{ feet}}{3 \text{ rolls}} = \dfrac{204 \text{ feet}}{9 \text{ rolls}}$

37. $\dfrac{286 \text{ gallons}}{12 \text{ acres}} = \dfrac{429 \text{ gallons}}{18 \text{ acres}}$

38. $\dfrac{166 \text{ gallons}}{14 \text{ acres}} = \dfrac{249 \text{ gallons}}{21 \text{ acres}}$

39. $\dfrac{82 \text{ miles}}{35 \text{ dollars}} = \dfrac{123 \text{ miles}}{53 \text{ dollars}}$

40. $\dfrac{156 \text{ feet}}{62 \text{ cents}} = \dfrac{234 \text{ feet}}{92 \text{ cents}}$

41. In Accounting I, section A, there are 18 female students and 14 male students. In Accounting I, section B, there are 27 female students and 21 male students. Is the ratio of female students to male students the same in each section of Accounting I?

42. In General Psychology, section C, there are 25 male students and 35 female students. In section B there are 30 male students and 42 female students. Is the ratio of male students to female students the same in each section of General Psychology?

43. A car traveled 550 miles in 15 hours. A bus traveled 230 miles in 6 hours. Did they travel at the same rate?

44. A machine folds 730 boxes in 6 hours. Another machine folds 1090 boxes in 9 hours. Do they fold boxes at the same rate?

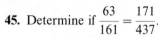

? To Think About

45. Determine if $\dfrac{63}{161} = \dfrac{171}{437}$.

    (a) By reducing each side to lowest terms.
    (b) By using the equality test for fractions.
    (c) Which method was faster? Why?

46. Determine if $\dfrac{169}{221} = \dfrac{247}{323}$.

    (a) By reducing each side to lowest terms
    (b) By using the equality test for fractions.
    (c) Which method was faster? Why?

47. Determine if $\dfrac{0.24}{0.84} = \dfrac{15.3}{53.55}$.

48. Determine if $\dfrac{4.8}{20.8} = \dfrac{8.49}{36.79}$.

Cumulative Review Problems

49. $9.6 + 7.8 + 2.56$
$+ 3.004 + 0.1765$

50. 2.83×5.002

51. $\begin{aligned} &29{,}366.215 \\ -\,&28{,}963.807 \end{aligned}$

52. $7.13\overline{)1.83954}$

 Calculator Problems

Determine if the following proportions are true or false.

53. $\dfrac{12.6}{25.2} = \dfrac{143.1}{286.2}$

54. $\dfrac{285}{305} = \dfrac{19.8}{22.6}$

For Extra Practice Examples and Exercises, turn to page 237.

Solutions to Odd-Numbered Practice Problems

1. 6 is to 8 as 9 is to 12.

$$\frac{6}{8} = \frac{9}{12}$$

3. (a) $\dfrac{2.4}{3} \overset{?}{=} \dfrac{12}{15}$

$3 \times 12 = 36$
↑
products equal
↓
$2.4 \times 15 = 36$

Thus $\dfrac{2.4}{3} = \dfrac{12}{15}$ is true.

(b) $\dfrac{2\frac{1}{3}}{6} \overset{?}{=} \dfrac{14}{38}$

$2\dfrac{1}{3} \times 38 = \dfrac{7}{3} \times \dfrac{38}{1} = \dfrac{266}{3} = 88\dfrac{2}{3}$

$6 \times 14 = 84$

products not equal

$2\dfrac{1}{3} \times 38 = 88\dfrac{2}{3}$

Thus $\dfrac{2\frac{1}{3}}{6} = \dfrac{14}{38}$ is false.

Answers to Even-Numbered Practice Problems

2. (a) True   **(b)** False   **4.** No. The two rates are not equal.

4.3 SOLVING PROPORTIONS

In some proportions we may know three quantities and wish to know the fourth, which is an unknown quantity. For example, an experiment may call for 5 grams of salt for 8 liters of water, which is the ratio

$$\frac{5 \text{ grams}}{8 \text{ liters}}$$

Suppose that we want to expand the experiment to 12 liters of water. How many grams of salt would we need to keep the proportion constant? We set up the proportion

$$\frac{5 \text{ grams}}{8 \text{ liters}} = \frac{n \text{ grams}}{12 \text{ liters}}$$

How do we solve for *n*? (See Example 6.) In this section we learn how to handle the numbers in a proportion to solve for the unknown value.

1 Consider this expression: "3 times a number yields 15. What is the number?" We could write this as

$$3 \times \boxed{?} = 15$$

and guess that the number $\boxed{?} = 5$. There is a better way of solving this problem, a way that eliminates the guesswork. Let the letter *n* represent the unknown number. We write

$$3 \times n = 15$$

This is called an **equation**. An equation has an equals sign. We want to find the number *n* in this equation without guessing. We will not change the value of *n* in the equation if we divide both sides of the equation by 3. Thus if

$$3 \times n = 15$$

we can say

$$\frac{3 \times n}{3} = \frac{15}{3}$$

which is

$$\frac{3}{3} \times n = 5$$

or

$$1 \times n = 5$$

Since 1 × any number is the same number, we know that *n* = 5. Any equation of the form *a* × *n* = *b* can be solved in this way. We divide both sides of an equation of the form *a* × *n* = *b* by the number that is multiplied by *n*. (The method will not work for ¬3 + *n* = 15, since the 3 is added to *n* and not multiplied by *n*.)

Example 1 Solve each equation to find the value of *n*.

(a) $16 \times n = 80$ **(b)** $24 \times n = 240$

(a) $16 \times n = 80$

$$\frac{16 \times n}{16} = \frac{80}{16}$$ Dividing each side by 16.

$$n = 5$$ Because $16 \div 16 = 1$ and $80 \div 16 = 5$.

(b) $24 \times n = 240$

$$\frac{24 \times n}{24} = \frac{240}{24}$$ Dividing each side by 24.

$$n = 10$$ Because $24 \div 24 = 1$ and $240 \div 24 = 10$. ∎

Practice Problem 1   Solve each equation to find the value of n.

(a) $5 \times n = 45$ **(b)** $7 \times n = 84$ ■

The same procedure is followed if the variable n is on the right side of the equation.

Example 2   Solve each equation to find the value of n.

(a) $66 = 11 \times n$ **(b)** $143 = 13 \times n$

(a) $66 = 11 \times n$ **(b)** $143 = 13 \times n$

$$\frac{66}{11} = \frac{11 \times n}{11}$$ Dividing each side by 11. $$\frac{143}{13} = \frac{13 \times n}{13}$$ Dividing each side by 13.

$6 = n$ Simplifying. $11 = n$ Simplifying.   ■

Practice Problem 2   Solve each equation to find the value of n.

(a) $108 = 9 \times n$ **(b)** $210 = 14 \times n$ ■

The numbers in the equations are not always whole numbers, and the answer to an equation is not always a whole number.

Example 3   Solve each equation to find the value of n.

(a) $16 \times n = 56$ **(b)** $18.2 = 2.6 \times n$

(a) $16 \times n = 56$ **(b)** $18.2 = 2.6 \times n$

$$\frac{16 \times n}{16} = \frac{56}{16}$$ Dividing each side by 16. $$\frac{18.2}{2.6} = \frac{2.6 \times n}{2.6}$$ Dividing each side by 2.6.

$n = 3.5$

$$\begin{array}{r} 3.5 \\ 16 \overline{)\ 56.0} \\ 48 \\ \hline 80 \\ 80 \\ \hline 0 \end{array}$$

$7 = n$

$$\begin{array}{r} 7. \\ 2.6_\wedge \overline{)\ 18.2_\wedge} \\ 18\ 2 \\ \hline 0 \end{array}$$ ■

Practice Problem 3   Solve each equation to find the value of n.

(a) $15 \times n = 63$ **(b)** $39.2 = 5.6 \times n$ ■

2 Find the Missing Number in a Proportion

We can write a proportion in which three numbers are known and one is unknown using the variable n as the unknown quantity. We work the equation to get it into the form $a \times n = b$ and solve for n.

Suppose that we want to know the value of n in the proportion

$$\frac{5}{12} = \frac{n}{144}$$

We multiply $5 \times 144 = 12 \times n$. This is sometimes called forming the "cross product" or "cross-multiplying." Now we have

$$720 = 12 \times n$$

Next we divide both sides by 12.

$$\frac{720}{12} = \frac{12 \times n}{12}$$

$$60 = n$$

Can we check to see if this is correct? Do we have a true proportion?

$$\frac{5}{12} \overset{?}{=} \frac{60}{144}$$

$$\frac{5}{12} \diagdown \frac{60}{144} \qquad \begin{array}{l} 12 \times 60 = 720 \leftarrow \\ \\ 5 \times 144 = 720 \leftarrow \end{array} \Bigg] \text{products equal}$$

Thus $\dfrac{5}{12} = \dfrac{60}{144}$ is true. We have checked our answer.

To Solve for a Missing Number in a Proportion

1. Find the two diagonal products (also called "cross-multiplying" or "forming a cross product").
2. Divide each side of the equation by the number multiplied by n.
3. Simplify the result.
4. Check your answer.

Example 4 Find the value of n in $\dfrac{25}{4} = \dfrac{n}{12}$.

$$25 \times 12 = 4 \times n \qquad \text{Finding the two diagonal products.}$$

$$300 = 4 \times n$$

$$\frac{300}{4} = \frac{4 \times n}{4} \qquad \text{Dividing each side by 4.}$$

$$75 = n \qquad \text{Simplifying.}$$

Check whether the proportion is true.

$$\frac{25}{4} \overset{?}{=} \frac{75}{12}$$

$$25 \times 12 \overset{?}{=} 4 \times 75$$

$$300 = 300 \checkmark$$

The proportion is true. The answer $n = 75$ is correct. ∎

Practice Problem 4 Find the value of n in $\dfrac{24}{n} = \dfrac{3}{7}$.

The answer to the problem may not be a whole number. ∎

Example 5 Find the value of n in $\dfrac{125}{2} = \dfrac{150}{n}$.

$$125 \times n = 2 \times 150 \qquad \text{Finding the two}$$
$$\text{diagonal products.}$$
$$125 \times n = 300$$

$$\frac{125 \times n}{125} = \frac{300}{125} \qquad \text{Dividing each side by 125.}$$

$$n = 2.4$$

$$\begin{array}{r} 2.4 \\ 125 \overline{)\ 300.0} \\ 250 \\ \hline 500 \\ 500 \\ \hline 0 \end{array}$$

$$Check: \quad \frac{125}{2} \overset{?}{=} \frac{150}{2.4}$$

$$125 \times 2.4 \overset{?}{=} 2 \times 150$$

$$300 = 300 \checkmark \quad ∎$$

Practice Problem 5 Find the value of n in $\dfrac{176}{4} = \dfrac{286}{n}$. ■

In some applied situations it is helpful to keep the labels of the units involved in the proportion.

Example 6 Find the value of n in $\dfrac{n \text{ grams}}{12 \text{ liters}} = \dfrac{5 \text{ grams}}{8 \text{ liters}}$.

$$8 \times n = 12 \times 5$$

$$8 \times n = 60$$

$$\dfrac{8 \times n}{8} = \dfrac{60}{8}$$

$$n = 7.5$$

The answer is 7.5 grams.   *Check:* $\dfrac{7.5 \text{ grams}}{12 \text{ liters}} = \dfrac{5 \text{ grams}}{8 \text{ liters}}$

$$7.5 \times 8 \overset{?}{=} 12 \times 5$$

$$60 = 60 \checkmark \quad ■$$

Practice Problem 6 Find the value of n in $\dfrac{80 \text{ dollars}}{5 \text{ tons}} = \dfrac{n \text{ dollars}}{6 \text{ tons}}$. ■

?

To Think About

Suppose that the proportion involved all fractional quantities. Could you still follow all the steps? Consider the case of finding n when

$$\frac{n}{3\frac{1}{4}} = \frac{5\frac{1}{6}}{2\frac{1}{3}}$$

We would have

$$2\frac{1}{3} \times n = 5\frac{1}{6} \times 3\frac{1}{4}$$

This can be written as

$$\frac{7}{3} \times n = \frac{31}{6} \times \frac{13}{4}$$

$$\boxed{\frac{7}{3} \times n = \frac{403}{24}} \qquad \text{equation (1)}$$

Now we divide each side of equation (1) by $\dfrac{7}{3}$. Do you see why?

$$\frac{\frac{7}{3} \times n}{\frac{7}{3}} = \frac{\frac{403}{24}}{\frac{7}{3}}$$

$$n = \frac{\frac{403}{24}}{\frac{7}{3}}$$

Be careful here. The right-hand side means $\frac{403}{24} \div \frac{7}{3}$, which we evaluate by *inverting* the second fraction and multiplying:

$$\frac{403}{\overset{}{\underset{8}{24}}} \times \frac{\overset{1}{3}}{7} = \frac{403}{56}$$

Thus $n = \frac{403}{56}$ or $7\frac{11}{56}$. Think about all the steps to solving this problem. Can you follow them? There is another way to do the problem. We could multiply each side of equation (1) by $\frac{3}{7}$. Do you see that this would yield the same results? Now then try problems 59–62. ■

EXERCISES 4.3

Solve each equation to find the value of n.

1. $5 \times n = 40$ **2.** $7 \times n = 56$ **3.** $84 = 2 \times n$ **4.** $69 = 3 \times n$ **5.** $6 \times n = 96$

6. $8 \times n = 136$ **7.** $117 = 9 \times n$ **8.** $115 = 5 \times n$ **9.** $7 \times n = 182$ **10.** $6 \times n = 144$

11. $3 \times n = 16.8$ **12.** $2 \times n = 19.6$ **13.** $2.6 \times n = 13$ **14.** $3.2 \times n = 48$ **15.** $40.6 = 5.8 \times n$

16. $56.7 = 6.3 \times n$ **17.** $260 \times n = 1430$ **18.** $180 \times n = 936$

Find the value of n in each proportion. Check your answer.

19. $\frac{n}{25} = \frac{4}{5}$ **20.** $\frac{n}{28} = \frac{3}{7}$ **21.** $\frac{6}{n} = \frac{3}{8}$ **22.** $\frac{4}{n} = \frac{2}{7}$ **23.** $\frac{12}{40} = \frac{n}{25}$ **24.** $\frac{18}{8} = \frac{n}{10}$ **25.** $\frac{25}{100} = \frac{8}{n}$

26. $\frac{9}{150} = \frac{6}{n}$ **27.** $\frac{n}{32} = \frac{3}{4}$ **28.** $\frac{n}{8} = \frac{225}{3}$ **29.** $\frac{n}{9} = \frac{49}{63}$ **30.** $\frac{n}{42} = \frac{5}{6}$ **31.** $\frac{18}{n} = \frac{3}{11}$ **32.** $\frac{16}{n} = \frac{2}{7}$

Find the value of n in each proportion. Round your answer to the nearest tenth if necessary.

33. $\frac{n}{5} = \frac{7}{4}$ **34.** $\frac{n}{8} = \frac{6}{5}$ **35.** $\frac{12}{n} = \frac{3}{5}$ **36.** $\frac{21}{n} = \frac{2}{3}$ **37.** $\frac{35}{n} = \frac{7}{5}$

38. $\frac{35}{n} = \frac{4}{3}$ **39.** $\frac{9}{26} = \frac{n}{52}$ **40.** $\frac{12}{15} = \frac{n}{25}$ **41.** $\frac{12}{8} = \frac{21}{n}$ **42.** $\frac{15}{12} = \frac{10}{n}$

43. $\dfrac{n}{24} = \dfrac{5.5}{1}$

44. $\dfrac{n}{16} = \dfrac{6.5}{1}$

45. $\dfrac{2.5}{n} = \dfrac{0.5}{10}$

46. $\dfrac{8}{n} = \dfrac{0.3}{1.5}$

47. $\dfrac{3}{4} = \dfrac{n}{3.8}$

48. $\dfrac{7}{8} = \dfrac{n}{4.2}$

49. $\dfrac{12.5}{16} = \dfrac{n}{12}$

50. $\dfrac{13.8}{15} = \dfrac{n}{6}$

51. $\dfrac{16\frac{2}{3}}{100} = \dfrac{3}{n}$

52. $\dfrac{22\frac{2}{9}}{100} = \dfrac{2}{n}$

Find the value of n. Round to the nearest hundredth if necessary.

53. $\dfrac{n \text{ grams}}{10 \text{ liters}} = \dfrac{7 \text{ grams}}{25 \text{ liters}}$

54. $\dfrac{n \text{ grams}}{14 \text{ liters}} = \dfrac{8 \text{ grams}}{22 \text{ liters}}$

55. $\dfrac{76 \text{ dollars}}{5 \text{ tons}} = \dfrac{n \text{ dollars}}{8 \text{ tons}}$

56. $\dfrac{88 \text{ dollars}}{10 \text{ tons}} = \dfrac{n \text{ dollars}}{7 \text{ tons}}$

57. $\dfrac{50 \text{ gallons}}{12 \text{ acres}} = \dfrac{36 \text{ gallons}}{n \text{ acres}}$

58. $\dfrac{70 \text{ gallons}}{25 \text{ acres}} = \dfrac{80 \text{ gallons}}{n \text{ acres}}$

To Think About

Study the "To Think About" example in the text. See if you can find n in each problem. Express n as a mixed number.

59. $\dfrac{n}{2\frac{1}{3}} = \dfrac{4\frac{5}{6}}{3\frac{1}{9}}$

60. $\dfrac{n}{7\frac{1}{4}} = \dfrac{2\frac{1}{5}}{4\frac{1}{8}}$

61. $\dfrac{8\frac{1}{6}}{n} = \dfrac{5\frac{1}{2}}{7\frac{1}{3}}$

62. $\dfrac{9\frac{3}{4}}{n} = \dfrac{8\frac{1}{2}}{4\frac{1}{3}}$

Find n. Round to the nearest hundredth if necessary.

63. $\dfrac{3.8}{1.56} = \dfrac{n}{7.2}$

64. $\dfrac{14.6}{3.8} = \dfrac{5.7}{n}$

Cumulative Review Problems

Evaluate by doing each operation in the proper order.

65. $4^3 + 20 \div 5 + 6 \times 3 - 5 \times 2$

66. $(1.6)^2 - 0.12 \times 3.5 + 36.8 \div 2.5$

Calculator Problems

Find the value of n. Round your answer to the nearest tenth if necessary.

67. $\dfrac{12.223}{25.12} = \dfrac{n}{50.2}$

68. $\dfrac{3055}{532} = \dfrac{750}{n}$

For Extra Practice Examples and Exercises, turn to page 238.

1. (a) $5 \times n = 45$ **(b)** $7 \times n = 84$ **3. (a)** $15 \times n = 63$ **(b)** $39.2 = 5.6 \times n$ **5.** $\dfrac{176}{4} = \dfrac{286}{n}$

$\dfrac{5 \times n}{5} = \dfrac{45}{5}$ $\dfrac{7 \times n}{7} = \dfrac{84}{7}$ $\dfrac{15 \times n}{15} = \dfrac{63}{15}$ $\dfrac{39.2}{5.6} = \dfrac{5.6 \times n}{5.6}$ $176 \times n = 286 \times 4 = 1144$

$n = 9$ $n = 12$ $n = 4.2$ $7 = n$ $\dfrac{176 \times n}{176} = \dfrac{1144}{176}$

$n = 6.5$

2. (a) $12 = n$ **(b)** $15 = n$ **4.** $n = 56$ **6.** $n = \$96$

4.4 APPLICATIONS OF PROPORTIONS

☐ After studying this section,
you will be able to:

1 *Solve applied problems using proportions*

The solutions to many problems in proportions rely on assumptions. For example, we may assume that a rate that holds true today is the same rate that existed in the past or that will continue to exist in the future. Consider this problem (Exercises 4.4, problem 1): "A discount store sells microwave ovens. For every 120 ovens sold, 7 are returned as defective. If the store sells 1800 ovens this month, approximately how many defective ones will be returned?

We arrive at the rate of defective ovens—7 per 120 sold—by past measurement. We *assume* that the conditions of manufacture that existed in the past continue to exist, so we *assume* that the rate of defective ovens holds steady. This assumption allows us to set $\dfrac{7}{20}$ equal to $\dfrac{n}{1800}$ to get the answer to the problem. Applying this technique in working proportions allows us to solve many practical problems.

1 Example 1 A recent citywide survey discovered that 37 pairs of eyeglasses in a sample of 120 pairs of eyeglasses were defective. If this rate remains the same each year, how many of the 36,000 pairs of eyeglasses made in the city each year are defective?

We will use the letter n to represent the number of defective eyeglasses in the city.

We compare the   to   the total number in
sample   the city

$$\dfrac{37 \text{ defective pairs}}{120 \text{ total pairs of eyeglasses}} = \dfrac{n \text{ defective pairs}}{36{,}000 \text{ total pairs of eyeglasses}}$$

Both rates refer to the number of glasses made in one year.

$37 \times 36{,}000 = 120 \times n$ Forming the cross product.

$1{,}332{,}000 = 120 \times n$ Simplifying.

$\dfrac{1{,}332{,}000}{120} = \dfrac{120 \times n}{120}$ Dividing each side by 120.

$11{,}100 = n$

Thus if the rate of defective eyeglasses holds steady, there are about 11,100 defective pairs of eyeglasses made in the city each year. ∎

Practice Problem 1 An automobile assembly line produced 243 engines yesterday of which 27 were defective. If the same rate is true each day, how many of the 4131 engines produced this month would be defective? ■

Example 2 A survey conducted on a college campus showed that approximately 3 out of every 11 students were left-handed. If there are 2310 students on campus, how many are left-handed?

Let n represent the number of left-handed students.

$$\underbrace{\text{We compare the}}_{\text{measured rate}} \quad \text{to} \quad \underbrace{\text{the campus wide}}_{\text{situation}}$$

$$\frac{3 \text{ left-handed students in survey}}{11 \text{ students in survey}} = \frac{n \text{ left-handed students on campus}}{2310 \text{ total students on campus}}$$

We assume that the measured rate may be applied to the entire campus.

$$3 \times 2310 = 11 \times n \qquad \text{Forming the cross product.}$$

$$6930 = 11 \times n \qquad \text{Simplifying.}$$

$$\frac{6930}{11} = \frac{11 \times n}{11} \qquad \text{Dividing each side by 11.}$$

$$630 = n$$

Thus there are approximately 630 left-handed students on campus. ■

Practice Problem 2 A survey conducted on campus showed that 5 out of every 19 students had quit smoking during the last five years. Approximately how many of the 7410 students now on campus quit smoking in the last five years? ■

To Think About

Does it matter how the fractions are set up? In Example 2 we set up the problem $\frac{3}{11} = \frac{n}{2310}$ to find $n = 630$. Could we also get the same answer setting up the equation

$$\frac{3}{n} = \frac{11}{2310} \qquad \text{or} \qquad \frac{2310}{11} = \frac{n}{3}?$$

The answer is yes. There is no restriction on how to set up the problem as long as the units—the labels—are in correctly corresponding positions. It would be correct to set up the problem in the form

$$\frac{\text{left-handed students in survey}}{\text{left-handed students on campus}} = \frac{\text{total students in survey}}{\text{total students on campus}}$$

or

$$\frac{\text{total students on campus}}{\text{total students in survey}} = \frac{\text{left-handed students on campus}}{\text{left-handed students in survey}}$$

But we **cannot use an equation** when we write

$$\frac{\text{left-handed students}}{\text{total students}} \neq \frac{\text{total students}}{\text{left-handed students}}$$

is *not* correct. Do you see why? See problems 25 and 26 for further examination of this idea. ■

Example 3 Ted's car can go 245 miles on 7 gallons of gas. Ted wants to take a trip of 455 miles. Approximately how many gallons of gas will this take?

Let $n =$ the unknown number of gallons.

$$\frac{245 \text{ miles}}{7 \text{ gallons}} = \frac{455 \text{ miles}}{n \text{ gallons}}$$

$245 \times n = 7 \times 455$ Forming the cross product.

$245 \times n = 3185$ Simplifying.

$$\frac{245 \times n}{245} = \frac{3185}{245}$$

$$n = 13$$

Ted will need approximately 13 gallons of gas for the trip.   ■

Practice Problem 3   Cindy's car travels 234 miles on 9 gallons of gas. How many gallons of gas will Cindy need to take a 312-mile trip?   ■

Example 4   In a certain gear, Alice's 10-speed bicycle has a gear ratio of 3 revolutions of the pedal for every 2 revolutions of the bicycle wheel. If her bicycle wheel is turning at 65 revolutions per minute, how many times must she pedal in 1 minute?

Let $n =$ the number of revolutions of the pedal.

$$\frac{3 \text{ revolutions of the pedal}}{2 \text{ revolutions of the wheel}} = \frac{n \text{ revolutions of the pedal}}{65 \text{ revolutions of the wheel}}$$

$3 \times 65 = 2 \times n$ Cross-multiplying.

$195 = 2 \times n$ Simplifying.

$$\frac{195}{2} = \frac{2 \times n}{2}$$ Dividing both sides by 2.

$$97.5 = n$$

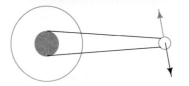

Alice will pedal at the rate of 97.5 revolutions in 1 minute.   ■

Practice Problem 4   Alicia must pedal at 80 revolutions per minute to ride her bicycle at 16 miles per hour. If she pedals at 90 revolutions per minute, how fast will she be riding?   ■

1. A discount store sells microwave ovens. For every 120 ovens sold, 7 are returned as defective. If the store sells 1800 ovens this month, approximately how many defective ones will be returned?

2. An automobile dealership has found that for every 140 cars sold, 23 will be brought back to the dealer for major repairs. If the dealership sells 980 cars this year, approximately how many cars will be brought back for major repairs?

3. A campus student/faculty ratio is 25 to 3. How many faculty members should the school have to maintain that ratio for a student body of 2575?

4. As a safety measure, the island ferry is required to have a ratio of 17 life jackets for every 14 passengers. If the ferry has 663 life jackets, what is the maximum number of passengers it can carry?

5. Linda's car uses 1.5 quarts of oil for every 900 miles she drives. She drove 6300 miles last year. At that rate, how many quarts of oil did she use?

6. Jeff's car uses 2.5 quarts of oil for every 2100 miles he drives. He drove 14,700 miles last year. At that rate, how many quarts of oil did he use last year?

7. Melinda traveled to Europe last year. The exchange rate was 11 Swiss francs for every 5 American dollars. Melinda exchanged 280 American dollars. How many Swiss francs did she receive?

8. When Douglas flew to London, the exchange rate was 7 British pounds for every 15 American dollars. Douglas exchanged 255 American dollars. How many British pounds did he receive?

9. On her bike, Mary Ellen pedals at 70 revolutions per minute to travel at 13 miles per hour. If she pedals at 80 revolutions per minute, how fast will she travel? Round your answer to the nearest tenth.

10. On his bike, Wally pedals at 65 revolutions per minute to travel at 12 miles per hour. If he pedals at 75 revolutions per minute, how fast will he travel? Round your answer to the nearest tenth.

(In problems 11 and 12, the shadow of one object is in the same ratio to the height of that object as a second shadow is to a second object if we look at the same moment, when the sun's rays cast the shadows.)

11. A man who is 6 feet tall casts a 5-foot shadow. At the same time, an office building casts a shadow of 143 feet. How tall is the building? Round your answer to the nearest tenth.

12. A woman who is 5 feet tall casts a shadow that is 7 feet long. At the same time an office building casts a shadow of 165 feet. How tall is the building?

13. Melissa can type the final copy of a three-page English paper in 22 minutes. How long will it take her to type a 14-page paper? Round your answer to the nearest minute.

14. Marcia can type the final copy of a four-page psychology paper in 27 minutes. How long will it take her to type a 15-page paper? Round your answer to the nearest minute.

15. The scale on a road map is as follows: 2 inches on the map represents 23 miles of actual distance. Two cities are 6.5 inches apart on the map. What is the approximate distance in miles between the two cities? Round your answer to the nearest mile.

16. The scale on a road map is as follows: 3 inches on the map represents 31 miles of actual distance. Two cities are 8.6 inches apart on a map. What is the approximate distance in miles between the two cities? Round your answer to the nearest mile.

17. The amount of electrical resistance in a wire is measured in ohms. A 16-foot wire has 0.4 ohm of resistance. How much resistance does a 80-foot length of the same type of wire have?

18. A 19-foot wire has 0.5 ohm of resistance. How much resistance would be in the same type of wire 95 feet long?

19. A recipe that feeds 14 people uses $2\frac{1}{2}$ cups of flour. How many cups of flour are needed for the same recipe prepared for 35 people?

20. Johnny's famous mountain french toast recipe calls for three eggs for every eight slices of french toast. He brought two dozen eggs on the retreat. Using the same ratio, how many slices of french toast can he make?

21. A basketball team won 11 of its first 16 games. To keep the same rate of winning, how many games will the team need to win in an 80-game season?

22. Coach Jones's basketball team has won seven out of every eight games over the last several years. The team has played 688 times. How many times have they won?

23. A baseball pitcher gave up 42 earned runs in 280 innings of pitching. At that rate, how many runs would he give up in a 9-inning game? Round your answer to the nearest hundredth. (This decimal is called the pitcher's **earned run average**.)

24. A baseball pitcher gave up 44 earned runs in 270 innings of pitching. At that rate, how many runs would he give up in a 9-inning game? Round your answer to the nearest hundredth. (This decimal is called the pitcher's **earned run average**.)

? To Think About

The rate of speed of 60 miles per hour is equivalent to the speed 88 feet per second. A car is traveling at 49 miles per hour. Several students attempt to solve the problem of how many feet per second the car is traveling.

Student A uses the proportion: $\dfrac{49}{60} = \dfrac{88}{n}$;

Student B uses the proportion: $\dfrac{60}{88} = \dfrac{49}{n}$

25. Is student A right? Why?

26. Is student B right? Why?

Student C uses the proportion: $\dfrac{n}{88} = \dfrac{60}{49}$;

Student D uses the proportion: $\dfrac{60}{88} = \dfrac{n}{49}$

27. Is student C right? Why?

28. Is student D right? Why?

29. In Deerfield, the average farm has a yield of 325 bushels of wheat for every 4 acres. Hector Ramonson has 22 acres of good wheat fields. The crop this year should bring $5.80 for each bushel of wheat. How much money might Hector plan to get from the sale of his wheat crop?

30. A paint manufacturer suggests 2 gallons of paint for every 750 square feet of wall. A painter wishes to paint 7875 square feet of wall in an office building with paint that costs $8.50 per gallon. How much will the painter spend for the paint?

31. Peter invested in 255 shares of stock. The company declared a stock split; each stock investor received five new shares to replace every three shares previously owned. *After* the split the stock was worth $188.00 per share. How much was Peter's stock worth after the split?

32. A 40-foot roll of fence wire weighs 37 pounds. The same type of fence wire comes in 100-foot rolls. Frank put 13 of the 100-foot rolls in his pickup truck. How much did his cargo weigh?

Cumulative Review Problems

33. Round to the nearest hundred: 56,179.

34. Round to the nearest ten thousandth: 196,379,910.

35. Round to the nearest tenth: 56.148.

36. Round to the nearest hundredth: 1.96341.

37. Round to the nearest ten-thousandth: 0.07615382.

 Calculator Problems

38. An auto parts store has found that 4 out of every 250 parts are defective. If the store owner purchased 3000 parts, estimate how many would be defective.

39. It has been found that 22 out of every 100 people mailed a survey will respond. If 15,500 are mailed a survey, estimate how many will respond.

For Extra Practice Examples and Exercises, turn to page 238.

Solutions to Odd-Numbered Practice Problems

1. $\dfrac{27 \text{ defective engines}}{243 \text{ engines produced}} = \dfrac{n \text{ defective engines}}{4131 \text{ engines produced}}$

$$27 \times 4131 = 243 \times n$$
$$111{,}537 = 243 \times n$$
$$\frac{111{,}537}{243} = \frac{243 \times n}{243}$$
$$459 = n$$

Thus we estimate that 459 engines would be defective. (Fortunately, most defects are minor and can be removed or remedied before the engine leaves the factory.)

3. $\dfrac{9 \text{ gallons of gas}}{234 \text{ miles traveled}} = \dfrac{n \text{ gallons of gas}}{312 \text{ miles traveled}}$

$$9 \times 312 = 234 \times n$$
$$2808 = 234 \times n$$
$$\frac{2808}{234} = \frac{234 \times n}{234}$$
$$12 = n$$

She would need 12 gallons of gas.

Answers to Even-Numbered Practice Problems

2. Approximately 1950   **4.** 18 miles per hour

Section 4.1

Write each ratio in simplest form. Express your answer as a fraction.

Example   8 feet to 12 feet or 8:12

$$\frac{8}{12} = \frac{2}{3}$$

1. 18:45

2. 50:70

3. 9 yards to 21 yards

4. 161 inches to 5 inches

5. 42 kilograms to 36 kilograms

Example   120 people participated in a taste test. Brand X was chosen by 42 people, brand Y by 57 people, and brand Z by 21 people. What was the ratio of people who chose **(a)** Brand Z to brand X **(b)** Brand Y to the total participants?

(a) $\frac{21}{42} = \frac{1}{2}$ **(b)** $\frac{57}{120}$

JILL'S BUDGET

Total Expenses	Rent	Utilities	Gasoline	Groceries	Car Payment
$1440	$900	$75	$100	$155	$210

State the ratio of each of the following.

6. Car payment to rent

7. Utilities to gasoline

8. Groceries to car payment

9. Rent to total expenses

10. Car payment to total expenses

Write as a rate in simplest form.

Example $230 for 45 pounds

$$\frac{230 \text{ dollars}}{45 \text{ pounds}} = \frac{46 \text{ dollars}}{9 \text{ pounds}}$$

11. $180 for 15 pieces of china

12. 75 pounds for 22 people

13. 5135 revolutions for every 20 miles

14. $195 for 25 square yards of carpet

15. 330 gallons of water for every 700 square feet of garden

Write as a unit rate. Round to the nearest tenth when necessary.

Example   Earn $456 in 38 hours.

$$\frac{456}{38} = 38\overline{)456}^{\,12}$$

$$\frac{\$456 \text{ dollars}}{38 \text{ hours}} = \frac{\$12 \text{ dollars}}{1 \text{ hour}} \text{ or } \$12 \text{ per hour}$$

16. Earn $278 in 40 hours.

17. Travel 302 miles on 17 gallons of gas.

18. 2150 words on 10 pages.

19. Travel 3155 kilometers in 12 hours.

20. $6250 was spent for 690 shares of stock. Find the cost per share to the nearest cent.

Section 4.2

Write a proportion.

Example   52 is to 38 as 26 is to 19.

$$\frac{52}{38} = \frac{26}{19}$$

1. 3 is to 21 as 1 is to 7

2. 50 is to 55 as 10 is to 11

3. 35 is to 95 as 7 is to 19

4. 27 is to 63 as 3 is to 7

5. 6.5 is to 10 as 13 is to 20

Determine if each proportion is true or false.

Example $\frac{15}{42} = \frac{5}{18}$

$$\frac{15}{42} = \frac{5}{18} \qquad \begin{array}{l} 15 \times 18 = 270 \\ 42 \times 5 = 210 \end{array}$$

270 does not equal 210, so 154/42 does not equal 5/18.

6. $\frac{16}{10} = \frac{32}{20}$ **7.** $\frac{8}{21} = \frac{16}{40}$ **8.** $\frac{47}{35} = \frac{75}{120}$

9. $\frac{6.6}{5.7} = \frac{19.8}{17.1}$ **10.** $\frac{3\frac{1}{2}}{5} = \frac{10\frac{1}{2}}{15}$

Example $\frac{85.5 \text{ miles}}{45 \text{ dollars}} = \frac{171 \text{ miles}}{90 \text{ dollars}}$

$$85.5 \times 90 = 7695$$

$$45 \times 171 = 7695$$

These are equal, so the rates are equal.

11. $\frac{160 \text{ feet}}{6.5 \text{ rolls}} = \frac{325 \text{ feet}}{20 \text{ rolls}}$

12. $\frac{351 \text{ gallons}}{19 \text{ acres}} = \frac{702 \text{ gallons}}{38 \text{ acres}}$

13. $\frac{55 \text{ gallons}}{13 \text{ acres}} = \frac{275 \text{ gallons}}{65 \text{ acres}}$

14. $\frac{75 \text{ miles}}{32 \text{ dollars}} = \frac{225 \text{ miles}}{96 \text{ dollars}}$

15. $\frac{145.5 \text{ feet}}{55 \text{ cents}} = \frac{291 \text{ feet}}{165 \text{ cents}}$

Section 4.3

Solve each equation to find the value of n.

Example $75 = 15 \times n$ $\dfrac{75}{15} = \dfrac{15 \times n}{15}$

$5 = n$

1. $80 = 10 \times n$

2. $108 = 9 \times n$

3. $13 \times n = 91$

4. $23 \times n = 184$

5. $16 \times n = 80$

Solve each equation to find the value of n to the nearest tenth.

Example $3.7 \times n = 11.1$

$$\frac{3.7 \times n}{3.7} = \frac{11.1}{3.7}$$

$$n = \frac{11.1}{3.7}, \quad n = 3$$

6. $5.6 \times n = 11.2$

7. $7.3 \times n = 30.4$

8. $13.5 = 5 \times n$

9. $21.8 = 2 \times n$

10. $8 \times n = 108.8$

Find the value of n in each proportion. Check your answer. Round your answer to the nearest tenth if necessary.

Example $\dfrac{25}{2} = \dfrac{40}{n}$ $25 \times n = 2 \times 40$

$25 \times n = 80$

$$\frac{25 \times n}{25} = \frac{80}{25}, \quad n = \frac{80}{25} = 3.2$$

11. $\dfrac{46}{115} = \dfrac{n}{5}$

12. $\dfrac{20}{3} = \dfrac{30}{n}$

13. $\dfrac{n}{36} = \dfrac{3.3}{9}$

14. $\dfrac{3}{4} = \dfrac{n}{25}$

15. $\dfrac{n}{28} = \dfrac{2.5}{12}$

Find the value. Round to the nearest hundredth if necessary.

Example $\dfrac{n \text{ grams}}{20 \text{ liters}} = \dfrac{8 \text{ grams}}{35 \text{ liters}}$

$$n \times 35 = 20 \times 8$$

$$n \times 35 = 160$$

$$\frac{n \times 35}{35} = \frac{160}{35}$$

$$n = 4.57$$

16. $\dfrac{n \text{ grams}}{32 \text{ liters}} = \dfrac{6 \text{ grams}}{30 \text{ liters}}$

17. $\dfrac{n \text{ grams}}{21 \text{ liters}} = \dfrac{8 \text{ grams}}{42 \text{ liters}}$

18. $\dfrac{93 \text{ dollars}}{6 \text{ tons}} = \dfrac{n \text{ dollars}}{5 \text{ tons}}$

19. $\dfrac{33 \text{ dollars}}{12 \text{ tons}} = \dfrac{n \text{ dollars}}{8 \text{ tons}}$

20. $\dfrac{60 \text{ gallons}}{13 \text{ acres}} = \dfrac{21 \text{ gallons}}{n \text{ acres}}$

Section 4.4

Example   A campus has a ratio of 60 students for every eight faculty members. If the student body is 3525, how many faculty should there be to maintain that rate?

$$\frac{60 \text{ students}}{8 \text{ faculty}} = \frac{3525 \text{ students}}{n \text{ faculty}}$$

$\dfrac{60}{8} = \dfrac{3525}{n}$ $60 \times n = 3525 \times 8$

$60 \times n = 28{,}200$

$$\frac{60 \times n}{60} = \frac{28{,}200}{60}$$

$$n = 470 \text{ faculty}$$

1. A campus has a ratio of 50 students for every six faculty. If the student body is 3250 students, how many faculty should there be to maintain that rate?

2. A factory needs eight workers to make 125 products per day. If the factory wants to make 750 products, how many workers are needed?

3. The college has 1750 students with 125 faculty. At this ratio how many students should they accept if they increase their faculty to 140?

4. Jason traveled to Europe last year. The exchange rate was 12 Swiss francs for every $5 American. Jason exchanged $370 American. How many Swiss francs was that worth?

5. Daniel exchanged $350 American at an exchange rate of 11 British pounds for every $20 American. How many British pounds was that worth?

Example   A man who is 6 feet tall casts a 5-foot shadow. At the same time, an office building casts a shadow 125 feet. How tall is the building? Round your answer to the nearest tenth.

$$\frac{6\text{-foot man}}{5\text{-foot shadow}} = \frac{n\text{-foot building}}{125\text{-foot shadow}}$$

$\dfrac{6}{5} = \dfrac{n}{125}$ $6 \times 125 = 5 \times n$

$750 = 5 \times n$

$$\frac{750}{5} = \frac{5 \times n}{5}$$

$$150 = n$$

6. A man who is 6 feet tall casts a 7-foot shadow. At the same time, an office building casts a shadow of 105 feet. How tall is the building? Round your answer to the nearest tenth.

7. In the setting sun a woman who is 5 feet tall casts a shadow that is 13 feet long. At the same time an office building casts a shadow of 117 feet. How tall is the building? Round your answer to the nearest tenth.

8. Sally can type the final copy of a five-page history paper in 55 minutes. How long will it take her to type a 20-page paper? Round your answer to the nearest minute.

9. Don can type the final copy of a seven-page term paper in 49 minutes. How long will it take him to type an 18-page paper? Round your answer to the nearest minute.

10. Ted owns a chain of fast-food restaurants. He became a millionaire. He calculated that for every $5 invested in the food chain he made $1020. In seven years how much did he invest in order to earn $193,800?

Example The scale on a road map is 2 inches on the map represents 35 miles of actual distance. Two cities are 9.2 inches apart. What is the approximate distance in miles between the two cities? Round your answer to the nearest mile.

$$\frac{2 \text{ inches}}{35 \text{ miles}} = \frac{9.2 \text{ inches}}{n \text{ miles}}$$

$$2 \times n = 35 \times 9.2$$

$$2 \times n = 322$$

$$\frac{2 \times n}{2} = \frac{322}{2}$$

$$n = 161 \text{ miles}$$

11. The scale on a road map is 3 inches, which represents 45 miles of actual distance. Two cities are 8.5 inches apart on a map. What is the approximate distance in miles between the two cities? Round your answer to the nearest mile.

12. The scale on a road map is 1 inch, which represents 20 miles actual distance. Two cities are 7.2 inches apart on the map. What is the approximate distance in miles between the two cities? Round your answer to the nearest mile.

13. A recipe that feeds 15 people uses $3\frac{1}{2}$ cups of flour. How many cups of flour are needed for the same recipe prepared for 45 people?

14. Sandy is making a cake. The recipe calls for two eggs for each cake. If Sandy wants to make 12 cakes, how many eggs will she need?

15. A basketball team won 12 of it first 16 games. To keep at this same rate of winning, how many games will the team need to win in a 64-game season?

CHAPTER ORGANIZER

Topic	Procedure	Examples
Forming a ratio, p. 213	A *ratio* is the comparison of two quantities that have the same units. A ratio is usually expressed as a fraction. The fractions should be in reduced form.	1. Find the ratio of 20 books to 35 books. $$\frac{20}{35} = \frac{4}{7}$$ 2. Find the ratio in simplest form of 88:99. $$\frac{88}{99} = \frac{8}{9}$$ 3. Bob earns $250 each week, but $15 is withheld for medical insurance. Find the ratio of medical insurance to total pay. $$\frac{\$15}{\$250} = \frac{3}{50}$$
Forming a rate, p. 214	A *rate* is a comparison of two quantities that have different units. A rate is usually expressed as a fraction in reduced form.	A college has 2520 students with 154 faculty. What is the rate of students to faculty? $$\frac{2520 \text{ students}}{154 \text{ faculty}} = \frac{180 \text{ students}}{11 \text{ faculty}}$$
Forming a unit rate, p. 214	A *unit rate* is a rate with a denominator of 1. Divide the denominator into the numerator to obtain the unit rate.	A car traveled 416 miles in 8 hours. Find the unit rate. $$\frac{416 \text{ miles}}{8 \text{ hours}} = 52 \text{ miles/hour}$$ Bob spread fertilizer over 1870 square feet with 50 pounds of fertilizer. Find the unit rate of square feet per pound. $$\frac{1870 \text{ square feet}}{50 \text{ pounds}} = 37.4 \text{ square feet/pound}$$
Writing proportions, p. 219	A *proportion* is a statement that two rates or two ratios are equal. A proportion statement a is to b as c is to d can be written $$\frac{a}{b} = \frac{c}{d}$$	Write a proportion for 17 is to 34 as 13 is to 26. $$\frac{17}{34} = \frac{13}{26}$$

Topic	Procedure	Examples
Determining if proportions are true or false, p. 220	For any two fractions where a, b, c, d are whole numbers and $b \neq 0$, $d \neq 0$, then: $$\frac{a}{b} = \frac{c}{d} \text{ if and only if } a \times d = b \times c.$$	1. Is the proportion $\frac{7}{56} = \frac{3}{24}$ true or false? $$7 \times 24 \stackrel{?}{=} 56 \times 3$$ $$168 = 168 \checkmark$$ The proportion is *true*. 2. Is the proportion $$\frac{64 \text{ gallons}}{5 \text{ acres}} = \frac{89 \text{ gallons}}{7 \text{ acres}}$$ true or false? $$64 \times 7 \stackrel{?}{=} 5 \times 89$$ $$448 \neq 445$$ The proportion is *false*.
Solving a proportion, p. 225	To solve a proportion where one value n is not known: 1. Cross-multiply. 2. Divide both sides of the equation by the number multiplied by n.	1. Solve for n: $$\frac{17}{n} = \frac{51}{9}$$ $17 \times 9 = 51 \times n$ Cross-multiplying. $153 = 51 \times n$ Simplifying. $\frac{153}{51} = \frac{51 \times n}{51}$ Dividing by 51. $3 = n$
Solving applied problems, p. 231	1. Write a proportion with n representing the unknown value. 2. Solve the proportion.	Bob purchased eight notebooks for \$19. How much would 14 notebooks cost? $$\frac{8 \text{ notebooks}}{\$19} = \frac{14 \text{ notebooks}}{n}$$ $8 \times n = 19 \times 14$ $8 \times n = 266$ $\frac{8 \times n}{8} = \frac{266}{8}$ $n = 33.25$ The 14 notebooks would cost \$33.25.

REVIEW PROBLEMS CHAPTER 4

4.1 *Write each ratio in simplest form. Express your answer as a fraction.*

1. 88:40

2. 26:39

3. 28:35

4. 250:475

5. 50 to 124

6. 27 to 81

7. 156 to 441

8. 280 to 343

9. 26 tons to 65 tons

10. 34 tons to 170 tons

11. 150 kilograms to 200 kilograms

12. 115 grams to 130 grams

Bob earns $215 per week and has $35 per week withheld for federal taxes and $20 per week withheld for state taxes.

13. Write the ratio of federal taxes withheld to earned income.

14. Write the ratio of state taxes withheld to earned income.

Write as a rate in simplest form.

15. 10 gallons of water for every 18 people

16. 256 revolutions every 12 minutes

17. 188 vibrations every 16 seconds

18. 12 cups of flour for every 38 people

Write as a unit rate. Round to the nearest tenth when necessary.

19. $2125 was paid for 125 shares of stock. Find the cost per share.

20. $1785 was paid for 105 chairs. Find the cost per chair.

21. $742.50 was spent for 55 square yards of carpet. Find the cost per square yard.

22. The baseball boosters club spent $768.80 for 62 tickets to the ball game. Find the cost per ticket.

4.2 *Write as a proportion.*

23. 12 is to 48 as 7 is to 28

24. 10 is to 21 as 30 is to 63

25. 7.5 is to 45 as 22.5 is to 135

26. 8.6 is to 43 as 17.2 is to 86

27. 136 is to 17 as 408 is to 51

28. 117 is to 61 as 351 is to 183

Determine if each proportion is true or false.

29. $\dfrac{16}{48} = \dfrac{2}{12}$

30. $\dfrac{20}{25} = \dfrac{8}{10}$

31. $\dfrac{24}{20} = \dfrac{18}{15}$

32. $\dfrac{84}{48} = \dfrac{14}{8}$

33. $\dfrac{37}{33} = \dfrac{22}{19}$

34. $\dfrac{15}{18} = \dfrac{18}{22}$

35. $\dfrac{84 \text{ miles}}{7 \text{ gallons}} = \dfrac{108 \text{ miles}}{9 \text{ gallons}}$

36. $\dfrac{156 \text{ revolutions}}{6 \text{ minutes}} = \dfrac{182 \text{ revolutions}}{7 \text{ minutes}}$

37. $\dfrac{1.6 \text{ pounds}}{32 \text{ feet}} = \dfrac{4.8 \text{ pounds}}{96 \text{ feet}}$

38. $\dfrac{3.9 \text{ pounds}}{45 \text{ feet}} = \dfrac{7.8 \text{ pounds}}{90 \text{ feet}}$

4.3 *Solve for n in each equation.*

39. $7 \times n = 161$

40. $8 \times n = 256$

41. $558 = 18 \times n$

42. $663 = 13 \times n$

Solve each proportion. If necessary, round to the nearest tenth.

43. $\dfrac{3}{11} = \dfrac{9}{n}$

44. $\dfrac{2}{7} = \dfrac{12}{n}$

45. $\dfrac{n}{28} = \dfrac{6}{24}$

46. $\dfrac{n}{28} = \dfrac{15}{20}$

47. $\dfrac{3}{7} = \dfrac{n}{9}$

48. $\dfrac{4}{9} = \dfrac{n}{8}$

49. $\dfrac{54}{72} = \dfrac{n}{4}$

50. $\dfrac{45}{135} = \dfrac{n}{3}$

51. $\dfrac{6}{n} = \dfrac{2}{29}$

52. $\dfrac{8}{n} = \dfrac{2}{81}$

53. $\dfrac{25}{7} = \dfrac{60}{n}$

54. $\dfrac{60}{10} = \dfrac{31}{n}$

55. $\dfrac{35 \text{ miles}}{28 \text{ gallons}} = \dfrac{15 \text{ miles}}{n \text{ gallons}}$

56. $\dfrac{8 \text{ defective parts}}{100 \text{ perfect parts}} = \dfrac{44 \text{ defective parts}}{n \text{ perfect parts}}$

57. $\dfrac{7 \text{ tons}}{5.5 \text{ horsepower}} = \dfrac{16 \text{ tons}}{n \text{ horsepower}}$

58. $\dfrac{27 \text{ feet}}{4 \text{ quarts}} = \dfrac{30.5 \text{ feet}}{n \text{ quarts}}$

4.4   *Solve each problem by using a proportion. Round your answer to the nearest hundredth if necessary.*

59. The school volunteers used 3 gallons of paint to paint two rooms. How many gallons would they need to paint 10 rooms of the same size?

60. Fred paid $77 for three chairs. How much would he pay for nine chairs?

61. The hospital has 84 patients for seven registered nurses. To keep this ratio, how many registered nurses would be needed for 108 patients?

62. The college has 1680 students with 100 faculty. To keep this ratio, how many students should they have if they increase the faculty to 130?

63. When Marguerite traveled, the rate of francs to American dollars was 24 French francs to 5 American dollars. How many French francs did Marguerite receive for 420 American dollars?

64. The food commons recommends 7.5 pounds of cold cuts for every 18 people at the reception. The men's glee club is having a reception for 120 people. How many pounds of cold cuts should they order?

65. Two cities that are actually 225 miles apart appear 3 inches apart on a map. If two other cities appear 8 inches apart on the map, in actuality, how many miles apart are they?

66. If Melissa pedals her bicycle at 84 revolutions per minute, she travels at 14 miles per hour. How fast does she go if she pedals at 96 revolutions per minute?

67. Michael conducted a science experiment and found that sound travels in air at 34,720 feet in 31 seconds. How many feet would sound travel in 60 seconds?

68. After college, Roberto began a chain of Pizza Villas in his home state. He eventually became a millionaire. He told his friends that for every $3 he invested, he earned $985 in 10 years. How much did he invest to earn $1 million?

69. In the setting sun, a 6-foot man has a shadow 16 feet long. At the same time, how tall is a building that has a shadow of 320 feet?

70. During the first 680 miles of a trip, Cindy and Melinda used 26 gallons of gas. They need to travel 200 more miles. Assuming that the car has the same rate of gas consumption:
  (a) How many more gallons of gas will they need?
  (b) If gas is $1.15 per gallon, what will fuel cost them for the last 200 miles?

Write each ratio *in simplest form.*

1. 33 to 44

2. 13 to 117

3. $150 to $175

4. 84 hours to 46 hours

To manufacture a certain type of television set costs $48 in labor, $36 in parts, $18 in packaging and shipping, and $9 in research and development.

5. Find the ratio of labor cost to parts cost.

6. Find the ratio of packaging and shipping cost to total cost.

7. Find the ratio of research and development costs to total cost.

Write each rate *in simplest form.*

8. $134 for six chairs

9. 3860 students for 185 faculty

10. 138 kilometers on 4 liters

11. 24 pounds for 16 people

12. 3150 revolutions for every 20 miles

1. _____

2. _____

3. _____

4. _____

5. _____

6. _____

7. _____

8. _____

9. _____

10. _____

11. _____

12. _____

13. _____
14. _____
15. _____
16. _____
17. _____
18. _____
19. _____
20. _____
21. _____
22. _____
23. _____
24. _____

Write as a unit rate. *Round to the nearest hundredth if necessary.*

13. 1246 words on seven pages

14. Travel 156 miles on 8 gallons of gas

15. $1685 was spent for 65 shares of stock

Write as a proportion.

16. 11 is to 7 as 55 is to 35

17. 27.5 is to 33 as 10 is to 12

18. 34 is to 30 as 51 is to 45

19. 238 is to 357 as 38 is to 57

Determine if each proportion is true or false.

20. $\dfrac{60}{100} = \dfrac{12}{25}$

21. $\dfrac{625}{15} = \dfrac{6875}{165}$

22. $\dfrac{45 \text{ dollars}}{135 \text{ pens}} = \dfrac{3 \text{ dollars}}{9 \text{ pens}}$

23. $\dfrac{0.7 \text{ meter}}{9.8 \text{ kilograms}} = \dfrac{3.6 \text{ meters}}{50 \text{ kilograms}}$

24. $\dfrac{204 \text{ gallons}}{12 \text{ acres}} = \dfrac{323 \text{ gallons}}{19 \text{ acres}}$

Solve for n in each equation.

1. $17 \times n = 136$

2. $34 = 4 \times n$

Solve for n in each proportion. Round to the nearest tenth if necessary.

3. $\dfrac{15}{10} = \dfrac{n}{2}$

4. $\dfrac{9}{n} = \dfrac{25}{135}$

5. $\dfrac{18}{54} = \dfrac{54}{n}$

6. $\dfrac{n}{42} = \dfrac{60}{28}$

7. $\dfrac{24}{n} = \dfrac{1.6}{3}$

8. $\dfrac{3.8}{80} = \dfrac{19}{n}$

9. $\dfrac{12 \text{ trees}}{\$35} = \dfrac{18 \text{ trees}}{\$n}$

10. $\dfrac{156 \text{ miles}}{3 \text{ hours}} = \dfrac{n \text{ miles}}{57 \text{ hours}}$

1. _____

2. _____

3. _____

4. _____

5. _____

6. _____

7. _____

8. _____

9. _____

10. _____

11. _____

12. _____

13. _____

14. _____

15. _____

16. _____

Solve each problem by using a proportion. Round to the nearest tenth if necessary.

11. Henry's car went 120 miles on 5 gallons of gas. How much gas would he need to travel 408 miles?

12. An 8-ounce jar of instant coffee contains 42 servings of coffee. How many servings are in a 12-ounce jar of instant coffee?

13. Wally purchased 56 shares of stock for $700. How many shares of stock could he have purchased with $1000?

14. New cars made at the Smithville plant have been discovered to have some defects. For every 810 cars manufactured, seven have steering defects. If the plant manufactured 2430 cars this month, how many may have steering defects?

15. It was found that 19 out of every 95 students at Western University are left-handed. If there are 4845 students at the university, how many are left-handed.

16. Two cities, A and B, are 530 miles apart, but on a road map they measure only 4 inches apart. On the same map two cities, C and D, are 10 inches apart. Approximately how many miles apart are they?

Write as a ratio *in simplest form. Express your answer as a fraction.*

1. 12:51

2. 60 to 175

Write as a rate *in simplest form. Express your answer as a fraction.*

3. 808 miles per 24 gallons

4. 2400 square feet per 35 pounds

Write as a unit rate. *Round to the nearest hundredth when necessary.*

5. 17 tons in five days

6. $36.96 for 7 hours

7. 5200 feet per 18 telephone poles

8. $9364 for 110 shares of stock

Write as a proportion.

9. 19 is to 31 as 57 is to 93

10. 12 is to 17 as 18 is to 25.5

11. 420 miles is to 18 gallons as 350 miles is to 15 gallons

12. 3 tablespoons of flour is to 16 people as 9 tablespoons of flour is to 48 people

Determine if each proportion is true or false.

13. $\dfrac{51}{24} = \dfrac{34}{16}$

14. $\dfrac{9.2}{10} = \dfrac{46}{50}$

15. $\dfrac{16 \text{ smokers}}{23 \text{ nonsmokers}} = \dfrac{80 \text{ smokers}}{115 \text{ nonsmokers}}$

16. $\dfrac{\$0.74}{16 \text{ ounces}} = \dfrac{\$1.85}{40 \text{ ounces}}$

1. _____

2. _____

3. _____

4. _____

5. _____

6. _____

7. _____

8. _____

9. _____

10. _____

11. _____

12. _____

13. _____

14. _____

15. _____

16. _____

17. _____

18. _____

19. _____

20. _____

21. _____

22. _____

23. _____

24. _____

25. _____

26. _____

27. _____

28. _____

29. _____

30. _____

Solve each proportion. If necessary, round to the nearest tenth.

17. $\dfrac{n}{10} = \dfrac{3}{5}$

18. $\dfrac{9}{2} = \dfrac{72}{n}$

19. $\dfrac{44}{n} = \dfrac{11}{2}$

20. $\dfrac{5.6}{11} = \dfrac{n}{66}$

21. $\dfrac{45 \text{ women}}{25 \text{ men}} = \dfrac{n \text{ women}}{20 \text{ men}}$

22. $\dfrac{3.5 \text{ ounces}}{4.2 \text{ grams}} = \dfrac{14 \text{ ounces}}{n \text{ grams}}$

23. $\dfrac{n \text{ inches of snow}}{7 \text{ inches of rain}} = \dfrac{12 \text{ inches of snow}}{1.4 \text{ inches of rain}}$

24. $\dfrac{7 \text{ pounds of bananas}}{\$n} = \dfrac{3 \text{ pounds of bananas}}{\$0.55}$

Solve each problem by using a proportion. Round your answer to the nearest hundredth if necessary.

25. Bob's recipe for pancakes that calls for three eggs will feed seven people. If he wants to feed 21 people, how many eggs will he need?

26. A steel cable 42 feet long weighs 130 pounds. How much will 20 feet of this cable weigh?

27. If 3 inches on a map represents 84 miles, what distance does 7 inches represent?

28. John and Nancy found it would cost $210 per year to fertilize their front lawn of 3000 square feet. How much would it cost to fertilize 5000 square feet?

29. If Jenny's car uses 1.5 quarts of oil every 2000 miles, how many quarts will it use in 8000 miles?

30. Stephen traveled 530 kilometers in 6 hours. At this rate, how far could he go in 8 hours?

53. Suppose that we want to change 36% to 0.36 by moving the decimal point two places to the left and dropping the % symbol. Explain the steps to show what is really involved in changing 36% to 0.36. Why does the rule work?

54. Suppose that we want to change 10.65 to 1065%. Give a complete explanation of the steps.

Take the value in each case and (a) write it as a decimal, (b) as a fraction with a denominator of a power of 10, and (c) as a reduced fraction.

55. 55562%

56. 80738%

Cumulative Review Problems

Write as a fraction in reduced form.

57. 0.56

58. 0.72

Write as a decimal.

59. $\dfrac{11}{16}$

60. $\dfrac{15}{16}$

For Extra Practice Examples and Exercises, turn to page 284.

Solutions to Odd-Numbered Practice Problems

1. (a) $\dfrac{51}{100} = 51\%$ **(b)** $\dfrac{68}{100} = 68\%$ **(c)** $\dfrac{7}{100} = 7\%$ **(d)** $\dfrac{26}{100} = 26\%$ **3. (a)** $\dfrac{0.5}{100} = 0.5\%$ **(b)** $\dfrac{0.06}{100} = 0.06\%$ **(c)** $\dfrac{0.003}{100} = 0.003\%$
5. (a) $80.6\% = 0.806$ **(b)** $2.5\% = 0.025$ **(c)** $0.29\% = 0.0029$ **(d)** $231\% = 2.31$

Answers to Even-Numbered Practice Problems

2. (a) 238% **(b)** 505% **(c)** This year the number of students who tried out was 121% of the number who tried out last year.
4. (a) 0.47 **(b)** 0.02 **(c)** 0.90 or 0.9 **6. (a)** 78% **(b)** 2% **(c)** 507% **(d)** 2.9% **(e)** 0.6%

5.2 CONVERSIONS BETWEEN PERCENTS, DECIMALS, AND FRACTIONS

☐ After studying this section, you will be able to:

1 Convert a percent to a fraction or a mixed number

2 Convert a fraction to a percent

3 Complete an equivalency chart of fractions, decimals, and percents

1 Convert a Percent to a Fraction

When we change a percent to a common fraction we remove the percent symbol and write the number over 100. The resulting fraction is reduced if possible.

Example 1 Write as a fraction in simplest form.

(a) 37%

(b) 75%

(c) 2%

(a) $37\% = \dfrac{37}{100}$

(b) $75\% = \dfrac{75}{100} = \dfrac{3}{4}$

(c) $2\% = \dfrac{2}{100} = \dfrac{1}{50}$ ∎

Practice Problem 1 Write as a fraction in simplest form.
(a) 71%

(b) 25%

(c) 8% ∎

In some cases it may be necessary to write the percent as a decimal before you write it as a fraction in simplest form.

Example 2 Write as a fraction in simplest form: 43.5%.

$$43.5\% = .435 \qquad \text{Changing \% to a decimal.}$$

$$= \frac{435}{1000} \qquad \text{Changing a decimal to a fraction.}$$

$$= \frac{87}{200} \qquad \text{Reducing the fraction.} \quad \blacksquare$$

Practice Problem 2 Write as a fraction in simplest form: 28.5%. ■

Example 3 Write each percent as a fraction in simplest form.

(a) 5.8%   **(b)** 36.75%

(a) $5.8\% = 0.058 = \dfrac{58}{1000} = \dfrac{29}{500}$

(b) $36.75\% = 0.3675 = \dfrac{3675}{10000} = \dfrac{147}{400}$ ■

Practice Problem 3 Write each percent as a fraction in simplest form.
(a) 8.4%   **(b)** 55.25% ■

If the percent is greater than 100%, the reduced fraction is usually changed to a mixed number.

Example 4 Write each percent as a mixed number.

(a) 225%   **(b)** 138%

(a) $225\% = 2.25 = 2\dfrac{25}{100} = 2\dfrac{1}{4}$ **(b)** $138\% = 1.38 = 1\dfrac{38}{100} = 1\dfrac{19}{50}$ ■

Practice Problem 4 Write each percent as a mixed number.
(a) 170%   **(b)** 288% ■

Sometimes we meet percentages that are not whole numbers, such as 9% or 10%, but that are some fractional value, such as $9\dfrac{1}{12}\%$ or $9\dfrac{3}{8}\%$. Extra steps will be needed to write this type of percent as a reduced fraction.

Example 5 Convert $9\dfrac{3}{8}\%$ to fraction form.

$$9\dfrac{3}{8}\% = \frac{9\dfrac{3}{8}}{100} \qquad \text{Converting the percent to a fraction.}$$

$$= 9\dfrac{3}{8} \div \dfrac{100}{1} \qquad \text{Writing division horizontally.}$$

$$= \dfrac{75}{8} \div \dfrac{100}{1} \qquad \text{Writing } 9\dfrac{3}{8} \text{ as a mixed fraction.}$$

$$= \dfrac{75}{8} \times \dfrac{1}{100} \qquad \text{Using the definition of division of fractions.}$$

$$= \dfrac{\overset{3}{\cancel{75}}}{8} \times \dfrac{1}{\underset{4}{\cancel{100}}} \qquad \text{Simplifying.}$$

$$= \dfrac{3}{32} \qquad \text{Reducing the fraction.} \quad \blacksquare$$

Practice Problem 5 Convert $7\frac{5}{8}\%$ to fraction form. ■

Some percents occur very often, especially with regard to money. Here are some common equivalencies that you may already know. If not, it is useful to memorize them:

$$25\% = \frac{1}{4} \qquad 33\frac{1}{3}\% = \frac{1}{3} \qquad 10\% = \frac{1}{10}$$

$$50\% = \frac{1}{2} \qquad 66\frac{2}{3}\% = \frac{2}{3}$$

$$75\% = \frac{3}{4}$$

For example, with respect to one dollar:

25% means 25 cents (one quarter).
50% means 50 cents (one half-dollar).
10% means 10 cents (one dime).

▨ Converting a Fraction to a Percent

A convenient way to change a fraction to a percent is to write the fraction in decimal form first, then convert the decimal to a percent.

Example 6 Write $\frac{3}{8}$ as a percent.

We see that $\frac{3}{8} = 0.375$ by dividing out $3 \div 8$.

$$
\begin{array}{r}
.375 \\
8\overline{\smash{)}3.000} \\
\underline{24} \\
60 \\
\underline{56} \\
40 \\
\underline{40} \\
0
\end{array}
$$

Thus $\frac{3}{8} = 0.375 = 37.5\%$. ■

Practice Problem 6 Write $\frac{5}{8}$ as a percent. ■

Example 7 Write each fraction as a percent.

(a) $\frac{4}{5}$ 　　　　　 **(b)** $\frac{7}{40}$ 　　　　　 **(c)** $\frac{39}{50}$

(a) $\frac{4}{5} = 0.8 = 80\%$ 　　　　　 **(b)** $\frac{7}{40} = 0.175 = 17.5\%$

(c) $\frac{39}{50} = 0.78 = 78\%$ ■

Practice Problem 7 Write each fraction as a percent.

(a) $\frac{3}{5}$ 　　　　　 **(b)** $\frac{21}{25}$ 　　　　　 **(c)** $\frac{7}{16}$ ■

Changing some fractions to decimal form results in an infinitely repeating decimal. In such cases, we usually round to the nearest hundredth of a percent.

Example 8   Write each fraction as a percent. Round to the nearest hundredth of a percent.

(a) $\dfrac{1}{6}$ (b) $\dfrac{15}{33}$

(a) $\dfrac{1}{6} = 0.16666\ldots$ by dividing out $1 \div 6$:

$$
\begin{array}{r}
0.166 \\
6\overline{)\,1.000} \\
6 \\
\overline{40} \\
36 \\
\overline{40}
\end{array}
$$

This repeating pattern could be written as $0.1\overline{6}$. If we round the decimal to the nearest ten thousandth, we have $\dfrac{1}{6} = 0.1667$. If we change this to a percent, we have

$$\dfrac{1}{6} = 16.67\%$$

This is correct to the nearest hundredth of a percent.

(b) $\dfrac{15}{33} = 0.45454545\ldots$ by dividing out $15 \div 33$:

$$
\begin{array}{r}
0.4545 \\
33\overline{)\,15.0000} \\
132 \\
\overline{180} \\
165 \\
\overline{150} \\
132 \\
\overline{180} \\
165 \\
\overline{15}
\end{array}
$$

This can be written as $0.\overline{45}$. If we round to the nearest ten thousandth, we have

$$\dfrac{15}{33} = 0.4545 = 45.45\%$$

This rounded value is correct to the nearest hundredth of a percent.   ■

Practice Problem 8   Write each fraction as a percent. Round to the nearest hundredth of a percent.

(a) $\dfrac{7}{9}$ (b) $\dfrac{19}{30}$ ■

In some cases the percent is left with a fraction.

Example 9   Express $\dfrac{11}{12}$ as a percent containing a fraction.

We terminate the division after two steps, and use the remainder in fraction form.

$$\begin{array}{r} 0.91 \\ 12\overline{)11.00} \\ \underline{108} \\ 20 \\ \underline{12} \\ 8 \end{array}$$

This division tells us that we can write

$$\frac{11}{12} = 0.91\frac{8}{12} = 0.91\frac{2}{3}$$

Notice that we have a decimal with a fraction. Now we must express this decimal as a percent. This is the same as "moving the decimal point over two places":

$$0.91\frac{2}{3} = 91\frac{2}{3}\% \quad \blacksquare$$

Practice Problem 9 Express $\dfrac{7}{12}$ as a percent containing a fraction. $\blacksquare$

3 We have seen so far that a fraction, a decimal, and a percent are three different forms (notations) for the same quantity. Now we summarize by completing a chart (a table of equivalent notations) in which one, or another, of these forms is given. Our task is to write the other, equivalent forms in the open boxes.

Example 10 Complete the following table of equivalent notations. Round decimals to the nearest ten thousandth. Round percents to the nearest hundredth of a percent.

Notice that you may have to go from fraction form to decimal form before you write the percent form. The pattern is

> Fraction

is changed to

> Decimal

is changed to

> Percent

Fraction	Decimal	Percent
$\dfrac{11}{16}$	0.6875	68.75%
$\dfrac{53}{200}$	0.265	26.5%
$\dfrac{43}{250}$	0.172	$17\frac{1}{5}\%$

$\blacksquare$

Practice Problem 10 Complete the following table of equivalent notations. Round decimals to the nearest ten thousandth. Round percents to the nearest hundredth of a percent. $\blacksquare$

Fraction	Decimal	Percent
$\frac{23}{99}$		
	0.516	
		$38\frac{4}{5}\%$

■

? To Think About

Sometimes fractions are converted to a percent by using a proportion. Can you see how this could be done? Write the proportion $\frac{7}{8} = \frac{n}{100}$.

$$7 \times 100 = 8 \times n \qquad \text{Crossing-multiplying.}$$
$$700 = 8 \times n \qquad \text{Simplifying.}$$
$$\frac{700}{8} = \frac{8 \times n}{8} \qquad \text{Dividing each side by 8.}$$
$$87.5 = n \qquad \text{Simplifying.}$$

Thus $\frac{7}{8} = \frac{87.5}{100} = 87.5\%$. You will have an opportunity to try this approach in problems 75 and 76. ■

EXERCISES 5.2

Write each percent as a fraction or as a mixed number.

1. 44% **2.** 58% **3.** 7% **4.** 9% **5.** 21% **6.** 19% **7.** 55% **8.** 35%

9. 75% **10.** 25% **11.** 20% **12.** 10% **13.** 14.5% **14.** 18.5% **15.** 17.6% **16.** 78.4%

17. 64.8% **18.** 14.6% **19.** 71.25% **20.** 38.75% **21.** 176% **22.** 228% **23.** 340% **24.** 420%

25. $8\frac{1}{4}\%$ **26.** $6\frac{3}{4}\%$ **27.** $2\frac{1}{6}\%$ **28.** $3\frac{5}{6}\%$ **29.** $12\frac{3}{8}\%$ **30.** $15\frac{5}{8}\%$

31. In a recent city election for Mayor, $58\frac{1}{2}\%$ of the voters voted Republican. What *fraction* of the voters in the city voted Republican?

32. During an inspection $17\frac{3}{4}\%$ of the labels printed had the wrong colors. What *fraction* of the labels was in the wrong color?

Write each fraction as a percent. Round to the nearest hundredth of a percent.

33. $\dfrac{3}{4}$ **34.** $\dfrac{1}{4}$ **35.** $\dfrac{1}{3}$ **36.** $\dfrac{2}{3}$ **37.** $\dfrac{5}{16}$ **38.** $\dfrac{7}{16}$ **39.** $\dfrac{7}{25}$ **40.** $\dfrac{17}{25}$

41. $\dfrac{11}{40}$ **42.** $\dfrac{9}{40}$ **43.** $\dfrac{7}{12}$ **44.** $\dfrac{5}{12}$ **45.** $\dfrac{18}{5}$ **46.** $\dfrac{7}{4}$ **47.** $2\dfrac{5}{6}$ **48.** $3\dfrac{1}{6}$

49. $4\dfrac{1}{8}$ **50.** $2\dfrac{5}{8}$ **51.** $\dfrac{3}{7}$ **52.** $\dfrac{6}{7}$ **53.** $\dfrac{15}{16}$ **54.** $\dfrac{11}{16}$ **55.** $\dfrac{19}{50}$ **56.** $\dfrac{31}{50}$

Express each fraction as a percent containing a fraction. (Study Example 9 for further help.)

57. $\dfrac{5}{6}$ **58.** $\dfrac{1}{6}$ **59.** $\dfrac{11}{16}$ **60.** $\dfrac{13}{16}$

Complete the following table of equivalent notations. Round decimals to the nearest ten thousandth. Round percents to the nearest hundredth of a percent.

	Fraction	Decimal	Percent
61.	$\dfrac{5}{12}$		
63.		0.06	
65.			40%
67.		0.345	
69.	$\dfrac{3}{200}$		
71.	$\dfrac{5}{9}$		
73.			$3\dfrac{1}{8}\%$

	Fraction	Decimal	Percent
62.	$\dfrac{7}{12}$		
64.		0.15	
66.			60%
68.		0.625	
70.	$\dfrac{7}{200}$		
72.	$\dfrac{3}{8}$		
74.			$2\dfrac{5}{8}\%$

? **To Think About**

Convert each fraction to a percent by using a proportion.

75. $\dfrac{123}{800}$ **76.** $\dfrac{417}{600}$

77. Write $28\dfrac{15}{16}\%$ as a fraction.   **78.** Write $18\dfrac{7}{12}\%$ as a fraction.

Solve for n in each problem.

79. $\dfrac{15}{n} = \dfrac{8}{3}$ **80.** $\dfrac{32}{12} = \dfrac{n}{3}$ **81.** $\dfrac{n}{11} = \dfrac{32}{4}$ **82.** $\dfrac{1}{3} = \dfrac{32}{n}$

For Extra Practice Examples and Exercises, turn to page 284.

Solutions to Odd-Numbered Practice Problems

1. (a) $71\% = \dfrac{71}{100}$ **(b)** $25\% = \dfrac{25}{100} = \dfrac{1}{4}$ **(c)** $8\% = \dfrac{8}{100} = \dfrac{2}{25}$

3. (a) $8.4\% = 0.084 = \dfrac{84}{1000} = \dfrac{21}{250}$ **(b)** $55.25\% = 0.5525 = \dfrac{5525}{10,000} = \dfrac{221}{400}$

5. $7\dfrac{5}{8}\% = 7\dfrac{5}{8} \div 100$ **7. (a)** $\dfrac{3}{5} = 0.60 = 60\%$ **9.** $\dfrac{7}{12}$ If we divide

$\quad = \dfrac{61}{8} \times \dfrac{1}{100}$ **(b)** $\dfrac{21}{25} = 0.84 = 84\%$

$\quad = \dfrac{61}{800}$ **(c)** $\dfrac{7}{16} = 0.4375 = 43.75\%$

$$\begin{array}{r} .58 \\ 12\,\overline{)\,7.00} \\ 60 \\ \hline 100 \\ 96 \\ \hline 4 \end{array}$$

we see that $\dfrac{7}{12} = 0.58\dfrac{4}{12} = 0.58\dfrac{1}{3}$. Thus $\dfrac{7}{12} = 58\dfrac{1}{3}\%$.

Answers to Even-Numbered Practice Problems

2. $\dfrac{57}{200}$ **4. (a)** $1\dfrac{7}{10}$ **(b)** $2\dfrac{22}{25}$ **6.** 62.5% **8. (a)** 77.78% **(b)** 63.33% **10.**

Fraction	Decimal	Percent
$\dfrac{23}{99}$	0.2323	23.23%
$\dfrac{129}{250}$	0.516	51.6%
$\dfrac{97}{250}$	0.388	$38\dfrac{4}{5}\%$

☐ After studying this section, you will be able to:

1 *Translate a percent problem into an equation*

2 *Solve a percent problem by solving an equation*

5.3A SOLVING PERCENT PROBLEMS BY TRANSLATING TO EQUATIONS

Like music, mathematics is a universal language. People from various countries and cultures can communicate certain ideas in this language. For example, a Russian-language mathematics book has much the same appearance as an English-language one, and it's not that hard to see mathematical phrases in one language that can be directly translated into the other.

In English-language problems, like the ones we look at in this section, we can translate from English words to mathematical symbols and back again. After we have the mathematical symbols arranged in an *equation*, we solve the equation. When we find those values that make the math equation true, we have also found the answer to our original word problem. For example,

A video recorder listed for $495. Last week the store had a 30% discount, and Wally bought the video recorder. What was the discount? How much did Wally pay for the video recorder?

We translate the word problem to a mathematical equation:

$$0.30 \times 495 = n$$

where the letter n represents the discount. By solving mathematically, we find that the discount was $148.50 and that Wally paid $346.50 for the recorder, which listed for almost $500. By *translating* English words to mathematical equations we can solve many percent problems (as well as other types).

❶ Translate a Percent Problem into an Equation

Robert has to pay a 5% sales tax on a purchase of $19.00. What amount of tax does he have to pay? To solve a percent problem, express it as an equation with an unknown quantity. We use the letter n to represent the number we do not know. The following table is helpful when translating from a percent problem to an equation.

English Word	Mathematical Symbol
of	Any multiplication symbol usually $\times$ or ( )
is	$=$
what	Any letter, usually n
find	$n =$

In Examples 1–6, we are practicing how to translate to an equation. Please do not solve the problem. Translate to an equation only.

Example 1 Translate to an equation.

What is 5% of 19.00?

$$n = 5\% \times 19.00 \quad \blacksquare$$

Practice Problem 1 Translate to an equation: What is 26% of 35? ■

Example 2 Translate to an equation.

142% of 35 is what?

$$142\% \times 35 = n \quad \blacksquare$$

Practice Problem 2 Translate to an equation: 155% of 20 is what? ■

Example 3 Translate to an equation.

Find 0.6% of 400.

Notice here that the words "what is" are missing. The word "find" in mathematics is equivalent to "what is."

Find 0.6% of 400

$$n = 0.6\% \times 400 \quad \blacksquare$$

Practice Problem 3 Translate to an equation: Find 0.8% of 350. ■

Example 4 Translate to an equation.

(a) 35% of what is 60?

(b) 7.2 is 120% of what?

(a) 35% of what is 60?

$$35\% \times n = 60$$

(b) 7.2 is 120% of what?

$$7.2 = 120\% \times n \quad \blacksquare$$

Practice Problem 4 Translate to an equation.

(a) 58% of what is 400? **(b)** 9.1 is 135% of what? ■

Example 5 Translate to an equation.

$$\underbrace{\text{What percent}}_{n} \text{ of } \overbrace{50 \text{ is } 10}^{\downarrow\ \downarrow\ \downarrow\ \downarrow} $$

$$n \qquad \times\ 50\ =\ 10$$

We see here that the words "what percent" are both represented by the letter n. ■

Practice Problem 5 Translate to an equation: What percent of 250 is 36? ■

Example 6 Translate to an equation.

(a) 30 is what percent of 16? **(b)** What percent of 3000 is 2.6?

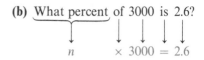

(a) 30 is $\underbrace{\text{what percent}}$ of 16? **(b)** $\underbrace{\text{What percent}}$ of 3000 is 2.6?

$$30\ =\qquad n\qquad \times\ 16 \qquad\qquad n\qquad \times\ 3000\ =\ 2.6 \quad■$$

Practice Problem 6 Translate to an equation.

(a) 50 is what percent of 20? **(b)** What percent of 2000 is 4.5? ■

◻ Solve a Percent Problem by Solving an Equation

The percent problems we have translated represent three different types. Consider the equation $60 = 20\%$ of 300, which we write as $60 = 20\% \times 300$. In general we can describe this type of problem as being of the form

$$\boxed{\text{amount} = \text{percent} \times \text{base}}$$

Any of the three quantities—amount, percent, or base—may be unknown.

1. When *we do not know the amount* we have an equation like

$$n = 20\% \times 300$$

2. When *we do not know the base* we have an equation like

$$60 = 20\% \times n$$

3. When *we do not know the percent* we have an equation like

$$60 = n \times 300$$

We will study an example of each type separately. It is not necessary to memorize the three types, but it is helpful to look carefully at each example. In each example, do the actual computation in a way that is easiest for you. This may be using a pencil and paper, using a calculator, or in some cases doing the problem mentally.

Solving Percent Problems When We Do Not Know the Amount

Example 7 What is 45% of 590?

$$\overset{\downarrow\ \downarrow\ \downarrow\ \downarrow\ \downarrow}{n\ \ =\ 45\%\ \times\ 590} \qquad \text{Translating to an equation.}$$

$$n = (0.45)(590) \qquad \text{Changing the percent to decimal form and multiplying.}$$

$$n = 265.5 \qquad \text{Multiplying } 0.45 \times 590. \quad ■$$

Practice Problem 7 What is 82% of 350? ■

Example 8

(a) 3.8% of 600 is what?   **(b)** Find 160% of 500.

(a) 3.8% of 600 is what?

$$3.8\% \times 600 = n$$ Translating to an equation.

$$(0.038)(600) = n$$ Changing the percent to decimal form and multiplying.

$$22.8 = n$$ Multiplying 0.038 × 600.

(b) Find 160% of 500   When you translate, remember that the word "find" is equivalent to "what" is."

$$n = 160\% \times 500$$

$$n = (1.60)(500)$$ Changing the percent to decimal form and multiplying.

$$n = 800$$ Multiplying 1.6 × 500. ■

Practice Problem 8

(a) 5.2% of 800 is what?   **(b)** Find 230% of 400. ■

Solving Percent Problems When We Do Not Know the Base

If a number is multiplied by the letter n this can be indicated by a multiplication sign, a parentheses, a dot, or by placing the number in front of the letter. Thus $3 \times n = 3(n) = 3 \cdot n = 3n$. In this section we encounter equations like $3n = 9$ and $0.5n = 20$. To solve these equations we use the procedures developed in Chapter 4. We divide each side by the number multiplied by n.

Example 9 90% of what is 342?

$$90\% \times n = 342$$ Translating to an equation.

$$0.9n = 342$$ Changing the percent to a decimal and multiplying by n. Recall that 0.90 = 0.9.

$$\frac{0.9n}{0.9} = \frac{342}{0.9}$$ Dividing each side of the equation by 0.9.

$$n = 380$$ Dividing 342 ÷ 0.9. ■

Practice Problem 9 45% of what is 162? ■

Example 10 12 is 0.6% of what?

$$12 = 0.6\% \times n$$ Translating to an equation.

$$12 = 0.006n$$ Changing 0.6% to decimal form and multiplying by n.

$$\frac{12}{0.006} = \frac{0.006n}{0.006}$$ Dividing each side of the equation by 0.006.

$$2000 = n$$ Dividing 12 ÷ 0.006. ■

Practice Problem 10 32 is 0.4% of what? ■

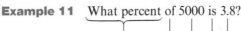

Example 11   What percent of 5000 is 3.8?

$$n \quad \times 5000 = 3.8 \qquad \text{Translating to an equation.}$$

$$5000n = 3.8 \qquad \text{(Multiplication is commutative.} \\ n \times 5000 = 5000 \times n.)$$

$$\frac{5000n}{5000} = \frac{3.8}{5000} \qquad \text{Dividing each side by 5000.}$$

$$n = 0.00076 \qquad \text{Dividing } 3.8 \div 5000.$$

$$n = 0.076\% \qquad \text{Expressing the decimal as a percent.} \quad \blacksquare$$

Practice Problem 11   What percent of 9000 is 4.5?   ∎

Example 12   90 is what percent of 20?

$$90 = \quad n \quad \times 20 \qquad \text{Translating to an equation.}$$

$$90 = 20n \qquad \text{(Multiplication is commutative.} \\ n \times 20 = 20 \times n.)$$

$$\frac{90}{20} = \frac{20n}{20} \qquad \text{Dividing each side by 20.}$$

$$4.5 = n \qquad \text{Dividing } 90 \div 20.$$

$$450\% = n \qquad \text{Expressing the decimal as a percent.} \quad \blacksquare$$

Practice Problem 12   198 is what percent of 33?   ∎

EXERCISES 5.3A

Translate into a mathematical equation. Use the letter n for the unknown quantity. Do not solve.

1. What is 38% of 500?
2. What is 96% of 700?
3. 75% of what is 9?
4. 85% of what is 51?

5. 17 is what percent of 85?
6. 36 is what percent of 144?
7. Find 128% of 4000.
8. Find 210% of 350.

9. What percent of 400 is 15?
10. What percent of 600 is 528?
11. 156 is 130% of what?
12. 200 is 160% of what?

Solve each percent problem for which we do not know the amount.

13. What is 60% of 250?
14. What is 80% of 175?
15. Find 152% of 600.
16. Find 136% of 500.

Solve each percent problem for which we do not know the base.

17. 26% of what is 312?
18. 32% of what is 200?
19. 52 is 4% of what?
20. 36 is 6% of what?

Solve each percent problem for which we do not know the percent.

21. What percent of 30 is 18?

22. What percent of 40 is 26?

23. 56 is what percent of 200?

24. 45 is what percent of 300?

Solve each percent problem. All the various types are represented.

25. 18% of 280 is what?

26. 12% of 260 is what?

27. 150% of what is 102?

28. 140% of what is 126?

29. 84 is what percent of 700?

30. 72 is what percent of 900?

31. Find 0.4% of 820.

32. Find 0.3% of 540.

33. What percent of 35 is 22.4?

34. What percent of 45 is 16.2?

35. 89 is 20% of what?

36. 63 is 60% of what?

37. 77 is what percent of 140?

38. 75 is what percent of 120?

39. What is 16.5% of 240?

40. What is 18.5% of 360?

41. Find 0.06% of 2400.

42. Find 0.03% of 3200.

? **To Think About**

43. What is $128\frac{1}{4}$% of 368?

44. $150\frac{1}{2}$% of what number is 10,535?

45. Find 12% of 30% of $1600.

46. Find 90% of 15% of 2700.

Cumulative Review Problems

Multiply.

47.   1.36
      × 1.8
      ‾‾‾‾‾

48.   2.04
      × 7.3
      ‾‾‾‾‾

Divide.

49. $0.06\overline{)170.04}$

50. $0.08\overline{)233.36}$

For Extra Practice Examples and Exercises, turn to page 285.

Solutions to Odd-Numbered Practice Problems

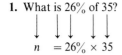

1. What is 26% of 35?

$n = 26\% \times 35$

3. Find 0.8% of 350

$n = 0.8\% \times 350$

5. What percent of 250 is 36?

$n \times 250 = 36$

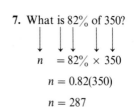

7. What is 82% of 350?

$n = 82\% \times 350$

$n = 0.82(350)$

$n = 287$

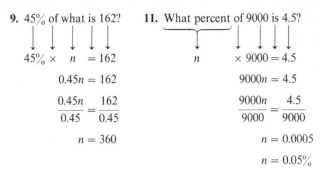

9. 45% of what is 162?

$$45\% \times \quad n \quad = 162$$

$$0.45n = 162$$

$$\frac{0.45n}{0.45} = \frac{162}{0.45}$$

$$n = 360$$

11. What percent of 9000 is 4.5?

$$n \quad \times 9000 = 4.5$$

$$9000n = 4.5$$

$$\frac{9000n}{9000} = \frac{4.5}{9000}$$

$$n = 0.0005$$

$$n = 0.05\%$$

Answers to Even-Numbered Practice Problems

2. $155\% \times 20 = n$ **4. (a)** $58\% \times n = 400$ **(b)** $91 = 135\% \times n$ **6. (a)** $50 = n \times 20$ **(b)** $n \times 2000 = 4.5$
8. (a) 41.6 **(b)** 920 **10.** 8000 **12.** 600%

After studying this section, you will be able to:

1 *Identify the parts of the percent proportion*

2 *Use the percent proportion to solve percent problems*

5.3B SOLVING PERCENT PROBLEMS BY USING PROPORTIONS (OPTIONAL)

We know there is usually more than one way to travel from point A to point B. In the construction industry, there is more than one way to make a building. In the restaurant business there is more than one way to prepare a salad. And in problem solving, there can be more than one way to reach a solution.

In this section we look at the same types of percent problems that we have solved by the method of translating to an equation. But now we will use another method, the method of percent proportion. The two methods work equally well. Both enable you to solve percent problems. Sometimes it's helpful to know two ways to travel from "point A," a problem, to "point B," its solution. Here we show you a second way for solving percent problems.

1 Identifying the Parts of the Percent Proportion

Consider the following relationship:

$$\frac{19}{25} = 76\%$$

This can be written as

$$\frac{19}{25} = \frac{76}{100}$$

In general, we can describe this type of problem by the percent proportion

$$\frac{a}{b} = \frac{p}{100} \qquad \text{where } a = \text{the amount}$$
$$b = \text{the base}$$
$$p = \text{the percent number}$$

To use this equation effectively, we need to be able to find the parts a, b, and p when they are included in a mathematical problem statement. The easiest number to find is the percent number. If the number is not given, we use the letter p (a variable) to represent that number.

Example 1 Find the percent number p in each of the following.

(a) Find 16% of 370.
(c) What percent of 18 is 4.5?

(b) 28% of what is 25?

(a) Find 16% of 370.

The value of p is 16.

(b) 28% of what number is 25?

The value of p is 28.

(c) What percent of 18 is 4.5?

p

We let p represent the unknown percent number. ■

Practice Problem 1 Find the percent number p.
(a) Find 83% of 460. **(b)** 18% of what number is 90?
(c) What percent of 64 is 8? ■

The base b is the entire quantity or the total involved. The number that is the base usually appears after the word "of." The amount a is the part being compared to the whole.

Example 2 Identify the base b and the amount a.

(a) 20% of 320 is 64. **(b)** 12 is 60% of what?

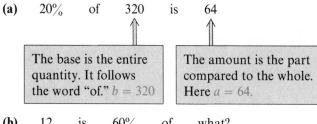

(a)   20%   of   320   is   64

| The base is the entire quantity. It follows the word "of." $b = 320$ | The amount is the part compared to the whole. Here $a = 64$. |

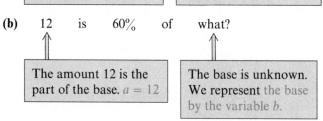

(b)   12   is   60%   of   what?

| The amount 12 is the part of the base. $a = 12$ | The base is unknown. We represent the base by the variable b. | ■

Practice Problem 2 Identify the base b and the amount a.
(a) 30% of 52 is 15.6. **(b)** 170 is 85% of what? ■

When identifying p, b, and a in a problem, it is easiest to identify p and b first. The remaining quantity or variable is a.

Example 3 Find p, b, and a.

(a) What is 52% of 300? **(b)** What percent of 30 is 18?

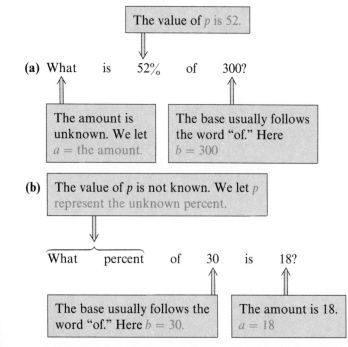

The value of p is 52.

(a) What   is   52%   of   300?

| The amount is unknown. We let a = the amount. | The base usually follows the word "of." Here $b = 300$ |

(b) The value of p is not known. We let p represent the unknown percent.

What   percent   of   30   is   18?

| The base usually follows the word "of." Here $b = 30$. | The amount is 18. $a = 18$ | ■

Practice Problem 3 Find p, b, and a.
(a) What is 18% of 240?   **(b)** What percent of 64 is 4?   ■

2 When we solve the percent proportion we will have enough information to state the numerical value for two of the three variables a, b, p in the equation

$$\frac{a}{b} = \frac{p}{100}$$

We first identify those two values, and then substitute those values into the equation. We will use our skills for solving proportions from Chapter 4.

Example 4 What is 16% of 380?

The percent $p = 16$. The base $b = 380$. The amount is unknown. We use the variable a. Thus

$$\frac{a}{b} = \frac{p}{100} \qquad \text{becomes} \qquad \frac{a}{380} = \frac{16}{100}$$

If we reduce the right-hand fraction, we have

$$\frac{a}{380} = \frac{4}{25}$$

$$25a = (4)(380) \qquad \text{Using cross-multiplication.}$$

$$25a = 1520 \qquad \text{Simplifying.}$$

$$\frac{25a}{25} = \frac{1520}{25} \qquad \boxed{\text{Dividing each side by 25.}}$$

$$a = 60.8 \qquad \text{Result of } 1520 \div 25.$$

Thus 16% of 380 is 60.8.   ■

Practice Problem 4 What is 32% of 550?   ■

Example 5 85% of what is 221?

The percent $p = 85$. The base is unknown. We use the variable b. The amount a is 221. Thus

$$\frac{a}{b} = \frac{p}{100} \qquad \text{becomes} \qquad \frac{221}{b} = \frac{85}{100}$$

If we reduce the right-hand fraction, we have

$$\frac{221}{b} = \frac{17}{20}$$

$$(221)(20) = 17b \qquad \text{Using cross multiplication.}$$

$$4420 = 17b \qquad \text{Simplifying.}$$

$$\frac{4420}{17} = \frac{17b}{17} \qquad \boxed{\text{Dividing each side by 17.}}$$

$$260 = b. \qquad \text{Result of } 4420 \div 17.$$

Thus 85% of 260 is 221.   ■

Practice Problem 5 68% of what is 476?   ■

Example 6   What percent of 4000 is 160?

The percent is unknown. We use the variable p. The base b is 4000. The amount a is 160. Thus

$$\frac{a}{b} = \frac{p}{100} \qquad \text{becomes} \qquad \frac{160}{4000} = \frac{p}{100}$$

If we reduce the right-hand fraction, we have

$$\frac{1}{25} = \frac{p}{100}$$

$$100 = 25p \qquad \text{Using cross-multiplication.}$$

$$\frac{100}{25} = \frac{25p}{25} \qquad \boxed{\text{Dividing each side by 25.}}$$

$$4 = p \qquad \text{Result of } 100 \div 25.$$

Thus 4% of 4000 is 160.   ■

Practice Problem 6   What percent of 3500 is 105?   ■

DEVELOPING YOUR STUDY SKILLS

TAKING NOTES IN CLASS

An important part of mathematics studying is taking notes. In order to take meaningful notes, you must be an active listener. Keep your mind on what the instructor is saying, and be ready with questions whenever you do not understand something.

If you have previewed the lesson material, you will be prepared to take good notes. The important concepts will seem somewhat familiar. You will have a better idea of what needs to be written down. If you frantically try to write all that the instructor says or copy all the examples done in class, you may find your notes to be nearly worthless when you are home alone. You may find that you are unable to make sense of what you have written.

Write down *important* ideas and examples as the instructor lectures, making sure that you are listening and following the logic. Include any helpful hints or suggestions that your instructor gives you or refers to in your text. You will be amazed at how easily these are forgotten if they are not written down.

Successful notetaking requires active listening and processing. Stay alert in class. You will realize the advantages of taking your own notes over copying those of someone else.

EXERCISES 5.3B

Identify p, b, and a. Do not solve for the unknown.

	p	b	a
1. 75% of 660 is 495.	_____	_____	_____
2. 65% of 820 is 532.	_____	_____	_____
3. What is 42% of 400?	_____	_____	_____
4. What is 56% of 600?	_____	_____	_____

Identify p, b, and a. Do not solve for the unknown.

	p	b	a
5. 49% of what is 2450?	———	———	———
6. 38% of what is 2280?	———	———	———
7. 30 is what percent of 50?	———	———	———
8. 70 is what percent of 350?	———	———	———
9. What percent of 25 is 10?	———	———	———
10. What percent of 36 is 9?	———	———	———
11. 400 is 160% of what?	———	———	———
12. 900 is 225% of what?	———	———	———

Solve each problem by using the percent proportion

$$\frac{a}{b} = \frac{p}{100}$$

13. 26% of 350 is what?

14. 42% of 450 is what?

15. 190% of what is 570?

16. 170% of what is 510?

17. 82 is what percent of 500?

18. 75 is what percent of 600?

19. Find 0.7% of 520.

20. Find 0.4% of 650.

21. What percent of 66 is 16.5?

22. What percent of 49 is 34.3?

23. 54 is 30% of what?

24. 120 is 40% of what?

25. 94.6 is what percent of 220?

26. 85.8 is what percent of 260?

27. What is 12.5% of 380?

28. What is 20.5% of 320?

29. Find 0.05% of 5600.

30. Find 0.04% of 8700.

? **To Think About**

31. What is $19\frac{1}{4}$% of 798?

32. $140\frac{1}{2}$% of what number is 10,397?

33. Find 18% of 20% of $3300.

34. Find 42% of 16% of $5500.

Solve each percent problem. Round to the nearest hundredth.

35. What percent of 4550 is 720?

36. What is 16.25% of 65,250?

37. 4.25% of 256.75 is what number?

38. 2,760 is 5.5% of what number?

For Extra Practice Examples and Exercises, turn to page 285.

Solutions to Odd-Numbered Practice Problems

1. (a) Find 83% of 460. **(b)** 18% of what number is 90? **(c)** What percent of 64 is 8?
$p = 83$ $p = 18$ The percent is unknown. Use the variable p.
3. (a) What is 18% of 240?   **(b)** What percent of 64 is 4?
   Percent $p = 18$ Percent is unknown; use the variable p.
   Base $b = 240$ Base $b = 64$
   Amount is unknown; use the variable a.   Amount $a = 4$
5. 68% of what is 476?
   Percent $p = 68$
   Base is unknown; use base $= b$.
   Amount $a = 476$

$$\frac{a}{b} = \frac{p}{100} \quad \text{becomes} \quad \frac{476}{b} = \frac{68}{100}$$

If we reduce the right-hand fraction, we have

$$\frac{476}{b} = \frac{17}{25}$$

$(476)(25) = 17b$	Using cross multiplication.
$11{,}900 = 17b$	Simplifying.
$\dfrac{11{,}900}{17} = \dfrac{17b}{17}$	Dividing each side by 17.
$700 = b$	Result of $11{,}900 \div 17$.

Thus 68% of 700 is 476.

Answers to Even-Numbered Practice Problems

2. (a) $b = 52$, $a = 15.6$ **(b)** Base $= b$, $a = 170$ **4.** 176   **6.** 3%

5.4 SOLVING APPLIED PERCENT PROBLEMS

☐ After studying this section, you will be able to:

Percent problems occur so frequently in everyday life that we can group them by category (commission, discount, interest, etc.) and look at each kind separately, as we do in this section.

1 *Solve general percent problems*

Many percent problems deal with money. *All* percent problems can be tackled using the five-step method of (1) reading and visualizing the problem, (2) writing down the information, (3) making an estimate of the answer, (4) using a mathematical method, such as an equation or proportion to solve for the unknown quantity, and (5) checking to see if the answer is reasonable.

2 *Solve percent problems involving sales tax*

3 *Solve commission problems*

4 *Solve problems where percents are added*

For example (Exercises 5.4, problem 4), we may have the problem, "In the last six years, the college basketball team won 154 out of 280 games. What percent of the games did they win?"

5 *Solve percent of increase/decrease problems*

6 *Solve discount problems*

7 *Solve simple-interest problems*

Reading, visualizing, writing down:

154 games won   280 games

- *Estimating:* The percent should be greater than 50% since if the team had won 140 out of 280 games it would have won 50% of its games.
- *Solving and checking for reasonableness:*

$$\frac{154}{280} = 0.55 = 55\%$$

The answer is reasonable.

🔳 General Percent Problems

Read through the problem carefully. It is often helpful to draw a sketch or diagram to help in understanding the problem.

Example 1 Of all the computers manufactured last month, an inspector found 18 that were defective. This is 2.5% of all the computers manufactured last month. We need to take 2.5% of an unknown number of computers. Let n = the number of unknown computers.

$$2.5\% \times n = 18$$

$$0.025 \times n = 18$$

Divide each side by 0.025.

$$\frac{0.025 \times n}{0.025} = \frac{18}{0.025}$$

$$n = 720$$

```
         720.
0.025 ) 18.000
        175
         50
         50
          0
```

A total of 720 computers were manufactured last month.

Is this answer reasonable? Let's estimate the answer. Since 2.5% is close to 3% we could estimate 3% of 700 is about 21. This is close to 18. Our answer is reasonable. ∎

Practice Problem 1 4800 people, or 12% of all passengers holding tickets for American Airlines flights in one month, did not show up for their flight. How many people held tickets that month? ∎

🔳 Percent Problems Involving Sales Tax

A sales tax is obtained by multiplying the sales tax rate by the purchase price:

> sales tax = sales tax rate × purchase price

Example 2 How much sales tax will you pay on a $299.00 color television set if the sales tax is 5%?

Sales tax = sales tax rate × purchase price   Sales tax = 0.05 × $299.00 = $14.95.

The sales tax is $14.95.

Is this reasonable? $299 is close to $300. 5% of 300 is $15. This is close to our answer, so we feel that our answer is reasonable. ∎

Practice Problem 2 A salesperson rented a hotel room for $62.30 per night. The tax in his state is 8%. What tax does he pay for one night at the hotel? ■

3 Commission Problems

When you are a salesperson, all that you get paid may be in part or in total a certain percentage of the sales you make. The amount of money you get that is a percentage of the value of your sales is called your **commission**. It is calculated by multiplying the percentage (called the **commission rate**) by the value of the sales.

$$\text{Commission} = \text{commission rate} \times \text{value of sales}$$

Example 3 A salesperson has a commission rate of 17%. She sells $32,500 worth of goods in a department store in two months. What is her commission?

$$\text{Commission} = \text{commission rate} \times \text{value of sales}$$
$$\text{Commission} = 17\% \times \$32,500$$
$$= 0.17 \times 32,500$$
$$= 5525$$

Her commission is $5525.00.
    Does this answer seem reasonable? ■

Practice Problem 3 A real estate salesperson earns a commission rate of 6% when he sells a $156,000 home. What is his commission? ■

4 Problems Where Percents Are Added

Percents of the *same quantity* can be added. For example, 50% of your salary added to 20% of your salary = 70% of your salary. 100% of your cost added to 15% of your cost = 115% of your cost. Problems like this are often called markup problems. If we add 15% of the cost of an item to the original cost, the markup is 15%. We will use this idea of adding percents in some applied situations.

Example 4 Walter and MaryAnn are going out to a restaurant. They have a limit of $63.25 to spend for the evening. They want to tip the waitress 15% of the cost of the meal. How much money can they afford to spend on the meal itself?

Visualize the situation.

$$\boxed{\begin{array}{c}\text{Cost}\\\text{of}\\\text{meal}\end{array}} + \boxed{\begin{array}{c}\text{tip}\\\text{of}\\15\%\end{array}} = \boxed{\$63.25}$$

Let $n =$ the cost of the meal. 15% of the cost = the amount of the tip. We want to add the percents of the meal.

$$\boxed{\begin{array}{c}\text{Cost of}\\\text{meal } n\end{array}} + \boxed{\begin{array}{c}\text{tip of } 15\%\\\text{of the cost}\end{array}} = \boxed{\$63.25}$$

$$100\% \text{ of } n + \quad 15\% \text{ of } n \quad = \quad \$63.25$$

Note that 100% of n added to 15% of n is 115% of n.

115% of $n = \$63.25$

$1.15 \times n = 63.25$

$\dfrac{1.15 \times n}{1.15} = \dfrac{63.25}{1.15}$ Dividing both sides by 1.15.

$$\begin{array}{r}55.\\1.15_\wedge \overline{)63.25_\wedge}\\575\\\overline{575}\\575\\\overline{0}\quad n=55\end{array}$$

They can spend up to $55.00 on the meal itself.
    Does this answer seem reasonable? ■

Practice Problem 4 Sue and Sam have $46.00 to spend at a restaurant, including a 15% tip. How much can they spend on the meal itself? ∎

5 Percent of Increase/Decrease Problems

We sometimes need to find the percent that something increases or decreases. If a car costs $7000 and the price decreases $1750, we say that the percent decrease is $\frac{1750}{7000} = 0.25 = 25\%$.

$$\text{Percent of decrease} = \frac{\text{amount of decrease}}{\text{original amount}}$$

Similarly, if a population of 12,000 people increases by 1920 people, we say that the percent increase is $\frac{1920}{12,000} = 0.16 = 16\%$.

$$\text{Percent of increase} = \frac{\text{amount of increase}}{\text{original amount}}$$

Example 5 The population of Center City increased from 50,000 to 59,500. What was the percent of increase?

Amount of increase:

$$\begin{array}{r} 59,500 \\ - 50,000 \\ \hline 9,500 \end{array}$$

$$\text{Percent of increase} = \frac{\text{amount of increase}}{\text{original amount}} = \frac{9500}{50,000}$$
$$= 0.19 = 19\%$$

The percent increase is 19%.
Does this answer seem reasonable? ∎

Practice Problem 5 A new car is bought for $15,000. A year later the value has decreased to $10,500. What is the percent of decrease? ∎

Buy Furniture at Sozio's

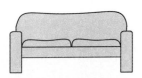

35% DISCOUNT ON
ALL FURNITURE THIS WEEK.

6 Discount Problems

Frequently, we see signs urging us to buy during a sale when the list price is discounted by a certain percent.

The amount of a discount is a product of the discount rate and the list price.

$$\text{Discount} = \text{discount rate} \times \text{list price}$$

Example 6 Jeff purchased a compact disc player on sale at 35% discount. The list price was $430.00.

(a) What was the amount of the discount?
(b) How much did Jeff pay for the CD player?

(a) Discount = discount rate × list price
$$= 35\% \times 430$$
$$= 0.35 \times 430$$
$$= 150.5$$

The discount is $150.50.

(b) We subtract the discount from the list price to get the selling price.

$$\begin{array}{ll} \$430.00 & \text{list price} \\ - \ \ 150.50 & \text{discount} \\ \hline \$279.50 & \text{selling price} \end{array}$$

Jeff paid $279.50 for the CD player.

Does this seem like a reasonable answer?

This problem can be solved in other ways. When a product has a 35% discount we could say that the sale price is 35% off the normal price. Or we could say that what we pay for the product is 65% (100% − 35%) of the normal price. ■

Practice Problem 6   Betty bought a car that lists for $13,600 at a 7% discount.
(a) What was the discount?   **(b)** What did she pay for the car?   ■

7 Simple-Interest Problems

Interest is money paid for the use of money. If you deposit money in a bank, the bank uses that money and pays you interest. If you borrow money, you pay the bank interest for the use of that money. The *principal* is the amount deposited or borrowed. Interest is usually expressed as a percent rate of the principal. The interest rate is assumed to be per year, unless otherwise stated. The formula often used in business to compute interest is

$$\text{Interest} = \text{principal} \times \text{rate} \times \text{time}$$
$$I = P \times R \times T$$

Example 7   Find the interest on a loan of $7500 borrowed at 13% for one year.

$$I = P \times R \times T$$

$P = \text{principal} = \$7500 \qquad R = \text{rate} = 13\% \qquad T = \text{time} = 1 \text{ year}$
$I = 7500 \times 13\% \times 1 = 7500 \times 0.13 = 975$

The interest is $975.00.
        Does this seem like a reasonable answer?   ■

Practice Problem 7   Find the interest on a loan of $5600 borrowed at 12% for one year.   ■

        Our formula is based on a yearly interest rate. Time periods of more than one year or a fractional part of a year are sometimes needed.

Example 8   Find the interest on a loan of $2500 that is borrowed at 9% for

(a) 3 years   **(b)** $\dfrac{1}{4}$ year $I = P \times R \times T$
                                                        $P = \$2500 \qquad R = 9\%$

(a) When $T = 3$ years we have

$$I = 2500 \times 0.09 \times 3 = 225 \times 3 = 675$$

The interest for 3 years is $675.

(b) When $T = \dfrac{1}{4}$ year we have

$$I = 2500 \times 0.09 \times \frac{1}{4}$$

$$= 225 \times \frac{1}{4}$$

$$= \frac{225}{4}$$

$$= 56.25$$

The interest for $\dfrac{1}{4}$ year is $56.25.

        Do these answers seem reasonable?   ■

Practice Problem 8   Find the interest on a loan of $1800 that is borrowed at 11% for

(a) 4 years   **(b)** $\dfrac{1}{2}$ year   ■

EXERCISES 5.4

Solve each percent problem. If necessary, round your answer to the nearest hundredth.

1. A student answered 36 of 45 questions correctly on a test. What percent of the student's answers were correct?

2. A total of 52 out of 65 students in the class are freshmen. What percent of the students are freshmen?

3. In a recent New York state survey, 66 out of 300 drivers interviewed said they did not regularly use seat belts. What percent of the drivers do not use seat belts?

4. In the last six years, the college basketball team won 154 of 280 games. What percent of the games did they win?

(Hint: In problems 5–8, you are finding the percent of a given value.)

5. How much tax will Harriet pay on a purchase of $126.00 if the sales tax rate is 6%?

6. How much tax will Stephen pay on a purchase of $134.00 if the sales tax rate is 6%?

7. Fred's home state has adopted a luxury tax rate of 8% on hotels. What is the tax on a room that costs $55.50 per night?

8. When Alice traveled on vacation, she paid a tax of 8% on her hotel bills. How much tax did she pay on a room that cost $63.50?

(Hint: In problems 9–12, you are given a percent to find an unknown number.)

9. Vijay now earns $5.64 per hour. This is 120% of what he earned last year. What did he earn per hour last year?

10. Sally now earns $6.50 per hour. This is 130% of what she earned last year. What did she earn per hour last year?

11. Defects were found in 72 parts manufactured last week by Ace Industries. This was 1.2% of the total number of parts manufactured last week. How many parts were manufactured last week?

12. A major airline had 99 flights last month that arrived over six hours late at their destination. This was 1.8% of their total flights. How many flights did this airline have last month?

The following problems represent a variety of all types of percent problems.

13. The caterer estimates at a banquet that 85% of the people will request coffee. If the banquet has 340 guests, how many people are expected to request coffee?

14. An insurance salesperson sold $150,000 worth of life insurance last week. He receives a commission of 0.5% of the worth of life insurance he sells. What commission did he earn last week?

15. Bobby purchased some new Reebok sneakers. The sales tax in his state is 5%. He paid $2.12 in tax. What was the sales price of the sneakers?

16. Maha received a raise of 7% of her monthly salary. The amount of her raise was $35 per month. What was her monthly salary before the raise?

17. Sam and Jody have a combined income of $1240 per month. Their food budget is $310 per month. What percent of their budget is allotted for food?

18. Michael has $21.60 of his weekly paycheck withheld for federal income tax. He earns $180 weekly. What percent of his income is withheld for federal income tax?

19. Dr. Vingarten discovered that he spent 36% of his income last year on federal and state taxes. He spent $21,600 on federal and state taxes. What was his income last year?

20. The college raised 15% of its operating income from alumni donations. Last year the alumni donations were $1,500,000. What was the college's operating income last year?

21. In a town of 18,000 eligible workers, the unemployment rate is 4.5%. How many eligible workers are unemployed?

22. Juan traveled 22,400 miles in his car last year. As a salesperson, 65% of his mileage is for business purposes. How many miles did he travel on business last year?

23. Jeff is taking his date to a restaurant. He has $40.25 to spend. He wants to tip the waitress 15% of the cost of the meal. How much money can he afford to spend on the meal itself?

24. Belinda asked Martin out to dinner. She has $47.50 to spend. She wants to tip the waitress 15% of the cost of the meal. How much money can she afford to spend on the meal itself?

25. The population of the town increased from 13,000 to 14,950. What was the percent of increase?

26. The cost of a certain Chevrolet Camaro increased from $16,000 last year to $17,280 this year. What was the percent of increase?

27. A television set originally priced at $495 was sold for $297. What was the percent of decrease?

28. Tay weighed 280 pounds two years ago. After careful supervision at a weight loss center, he reduced his weight to 196 pounds. What was the percent of decrease in his weight?

29. The sales tax rate in Barbara's home state is 7%. The purchase price of Barbara's new coat is $110.00.
    (a) What is the sales tax?
    (b) What is the final price of the coat?

30. The sales tax rate in Samuel's home state is 6%. The purchase price of Samuel's new blazer is $149.00.
    (a) What is the sales tax?
    (b) What is the final price of the blazer?

31. Carolyn bought a new pickup truck. The list price was $16,380, but the dealer gave her a discount of 4%.
    (a) What was the discount?
    (b) How much did she pay for the truck?

32. Russell bought a car that was used as a demonstrator. The list price was $14,500, but the dealer gave him a discount of 9%.
    (a) What was the discount?
    (b) How much did he pay for the car?

33. A video recorder listed for $495. Last week the store had a 30% discount, and Wally bought the video recorder.
    (a) What was the discount?
    (b) How much did Wally pay for the video recorder?

34. Joan and Bob looked at a new living room set that had a list price of $2200. They purchased it during a sale when the discount was 18%.
    (a) What was the discount?
    (b) How much did they pay for the living room set?

35. Lavelle's Department Store offers its salespersons a commission of 14%. Nicole sold $150,000 worth of merchandise last week. What commission did she earn?

36. Fidelity Mutual Insurance offers its salespersons a commission of 0.8%. In three months, Yvonne sold $890,600 worth of life insurance. What was her commission for those three months?

37. A real estate agent sold a total of $1,360,000 in property and earned $40,800 in commissions. What commission rate did she earn?

38. Adam sold encyclopedias last summer to raise tuition money. He sold $37,500 worth of encyclopedias and was paid $3000 in commissions. What commission rate did he earn?

The following four problems refer to simple interest.

39. Jino took out an education loan of $2500 at an interest rate of 8% for one year. He then paid back the loan.
    (a) How much interest did he accrue?
    (b) How much did it cost to pay off the loan totally?

40. Natasha had $1600 in the bank for one year. Her savings account had an interest rate of 6%. She then withdrew all her money.
    (a) How much interest did she earn?
    (b) How much money did she withdraw from the bank?

41. Wendy placed $2300 in her savings account for two years. Her savings account earns 5.5% interest. She then withdrew all her money.
   (a) How much interest did she earn?
   (b) How much money did she withdraw from the bank?
   (c) If she had only deposited the money for six months, how much interest would she have earned?

42. Carl took out a $2900 loan at Friendly Finance Company. The company charges 18% interest. At the end of two years, Carl paid off the loan.
   (a) How much interest did Carl have to pay?
   (b) How much did it cost to pay off the loan totally?
   (c) If he had only taken out the loan for three months, how much interest would he have had to pay?

 To Think About

43. The college bought a new dump truck that listed for $29,340.00. The dealer gave the college 5% discount on the list price. The college had to pay a 3% sales tax on the purchase price of the truck. How much did it cost the college to buy the truck? Round your answer to the nearest cent.

44. A farmer sold 100 pounds of potatoes for $10.00. The shipper charged $4.00 and brought them to a wholesaler. The wholesaler increased the price 25% and sold them to a grocery store. The grocery store increased the price by 18%. One customer bought them. He tipped the bag boy 5% of his total purchase price. Including the tip, how much did the customer pay? Round your answer to the nearest cent.

Cumulative Review Problems

45. Round to the nearest thousand: 1,698,481.

46. Round to the nearest hundred: 2,452,399.

47. Round to the nearest hundredth: 1.63474.

48. Round to the nearest thousandth: 0.793468.

49. Round to the nearest ten thousandth: 0.055613.

50. Round to the nearest ten thousandth: 0.079152.

 Calculator Problems

Round all answers to the nearest cent in problems 51–53.

51. How much sales tax will you pay on a car that cost $16,285.50 if the sales tax is 6%?

52. The Billardis purchased a living room and dining room set on sale at 35% discount. The list price was $4,250.
   (a) What was the discount?
   (b) How much did the Billardis pay for the furniture?

53. Find the interest on a loan of $26,750 that is borrowed at 8.75% for three years.

For Extra Practice Examples and Exercises, turn to page 285.

Solutions to Odd-Numbered Practice Problems

1. Let n = number of people with reserved airline tickets.

$$12\% \text{ of } n = 4800$$
$$0.12 \times n = 4800$$
$$\frac{0.12 \times n}{0.12} = \frac{4800}{0.12} \quad n = 40,000$$

40,000 people had airline tickets that month.

3. Commission = commission rate × value of sales

$$\text{Commission} = 6\% \times \$156,000$$
$$= 0.06 \times 156,000$$
$$= 9360$$

His commission is $9360.

5.

$$\begin{array}{r} 15{,}000 \\ -\ 10{,}500 \\ \hline 4{,}500 \end{array}$$ the amount of decrease

$$\text{Percent of decrease} = \frac{\text{amount of decrease}}{\text{original amount}} = \frac{\$4500}{\$15{,}000}$$
$$= 0.30 = 30\%$$

The rate of decrease is 30%.

7. $I = P \times R \times T$
$$P = \$5600 \qquad R = 12\% \qquad T = 1 \text{ year}$$
$$I = 5600 \times 12\% \times 1$$
$$= 5600 \times 0.12$$
$$= 672$$

The interest is $672 for 1 year.

Answers to Even-Numbered Practice Problems

2. $4.984 rounded to the nearest hundred is $4.98. **4.** $40.00 **6. (a)** $952 **(b)** $12,648 **8. (a)** $792 **(b)** $99

EXTRA PRACTICE: EXAMPLES AND EXERCISES

Section 5.1

Write each as a percent.

Example $\dfrac{0.8}{100}$

$$\frac{0.8}{100} = 0.8\%$$

1. $\dfrac{0.7}{100}$ **2.** $\dfrac{2.3}{100}$ **3.** $\dfrac{9}{100}$

4. $\dfrac{275}{100}$ **5.** $\dfrac{355}{100}$

Example 13 out of 100 apples were spoiled.

$$\frac{13}{100} = 13\% \text{ of the apples were spoiled}$$

6. 17 out of 100 radios were defective

7. 87 out of 100 new car owners were satisfied with their purchase

8. 59 out of 100 high school students have decided on a major

9. 2 out of 100 employees were late for work

10. 69 out of 100 college students participated in college activities

Write each percent as a decimal.

Example 3.59%

0.0359   Move the decimal point two places to the *left*.

11. 5.89% **12.** 18.75% **13.** 28%

14. 7% **15.** 355%

Write each decimal as a percent.

Example 0.0033

0.0033 = 0.33%   Move the decimal point two places to the *right*.

16. 0.0122 **17.** 0.0056 **18.** 0.2

19. 1.25 **20.** 3.95

Section 5.2

Write each percent as a fraction or mixed number in simplest form.

Example 20.75%

$$\frac{20.75}{100} \times \frac{100}{100} \qquad \text{This multiplication removes the decimal.}$$

$$\frac{2075}{10{,}000} = \frac{83}{400} \qquad \begin{array}{l}\text{Reducing by dividing numerator and}\\ \text{denominator by 25.}\end{array}$$

$$20.75\% = \frac{83}{400}$$

1. 35.75% **2.** 45.75% **3.** 52%
4. 86% **5.** 165%

Example $8\dfrac{2}{5}\%$

$$\frac{8\frac{2}{5}}{100} = 8\frac{2}{5} \div \frac{100}{1} = 8\frac{2}{5} \times \frac{1}{100} = \frac{42}{5} \times \frac{1}{100} = \frac{21}{5} \times \frac{1}{50} = \frac{21}{250}$$

6. $6\dfrac{3}{4}\%$ **7.** $7\dfrac{1}{2}\%$ **8.** $15\dfrac{7}{8}\%$

9. $2\dfrac{1}{4}\%$ **10.** $9\dfrac{4}{5}\%$

Write each fraction as a percent. Round to the nearest hundredth of a percent.

Example $\dfrac{5}{8}$

$$\frac{5}{8} \quad 8\,\overline{)5.00}^{\,0.625} \quad \frac{5}{8} = 0.625 = 62.5\% \quad \begin{array}{l}\text{Move the decimal point}\\ \text{two places to the right.}\end{array}$$

11. $\dfrac{3}{4}$ **12.** $\dfrac{7}{8}$ **13.** $\dfrac{9}{16}$

14. $\dfrac{21}{50}$ **15.** $\dfrac{20}{7}$

Write each fraction as a percent. Round to the nearest hundredth of a percent.

Example $\dfrac{2}{15}$

$\dfrac{2}{15} = 0.13333\ldots$ or $0.1\overline{3}$

$0.13333\ldots$ rounded to the nearest ten
thousandth $= 0.1333$
$0.1333 = 13.33\%$ which is correct to the nearest
hundredth of a percent

16. $\dfrac{5}{6}$

17. $\dfrac{2}{3}$

18. $\dfrac{8}{9}$

19. $\dfrac{8}{11}$

20. $2\dfrac{1}{6}$

Section 5.3

Solve each percent problem. Round to the nearest hundredth of a percent.

Example   22 is what percent of 85?

22 is what percent of 85 ⟵ this is the denominator
⤷ this is the numerator

$\dfrac{22}{85} = 0.2588$ rounded to nearest ten thousandth

$0.2588 = 25.88\%$

1. 15 is what percent of 38?
2. 27 is what percent of 54?
3. What percent of 134 is 85?
4. What percent of 345 is 5?
5. What percent is 45 of 122?

Solve each problem. Round to the nearest hundredth of a percent.

Example   What is 18% of 92?

What is 18% of 92?
⤷ this tells us that 0.18×92
$0.18 \times 92 = 16.56$

6. What is 29% of 76?
7. What is 77% of 31?
8. 42% of 165 is what number?
9. 65% of 210 is what number?
10. Find 123% of 470.

Solve each percent problem. Round to the nearest hundredth.

Example   55% of what number is 58?

55% of $n = 58$

$0.55 \times n = 58 \Rightarrow \dfrac{0.55 \times n}{0.55} = \dfrac{58}{0.55} \Rightarrow n = \dfrac{58}{0.55} = 105.45$

11. 34% of what number is 66?
12. 120% of what number is 156?
13. 75 is 16% of what number?
14. 90 is 25% of what number?
15. 6 is 0.05% of what number?

Solve each percent problem. Round to the nearest hundredth.

Example
(a) 60% of what number is 25?
(b) 0.75% of 250 is what number?
(a) 0.60 of $n = 25$
$0.60 \times n = 25$

$\dfrac{25}{0.60} = 41.67$

(b) $0.75\% \times 250 = 0.0075 \times 250 = 1.875 \approx 1.88$

16. 75% of what number is 220?
17. What is 15% of 130?
18. 0.2% of 210 is what number?
19. What percent of 80 is 20?
20. 10 is what percent of 240?

Section 5.4

Round all answers to the nearest hundredth.

Example   A lawyer discovered that he spent 38% of his income last year on federal and state taxes. He spent $28,500 on federal and state taxes. What was his income last year?

$28,500 is 38% of what?

$\$28{,}500 = 0.38 \times n$

$\dfrac{28{,}500}{0.38} = \dfrac{0.38 \times n}{0.38}$

$75{,}000 = n$

1. The caterer estimates that 80% of the people at a banquet will request coffee. If the banquet has 550 guests, how many people are expected to request coffee?

2. Jean purchased a new watch. The sales tax in her state is 6%. She paid $3.90 in tax. What was the sales price of the watch?

3. Michael received a raise of $7\dfrac{1}{2}\%$ of his monthly salary. The amount of his raise was $42.75 per month. What was his monthly salary before the raise?

4. Susan and Matt have a combined income of $1950 per month. Their food budget is $350 per month. What percent of their budget is allotted for food?

5. Nick has $27.50 of his weekly paycheck withheld for federal income tax. He earns $220 weekly. What percent of his income is withheld for federal income tax?

Example William earns a commission at the rate of 8.5%. Last month he sold $15,000 worth of goods. How much commission did he make?

$$\text{Commission} = \text{commission rate} \times \text{value of sales}$$
$$= \quad 0.085 \quad \times \quad 15{,}000$$
$$= \$1275$$

6. An insurance salesman sold $220,000 worth of life insurance last week. He receives a commission of 0.7% of worth of life insurance he sells. What commission did he earn last week?

7. C & J Radio Shop offers its salespersons a commission of 12%. Sam sold $175,000 worth of merchandise last year. What commission did she earn?

8. A-Star Insurance offers its salespersons a commission of 0.9%. In three months, Les sold $960,500 worth of life insurance. What was his commission for those three months?

9. A real estate agent sold a total of $1,750,000 in property and earned $43,750 in commissions. What commission rate did she earn?

10. Jennifer sold $42,700 worth of encyclopedias. She was paid $3843 in commissions. What commission rate did she earn?

Example Carlos bought a car that lists for $16,500 at a 5% discount.
(a) What was the discount?
(b) What did he pay for the car?
(a) Discount = discount rate × list price
$$= \quad 0.05 \quad \times \quad 16{,}500$$
$$= \$825$$
(b) $16,500   list price
  −   825   discount
    $15,675   selling price

11. Janice bought a new 4 × 4 truck. The list price was $18,950, but the dealer gave her a discount of 3%.
(a) What was the discount?
(b) How much did she pay for the truck?

12. Margie bought a car that was used as a demonstrator. The list price was $13,850, but the dealer gave her a discount of 8%.
(a) What was the discount?
(b) How much did he pay for the car?

13. A color television set listed for $695. Last week the store had a 25% discount, and Sandy bought the color set.
(a) What was the discount?
(b) How much did Sandy pay for the set?

14. Jenea and Allen looked at a new dining room set that had a list price of $1900. They purchased it during a sale when the discount was 15%.
(a) What was the discount?
(b) How much did they pay for the dining room set?

15. Mr. Sanson purchased a computer at a 30% discount. The list price was $2150.
(a) What was the discount?
(b) What did he pay for the computer?

Example Joseph placed $2900 in his savings account for two years. His account earns 7.5% interest. He then withdrew all his money.
(a) How much interest did he earn?
(b) How much money did he withdraw from the bank?
(c) If he only deposited the money for three months, how much interest would he earn?

(a) $I = P \times R \times T$
$$= \$2900 \times 0.075 \times 2$$
$$= \$435$$
(b) He withdrew the $2900 plus the interest earned.
$$2900 + 435 = \$3335$$
(c) $I = P \times R \times T$
$$= 2900 \times 0.075 \times \frac{1}{4} \longleftarrow 3 \text{ months} = \frac{3}{12} = \frac{1}{4} \text{ of a year}$$
$$= \$54.375 \qquad \text{(time } must \text{ be in years)}$$
$$= \$54.38 \text{ rounded}$$

16. Art took out an education loan of $5000 at an interest rate of 7% for one year. He than paid back the loan.
(a) How much interest did he accrue?
(b) How much did it cost to pay off the loan totally?

17. Bob had $1200 in the bank for one year. His savings account had an interest rate of 6.5%. He then withdrew all his money.
(a) How much interest did he earn?
(b) How much money did he withdraw from the bank?

18. Natalie placed $3200 in her savings account for three years. Her savings account earns 7.5% interest. She then withdrew all her money.
(a) How much interest did she earn?
(b) How much money did she withdraw from the bank?
(c) If she had only deposited the money for six months, how much interest would she have earned?

19. Curt took out a $3500 loan at a finance company. The company charges 18% interest. At the end of two years Curt paid off the loan.
(a) How much interest did Curt have to pay?
(b) How much did it cost to totally pay off the loan?
(c) If he had only taken out the loan for four months, how much interest would he have had to pay?

20. Mr. Jensen invested $6500 in mutual funds earning 11% simple interest.
(a) How much interest will he earn in three months?
(b) How much interest will he earn in two years?

CHAPTER ORGANIZER

Topic	Procedure	Examples
Converting a decimal to a percent, p. 255	1. Move the decimal point two places to the right. 2. Add the percent sign.	$0.19 = 19\%$ $0.516 = 51.6\%$ $0.04 = 4\%$ $1.53 = 153\%$ $0.006 = 0.6\%$
Converting a fraction with a denominator of 100 to a percent, p. 253	1. Use the numerator only. 2. Add the percent sign.	$\dfrac{29}{100} = 29\%$ $\dfrac{5.6}{100} = 5.6\%$ $\dfrac{3}{100} = 3\%$ $\dfrac{7\frac{1}{3}}{100} = 7\frac{1}{3}\%$ $\dfrac{231}{100} = 231\%$
Changing a fraction (whose denominator is not 100) to a percent, p. 259	1. Divide the numerator into the denominator and obtain a decimal. 2. Change the decimal to a percent.	$\dfrac{13}{50} = 0.26 = 26\%$ $\dfrac{1}{20} = 0.05 = 5\%$ $\dfrac{3}{800} = 0.00375 = 0.375\%$ $\dfrac{312}{200} = 1.56 = 156\%$
Changing a mixed number to a percent, p. 258	1. Change the mixed number to decimal form. 2. Transform the decimal to an equivalent percent.	$3\frac{1}{4} = 3.25 = 325\%$ $6\frac{4}{5} = 6.8 = 680\%$ $7\frac{3}{8} = 7.375 = 737.5\%$
Changing a percent to a decimal, p. 255	1. Move the decimal point two places to the left. 2. Drop the percent sign.	$49\% = 0.49$ $2\% = 0.02$ $0.5\% = 0.005$ $196\% = 1.96$ $1.36\% = 0.0136$
Changing a percent to a fraction, p. 257	1. Change the percent to a fraction with a denominator of 100. 2. If the numerator does not contain a decimal or a fraction, reduce the fraction if possible. 3. If the percent contains a decimal, first change the percent to a decimal. Then write it as a fraction, and reduce the fraction. 4. If the numerator contains a fraction, change the numerator into an improper fraction. Then simplify by the "invert and multiply" rule.	$25\% = \dfrac{25}{100} = \dfrac{1}{4}$ $2.72\% = 0.0272$ $38\% = \dfrac{38}{100} = \dfrac{19}{50}$ $= \dfrac{272}{10,000} = \dfrac{17}{625}$ $130\% = \dfrac{130}{100} = \dfrac{13}{10}$ $7\frac{1}{8}\% = \dfrac{7\frac{1}{8}}{100}$ $5.8\% = 0.058$ $= 7\frac{1}{8} \div \dfrac{100}{1}$ $= \dfrac{58}{1000} = \dfrac{29}{500}$ $= \dfrac{57}{8} \times \dfrac{1}{100}$ $= \dfrac{57}{800}$
Solving percent problems by translating to equations p. 264	1. Translate by replacing: "of" by x "is" by $=$ "what" by n "find" by $n =$ 2. Solve the resulting equation.	**(a)** What is 3% of 56? $n = 3\% \times 56$ $n = (0.03)(56)$ $= 1.68$ **(b)** 16% of what is 208? $16\% \times n = 208$ $0.16n = 208$ $\dfrac{0.16n}{0.16} = \dfrac{208}{0.16}$ $n = 1300$

Topic	Procedure	Examples
Solving percent problems by using proportions (Optional), p. 270	1. Identify the parts of the percent proportion. a = the amount b = the base (the entire quantity involved; it usually appears after the word "of") p = the percent number 2. Solve the percent proportion $\dfrac{a}{b} = \dfrac{p}{100}$ with the values obtained in step 1.	(a) What is 28% of 420? The percent $p = 28$. The base $b = 420$. The amount a is unknown, we use the variable a. $$\frac{a}{b} = \frac{p}{100} \qquad \frac{a}{420} = \frac{28}{100}$$ If we reduce the right-hand side, we have $$\frac{a}{420} = \frac{7}{25}$$ $$25a = (7)(420)$$ $$25a = 2940$$ $$\frac{25a}{25} = \frac{2940}{25}$$ $$a = 117.6$$ (b) 64% of what is 320? The percent $p = 64$. The base is unknown. We use the variable b. The amount $a = 320$. $$\frac{a}{b} = \frac{p}{100} \qquad \frac{320}{b} = \frac{64}{100}$$ If we reduce the right-hand side, we have $$\frac{320}{b} = \frac{16}{25}$$ $$(320)(25) = 16b$$ $$8000 = 16b$$ $$\frac{8000}{16} = \frac{16b}{16}$$ $$500 = b$$

REVIEW PROBLEMS CHAPTER 5

5.1 *Write as a percent. Round to the nearest hundredth of a percent if necessary.*

1. 0.87

2. 0.55

3. 0.276

4. 0.521

5. 0.0713

6. 0.0608

7. 2.52

8. 4.37

9. 1.036

10. 1.052

11. 0.006

12. 0.002

13. 0.0029

14. 0.0053

15. $\dfrac{72}{100}$

16. $\dfrac{81}{100}$

17. $\dfrac{19.5}{100}$

18. $\dfrac{21.6}{100}$

19. $\dfrac{0.24}{100}$

20. $\dfrac{0.98}{100}$

21. $\dfrac{4\frac{1}{12}}{100}$

22. $\dfrac{3\frac{3}{8}}{100}$

23. $\dfrac{317}{100}$

24. $\dfrac{225}{100}$

5.2 *Change each fraction to a percent. Round to the nearest hundredth of a percent, when necessary.*

25. $\dfrac{16}{25}$ **26.** $\dfrac{22}{25}$ **27.** $\dfrac{18}{20}$ **28.** $\dfrac{13}{40}$ **29.** $\dfrac{5}{11}$ **30.** $\dfrac{4}{9}$ **31.** $2\dfrac{1}{4}$

32. $3\dfrac{3}{4}$ **33.** $4\dfrac{3}{7}$ **34.** $5\dfrac{2}{9}$ **35.** $\dfrac{152}{80}$ **36.** $\dfrac{165}{90}$ **37.** $\dfrac{3}{800}$ **38.** $\dfrac{2}{500}$

Change each percent to decimal form.

39. 0.2% **40.** 0.7% **41.** 21.9% **42.** 43.1% **43.** 166% **44.** 139% **45.** $32\dfrac{1}{8}\%$ **46.** $26\dfrac{3}{8}\%$

Change each percent to fractional form.

47. 82% **48.** 38% **49.** 185% **50.** 215% **51.** 16.4%

52. 30.5% **53.** $31\dfrac{1}{4}\%$ **54.** $3\dfrac{1}{8}\%$ **55.** 0.05% **56.** 0.06%

Complete the following chart to find equivalent forms.

	Fractional Notation	Decimal Notation	Percent Notation
57.	$\dfrac{3}{5}$		
58.	$\dfrac{7}{8}$		
59.			37.5%
60.			56.25%
61.		0.008	
62.		0.45	

5.3 *Solve each percent problem. Round to the nearest hundredth if necessary.*

63. What is 96% of 300?   **64.** What is 85% of 260?   **65.** 3 is 20% of what number?

66. 90 is 45% of what number?   **67.** 50 is what percent of 125?   **68.** 18 is what percent of 50?

69. Find 125% of 46.　　　　**70.** Find 118% of 60.　　　　**71.** 92% of what number is 147.2?

72. 68% of what number is 95.2?　　　　**73.** What percent of 70 is 14?　　　　**74.** What percent of 60 is 28?

75. 0.5% of 2600 is what number?　　　　**76.** 0.8% of 3500 is what number?

77. What percent of 28 is 130?　　　　**78.** What percent of 36 is 120?

5.4　*Solve each applied percent problem. Round your answer to the nearest hundredth if necessary.*

79. Professor Wonson found that 34% of his class is left-handed. He has 150 students in his class. How many are left-handed?

80. A Vermont truck dealer found that 64% of all the trucks he sold were four-wheel drive. If he sold 150 trucks, how many had four-wheel drive?

81. A total of six computers in a shipment of 144 computers were found to be defective. What percent were defective?

82. Rafael obtained 16 correct answers on a test containing 22 questions. What percent of his problems were correct?

83. Today Yvonne's car has 61% of the value that it had 2 years ago. Today it is worth $6832. What was it worth 2 years ago?

84. A charity organization spent 12% of its budget for administrative expenses. It spent $18,000 on administrative expenses. What was the total budget?

85. In a test of car tires three of 20 tires failed the test. What percent failed the test?

86. Moorehouse Industries received 600 applications and hired 45 of the applicants. What percent of the applicants obtained a job?

87. Roberta earns a commission at the rate of 7.5%. Last month she sold $16,000 worth of goods. How much commission did she make?

88. Gary purchased a car for $18,600. The sales tax in his state is 3%. What did he pay in sales tax?

89. The average temperature in Winchester during the last 10 years has been 45.8°. In the previous 10 years the average temperature was 44°. What percent of increase is this?

90. Sally invested $6000 in mutual funds earning 11% simple interest. How much interest will she earn in
(a) Six months?
(b) Two years?

91. Irene purchased a dining room set at a 20% discount. The list price was $1595.
(a) What was the discount?
(b) What did she pay for the set?

92. Joan and Michael budget 38% of their income for housing. They spend $684 per month for housing. What is their monthly income?

Write as a percent.

1. 0.46

2. 0.93

3. 0.023

4. 0.057

5. 4.82

6. 5.03

7. 0.002

8. 0.009

9. $\dfrac{17}{100}$

10. $\dfrac{29}{100}$

11. $\dfrac{28.6}{100}$

12. $\dfrac{16.4}{100}$

13. $\dfrac{7\frac{3}{8}}{100}$

14. $\dfrac{4\frac{3}{4}}{100}$

1. _____

2. _____

3. _____

4. _____

5. _____

6. _____

7. _____

8. _____

9. _____

10. _____

11. _____

12. _____

13. _____

14. _____

15. _____

16. _____

17. _____

18. _____

19. _____

20. _____

21. _____

22. _____

23. _____

24. _____

25. _____

26. _____

27. _____

28. _____

29. _____

30. _____

31. _____

32. _____

33. _____

34. _____

35. _____

36. _____

Change each fraction to a percent. Round to the nearest hundredth of a percent, when necessary.

15. $\dfrac{17}{20}$ **16.** $\dfrac{45}{90}$ **17.** $\dfrac{36}{20}$

18. $\dfrac{17}{16}$ **19.** $\dfrac{5}{7}$ **20.** $\dfrac{4}{7}$

21. $\dfrac{18}{56}$ **22.** $\dfrac{17}{43}$ **23.** $7\dfrac{3}{4}$

24. $9\dfrac{3}{5}$ **25.** $\dfrac{5}{800}$ **26.** $\dfrac{3}{500}$

Write each percent as a fraction in simplified form.

27. 84%   **28.** 35%   **29.** 77%

30. 51%   **31.** 165%   **32.** 210%

33. $4\dfrac{1}{8}\%$ **34.** $3\dfrac{1}{4}\%$

35. $21\dfrac{2}{3}\%$ **36.** $17\dfrac{1}{3}\%$

Solve the following percent problems. Round to the nearest hundredth if necessary.

1. What is 25% of 1300?

2. Find 35% of 420.

3. What percent is 120 of 135?

4. 70 is what percent of 120?

5. 22% of what number is 286?

6. 19.5 is 7% of what number?

7. 117 is 130% of what number?

8. 160% of what number is 240?

9. 7 out of 250 parts are defective. What percent is defective?

10. 11 out of 400 students did not register for class. What percent did not register?

1. _____

2. _____

3. _____

4. _____

5. _____

6. _____

7. _____

8. _____

9. _____

10. _____

11. _____

12. _____

13. _____

14. _____

15. _____

16. _____

11. Terry bought a new television set for $245.00. The sales tax rate is 5% in his state. How much sales tax did she pay?

12. Vida sells encyclopedias and earns a commission rate of 14%. She sold $168,000 worth of encyclopedias last year. What was her commission?

13. The population of West Townsend was 16,000 ten years ago. It is now 19,360. What was the rate of increase?

14. Shao purchased a sofa that was listed at $499 at a 24% discount.
    (a) What was his discount?
    (b) How much did he pay for the sofa?

15. Inez took out a loan of $3000 at 14% interest.
    (a) How much interest will she pay if she borrows the money for two years?
    (b) How much interest will she pay if she borrows the money for three months?

16. Marshall now earns $26,400 per year. This is 120% of what he earned last year. What was his annual salary last year?

Write as a percent. Round to the nearest hundredth of a percent if necessary.

1. 0.56

2. 0.03

3. 0.008

4. 0.127

5. 1.35

6. $\dfrac{83}{100}$

7. $\dfrac{5.6}{100}$

8. $\dfrac{2\frac{1}{6}}{100}$

Change each fraction to a percent. Round to the nearest hundredth of a percent if necessary.

9. $\dfrac{17}{40}$

10. $\dfrac{70}{350}$

11. $\dfrac{102}{85}$

12. $2\frac{1}{4}$

Write each decimal as a percent.

13. 0.1852

14. 2.136

Write each percent as a fraction in simplified form.

15. 165%

16. $9\frac{1}{4}\%$

1. _____

2. _____

3. _____

4. _____

5. _____

6. _____

7. _____

8. _____

9. _____

10. _____

11. _____

12. _____

13. _____

14. _____

15. _____

16. _____

17. _____

18. _____

19. _____

20. _____

21. _____

22. _____

23. _____

24. _____

25. _____

26. _____

27. _____

28. _____

29. _____

30. _____

Solve each percent problem. Round to the nearest hundredth if necessary.

17. What is 19% of 167?

18. 38.4 is 24% of what number?

19. What percent of 42 is 30?

20. Find 0.4% of 25,000.

21. 12% of what number is 600?

22. 156 is what percent of 300?

23. 126% of 470 is what number?

24. What percent is 8 of 450?

Solve each applied percent problem. Round to the nearest hundredth if necessary.

25. A real estate agent sells a house for $136,700. She gets a commission of 3% on the sale. What is her commission?

26. Julia and Charles bought a new dishwasher at a 22% discount. The list price was $428.
   (a) What was the discount?
   (b) How much did they pay for the dishwasher?

27. An inspector found that 85 out of 94 parts were not defective. What percent of the parts were not defective?

28. At Cedars University the number of hours a three-credit college course meets in a semester is 48 hours. Twenty years ago a three-credit course met for 53 hours. What is the percent of decrease?

29. A total of 5980 people voted in the city election. This was 46% of the registered voters. How many registered voters are in the city?

30. Wanda borrowed $2000 at a 14% simple rate of interest.
   (a) How much interest did she pay in six months?
   (b) How much interest did she pay in two years?

Approximately one-half of this test is based on Chapter 5 material. The remainder is based on material covered in Chapters 1–4.

Do each problem. Simplify your answer.

1. Add: 38
            196
         + 2007

2. Subtract: 23,007
                  − 14,563

3. Multiply: 126
               × 42

4. Divide: $36 \overline{)3204}$.

5. Add: $2\frac{1}{4} + 3\frac{1}{3}$.

6. Subtract: $\frac{11}{12} - \frac{5}{6}$.

7. Multiply: $3\frac{17}{36} \times \frac{21}{25}$.

8. Divide: $\frac{5}{12} \div 1\frac{3}{4}$.

9. Round to the nearest tenth: 5731.652.

10. Add: 5.6
              3.21
             18.3
          + 7.008

11. Multiply: 5.62
                × 0.3

12. Divide: $1.4 \overline{)0.5152}$.

13. Write as a rate in simplest form: 78 pounds to 130 square feet.

14. Is this proportion true or false?

$$\frac{20}{25} = \frac{300}{375}$$ Cross multiply

15. Solve the proportion: $\frac{8}{2.5} = \frac{n}{7.5}$.

16. A college has a ratio of 3 faculty for every 19 students. The student body presently has 4263 students. How many faculty are there? Round to the nearest whole number.

1. _____

2. _____

3. _____

4. _____

5. _____

6. _____

7. _____

8. _____

9. _____

10. _____

11. _____

12. _____

13. _____

14. _____

15. _____

16. _____

17. _____

18. _____

19. _____

20. _____

21. _____

22. _____

23. _____

24. _____

25. _____

26. _____

27. _____

28. _____

29. _____

30. _____

In problems 17–30, round to the nearest hundredth if necessary.

Write as a percent.

17. 0.023

18. $\dfrac{46.8}{100}$

19. 1.98

20. Change $\dfrac{3}{80}$ to a percent.

Write as a decimal.

21. 243%

22. $6\dfrac{3}{4}\%$

23. What percent of 214 is 38?

24. Find 1.7% of 6740.

25. 219 is 73% of what number?

26. 114% of 630 is what number?

27. Alice bought a new car. She got a 7% discount. The car listed for $9000. How much did she pay for the car?

28. A total of 896 freshmen were admitted to King Frederich College. This is 28% of the student body. How big is the student body?

29. The air pollution level in Centerville is 8.86. Ten years ago it was 7.96. What is the percent of increase of the air pollution level?

30. Fred borrowed $1600 for two years. He was charged for simple interest at an interest rate of 11%. How much interest did he pay?

Measurement

Constant calculations and measurements are made and monitored by air traffic controllers who ensure that each airplane and jet takes off and lands safely.

6

PRETEST CHAPTER 6

If you are familiar with the topics in this chapter, take this test now. Check your answers with those in the back of the book. If an answer was wrong or you couldn't do a problem, study the appropriate section of the chapter.

If you are not familiar with the topics in this chapter, don't take this test now. Instead, study the examples, work the practice problems, and then take the test.

This test will help you identify those concepts that you have mastered and those that need more study.

Section 6.1 Convert. When necessary, express your answer as a decimal rounded to the nearest hundredth.

1. 17 ft = _____ in.
2. 7 gal = _____ pt
3. 2 mi = _____ yd
4. 3.2 tons = _____ lb
5. 22 min = _____ sec
6. 6 gal = _____ qt

Section 6.2 Perform each conversion.

7. 5.32 km = _____ m
8. 46.8 m = _____ cm
9. 986 mm = _____ cm
10. 27 mm = _____ m
11. 5296 mm = _____ cm
12. 123 m = _____ km

Convert to meters and add.

13. 1.2 km + 192 m + 984 m
14. 3862 cm + 9342 mm + 46.3 m

Section 6.3 Perform each conversion.

15. 3.82 L = _____ mL
16. 3162 g = _____ kg
17. 56.3 kg = _____ t
18. 4.8 kL = _____ L
19. 568 mg = _____ g
20. 8.9 L = _____ cm³

Section 6.4 Perform each conversion. Round to the nearest hundredth when necessary.

21. 12 cm = _____ in.
22. 4.2 ft = _____ m
23. 96 km = _____ mi
24. 482 gal = _____ L
25. 1.4 oz = _____ g
26. 47 kg = _____ lb

Section 6.5 Write in scientific notation.

27. 486,000
28. 0.00002

Write in ordinary notation.

29. 5.93×10^{-3}
30. 2.6×10^8

Section 6.6 Solve each problem. Round your answer to the nearest hundredth if necessary.

31. Find the perimeter of the triangle at left. Express your answer in **feet**.
32. The radio reported the temperature today at 35°C. The record high temperature for this day is 98°F.
    (a) What was the Fahrenheit temperature today?
    (b) Did the temperature today set a new record?
33. Juanita traveled in Mexico for 2 hours at 95 kilometers per hour. She had a distance of 130 miles to travel. How far does she still need to travel? (Express your answer in **miles**.)
34. A pump is running 1.5 quarts per minute. How many gallons per hour is this?
35. One beaker contains 2.89×10^{15} atoms of radioactive iodine, and a second beaker contains 4.31×10^{15} atoms. What is the total number of atoms in both containers?
36. A garage charges $1.50 per hour. Bob parked in the garage for $1\frac{1}{4}$ days. What was his parking charge?

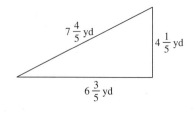

$7\frac{4}{5}$ yd

$4\frac{1}{5}$ yd

$6\frac{3}{5}$ yd

6.1 AMERICAN UNITS

After studying this section, you will be able to:

1 Know the basic units in the American system

2 Convert from one unit to another.

We often ask questions of measurement. How far is it to work? How much does this hold? What is the weight of this box? How long will it be until class is over? To answer these questions we need to agree on a unit of measure for each type of measurement.

At present there are two main systems of measurements, each with its own set of units: the metric system and the American system. Most countries in the world and most scientific workers use the metric system. In the United States, however, most measurements are made in American units.

The United States is using the metric system more frequently now and may eventually convert to metric units as the standard. But for now we need to be familiar with American units, which we cover in this section.

1 Here are the important relationships you should know in order to understand and use the American system of measurement. Your instructor may require you to memorize them.

Length
12 inches = 1 foot
3 feet = 1 yard
5280 feet = 1 mile
1760 yards = 1 mile

Time
60 seconds = 1 minute
60 minutes = 1 hour
24 hours = 1 day
7 days = 1 week

Note that time is measured in the same units in both metric and American systems.

Weight
16 ounces = 1 pound
2000 pounds = 1 ton

Volume
2 cups = 1 pint
2 pints = 1 quart
4 quarts = 1 gallon

We can choose to measure a given object—say, a bridge—using a small unit (an inch), a larger unit (a foot), or a still larger unit (a mile). We may say that the bridge spans 7920 inches, 660 feet, or an eighth of a mile. Although we probably would not express our measurement in terms of an inch, the bridge length is the same whatever unit of measurement we choose.

Notice that the smaller the measuring unit, the larger the number of those units in the final measurement. The inch is the smallest unit in our example, and the inch measurement has the greatest number of units (7920). Conversely, the mile is the largest unit, and it has the smallest number of units (an eighth equals 0.125). Whatever measuring system you use, and whatever you measure (length, volume, and so on), the smaller the unit you use, the larger the number of those units.

After studying the values in the length, time, weight, and volume tables, see if you can quickly do Example 1.

Example 1 Answer rapidly the following questions.

(a) How many inches in a foot?
(b) How many yards in a mile?
(c) How many seconds in a minute?
(d) How many hours in a day?
(e) How many pounds in a ton?
(f) How many cups in a pint?

(a) 12   **(b)** 1760   **(c)** 60   **(d)** 24   **(e)** 2000   **(f)** 2   ∎

Practice Problem 1 Answer rapidly the following questions.
(a) How many feet in a yard?
(b) How many feet in a mile?
(c) How many minutes in an hour?
(d) How many days in a week?
(e) How many ounces in a pound?
(f) How many pints in a quart?
(g) How many quarts in a gallon?   ∎

2 Converting One Measurement to Another

Every measurement has two parts: a number and a unit. There are several approaches people take in converting one measurement to another. The easiest to remember and use is the concept of multiplying a fraction by 1. When we multiply a value by 1, the value is not changed. Consider the relationship 12 inches = 1 foot. If we used that relationship to write a fraction of 1, we could write

$$1 = \frac{12 \text{ inches}}{1 \text{ foot}} = \frac{1 \text{ foot}}{12 \text{ inches}}$$

These two fractions are called **unit fractions**.

If we want to change 180 inches to a certain number of feet, we could multiply 180 inches by a unit fraction. We know that this will not change the value of the measurement, only the way it is written, in new units. This technique of changing units by the use of a unit fraction is often called "dimensional analysis." Since we start with inches, we select a unit fraction with inches in the denominator. Since we want to change to feet, we select a unit fraction with feet in the numerator. We treat the units as if they were factors, and "cancel." We divide the numerals as we ordinarily do.

$$180 \text{ inches} \times \frac{1 \text{ foot}}{12 \text{ inches}} = \frac{180 \text{ feet}}{12} = 15 \text{ feet}$$

It is important when multiplying by this unit fraction of 1 to select the correct unit fraction.

> When using unit fractions the unit we want to change to should be in the numerator of the unit fraction and in the answer. The unit we want to eliminate should be in our original problem and in the denominator of the unit fraction. We cancel this unit.

Example 2 Convert each measurement.

(a) 8800 yards to miles

(b) 26 feet to inches

(a) $8800 \text{ yards} \times \frac{1 \text{ mile}}{1760 \text{ yards}} = \frac{8800}{1760} \text{ miles} = 5 \text{ miles}$

(b) $26 \text{ feet} \times \frac{12 \text{ inches}}{1 \text{ foot}} = 26 \times 12 \text{ inches} = 312 \text{ inches}$ ■

Practice Problem 2 Convert each measurement.
(a) 15,840 feet to miles

(b) 17 feet to inches ■

Some conversions involve fractions or decimals.

Example 3 Convert each measurement.

(a) 26.48 miles to yards

(b) $3\frac{2}{3}$ feet to yards

(a) $26.48 \text{ miles} \times \frac{1760 \text{ yards}}{1 \text{ mile}} = 46{,}604.8 \text{ yards}$

(b) $3\frac{2}{3} \text{ feet} \times \frac{1 \text{ yard}}{3 \text{ feet}} = \frac{11}{3} \times \frac{1}{3} \text{ yard} = \frac{11}{9} \text{ yards} = 1\frac{2}{9} \text{ yards}$ ■

Practice Problem 3 Convert each measurement.

(a) 18.93 miles to feet

(b) $16\frac{1}{2}$ inches to yards ■

Example 4 Lynda's new car weighs 2.43 tons. How many pounds is that?

$$2.43 \text{ tons} \times \frac{2000 \text{ pounds}}{1 \text{ ton}} = 4860 \text{ pounds} \quad \blacksquare$$

Practice Problem 4 The package weighs 760.5 pounds. How many ounces does it weigh? ■

Example 5 The chemistry lab has 34 quarts of weak hydrochloric acid. How many gallons of this acid are in the lab? (Express your answer as a decimal.)

$$34 \text{ quarts} \times \frac{1 \text{ gallon}}{4 \text{ quarts}} = \frac{34}{4} \text{ gallons} = 8.5 \text{ gallons} \quad \blacksquare$$

Practice Problem 5 19 pints of milk is the same as how many quarts? (Express your answer as a decimal.) ■

Example 6 The printout on a computer program reads that a particular job took 144 seconds. How many minutes is that? (Express your answer as a decimal.)

$$144 \text{ seconds} \times \frac{1 \text{ minute}}{60 \text{ seconds}} = \frac{144}{60} \text{ minutes} = 2.4 \text{ minutes} \quad \blacksquare$$

Practice Problem 6 Joe's time card read "Hours worked today: 7.2." How many minutes are in 7.2 hours? ■

Example 7 The all-night Charlotte garage charges $1.50 per hour for parking both day and night. A businessman left his car there for $2\frac{1}{4}$ days. How much was he charged?

$2\frac{1}{4} \text{ days} = \frac{9}{4} \text{ days}$ Changing to an improper fraction so that multiplication can be done.

$\frac{9}{\overset{}{4}} \text{ days} \times \frac{\overset{6}{\cancel{24}} \text{ hours}}{1 \text{ day}} = 54 \text{ hours}$ Canceling the numeral 4 as well as the day unit.

$54 \text{ hours} \times \frac{1.50 \text{ dollars}}{1 \text{ hour}} = 81 \text{ dollars}$ Multiplying each hour by $1.50 per hour.

The businessman paid $81.00. ■

Practice Problem 7 A businesswoman parked her car at a garage for $1\frac{3}{4}$ days. The garage charges $1.50 per hour. How much did she pay to park the car? ■

To Think About

Where did people first come up with the idea of "multiply by a unit fraction"? What mathematical principles are really involved? Actually, multiplying by a unit fraction is the same as solving a proportion. Consider Example 5, where we changed 34 quarts to 8.5 gallons by multiplying:

$$34 \text{ quarts} \times \frac{1 \text{ gallon}}{4 \text{ quarts}} = \frac{34}{4} \text{ gallons} = 8.5 \text{ gallons}$$

What we were actually doing is setting up a proportion

1 gallon is to 4 quarts as n gallons is to 34 quarts

and solving:

$$\frac{1 \text{ gallon}}{4 \text{ quarts}} = \frac{n \text{ gallons}}{34 \text{ quarts}}$$

Cross-multiply:

$$1 \text{ gallon} \times 34 \text{ quarts} = 4 \text{ quarts} \times n \text{ gallons}$$

Divide both sides of the equation by 4 quarts:

$$\frac{1 \text{ gallon} \times 34 \cancel{\text{ quarts}}}{4 \cancel{\text{ quarts}}} = \frac{\cancel{4 \text{ quarts}} \times n \text{ gallons}}{\cancel{4 \text{ quarts}}}$$

$$1 \text{ gallon} \times \frac{34}{4} = n \text{ gallons}$$

$$8.5 \text{ gallons} = n \text{ gallons}$$

Thus the number of gallons is 8.5. Using proportions takes a little longer, so multiplying by a unit fraction is the more popular method. For more use of proportions in converting units, see problems 55 and 56. ■

EXERCISES 6.1

From memory, write down the equivalent values.

1. 1 foot = _____ inches

2. 1 yard = _____ feet

3. _____ yards = 1 mile

4. _____ feet = 1 mile

5. 1 ton = _____ pounds

6. 1 pound = _____ ounces

7. _____ quarts = 1 gallon

8. _____ cups = 1 pint

9. 1 quart = _____ pints

10. 1 day = _____ hours

11. _____ seconds = 1 minute

12. _____ minutes = 1 hour

Convert using unit fractions. When necessary, express your answer as a decimal.

13. 12 feet = _____ yards

14. 21 feet = _____ yards

15. 84 inches = _____ feet

16. 180 inches = _____ feet

17. 10,560 feet = _____ miles

18. 5280 yards = _____ miles

19. 7 miles = _____ yards

20. 6 miles = _____ feet

21. 9 feet = _____ inches

22. 7 feet = _____ inches

23. 41 yards = _____ feet

24. 36 yards = _____ feet

25. 75 inches = _____ feet

26. 87 inches = _____ feet

27. 176 ounces = _____ pounds

28. 144 ounces = _____ pounds

29. 13 tons = _____ pounds

30. 17 tons = _____ pounds

31. 2.25 pounds = _____ ounces

32. 4.25 pounds = _____ ounces

33. 7 gallons = _____ quarts

34. 5 gallons = _____ quarts

35. 18 pints = _____ quarts

36. 24 pints = _____ quarts

37. 31 pints = _____ cups

38. 27 pints = _____ cups

39. 8 gallons = _____ pints

40. 6 gallons = _____ pints

41. 5 weeks = _____ days

42. 7 weeks = _____ days

43. 660 minutes = _____ hours

44. 780 minutes = _____ hours

45. 11 days = _____ hours

46. 7 days = _____ hours

47. 70 minutes = _____ seconds

48. 50 minutes = _____ seconds

49. 18 hours = _____ seconds

50. 15 hours = _____ seconds

51. A businessman left his car in an all-night garage for $2\frac{1}{2}$ days. The garage charges $2.25 per hour. How much did he pay for parking?

52. A businesswoman left her car in an all-night garage for $1\frac{1}{2}$ days. The garage charges $1.80 per hour. How much did she pay for parking?

53. The supermarket charges $2.00 per pound for lean hamburger. Sally bought 23 ounces. How much did she pay?

54. The supermarket charges $1.60 per pound for regular hamburger. Wally bought 28 ounces. How much did he pay?

 To Think About

There are 6080 feet in a nautical mile and 5280 feet in a land mile. On sea or in the air, distance and speed are often measured in nautical miles. The ratio of regular miles to nautical miles is 38 to 33.

55. An ocean liner traveled 1000 nautical miles. That is equivalent to how many land miles? Round to the nearest whole mile.

56. An ocean liner traveled a distance equivalent to 1000 land miles. How many nautical miles did it travel? Round to the nearest whole mile.

57. A spacecraft travels 190,000 miles on its way to the moon. How many inches did it travel?

58. A space probe travels for 3 years to get to the planet Saturn. How many *hours* did the trip take? (Assume 365 days in 1 year.)

Cumulative Review Problems

Solve each proportion.

59. $\dfrac{3}{7} = \dfrac{n}{91}$

60. $\dfrac{12}{4.5} = \dfrac{16}{n}$

61. Round to the nearest tenth: 5.99317.

62. Round to the nearest hundredth: 52.196341.

 Calculator Problems

63. 3462.56 feet = _____ inches

64. 6755 yards = _____ feet

65. 65.62 pounds = _____ ounces

66. 120 tons = _____ pounds   **67.** 42 hours = _____ minutes

For Extra Practice Examples and Exercises, turn to page 336.

Solutions to Odd-Numbered Practice Problems

1. (a) 3 feet in 1 yard   **(b)** 5280 feet in 1 mile   **(c)** 60 minutes in 1 hour   **(d)** 7 days in 1 week   **(e)** 16 ounces in 1 pound
(f) 2 pints in 1 quart   **(g)** 4 quarts in 1 gallon

3. (a) $18.93 \text{ miles} \times \dfrac{5280 \text{ feet}}{1 \text{ mile}} = 99{,}950.4 \text{ feet}$ **(b)** $16\frac{1}{2} \text{ inches} \times \dfrac{1 \text{ yard}}{36 \text{ inches}} = 16\frac{1}{2} \times \frac{1}{36} \text{ yards} = \dfrac{\overset{11}{\cancel{33}}}{2} \times \dfrac{1}{\underset{12}{\cancel{36}}} \text{ yards} = \dfrac{11}{24} \text{ yards}$

5. $19 \text{ pints} \times \dfrac{1 \text{ quart}}{2 \text{ pints}} = \dfrac{19}{2} \text{ quarts} = 9.5 \text{ quarts}$

7. $1\frac{3}{4} \text{ days} \times \dfrac{24 \text{ hours}}{1 \text{ day}} = \dfrac{7}{\underset{1}{\cancel{4}}} \times \dfrac{\overset{6}{\cancel{24}}}{1} \text{ hours} = 42 \text{ hours time parked.}$ For cost, we multiply $42 \text{ hours} \times \dfrac{1.50 \text{ dollars}}{1 \text{ hour}} = 63.00 \text{ dollars.}$ She

paid $63.00.

Answers to Even-Numbered Practice Problems

2. (a) 3 miles   **(b)** 204 inches   **4.** 12,168 ounces   **6.** 432 minutes

---

6.2 METRIC MEASUREMENTS: LENGTH

After studying this section, you will be able to:

1 Understand the units of length in the metric system

2 Convert from one metric unit to another

The metric system of measurement is used in every industrialized nation of the world except the United States. Each year its use increases in the United States, so it is important to become familiar with the metric system. The metric system is a special system of making measurements designed for ease in calculating and in converting from one unit to another. In the metric system the meter is the basic unit of measurement for length. To measure lengths very much larger or smaller than the meter, the metric system uses units based on the meter and powers of 10.

1 In the metric system, the basic unit of length measurement is the meter. It is just slightly longer than a yard. To be precise, the meter is exactly 39.37 inches long.

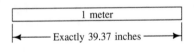

Many units of measurement are based on the meter. They are all obtained by multiplying or dividing the number of meter by a power of 10.
The metric system offers a simple way to write the word for a larger unit or a smaller unit of measurement. These smaller and larger units are formed by placing a metric prefix in front of the basic unit, the meter. The most commonly used prefixes are kilo-, centi-, and milli-. A longer list of metric prefixes and their meaning follows.

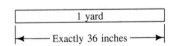

Prefix	Meaning
kilo-	thousand
hecto-	hundred
deka-	ten
deci-	tenth
centi-	hundredth
milli-	thousandth

For example, *kilo-* means thousand, so a *kilo*meter is a thousand meters. Similarly, *centi-* means one-hundredth, so a *centi*meter is one-hundredth of a meter. And *milli-* means one-thousandth, so a *milli*meter is one-thousandth of a meter.

The kilometer is used to measure distances much larger than meter. How far did you travel in a car? The centimeter is used to measure shorter lengths. What are the dimensions of this textbook? The millimeter is used to measure very small lengths. What is the width of the lead in a pencil?

2 How do we convert from one metric unit to another? For example, how do we change 5 kilometers into an equivalent number of meters? Let's look at how to express equal measurements using various metric units and powers of 10.

In the metric system, when we convert a measurement in one unit to an equivalent measurement in another unit we can first change the prefix. Then we can decide how to use the decimal point to arrive at the proper number. You will recall from Chapter 3 that when we multiply by 10 we move the decimal point one place to the right. When we divide by 10 we move the decimal point one place to the left. This knowledge helps us make metric conversions.

Changing From Larger Metric Units to Smaller Ones

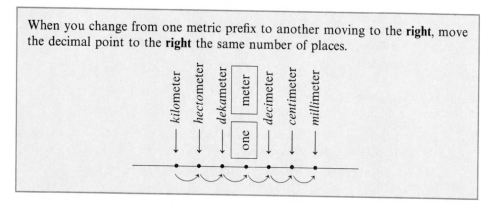

When you change from one metric prefix to another moving to the **right**, move the decimal point to the **right** the same number of places.

Thus 1 meter = 100 centimeters because we move two places to the right on the chart of prefixes and we also move the decimal point (1.00) two places to the right. Now let us examine the four most commonly used metric measurements of length and their abbreviations.

Commonly Used Metric Lengths

    1 kilometer (km) = 1000 meters

    1 meter (m) (the basic unit of length in the metric system)

    1 centimeter (cm) = 0.01 meter

    1 millimeter (mm) = 0.001 meter

Now let us examine how we can change a measurement in a larger unit to an equivalent measurement in terms of a smaller unit.

Example 1

(a) Change 5 kilometers to meters.   **(b)** Change 20 meters to centimeters.

(a) From kilometer to meter we move 3 units to the right, so we move the decimal point three places to the right.

        5 kilometers = 5.000 meters (move three places) = 5000 meters

(b) To go from meters to centimeters, we move two places to the right. Thus we move the decimal point two places to the right.

      20 meters = 20.00 centimeters (move two places) = 2000 centimeters   ■

Practice Problem 1

(a) Change 4 meters to centimeters.
(b) Change 30 centimeters to millimeters. ■

Changing From Smaller Metric Units to Larger Ones

When you change from one metric prefix to another moving to the **left**, move the decimal point to the **left** the same number of places.

kilometer   hectometer   dekameter   one | meter   decimeter   centimeter   millimeter

Let us now try this procedure to change some metric measurements made with smaller units into equivalent measures in terms of a larger unit.

Example 2

(a) Change 7 centimeters to meters.
(b) Change 56 millimeters to kilometers.

(a) To go from *centi*meters to meters, we move two places to the left. Thus we move the decimal point two places to the left.

$$7 \text{ centimeters} = 0.07 \text{ meter} \quad \text{(move two places to the left)}$$
$$= 0.07 \text{ meter}$$

(b) To go from *milli*meters to *kilo*meters, we move six places to the left. Thus we move the decimal point six places to the left.

$$56 \text{ millimeters} = 0.000056. \text{ kilometer} \quad \text{(move six places to the left)}$$
$$= 0.000056 \text{ kilometer} \quad ■$$

Practice Problem 2

(a) Change 3 *milli*meters to meters.
(b) Change 47 *centi*meters to *kilo*meters. ■

Thinking Metric: Having an Idea of How Large Something is in Metric Units

Long distances are customarily measured in kilometers. A kilometer is about 0.62 mile. It takes about 1.6 kilometers to make a mile. The following comparison is not, of course, drawn to actual size, but the relationship on size between the kilometer and the mile is accurate.

| 1 kilometer |

| 1 mile   (about 1.6 kilometers) |

Many small distances are measured in centimeters. A centimeter is about 0.394 inch. It takes about 2.54 centimeters to make an inch. You can get a good idea of their size by looking at a ruler measured in inches and centimeters.

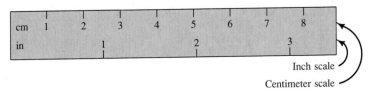

Try to visualize how many centimeters long or wide this book is. It is 21.5 centimeters wide and 27.8 centimeters long.

A millimeter is a very small unit of measurement, often used in manufacturing.

The thread end of a bolt may be 3 mm wide.

A paperclip is made of wire 0.8 mm thick.

The ruler shows that 10 mm = 1 cm. You can clearly see that 20 mm is equal to 2 cm, that 30 mm is equal to 3 cm, and so on.

| cm | 1 | 2 | 3 | 4 | 5 | 6 | 7 |
| mm | 10 | 20 | 30 | 40 | 50 | 60 | 70 |

Millimeter - centimeter ruler

To be sure our understanding of metric units is clear, let's try to select the "best" metric unit for measuring the length of an object.

Example 3 Bob measured the width of a doorway in his house. He wrote down "73." But what unit of measurement did he use?

(a) 73 kilometers   **(b)** 73 meters   **(c)** 73 centimeters

The logical choice is (c), 73 centimeters. The other two units would be much too long! A meter is close to a yard, and a doorway would not be 73 yards—equal to 219 feet—wide! A kilometer is even larger than a meter. ■

Practice Problem 3 Anthony measured the length of his car. He wrote down 3.8. Which unit of measurement did he mean?
(a) 3.8 kilometers   **(b)** 3.8 meters   **(c)** 3.8 centimeters ■

The most frequent metric conversions in length are done between kilometers, meters, centimeters, and millimeters. Their abbreviations are km, m, cm, and mm, and you should be able to use them correctly. No period is used after the letter or letters of the abbreviation.

Example 4 Convert each measurement.

(a) 364 m to km   **(b)** 982 cm to m   **(c)** 16 mm to cm   **(d)** 5.2 m to mm

In each of the first three cases, we move the decimal point to the left.
(a) 364 m = 0.364 km   (three places to left)
    = 0.364 km
(b) 982 cm = 9.82 m   (two places to left)
    = 9.82 m
(c) 16 mm = 1.6 cm   (one place to left)
    = 1.6 cm
Here we need to move the decimal point to the right.
(d) 5.2 m = 5200. mm   (three places to right)
    = 5200 mm ■

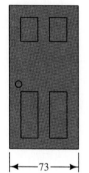

—73—

Practice Problem 4 Convert.

(a) 1527 m to km   **(b)** 375 cm to m

(c) 128 mm to cm   **(d)** 46 m to mm ■

The other metric units of length are the hectometer, the dekameter, and the decimeter. These are not used too frequently but it is good to have some understanding of their length and how it relates to the basic unit of one meter. A complete list of the metric lengths we have discussed appears in the following table:

Metric Lengths with Abbreviations

1 kilometer (km) = 1000 meters

1 hectometer (hm) = 100 meters

1 dekameter (dam) = 10 meters

1 meter (m)

1 decimeter (dm) = 0.1 meter

1 centimeter (cm) = 0.01 meter

1 millimeter (mm) = 0.001 meter

Example 5 Convert.

(a) 426 decimeters to kilometers   **(b)** 9.47 hectometers to meters

(a) We are converting from a smaller unit, dm, to a larger one, km. Therefore, there will be fewer units in our answer (426 will "get smaller"). We move the decimal point four places to the left.

$$426 \text{ dm} = 0.0426 \text{ km} \quad \text{(four places to left)}$$
$$= 0.0426 \text{ km}$$

(b) We are converting from a larger unit, hm, to a smaller one, m. Therefore, there will be more units in our answer (9.47 will "get larger"). We move the decimal point two places to the right.

$$9.47 \text{ hm} = 947. \text{ m} \quad \text{(two places to right)}$$
$$= 947 \text{ m} \quad ■$$

Practice Problem 5 Convert.

(a) 389 millimeters to dekameters   **(b)** 0.48 hectometer to centimeters ■

When several metric measurements are to be added, we change them to a convenient common unit.

Example 6 Add: 125 m + 1.8 km + 793 m.

First we change the kilometer measurement to a measurement in meters.

$$1.8 \text{ km} = 1800 \text{ m}$$

Then we add

$$
\begin{array}{r}
125 \text{ m} \\
1800 \text{ m} \\
+ \quad 793 \text{ m} \\
\hline
2718 \text{ m}
\end{array}
\quad ■
$$

Practice Problem 6 Add 782 cm + 2 m + 537 cm. ■

To Think About

Is the biggest length in the metric system a kilometer? Is the smallest length a millimeter? No. The system continues on for very large units and very small ones. Usually, only scientists and their staffs use these units. Each of these units is a power of 10^3 of the next smaller unit, as shown on the following chart.

> 1 gigameter = 1,000,000,000 meters
>
> 1 megameter = 1,000,000 meters
>
> 1 kilometer = 1000 meters
>
> 1 meter
>
> 1 millimeter = 0.001 meter
>
> 1 micrometer = 0.000001 meter
>
> 1 nanometer = 0.000000001 meter

For example, 26 megameters equals a length of 26,000,000 meters. A length of 31 micrometers equals a length 0.000031 meter.

A Challenge for You

Complete.

1. 18 megameters = _____ kilometers
2. 26 millimeters = _____ micrometers
3. 17 nanometers = _____ millimeters
4. 38 meters = _____ megameters ■

EXERCISES 6.2

Indicate if each statement is true or false.

1. 1 meter = 0.01 centimeter

2. 1 kilometer = 0.001 meter

3. 10 centimeters = 100 millimeters

4. 10 millimeters = 1 centimeter

5. An airport runway might be 2 kilometers long.

6. A man might be 2 meters tall.

7. A page of this book is about 2 centimeters thick.

8. The glass in a drinking glass is usually about 2 millimeters thick.

9. A kilometer is shorter than a mile.

10. A meter is longer than a yard.

The following conversions involve some of the units that are not used extensively. You should be able to perform each conversion, but it is not necessary to do it from memory.

11. 3 kilometers = _____ hectometers

12. 7 hectometers = _____ dekameters

13. 27 meters = _____ decimeters

14. 35 decimeters = _____ centimeters

15. 198 millimeters = _____ decimeters

16. 2.93 decimeters = _____ hectometers

17. 49.7 meters = _____ hectometers

18. 562 hectometers = _____ kilometers

19. 0.5236 hectometer = _____ centimeters

20. 0.7175 dekameter = _____ millimeters

The following conversions involve metric units that are very commonly used. You should be able to perform each conversion without any notes and without consulting the book.

21. 37 centimeters = _____ millimeters

22. 44 centimeters = _____ millimeters

23. 4.2 kilometers = _____ meters

24. 7.6 kilometers = _____ meters

25. 328 millimeters = _____ meter

26. 508 millimeters = _____ meter

27. 48.2 centimeters = _____ meter

28. 82.5 centimeters = _____ meter

29. 2 kilometers = _____ centimeters

30. 7 kilometers = _____ centimeters

31. 78,000 millimeters = _____ kilometer

32. 96,000 millimeters = _____ kilometer

33. 0.5386 kilometer = _____ millimeters

34. 0.3279 kilometer = _____ millimeters

Abbreviations are used in the following. Fill in the blanks with the appropriate value.

35. 96 mm = _____ cm = _____ m

36. 83 mm = _____ cm = _____ m

37. 3582 mm = _____ m = _____ km

38. 7812 mm = _____ m = _____ km

39. 0.32 cm = _____ m = _____ km

40. 0.81 cm = _____ m = _____ km

41. The Johnsons measured the length of their living room. Choose the most logical measurement.
(a) 7 km   (b) 7 m   (c) 7 cm

42. Wally measured the length of his college identification card. It was which of the following?
(a) 10 km   (b) 10 m   (c) 10 cm

43. Felipe measured the thickness of a plastic trash bag. Choose the most logical measurement.
(a) 1.5 km   (b) 1.5 m   (c) 1.5 mm

44. Central University has a modern campus. The director of housing measured the width of the campus. Choose the most logical measurement.
(a) 1.8 km   (b) 1.8 m   (c) 1.8 mm

Change to a convenient unit of measure and add.

45. 178 m + 3.2 km + 248 m

46. 798 m + 3.51 km + 661 m

47. 5.2 cm + 361 cm + 968 mm

48. 4.8 cm + 607 cm + 834 mm

49. 82 m + 471 cm + 0.32 km

50. 46 m + 986 cm + 0.884 km

51. Two metal sheets are 0.56 centimeters and 0.72 centi-meters thick. An insulating foil is placed between them that is 5.268 millimeters thick. When the three layers are placed together, what is the total thickness?

52. A plywood board is 2.2 centimeters thick. A layer of tar paper is 3.42 millimeters thick. A layer of false brick siding is 2.7 centimeters thick. A house wall consists of these three layers. How thick is the wall?

 To Think About

53. A space probe is being built to travel 3816 giga-meters. How many meters is that?

54. An electron microscope is focused on an object 423 nanometers long. How many meters is that?

55. One type of copper wire costs $12.82 per meter. An electrician needs one length that is 16.8 meters long and another length that is 980 centimeters long. How much will it cost her to buy the wire she needs?

56. The highway department determines that highway resurfacing costs $3560 per meter for a four-lane divided highway. How much will it be in total cost to resurface two sections of this highway, one that is 1.4 kilometers long and one that is 930 meters long?

Cumulative Review Problems

Write in words.

57. 38,516,243

58. 77,298,046

59. 0.2173

60. 0.1297

61. Write as a percent: 0.612.

62. Write as a decimal: 9.34%.

 Calculator Problem

Change to a convenient unit of measure and add.

63. 689 km + 159.2 m + 498 km + 76.2 km

For Extra Practice Examples and Exercises, turn to page 336.

Solutions to Odd-Numbered Practice Problems

1. (a) 4 meters = 4.00 centimeters   (two places to right) = 400 centimeters
   (b) 30 centimeters = 300 millimeters   (one place to right) = 300 millimeters
3. The car length would logically be choice **(b)**, 3.8 meters.
5. (a) 389 millimeters = 0.0389 dekameter   (four places to left)   **(b)** 0.48 hectometer = 4800 centimeters   (four places to right)

Answers to Even-Numbered Practice Problems

2. (a) 0.003 m   **(b)** 0.00047 km   **4. (a)** 1.527 km   **(b)** 3.75 m   **(c)** 12.8 cm   **(d)** 46,000 mm   **6.** 1519 cm

1 Convert between the various metric units of volume measurement

2 Convert between the various metric units of weight measurement

6.3 METRIC MEASUREMENTS: VOLUME AND WEIGHT

As the United States gradually "goes metric" we see more and more commonly sold products measured in metric units. Now soft drinks come in 1- or 2-liter bottles. Some gas stations now sell gasoline at a certain number of dollars per liter rather than at a rate per gallon. Very lightweight items such as medicines are measured in grams, and items that we ordinarily measure in pounds in the American system are measured in kilograms. (A kilogram is about 2 pounds.)

Try to get a feeling for the sizes of these metric units. As you convert measurements from one unit to another in this section, you'll be able to estimate your answers to see if they are reasonable.

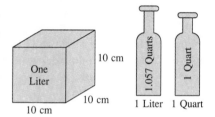

1 The volume measurement is based on the liter. A **liter** is defined as the volume of a box that is 10 centimeters on a side. Because volume equals length × width × height, 1 liter = 10 cm × 10 cm × 10 cm = 1000 cm³. (A cubic centimeter may be written as cc, so we sometimes see 1000 cc = 1 liter.) A liter is slightly larger than a quart. One liter of liquid is 1.057 quarts of that liquid.

The most common metric units of volume are the milliliter, the liter, and the kiloliter. Often a capital letter L is used to abbreviate liter.

> 1 kiloliter (kL) = 1000 liters
>
> 1 liter (L)
>
> 1 milliliter (mL) = 0.001 liter

We know that 1000 cc = 1 liter. Dividing each side of that equation by 1000, we get 1 cc = 1 mL.

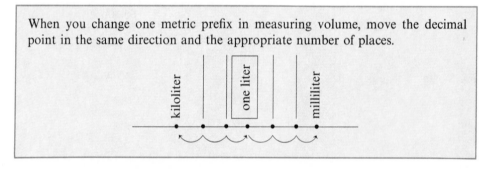

When you change one metric prefix in measuring volume, move the decimal point in the same direction and the appropriate number of places.

The prefixes for liter follow the pattern we have seen for meter. The *kilo-* is three places to the left of the liter, and the *milli-* is three places to the right. In converting among volume and weight measurements in the metric system, we follow the rules we learned in converting length measurements. For example, how many milliliters is 5 L? The prefix milli- is three places to the right of liter, so we move the decimal point three places to the right. Thus 5 L = 5000 mL.

Example 1 Convert.

(a) 3 L = _____ mL   (b) 24 kL = _____ L   (c) 0.084 L = _____ mL

(a) 3 L = 3000 mL   (b) 24 kL = 24,000 L
(c) 0.084 L = 84 mL   ■

Practice Problem 1 Convert.
(a) 5 L = _____ mL   (b) 84 kL = _____ L   (c) 0.732 L = _____ mL
                                                                          ■

As we convert from a smaller unit to the next larger unit (for example, from mL to L), we are moving three prefixes to the left, and we also move the decimal point three places to the left in the given quantity (2000 mL = 2 L).

Example 2   Convert.

(a) 26.4 mL = _____ L   **(b)** 5982 mL = _____ L   **(c)** 6.7 L = _____ kL

(a) 26.4 mL = 0.0264 L   **(b)** 5982 mL = 5.982 L
(c) 6.7 L = 0.0067 kL   ■

Practice Problem 2   Convert.
(a) 15.8 mL = _____ L   **(b)** 12340 mL = _____ L   **(c)** 86.3 L = _____ kL   ■

We have seen that the milliliter is the volume of a box 1 cm on each side and that 1 milliliter is equivalent to 1 cubic centimeter. We can write this as 1 mL = 1 cm³. (In medical fields the cubic centimeter is often abbreviated by the letters cc instead of cm³.)

Example 3   Convert.

(a) 26 mL = _____ cm³   **(b)** 0.82 L = _____ cm³

(a) 26 mL = 26 cm³
A milliliter and a cubic centimeter are equivalent.
(b) 0.82 L = 820 cm³
We use the same rule to convert a liter to a cubic centimeter as we do to convert a liter to a milliliter.   ■

Practice Problem 3   Convert.
(a) 396 mL = _____ cm³   **(b)** 0.096 L = _____ cm³   ■

Example 4   A special cleaning fluid to rinse test tubes in the chemistry lab costs $40.00 per liter. What is the cost per milliliter?

Since the milliliter is only $\frac{1}{1000}$ of a liter, its cost is only $\frac{1}{1000}$ of $40.00.

$$\frac{\$40.00}{1000} = \$0.040$$

Thus each milliliter costs $0.04, which is quite expensive.   ■

Practice Problem 4   A purified acid sells for $110 per liter. What does it cost per milliliter?   ■

2   In the metric system the basic unit of weight is the gram. A *gram* is the weight of the water in a box that is 1 centimeter on each side. To get an idea of how small a gram is, we note that two small paper clips weigh about 1 gram. A gram is only about 0.035 ounce.
   One kilogram is 1000 times larger than a gram. A kilogram is about 2.2 pounds. The most common measures of weight in the metric system are shown below.

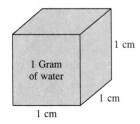

Common Metric Weight Measurements

1 metric ton (t) = 1,000,000 grams

1 kilogram (kg) = 1000 grams

1 gram (g)

1 milligram (mg) = 0.001 gram

As we convert from a larger unit to the next smaller unit on the chart (for example, from kg to g), we move three places to the right. This moves the decimal point three places to the right in the given quantity (4 kg = 4000 g).

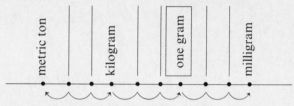

When you change one metric prefix in measuring weight, move the decimal point in the same direction and the appropriate number of places.

Example 5   Convert.

(a) 2 t = _____ kg   (b) 0.42 kg = _____ g   (c) 19.63 g = _____ mg

(a) 2 t = 2000 kg   (b) 0.42 kg = 420 g
(c) 19.63 g = 19,630 mg   ■

Practice Problem 5   Convert.
(a) 3.2 t = _____ kg   (b) 7.08 kg = _____ g   (c) 0.526 g = _____ mg   ■

As we convert from a smaller unit to the next larger unit on the chart (for example, from g to kg), we move three places to the left. This moves the decimal point three places to the left in the given quantity (6000 g = 6 kg).

Example 6   Convert.

(a) 283 kg = _____ t   (b) 14.628 g = _____ kg   (c) 7.98 mg = _____ g

(a) 283 kg = 0.283 t   (b) 14.628 g = 0.014628 kg
(c) 7.98 mg = 0.00798 mg   ■

Practice Problem 6   Convert.
(a) 59 kg = _____ t   (b) 6.152 g = _____ kg   (c) 28.3 mg = _____ g   ■

Example 7   A chemical that costs $0.03 per gram will cost what amount per kilogram?

Since there are 1000 grams in a kilogram, a chemical that costs $0.03 per gram would cost 1000 times as much per kilogram.

$$1000 \times \$0.03 = \$30.00$$

It would cost $30.00 per kilogram.   ■

Practice Problem 7   Coffee that costs $10.00 per kilogram will cost what amount per gram?   ■

Example 8   Select the most reasonable weight for Tammy's Toyota.

(a) 820 t   (b) 820 g   (c) 820 kg   (d) 820 mg

The most logical answer is (c) 820 kg. The other weight values are much too large or much too small. Since a kilogram is slightly more than 2 pounds, we see that this weight of 820 kg more closely tells the weight of a car.   ■

Practice Problem 8   Select the most reasonable weight for Hank, starting linebacker for the college football team.
(a) 120 kg   (b) 120 g   (c) 120 mg   ■

To Think About

?

When dealing with very small particles or atomic elements, scientists sometimes use units smaller than a gram.

Small Weight Measurements

1 milligram = 0.001 gram

1 microgram = 0.000001 gram

1 nanogram = 0.000000001 gram

1 picogram = 0.000000000001 gram

We could make the following conversions:

2.6 picograms = 0.0026 nanogram

29.7 micrograms = 0.0297 milligram

58 nanograms = 58000 picograms

58 nanograms = 0.058 microgram

Some exercises of this type appear in problems 61 and 62. ■

EXERCISES 6.3

Indicate if each statement is true or false.

1. 1 milliliter = 0.001 liter

2. 1 kiloliter = 1000 liters

3. Milk can be purchased in kiloliter jugs at the food store.

4. Small amounts of medicine are often measured in liters.

5. 1 kilogram = 1000 grams

6. 1 gram = 1000 milligrams

7. 1 metric ton = 100 grams

8. 0.1 gram = 10 milligrams

9. A nickel coin weighs about 5 grams.

10. A convenient size for a family purchase of hamburger is 2 kilograms of hamburger.

Perform each conversion.

11. 64 kL = _____ L

12. 37 kL = _____ L

13. 5.3 L = _____ mL

14. 4.2 L = _____ mL

15. 18.9 mL = _____ L

16. 31.5 mL = _____ L

17. 752 L = _____ kL

18. 493 L = _____ kL

19. 2.43 kL = _____ mL

20. 1.76 kL = _____ mL

21. 82 mL = _____ cm^3

22. 152 mL = _____ cm^3

23. 5261 mL = _____ kL

24. 28,156 mL = _____ kL

25. $74 \text{ L} = \underline{\hspace{1cm}} \text{ cm}^3$

26. $122 \text{ L} = \underline{\hspace{2cm}} \text{ cm}^3$

27. $162 \text{ g} = \underline{\hspace{1cm}} \text{ kg}$

28. $294 \text{ g} = \underline{\hspace{1cm}} \text{ kg}$

29. $27 \text{ mg} = \underline{\hspace{1cm}} \text{ g}$

30. $13 \text{ mg} = \underline{\hspace{1cm}} \text{ g}$

31. $6328 \text{ mg} = \underline{\hspace{1cm}} \text{ g}$

32. $986 \text{ mg} = \underline{\hspace{1cm}} \text{ g}$

33. $2.92 \text{ kg} = \underline{\hspace{1cm}} \text{ g}$

34. $14.6 \text{ kg} = \underline{\hspace{1cm}} \text{ g}$

35. $17 \text{ t} = \underline{\hspace{1cm}} \text{ kg}$

36. $36 \text{ t} = \underline{\hspace{1cm}} \text{ kg}$

37. $0.32 \text{ g} = \underline{\hspace{2cm}} \text{ kg}$

38. $0.78 \text{ g} = \underline{\hspace{2cm}} \text{ kg}$

39. $7896 \text{ g} = \underline{\hspace{2cm}} \text{ t}$

40. $12{,}315 \text{ g} = \underline{\hspace{2cm}} \text{ t}$

41. $5.9 \text{ kg} = \underline{\hspace{2cm}} \text{ mg}$

42. $0.83 \text{ kg} = \underline{\hspace{2cm}} \text{ mg}$

Fill in the blank with the appropriate value.

43. $7 \text{ mL} = \underline{\hspace{1cm}} \text{ L} = \underline{\hspace{2cm}} \text{ kL}$

44. $18 \text{ mL} = \underline{\hspace{1cm}} \text{ L} = \underline{\hspace{2cm}} \text{ kL}$

45. $128 \text{ cm}^3 = \underline{\hspace{1cm}} \text{ L} = \underline{\hspace{2cm}} \text{ kL}$

46. $199.8 \text{ cm}^3 = \underline{\hspace{1cm}} \text{ L} = \underline{\hspace{2cm}} \text{ kL}$

47. $522 \text{ mg} = \underline{\hspace{1cm}} \text{ g} = \underline{\hspace{2cm}} \text{ kg}$

48. $49 \text{ mg} = \underline{\hspace{1cm}} \text{ g} = \underline{\hspace{2cm}} \text{ kg}$

49. $6822 \text{ g} = \underline{\hspace{1cm}} \text{ kg} = \underline{\hspace{2cm}} \text{ t}$

50. $5096 \text{ g} = \underline{\hspace{1cm}} \text{ kg} = \underline{\hspace{2cm}} \text{ t}$

51. Alice bought a jar of apple juice at the store. Choose the most logical measurement.
(a) 0.32 kL
(b) 0.32 L
(c) 0.32 mL

52. A nurse gave an injection of insulin to a diabetic patient. Choose the most logical measurement.
(a) 4 kL
(b) 4 L
(c) 4 mL

53. A shipowner purchased a new oil tanker. Choose the most logical measurement.
(a) 4000 t
(b) 4000 kg
(c) 4000 g

54. Robert bought a new psychology textbook. Choose the most logical measurement.
(a) 0.49 *t*
(b) 0.49 kg
(c) 0.49 g

Find the convenient unit of measure and add.

55. $27 \text{ L} + 982 \text{ mL} + 13.6 \text{ L}$

56. $152 \text{ L} + 473 \text{ mL} + 77.3 \text{ L}$

57. $5 \text{ t} + 3.82 \text{ t} + 983 \text{ kg}$

58. $7.8 \text{ t} + 669 \text{ kg} + 5.23 \text{ t}$

59. 24 mg + 136 mg + 0.26 kg

60. 78 mg + 221 mg + 0.14 kg

⏺ To Think About

Round answers to the nearest tenth.

61. 5632 picograms = _____ micrograms

62. 0.076182 milligram = _____ nanograms

63. A doctor recommended that a patient take 5 grams of vitamin C every week.
  (a) In the month of September (30 days), how many *milligrams per day* should the patient have?
  (b) In a year which has 365 days, how many *kilograms* per year should the patient have?

64. A small commuter plane was loaded with the luggage of five people. The passengers' suitcases weighed 26.8 kilograms, 14.9 kilograms, 21.8 kilograms, 17,600 grams, and 9860 grams, respectively.
  (a) What was the total weight of the five suitcases?
  (b) The pilot said the maximum weight of cargo (suitcases) he could carry in his small plane was 19.5 kilograms for each of the five passengers. How much less than the maximum allowed weight was the weight of the five suitcases?

Cumulative Review Questions

65. 14 out of 70 is what percent?

66. What is 23% of 250?

67. What is 1.7% of $18,900?

68. A salesperson earns a commission of 8%. She sold furniture worth $8960. How much commission will she earn?

 Calculator Problem

69. .862 L + 223 mL + 3522 mL + 5.68 L + 2.90 L

For Extra Practice Examples and Exercises, turn to page 337.

Solutions to Odd-Numbered Practice Problems

1. (a) 5 L = 5000 mL **(b)** 84 kL = 84,000 L **(c)** 0.732 L = 732 mL
3. (a) 396 mL = 396 cm³ (because 1 milliliter = 1 cubic centimeter) **(b)** 0.096 L = 96 cm³
5. (a) 3.2 t = 3200 kg **(b)** 7.08 kg = 7080 g **(c)** 0.526 g = 526 mg

7. A gram is $\dfrac{1}{1000}$ of a kilogram. If the coffee costs $10.00 per kilogram, then 1 gram would be $\dfrac{1}{1000}$ of $10.

$$\frac{1}{1000} \times \$10 = \frac{\$10.00}{1000} = \$0.01$$

The coffee costs 1¢ per gram.

Answers to Even-Numbered Practice Problems

2. (a) 0.0158 L **(b)** 12.34 L **(c)** 0.0863 kL **4.** $0.11 per milliliter **6. (a)** 0.059 t **(b)** 0.006152 kg **(c)** 0.0283 g
8. 120 kg

6.4 CONVERSION OF UNITS [OPTIONAL]

So far we've seen how to convert units when working *within* either the American or metric system. If you're using the metric system, you can express a given measurement in terms of millimeters, centimeters, or meters.

Many people, however, work in *both* the metric and the American systems. If you study such fields as chemistry, electromechanical technology, business, x-ray technology, nursing, or computers, you will probably need to convert measurements between the two systems. We learn that skill in this section. To convert between American units and metric units, it is helpful to have equivalent values. The most commonly used equivalents are listed below. Most of these equivalents are approximate.

EQUIVALENT MEASURES

	American to Metric	Metric to American
Units of length	1 mile = 1.61 kilometers 1 yard = 0.914 meter 1 foot = 0.305 meter 1 inch = 2.54 centimeters	1 kilometer = 0.62 mile 1 meter = 3.28 feet 1 meter = 1.09 yards 1 centimeter = 0.394 inch
Units of volume	1 gallon = 3.79 liter 1 quart = 0.946 liter	1 liter = 0.264 gallon 1 liter = 1.06 quarts
Units of weight	1 pound = 0.454 kg 1 ounce = 28.35 g	1 kilogram = 2.2 pounds 1 gram = 0.0353 ounce

In each case, to convert from one unit to the other, multiply by 1 (a unit fraction). Choose the unit fraction from the equivalent measures table so that the unit in the denominator cancels with the unit in the numerator. Handling the numbers in problems of this type is best done on a calculator. You are encouraged to use a calculator in this section of the text.

1 To change 5 miles to kilometers we use the conversion table to make a unit fraction and multiply the unit fraction by 5 miles.

$$5 \text{ miles} \times \frac{1.61 \text{ kilometers}}{1 \text{ mile}} = 5 \times 1.61 \text{ km} = 8.05 \text{ kilometers}$$

Example 1

(a) Convert 3 feet to meters.

(b) Convert 6 inches to centimeters.

(a) $3 \text{ feet} \times \dfrac{0.305 \text{ meter}}{1 \text{ foot}} = 0.915 \text{ meter}$

(b) $6 \text{ inches} \times \dfrac{2.54 \text{ centimeters}}{1 \text{ inch}} = 15.24 \text{ centimeters}$ ■

Practice Problem 1

(a) Convert 7 feet to meters.

(b) Convert 4 inches to centimeters. ■

Unit abbreviations are quite common, so we will use them for the remainder of this section. We list them here for reference for your convenience.

American Measure (Alphabetical Order)	Standard Abbreviation
feet	ft
gallon	gal
inch	in.
mile	mi
ounce	oz
pound	lb
quart	qt
yard	yd

Metric Measure	Standard Abbreviation
centimeter	cm
gram	g
kilogram	kg
kilometer	km
liter	L
meter	m
millimeter	mm

Example 2

(a) Convert 26 m to yd.

(a) $26 \ \cancel{m} \times \dfrac{1.09 \ \text{yd}}{1 \ \cancel{m}} = 28.34 \ \text{yd}$

(b) Convert 1.9 km to mi.

(b) $1.9 \ \cancel{km} \times \dfrac{0.62 \ \text{mi}}{1 \ \cancel{km}} = 1.178 \ \text{mi}$ ■

Practice Problem 2

(a) Convert 17 m to yd.

(b) Convert 29.6 km to mi. ■

❷ Example 3

(a) Convert 14 gal to L.

(a) $14 \ \cancel{gal} \times \dfrac{3.79 \ \text{L}}{1 \ \cancel{gal}} = 53.06 \ \text{L}$

(b) Convert 2.5 L to qt.

(b) $2.5 \ \cancel{L} \times \dfrac{1.06 \ \text{qt}}{1 \ \cancel{L}} = 2.65 \ \text{qt}$ ■

Practice Problem 3

(a) Convert 26 gal to L.

(b) Convert 6.2 L to qt. ■

Example 4

(a) Convert 5.6 lb to kg.

(a) $5.6 \ \cancel{lb} \times \dfrac{0.454 \ \text{kg}}{1 \ \cancel{lb}} = 2.5424 \ \text{kg}$

(b) Convert 152 g to oz.

(b) $152 \ \cancel{g} \times \dfrac{0.0353 \ \text{oz}}{1 \ \cancel{g}} = 5.3656 \ \text{oz}$ ■

Practice Problem 4

(a) Convert 16 lb to kg.

(b) Convert 280 g to oz. ■

Some conversions will involve more than one step, because you will have to multiply by unit fractions more than once.

Example 5 Convert 235 cm to ft. Round your answer to the nearest hundredth of a foot.

Our first unit fraction converts centimeters to inches. Our second unit fraction converts inches to feet.

$$235 \ \cancel{cm} \times \dfrac{0.394 \ \cancel{in.}}{1 \ \cancel{cm}} \times \dfrac{1 \ \text{ft}}{12 \ \cancel{in.}} = \dfrac{92.59}{12} \ \text{ft}$$

$$= 7.72 \ \text{ft} \quad \text{(rounded to the nearest hundredth)} \quad ■$$

Practice Problem 5 Convert 180 cm to ft. ■

The same rules can be followed for a rate such as 50 miles per hour, which we write as $\dfrac{50 \ \text{miles}}{\text{hour}}$.

Example 6 Convert 100 km/hr to mi/hr.

$$\dfrac{100 \ \cancel{km}}{\text{hr}} \times \dfrac{0.62 \ \text{mi}}{1 \ \cancel{km}} = 62 \ \text{mi/hr}$$

Thus 100 km/hr per hour is approximately equal to 62 mi/hr. ■

Practice Problem 6 Convert 88 km/hr to mi/hr. (Round to the nearest hundredth.) ■

Example 7 A camera film that is 35 mm wide is how many inches wide?

We first convert from millimeters to centimeters by moving the decimal point in the number 35 one place to the left.

$$35 \ \text{mm} = 3.5 \ \text{cm}$$

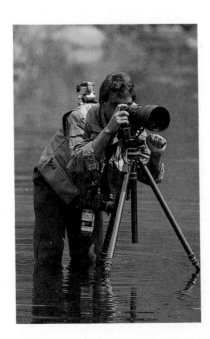

Then we convert using a unit fraction.

$$3.5 \text{ c\cancel{m}} \times \frac{0.394 \text{ in.}}{1 \text{ c\cancel{m}}} = 1.379 \text{ in.} \quad \blacksquare$$

Practice Problem 7 The city police use a 9-mm automatic pistol. If the pistol shoots a bullet 9 mm wide, how many inches wide is this? (Round to the nearest hundredth.) ■

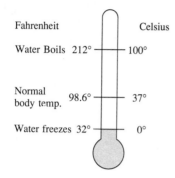

Fahrenheit · Celsius

Water Boils 212° — 100°

Normal body temp. 98.6° — 37°

Water freezes 32° — 0°

❸ Temperature

In the metric system, temperature is measured on the Celsius scale. Water boils at 100° (100°C) and freezes at 0° (0°C) on the Celsius scale. In the Fahrenheit system, water boils at 212° (212°F) and freezes at 32° (32°F). To convert Celsius to Fahrenheit, we can use the formula

$$F = 1.8 \times C + 32$$

Example 8 When the temperature is 35°C, what is the Fahrenheit reading?

$$F = 1.8 \times C + 32$$
$$= 1.8 \times 35 + 32$$
$$= 63 + 32$$
$$= 95$$

The temperature is 95°F. ■

Practice Problem 8 Convert 20°C to Fahrenheit temperature. ■

To convert Fahrenheit temperature to Celsius, we can use the formula

$$C = \frac{5 \times F - 160}{9}$$

where F is the number of Fahrenheit degrees and C is the number of Celsius degrees.

■

Example 9 When the temperature is 176°F, what is the Celsius reading?

$$C = \frac{5 \times F - 160}{9}$$
$$= \frac{5 \times 176 - 160}{9}$$
$$= \frac{880 - 160}{9} = \frac{720}{9} = 80$$

The temperature is 80°C. ■

Practice Problem 9 Convert 181°F to Celsius temperature. ■

?

To Think About

Conversion of Square Unit Measures Used in Area

4 yards

2 yards

Suppose we consider a rectangle that measured 2 yards wide by 4 yards long. The area would be 2 yards × 4 yards = 8 square yards. How could you change 8 square yards to square meters?

Suppose that we look at 1 square yard. Each side is 1 yard, which is equivalent to 0.9144 meter.

1 yard | 1 yard

0.9144 meter | 0.9144 meter

Area = 1 yard × 1 yard = 0.9144 meter × 0.9144 meter

Area = 1 square yard = 0.8361 square meter
(rounded to nearest ten thousandth)

Thus 1 yd² = 0.8361 m². Therefore,

$$8 \text{ yd}^2 \times \frac{0.8361 \text{ m}^2}{1 \text{ yd}^2} = 6.6888 \text{ m}^2$$

8 square yards = 6.6888 square meters. We will explore this further in problems 61 and 62. ■

EXERCISES 6.4

Perform each conversion. Round the answer to the nearest hundredth.

1. 7 ft to m

2. 11 ft to m

3. 9 in. to cm

4. 13 in. to cm

5. 14 m to yd

6. 18 m to yd

7. 26.5 m to yd

8. 29.3 m to yd

9. 13 km to mi

10. 17 km to mi

11. 24 yd to m

12. 31 yd to m

13. 82 mi to km

14. 68 mi to km

15. 25 m to ft

16. 35 m to ft

17. 17.5 cm to in.

18. 19.6 cm to in.

19. 200 m to yd

20. 350 m to yd

21. 7 gal to L

22. 5 gal to L

23. 280 gal to L

24. 140 gal to L

25. 23 qt to L

26. 28 qt to L

27. 19 L to gal

28. 15 L to gal

29. 4.5 L to qt

30. 6.5 L to qt

31. 32 lb to kg

32. 27 lb to kg

33. 7 oz to g

34. 9 oz to g

35. 18 kg to lb

36. 22 kg to lb

37. 126 g to oz

38. 186 g to oz

39. 166 cm to ft

40. 142 cm to ft

41. 16.5 ft to cm

42. 19.5 ft to cm

43. 50 km/hr to mi/hr

44. 60 km/hr to mi/hr

45. 60 mi/hr to km/hr

46. 40 mi/hr to km/hr

47. A wire that is 13 mm wide is how many inches wide?

48. A bolt that has 7 mm thread is how many inches wide?

49. 40°C to Fahrenheit

50. 60°C to Fahrenheit

51. 85°C to Fahrenheit

52. 105°C to Fahrenheit

53. 12°C to Fahrenheit

54. 21°C to Fahrenheit

55. 131°F to Celsius

56. 68°F to Celsius

57. 168°F to Celsius

58. 112°F to Celsius

59. 86°F to Celsius

60. 98°F to Celsius

 To Think About

Round your answer to four decimal places.

61. 28 square inches = ? square centimeters

62. 36 square meters = ? square yards

63. Change 48 feet per second to miles per hour.

64. Change 55 miles per hour to feet per second.

Cumulative Review Problems

Do each problem in the correct order.

65. $2^3 \times 6 - 4 + 3$

66. $5 + 2 - 3 + 5 \times 3^2$

67. $2^2 + 3^2 + 4^3 + 2 \times 7$

68. $5^2 + 4^2 + 3^2 + 3 \times 8$

 Calculator Problem

Make each conversion. Round your answer to the nearest hundredth.

69. 1926.33 L to gal

70. 152.3 in. to cm

71. 676.9 lb to kg

For Extra Practice Examples and Exercises, turn to page 337.

Solutions to Odd-Numbered Practice Problems

1. (a) $7 \text{ ft} \times \dfrac{0.305 \text{ m}}{1 \text{ ft}} = 2.135 \text{ m}$ **(b)** $4 \text{ in.} \times \dfrac{2.54 \text{ cm}}{1 \text{ in.}} = 10.16 \text{ cm}$

3. (a) $26 \text{ gal} \times \dfrac{3.79 \text{ L}}{1 \text{ gal}} = 98.54 \text{ L}$ **(b)** $6.2 \text{ L} \times \dfrac{1.06 \text{ qt}}{1 \text{ L}} = 6.572 \text{ qt}$ **5.** $180 \text{ cm} \times \dfrac{0.394 \text{ in.}}{1 \text{ cm}} \times \dfrac{1 \text{ ft}}{12 \text{ in.}} = 5.91 \text{ ft}$

7. $9 \text{ mm} = 0.9 \text{ cm}$ $0.9 \text{ cm} \times \dfrac{0.394 \text{ in.}}{\text{cm}} = 0.3546 \text{ in.}$ rounded to 0.35 in.

9. $181°F = C = \dfrac{5 \times 181 - 160}{9} = \dfrac{905 - 160}{9} = \dfrac{745}{9} = 82.7°C$ rounded to nearest tenth

---

6.5 SCIENTIFIC NOTATION

Scientists who frequently work with large or small measurements use a certain way to write numbers, called "scientific notation." Our usual way of writing a number, which we call "standard notation" or "standard form," expresses a quantity such as three million as

$$3,000,000$$

In scientific notation we write the same quantity as

$$3.0 \times 10^6$$

For a very small number, like two one-millionth, the standard form is

$$0.000002$$

The same quantity in scientific notation is

$$2.0 \times 10^{-6}$$

Notice that each number in scientific notation has two parts: (1) a number that is 1 or greater but less than 10, which is multiplied by (2) a power of 10. That power is an integer. In this section we learn how to go back and forth between standard and scientific notation.

1 Each of the positive whole numbers we have used has an opposite, negative number. If we place them on a number line, it would look like this:

These negative numbers are needed in order to write certain numbers in scientific notation.

We recall from Section 1.6 that

$$10^3 = 1000$$
$$10^2 = 100$$
$$10^1 = 10$$

At that time we introduced a definition that

$$\boxed{10^0 = 1}$$

If we now make a simple table of powers of 10, we can follow the pattern for 10^{-1}, 10^{-2}, 10^{-3}, and so on. We see that, going down the list, each number is one-tenth of the one above it.

$10^6 = 1,000,000$	$10^0 = 1$
$10^5 = 100,000$	$10^{-1} = 0.1$
$10^4 = 10,000$	$10^{-2} = 0.01$
$10^3 = 1,000$	$10^{-3} = 0.001$
$10^2 = 100$	$10^{-4} = 0.0001$
$10^1 = 10$	

After studying this section, you will be able to:

1 *Write numbers in scientific notation*

2 *Transform numbers in scientific notation to standard form (standard notation)*

Consider the following:

$$6700 = 6.7 \times 1000 = 6.7 \times 10^3$$

What we have done is to put a number in scientific notation.

> A positive number is written in scientific notation if it is in the form $a \times 10^n$, where a is a number greater than (or equal to) 1 and less than 10, and n is an integer.

Let us look at two more cases.

$$530 = 5.3 \times 100 \qquad\qquad 156{,}000 = 1.56 \times 100{,}000$$
$$= 5.3 \times 10^2 \qquad\qquad\qquad = 1.56 \times 10^5$$

These numbers are in scientific notation.

We want to look at a couple of cases where the power of 10 is negative. Consider

$$\frac{8.5}{100} = 8.5 \times \frac{1}{100} = 8.5 \times 10^{-2}$$

We can approach this a second way.

$$\frac{8.5}{100} = 0.085$$

Thus $0.085 = 8.5 \times 10^{-2}$.

In a similar fashion we could show that

$$0.76 = 7.6 \times 10^{-1}$$
$$0.0025 = 2.5 \times 10^{-3}$$
$$0.00088 = 8.8 \times 10^{-4}$$

Now that we see some examples of numbers in scientific notation, let us think through the steps of the next few examples. Exactly how can we put a number in scientific notation?

Example 1 Write 156 in scientific notation.

We want to determine the power of 10. We first change the given number to a number greater than or equal to 1 and less than 10. To determine that exponent, we look at the shift of the decimal point.

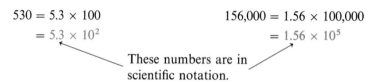

starting position of decimal point	ending position of decimal point

The decimal point moves $\boxed{2}$ places to the right. We put $\boxed{2}$ for the power of 10.

$$156 = 1.56 \times 10^2 \quad \blacksquare$$

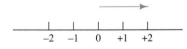

Practice Problem 1 Write 896 in scientific notation. ■

Example 2 Write in scientific notation.

(a) 57   (b) 9826   (c) 163,457

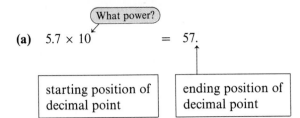

(a) 5.7×10 = 57.

The decimal point moves $\boxed{1}$ place to the right. We therefore put $\boxed{1}$ for the power of 10.

$$57 = 5.7 \times 10^1$$

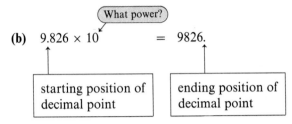

(b) 9.826×10 = 9826.

The decimal point moves $\boxed{3}$ places to the right. We therefore put $\boxed{3}$ for the power of 10.

$$9826 = 9.826 \times 10^3$$

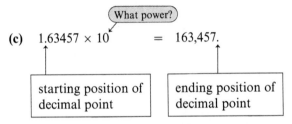

(c) 1.63457×10 = 163,457.

The decimal point moves $\boxed{5}$ places to the right. We therefore put $\boxed{5}$ for the power of 10.

$$163,457 = 1.63457 \times 10^5 \quad \blacksquare$$

Practice Problem 2   Write in scientific notation.
(a) 48   **(b)** 3729   **(c)** 506,936   ■

When changing to scientific notation, the extra zeros to the *right* of the final nonzero digit may be eliminated. This does not change the value of your answer.

Example 3   Write in scientific notation.

(a) 3100   **(b)** 700   **(c)** 4,500,000

(a) $3100 = 3.100 \times 10^3 = 3.1 \times 10^3$ **(b)** $700 = 7.00 \times 10^2 = 7 \times 10^2$
(c) $4,500,000 = 4.500000 \times 10^6 = 4.5 \times 10^6$ ■

Practice Problem 3   Write in scientific notation.
(a) 4600   **(b)** 900   **(c)** 3,800,000   ■

When we start with a number that is less than 1 and write it in scientific notation, we will get a result with 10 to a negative power. Think carefully through the steps of the following example.

Example 4   Write in scientific notation.

(a) 0.036   **(b)** 0.72   **(c)** 0.000589   **(d)** 0.008

(a) We change the given number to a number greater than or equal to 1 and less than 10. Thus we change 0.036 to 3.6.

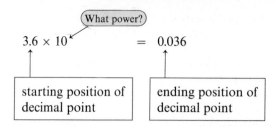

The decimal point moved in the negative direction 2 places. Since we moved the decimal point to the left we therefore use -2 for the power of 10.

$$0.036 = 3.6 \times 10^{-2}$$

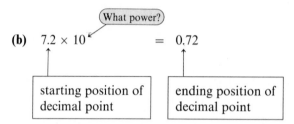

(b) 7.2×10 = 0.72

The decimal point moved in the negative direction 1 place. Because we moved the decimal point to the left, we use -1 for the power of 10.

$$0.72 = 7.2 \times 10^{-1}$$

Similarly, we find that

(c) $0.000589 = 5.89 \times 10^{-4}$ Moving the decimal point four places to the left.
(d) $0.008 = 8 \times 10^{-3}$ Moving the decimal point three places to the left.

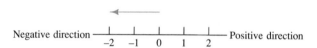

Practice Problem 4   Write in scientific notation.
(a) 0.076   **(b)** 0.982   **(c)** 0.000312   **(d)** 0.006 ■

2   Often, we are given a number in scientific notation and want to write it in standard notation—that is, we want to write it in what we consider "ordinary form." To do this, we reverse the process we've just used. If the number in scientific notation has a positive power of 10, we move the decimal point to the right to convert to standard notation.

To help in writing numbers in scientific notation, just remember these two concepts:

> **1.** A number that is larger than 10 will always have a positive exponent as the power of 10 when it is written in scientific notation.
> **2.** A number that is smaller than 1 will always have a negative exponent as a power of 10 when it is written in scientific notation.

Example 5   Write in standard notation.

(a) 8.32×10^2 **(b)** 5.8671×10^4

(a) $8.32 \times 10^2 = 832.$ Moving the decimal point two places to the right.
(b) $5.8671 \times 10^4 = 58,671.$ Moving the decimal point four places to the right. ■

Practice Problem 5   Write in standard notation.
(a) 3.08×10^2 $\qquad\qquad\qquad\qquad$ **(b)** 6.543×10^3 ■

In some cases we will need to add zeros as we move the decimal point to the right.

Example 6   Write in standard notation.

(a) 5.6×10^2 $\qquad\qquad$ **(b)** 9.8×10^5 $\qquad\qquad$ **(c)** 3×10^3

(a) $5.6 \times 10^2 = 560$ $\qquad$ Moving the decimal point two places to the right.
$\qquad\qquad\qquad\qquad\qquad\quad$ Add one zero.
(b) $9.8 \times 10^5 = 980,000$ $\qquad$ Moving the decimal point five places to the right.
$\qquad\qquad\qquad\qquad\qquad\qquad\quad$ Add four zeros.
(c) $3 \times 10^3 = 3000$ $\qquad$ Moving the decimal point three places to the right.
$\qquad\qquad\qquad\qquad\qquad\quad$ Add three zeros.   ■

Practice Problem 6   Write in standard notation.
(a) 8.5×10^2 $\qquad\qquad$ **(b)** 4.3×10^5 $\qquad\qquad$ **(c)** 6×10^4 ■

If the number in scientific notation has a negative power of 10, we move the decimal point to the left to convert to standard notation. In some cases we need to add zeros as we move the decimal point to the left.

Example 7   Write in standard notation.

(a) 7.16×10^{-1} $\qquad\qquad$ **(b)** 2.48×10^{-3} $\qquad\qquad$ **(c)** 1.2×10^{-4}

(a) $7.16 \times 10^{-1} = 0.716$ $\qquad$ Moving the decimal point one place to the left.
(b) $2.48 \times 10^{-3} = 0.00248$ $\qquad$ Moving the decimal point three places to the left.
(c) $1.2 \times 10^{-4} = 0.00012$ $\qquad$ Moving the decimal point four places to the left.   ■

Practice Problem 7   Write in standard notation.
(a) 8.56×10^{-1} $\qquad\qquad$ **(b)** 7.72×10^{-3} $\qquad\qquad$ **(c)** 2.6×10^{-5} ■

Numbers in scientific notation may be added or subtracted if they have the same power of 10. We add or subtract the decimal part and leave the power of 10 unchanged.

Example 8

(a) Add: 5.89×10^{20} miles $+ 3.04 \times 10^{20}$ miles.
(b) Subtract: 9.63×10^{17} pounds $- 2.98 \times 10^{17}$ pounds.

(a) $\qquad$ 5.89×10^{20} miles $\qquad\qquad$ **(b)** $\qquad$ 9.63×10^{17} pounds
$\qquad$ $\underline{+ 3.04 \times 10^{20}\text{ miles}}$ $\qquad\qquad\qquad$ $\underline{- 2.98 \times 10^{17}\text{ pounds}}$
$\qquad\qquad$ 8.93×10^{20} miles $\qquad\qquad\qquad\quad$ 6.65×10^{17} pounds   ■

Practice Problem 8
(a) Add: 6.85×10^{22} kilograms $+ 2.09 \times 10^{22}$ kilograms.
(b) Subtract: 8.04×10^{30} tons $- 6.98 \times 10^{30}$ tons.   ■

Sometimes it may be necessary to move the decimal point so that all the powers of 10 are the same.

Example 9   Add: $7.2 \times 10^6 + 5.2 \times 10^5$.

$$5.2 \times 10^5 = 520,000$$

But $520,000$ can be written as 0.52×10^6. Now we can add.

$$7.2\ \times 10^6$$
$$\underline{+\ 0.52 \times 10^6}$$
$$7.72 \times 10^6$$ ■

Practice Problem 9 Subtract: $4.36 \times 10^5 - 3.1 \times 10^4$. ■

To Think About

What would you do with the exponents when you multiply numbers in scientific notation? Look at this pattern:

$$200 \times 4000 = 800,000$$

In scientific notation this is equivalent to

$$2 \times 10^2 \times 4 \times 10^3 = 8 \times 10^5$$

To multiply two numbers in scientific notation, we *multiply the two numbers* multiplied by the power of 10. Then we create a new power of 10 by *adding the exponents of 10*. For example,

$$1.63 \times 10^4 \times 2.0 \times 10^3 = 1.63 \times 2.0 \times 10^7$$
$$= 3.26 \times 10^7$$

Another example is

$$5.3 \times 10^{10} \times 3.8 \times 10^{17} = 5.3 \times 3.8 \times 10^{27}$$
$$= 20.14 \times 10^{27}$$
$$= 2.014 \times 10^1 \times 10^{27}$$
$$= 2.014 \times 10^{28}$$

Additional examples of this type are found in problems 67 and 68. ■

EXERCISES 6.5

Write in scientific notation.

1. 26   **2.** 89   **3.** 137   **4.** 542   **5.** 7163   **6.** 4529

7. 120   **8.** 340   **9.** 500   **10.** 700   **11.** 26,300   **12.** 78,100

13. 199,000   **14.** 238,000   **15.** 1,710,000   **16.** 2,034,000   **17.** 12,000,000   **18.** 19,000,000

19. 0.67   **20.** 0.42   **21.** 0.398   **22.** 0.512   **23.** 0.00279   **24.** 0.00613

25. 0.4   **26.** 0.3   **27.** 0.0015   **28.** 0.0073   **29.** 0.000016   **30.** 0.000072

31. 0.00000531   **32.** 0.00000198   **33.** 0.0007   **34.** 0.00005

Write in standard notation.

35. 1.6×10^1 **36.** 2.8×10^2 **37.** 5.36×10^4 **38.** 2.19×10^4 **39.** 6.2×10^{-2}

40. 3.5×10^{-2} **41.** 5.6×10^{-5} **42.** 7.1×10^{-5} **43.** 8.5×10^{-4} **44.** 4.5×10^{-3}

45. 9×10^{11} **46.** 8×10^{10} **47.** 3×10^{-7} **48.** 4×10^{-6} **49.** 3.862×10^{-8}

50. 8.139×10^{-9} **51.** 4.6×10^{12} **52.** 3.8×10^{11} **53.** 6.721×10^{10} **54.** 4.039×10^{9}

Write each number in scientific notation.

55. In 1 year light will travel 5,878,000,000,000 miles.

56. An electron has a charge of 0.00000000048 electrostatic unit.

57. Yellow light has a wavelength of 0.00000059 meter.

58. The world's forests total 2,700,000,000 acres of wooded area.

Add.

59. 3.38×10^{7} dollars $+ 5.63 \times 10^{7}$ dollars

60. 8.17×10^{9} atoms $+ 2.76 \times 10^{9}$ atoms

61. 2.98×10^{12} kilometers $+ 5.07 \times 10^{12}$ kilometers

62. 3.15×10^{11} pounds $+ 4.98 \times 10^{11}$ pounds

Subtract.

63. 7.18×10^{15} miles $- 2.79 \times 10^{15}$ miles

64. 5.29×10^{12} acres $- 1.99 \times 10^{12}$ acres

65. 4×10^{8} feet $- 3.76 \times 10^{7}$ feet

66. 9×10^{10} meters $- 1.26 \times 10^{9}$ meters

 To Think About

Multiply, and leave your answer in scientific notation.

67. $9.2 \times 10^{20} \times 3.5 \times 10^{15}$

68. $4.3 \times 10^{16} \times 7.6 \times 10^{18}$

69. Write in scientific notation: 1,387,156,284,376,542.

70. Write in standard notation: 4.93×10^{-28}.

Cumulative Review Problems

Calculate.

71. 7.63×2.18 **72.** 5.92×1.98 **73.** $0.53 \overline{)0.13674}$ **74.** $0.42 \overline{)0.15204}$

Add or subtract.

75. $7.01 \times 10^{11} + 6.99 \times 10^{11} + 4 \times 10^{11} + 3.78 \times 10^{11}$ **76.** $5 \times 10^9 - 2.36 \times 10^9$

For Extra Practice Examples and Exercises, turn to page 337.

Solutions to Odd-Numbered Practice Problems

1. $896 = 8.96 \times 10^2$ **3. (a)** $4600 = 4.6 \times 10^3$ **(b)** $900 = 9 \times 10^2$ **(c)** $3{,}800{,}000 = 3.8 \times 10^6$
5. (a) $3.08 \times 10^2 = 308$ **(b)** $6.543 \times 10^3 = 6543$
7. (a) $8.56 \times 10^{-1} = 0.856$ **(b)** $7.72 \times 10^{-3} = 0.00772$ **(c)** $2.6 \times 10^{-5} = 0.000026$
9. $4.36 \times 10^5 = 436{,}000$. But $436{,}000$ is also 43.6×10^4, so

$$\begin{array}{r} 43.6 \times 10^4 \\ -\ \ 3.1 \times 10^4 \\ \hline 40.5 \times 10^4 \end{array} \quad \text{or} \quad 4.05 \times 10^5$$

Answers to Even-Numbered Practice Problems

2. (a) 4.8×10^1 **(b)** 3.729×10^3 **(c)** 5.06936×10^5 **4. (a)** 7.6×10^{-2} **(b)** 9.82×10^{-1} **(c)** 3.12×10^{-4} **(d)** 6×10^{-3}
6. (a) 850 **(b)** $430{,}000$ **(c)** $60{,}000$ **8. (a)** 8.94×10^{22} kg   **(b)** 1.06×10^{30} t

---

After studying this section,
you will be able to:

1 *Solve problems involving various metric and American measurements (conversion of one unit to another is often required)*

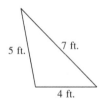
5 ft. 7 ft. 4 ft.

6.6 APPLIED PROBLEMS

The applied problems in this section require you to use your previously learned skills. Some of the applied problems also require you to read closely, visualize the problem, and decide what conversions will help you answer the question. Try to start each applied problem with a careful look at the units given, and then choose the appropriate system to which you will convert your measurements.

1 The perimeter of a triangle is the sum of the lengths of its three sides. The triangle at left has a perimeter of 16 feet.

A rectangle is a four-sided figure whose sides meet at right angles. The opposite sides of a rectangle are equal. The rectangle below has a perimeter of 22 meters. The perimeter of a rectangle is the sum of the lengths of all four sides. It is also the sum of double the width added to double the length.

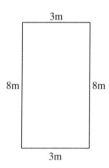
3m
8m   8m
3m

Example 1   Find the perimeter of the triangle. Express the answer in feet.

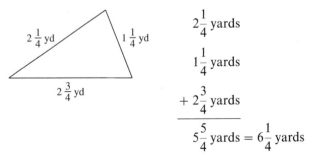

$2\frac{1}{4}$ yd $1\frac{1}{4}$ yd

$2\frac{3}{4}$ yd

$$\begin{array}{r} 2\frac{1}{4} \text{ yards} \\ 1\frac{1}{4} \text{ yards} \\ +\ 2\frac{3}{4} \text{ yards} \\ \hline 5\frac{5}{4} \text{ yards} = 6\frac{1}{4} \text{ yards} \end{array}$$

Now change yards to feet using 1 yard = 3 feet.

$$6\frac{1}{4} \text{ yards} \times \frac{3 \text{ feet}}{1 \text{ yard}} = \frac{25}{4} \text{ yards} \times \frac{3 \text{ feet}}{1 \text{ yard}} = \frac{75}{4} \text{ feet} = 18\frac{3}{4} \text{ feet} \quad \blacksquare$$

Practice Problem 1 Find the perimeter of the rectangle. Express the answer in feet. ■

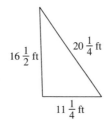

Example 2 How many 210-*liter* gasoline barrels can be filled from 5.04 *kiloliters* of gasoline?

$$5.04 \text{ kiloliters} = 5040 \text{ liters}$$

How many 210-liter barrels can be filled by 5040 liters? We divide

$$\frac{5040 \text{ liters}}{210 \text{ liter}} = 24$$

Thus we can fill 24 of the 210-liter barrels. ■

Practice Problem 2 A lab assistant must use 18.06 liters of solution to fill 42 jars. How many milliliters of a solvent will go into each one? ■

EXERCISES 6.6

Solve each.

1. Find the perimeter of the triangle. Express your answer in yards.

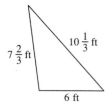

2. Find the perimeter of the triangle. Express your answer in yards.

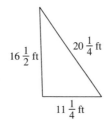

3. A farmer has 150 yards of fencing to fence in a triangular region. One side is to be 50 yards and a second side 60 yards. How much fencing will the farmer have left for the third side?

4. A farmer has 200 yards of fencing to fence in a triangular region. One side is to be 75 yards, and a second, 65 yards. How much fencing will the farmer have left for the third side?

5. A rectangular picture window measures 87 centimeters × 152 centimeters. Window insulation is placed along all four sides. The insulation costs $0.07 per centimeter. What will it cost to insulate the window?

6. A rectangular doorway measures 90 centimeters × 200 centimeters. Weatherstripping is placed on the top and the two sides. The weatherstripping costs $0.06 per centimeter. What does it cost to weatherstrip the door?

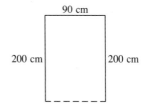

7. Forty full oil drums contain a total of 6.8 kiloliters of oil. How many *liters* of oil are in each drum?

8. A lab assistant must use 28 *liters* of solvent to fill 40 jars. How many *milliliters* of solvent will be in each jar?

9. A rope 17.8 *meters* long is cut into five equal pieces. How many *centimeters* long is each piece?

10. Bob found his trip to work was 12 *kilometers*. The last $\frac{1}{3}$ of the trip is in heavy traffic. For how many *meters* is he in heavy traffic?

11. How many 24-*milliliter* samples can be obtained from 3 *liters* of a chemical solution?

12. How many 150-liter oil barrels can be filled from 6 kiloliters of oil?

13. The temperature today in Winchester is 30°C. The record temperature for this date in Winchester is 89°F. By how many degrees Fahrenheit is the temperature today below the record?

14. The new computer from West Germany carries a sign "Do not operate this computer in temperatures above 80°F." The manual, however, specifies that the computer should not be operated above 25°C. By how many degrees Fahrenheit is the sign in error?

15. A French recipe for croissants calls for an oven temperature of 225°C. The cook set the oven for 425°F. By how many degrees Fahrenheit was the temperature off? Was the oven too cool or too hot?

16. A French cook wishes to bake potatoes in an oven at 195°C. The American assistant set the oven at 400°F. By how many degrees Fahrenheit was the oven temperature off? Was the oven too hot or too cold?

17. In his trip to Mexico, Juan traveled at 100 kilometers/hour for 4.5 hours. His total trip will be 300 miles long. How many miles does he still need to travel? (Round the answer to the nearest mile.)

18. In her trip to Mexico, Rosetta traveled at 85 kilometers/hour for 3.5 hours. Her total trip will be 210 miles. How many miles does she still need to travel? (Round the answer to the nearest mile.)

19. Denise traveled nonstop in Canada a distance of 170 miles in 3 hours on the highway. The speed limit is 100 kilometers/hour.
    (a) What was her average speed in kilometers per hour? (Round to the nearest kilometer per hour.)
    (b) Did she break the speed limit?

20. A small corporate jet travels at 600 kilometers/hour for 1.5 hours. The pilot said the plane would be on time if it met the goal of traveling at 350 miles per hour.
    (a) What was the jet's speed in miles per hour?
    (b) Did they arrive on time?

21. A septic tank is being emptied at the rate of 4 pints per minute. How many gallons per hour is this?

22. A swimming pool is being filled at the rate of 6 pints per minute. How many gallons per hour is this?

23. Trucks in Bob's state are taxed each year at $0.02 per pound. Bob's empty truck weighs 2.4 tons. What is his annual tax?

24. Trucks in Sam's home state are taxed each year at $0.03 per pound. Sam's empty truck weighs 1.8 tons. What is his annual tax?

25. The first leg of a journey by a space probe was 2.6×10^{12} kilometers. The second leg was 8.6×10^{12} kilometers. The final distance was 7.3×10^{12} kilometers. How far is the space probe from earth after traveling these three distances?

26. The distance that light travels in a year is 6×10^{12} miles. How far does light travel in seven years?

 To Think About

27. Kathleen's car gets 32 miles per gallon. Smithville is 468 kilometers away. If Kathleen buys gas at $1.10 per gallon and drives to Smithville, how much would the fuel used in the trip cost? Round your answer to the nearest cent.

28. Vince's truck gets 18 miles per gallon of diesel fuel. He drives a distance of 365 kilometers. His diesel fuel cost is $1.05 per gallon. How much does the fuel used in the trip cost him? Round your answer to the nearest cent.

Cumulative Review Problems

Solve for n.

29. $\dfrac{n}{16} = \dfrac{2}{50}$

30. $\dfrac{1.7}{n} = \dfrac{51}{6}$

 Calculator Problems

31. Jack took a summer-long trip and drove 18,963.2 kilometers. How many **miles** did he drive?

32. Find the perimeter of the triangle. Express your answer in **inches**.

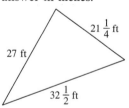

21 $\frac{1}{4}$ ft

27 ft

32 $\frac{1}{2}$ ft

For Extra Practice Examples and Exercises, turn to page 338.

Solution to Odd-Numbered Practice Problem

1. $8\dfrac{1}{3} + 2\dfrac{2}{3} + 8\dfrac{1}{3} + 2\dfrac{2}{3} = 20\dfrac{6}{3}$

The sum of all four sides is $20\dfrac{6}{3}$. $20\dfrac{6}{3} = 20 + 2 = 22$ yards. (You may also obtain the same perimeter by multiplying $2 \times$ width and adding that amount to $2 \times$ length.)

$$22 \text{ yards} \times \frac{3 \text{ feet}}{1 \text{ yard}} = 66 \text{ feet}$$

Answer to Even-Numbered Practice Problem

2. 430 ml

Section 6.1

Round to the nearest hundredth.

Example   Convert 2.55 miles to feet.

$$2.55 \text{ miles} \times \frac{5280 \text{ feet}}{1 \text{ mile}} = 2.55 \times 5280 \text{ feet} = 13{,}464 \text{ feet}$$

1. 15 feet = _____ yards
2. 72 yards = _____ miles
3. 6 miles = _____ yards
4. 12 feet = _____ inches
5. 70 inches = _____ feet

Example   Convert 22 pints to quarts.

$$22 \text{ pints} \times \frac{1 \text{ quart}}{2 \text{ pints}} = \frac{22}{2} \text{ quarts} = 11 \text{ quarts}$$

6. 180 ounces = _____ pounds
7. 8 tons = _____ ounces
8. 3.5 pounds = _____ ounces
9. 5 gallons = _____ quarts
10. 14 pints = _____ quarts

Example   Convert 3 days to hours.

$$3 \text{ days} \times \frac{24 \text{ hours}}{1 \text{ day}} = 3 \times 24 \text{ hours} = 72 \text{ hours}$$

11. 5 days = _____ hours
12. 3 weeks = _____ days
13. 540 minutes = _____ hours
14. 15 hours = _____ seconds
15. 56 days = _____ weeks

Example   John left his car in the airport garage for $3\frac{1}{2}$ days. The garage charges \$1.25 per hour. How much did he pay for parking?

$$3\frac{1}{2} = \frac{7}{2} \text{ days} \times \frac{24 \text{ hours}}{1 \text{ day}} = \frac{7}{2} \times 24 \text{ hours} = 84 \text{ hours}$$

$$84 \text{ hours} \times \frac{\$1.25}{\text{hour}} = 84 \times \$1.25 = \$105$$

16. Potato salad cost \$1.60 per pound. Bill bought 64 ounces of potato salad. How much did he pay?
17. A businesswoman left her car in an all-night garage for $2\frac{1}{2}$ days. The garage charges \$1.00 per hour. How much did she pay for parking?
18. The supermarket charges \$2.50 per pound for steak. Lisa bought 26 ounces of steak. How much did she pay?
19. The supermarket charges \$2.75 per pound for pork. Art bought 32 ounces of pork. How much did he pay?
20. John left his car in a long-term parking lot for three weeks. The parking fee is \$1.25 per day. How much did he pay for parking?

Section 6.2

Use this chart to do each conversion.

kilo—hecto—deka—meter—deci—centi—milli—

Example   25.4 centimeters = _____ meter

Move decimal point two places to the left; go from centi to meter.

$$25.4 \text{ centimeters} = 0.254 \text{ meter}$$

1. 38.6 centimeters _____ meter
2. 3 kilometers = _____ meters
3. 65,000 millimeters = _____ kilometers
4. 2 kilometers = _____ centimeters
5. 55 meters = _____ millimeters

Abbreviations are used in the following problems. Fill in the blanks with the appropriate value.

Example   5677 mm = _____ cm = _____ m

$$5677 \text{ millimeters} = 567.7 \text{ centimeters} = 5.677 \text{ meters}$$

6. 6899 mm = _____ cm = _____ m
7. 0.68 cm = _____ m = _____ km
8. 34 km = _____ m = _____ mm
9. 2334 cm = _____ m = _____ km
10. 4455 mm = _____ cm = _____ m

Choose the most logical measurement.

Example   The measurements of a patio

A meter would be the most logical measurement because a meter is a little longer than a yard.

11. The length of a credit card
12. The length of a kitchen
13. The thickness of a box
14. The length of a large park
15. The width of a window

Change to a convenient unit of measure and add.

Example   3.2 cm + 880 cm + 0.77 m

Change meters to centimeters and add. 0.77 m = 77 cm, so

$$\begin{array}{r} 77 \text{ cm} \\ 880 \text{ cm} \\ + \quad 3.2 \text{ cm} \\ \hline 960.2 \text{ cm} \end{array}$$

16. 36.8 cm + 4.6 cm + 0.89 m
17. 45 m + 2.3 cm + 498 cm
18. 879 m + 8.21 m + 9.33 km
19. 357 m + 4.79 km + 296 m
20. 8.6 cm + 490 cm + 335 mm

Section 6.3

Convert.

Example   677 L = _____ kL

  677 L = 0.677 kL   Move the decimal point three places
                                       to the left.

1. 233 L = _____ kL
2. 8.34 kL = _____ mL
3. 67.2 mL = _____ L
4. 5.6 L = _____ mL
5. 75 kL = _____ L

Example   85 L = _____ cm^3

    1 mL = 1 cm^3, so another way to state the above is

        85 L = _____ mL or cm^3

            85 L = 85,000 mL   or   85,000 cm^3

6. 66 L = _____ cm^3
7. 85 mL = _____ cm^3
8. 77,123 mL = _____ kL
9. 387 mL = _____ kL
10. 29 L = _____ cm^3

Example   4.88 g = _____ mg

  4.88 g = 4880 mg.   Move the decimal point three places
                                 to the right.

11. 6.99 kg = _____ mg
12. 8211 mg = _____ kg
13. 9 kg = _____ g
14. 15 mg = _____ g
15. 72 t = _____ kg

Find the convenient unit of measure and add.

Example   31 L + 863 mL + 12.8 L

        863 mL = 0.863 L   31.000 L
                                  0.863 L
                            + 12.800 L
                               44.663 L

16. 26 L + 557 mL + 13.7 L
17. 1223 mL + 12 L + 5.89 L
18. 4 t + 552 kg + 6.22 t
19. 102 kg + 9.1 t + 4.88 t
20. 44 mg + 167 mg + 0.77 kg

Section 6.4

Perform each conversion. Round your answer to the nearest hundredth.

Example   29 yd to m

    Use the fact that 1 yd = 0.914 m.

$$\frac{29 \text{ yd}}{1} \times \frac{0.914 \text{ m}}{1 \text{ yd}} = 29 \times 0.914 \text{ m}$$

                = 26.506 m   rounded = 26.51 m

1. 13 yd to m   2. 55 m to yd
3. 5 in. to cm   4. 44.2 in. to cm
5. 4 km to mi

Example   19 L to gal

    Use the fact that 1 gal = 3.79 L.

$$\frac{19 \text{ L}}{1} \times \frac{1 \text{ gal}}{3.79 \text{ L}} = 19 \times \frac{1 \text{ gal}}{3.79} = \frac{19 \text{ gal}}{3.79} = 5.01 \text{ gal}$$

6. 22 L to gal   7. 3.6 L to gal
8. 5 gal to L   9. 12 gal to L
10. 67 qt to L

Example   21.2 ft to cm

    Since 2.54 cm = 1 inch, first change feet to inches then inches to centimeters.

$$\frac{21.2 \text{ ft}}{1} \times \frac{12 \text{ in.}}{1 \text{ ft}} \times \frac{2.54 \text{ cm}}{1 \text{ in.}} = 21.2 \times 12 \times 2.54 \text{ cm} = 646.18 \text{ cm}$$

11. 13.5 ft to cm   12. 4.6 ft to cm
13. 177 cm to ft   14. 301 cm to ft
15. 31 ft to cm

Round to the nearest hundredth.

Example   129°F to Celsius

$$C = \frac{5 \times F - 160}{9}$$

$$= \frac{5 \times 129 - 160}{9} \qquad C = 645 - 160 = \frac{485}{9} = 53.89°C$$

16. 45°F to Celsius   17. 98°F to Celsius
18. 166°F to Celsius   19. 35°C to Fahrenheit
20. 88°C to Fahrenheit

Section 6.5

Write in scientific notation.

Example   23,900

  2.39×10^4 Move the decimal point four places to the left,
                so multiply by 10^4.

1. 445   2. 76,900
3. 130   4. 39.87
5. 988.23

Example   0.0098

  9.8×10^{-3} Move the decimal point to the right three
                 places, so multiply by 10^{-3}.

6. 0.012   7. 0.00456
8. 0.59   9. 0.000985
10. 0.00553

Write in standard notation.

Example

(a) 5.6×10^{-2} (b) 8×10^5

(a) 0.056   (b) 800,000

11. 3.8×10^{-3}

12. 89.2×10^{-1}

13. 0.055×10^2

14. 9×10^3

15. 1.2×10^{-5}

Section 6.6

Solve each problem.

Example John has 220 yards of fencing to fence in a triangular region. One side is to be 75 yards and a second side 55 yards. How much fencing will John have left for the third side?

$$
\begin{array}{ll}
75 & \text{fencing for the first side} \\
+ 55 & \text{fencing for second side} \\
\hline
130 & \\
\end{array}
$$

$$
\begin{array}{ll}
220 & \text{total fencing} \\
- 130 & \text{fencing for sides 1 and 2} \\
\hline
90 & \text{amount left for side 3} \\
\end{array}
$$

1. Find the perimeter of the triangle where the first side is $6\frac{2}{3}$ feet, the second side is 11 feet, and the third side is $8\frac{1}{3}$ feet. Express your answer in yards.

2. Find the perimeter of the triangle where the first side is $20\frac{1}{2}$ feet, the second side is $18\frac{1}{4}$ feet, and the third side is $12\frac{1}{2}$ feet. Express your answer in yards.

3. A farmer has 270 yards of fencing to fence in a triangular region. One side is to be 110 yards, and a second 80 yards. How much fencing will the farmer have left for the third side?

4. A farmer has 120 yards of fencing to fence in a triangular region. One side is to be 60 yards, and a second 30 yards. How much fencing will the farmer have left for the third side?

5. A rectangular picture window measures 67 centimeters by 132 centimeters. Window insulation is placed along all four sides. The insulation costs $0.09 per centimeter. What will it cost to insulate the window?

Example Mr. Snow found his trip to work was 15 *kilometers*. The last $\frac{1}{4}$ of the trip is in heavy traffic. For how many *meters* is he in heavy traffic?

$15 \text{ km} = 15,000 \text{ m}$ length of his trip in meters

$\frac{1}{4} \times 15,000 \text{ m} = 3750 \text{ meters}$ number of meters in heavy traffic

6. Forty full oil drums contain a *total* of 9.8 kiloliters of oil. How many liters of oil are in *each* drum?

7. A lab assistant must use 35 liters of solvent to fill 50 jars. How many milliliters of solvent will be in *each* jar?

8. A rope 16.4 meters long is cut into three equal pieces. How many centimeters long is each piece?

9. Bob found his trip to work was 16 kilometers. The last $\frac{1}{3}$ of the trip is in heavy traffic. For how many meters is he in heavy traffic?

10. How many 40 milliliter samples can be obtained from 5 liters of a chemical solution?

Example The temperature today in Springfield is 25°C. The record temperature for this date in Springfield is 92°F. By how many degrees Fahrenheit is the temperature today short of a record?

$F = 1.8 \times C + 32$

$= 1.8 \times 25 + 32$ be sure to multiply before adding

$= 45 + 32$

$= 77°F$ the temperature today in Springfield

$$
\begin{array}{ll}
92°F & \text{record temperature} \\
- 77°F & \text{today's temperature} \\
\hline
15°F & \text{the number of degrees short of the record} \\
\end{array}
$$

11. The temperature today in Los Angeles is 20°C. The record temperature for this date in Los Angeles is 88°F. By how many degrees Fahrenheit is the temperature today short of a record?

12. An electronic machine carries a sign "Do not operate this machine in temperatures above 85°F." The machine was made in France and the manual specifies that the machine should not be operated above 30°C. By how many degrees Fahrenheit is the sign in error?

13. A cake recipe calls for an oven temperature of 205°C. The cook set the oven for 420°F. By how many degrees Fahrenheit was the temperature off? Was the oven too cool or too hot?

14. The temperature today in New York is 92°F. How many degrees Celsius is this?

15. A cook wishes to bake a roast in an oven at 180°C. His assistant set the oven at 375°F. By how many degrees Fahrenheit was the oven temperature off? Was the oven too hot or too cold?

Convert. When necessary, express your answer as a decimal rounded to the nearest hundredth.

1. 1.6 tons = _____ lb

2. 19 ft = _____ in.

3. 21 gal = _____ qt

4. 36,960 ft = _____ mi

5. 3 cups = _____ qt

6. 1800 sec = _____ min

Perform each conversion. Do not round off.

7. 27.3 cm = _____ m

8. 9.2 km = _____ m

9. 46 mm = _____ cm

10. 9.88 cm = _____ m

11. 12.7 m = _____ cm

12. 0.936 cm = _____ mm

13. 46 L = _____ kL

14. 127 L = _____ mL

15. 28.9 mg = _____ g

16. 983 g = _____ kg

17. 0.92 L = _____ mL

18. 9.42 g = _____ mg

1. _____
2. _____
3. _____
4. _____
5. _____
6. _____
7. _____
8. _____
9. _____
10. _____
11. _____
12. _____
13. _____
14. _____
15. _____
16. _____
17. _____
18. _____

Perform each conversion. Round to the nearest hundredth when necessary.

19. 42 mi = _____ km

20. 1.78 yd = _____ m

21. 9 cm = _____ in.

22. 38 L = _____ gal

23. 7.3 kg = _____ lb

Write in scientific notation.

24. 96,379

25. 0.00004

Write in standard notation.

26. 4.58×10^{-7}

27. 6.7×10^6

Solve each problem. Round your answer to the nearest hundredth if necessary.

28. A rectangular picture frame measures 3 m × 7 m.
   (a) What is the perimeter of the picture frame in meters?
   (b) What is the perimeter of the picture frame in yards?

29. The temperature is 80°F today. Kristen's computer contains a warning not to operate above 35°C.
   (a) How many degrees Fahrenheit are there between the two temperatures?
   (b) Can she use her computer today?

30. One astronomer estimates there are 5.62×10^{23} stars in the sky. Another estimates there are 6.69×10^{23} stars in the sky. By how much do these estimates differ?

31. A pump is running at 5.5 quarts per minute. How many gallons per hour is this?

32. The speed limit on this Canadian road is 100 km/hr.
   (a) How far can Samuel travel at this speed limit in 3 hours?
   (b) If Samuel has to travel 200 miles, how much farther will he need to go after 3 hours of driving at 100 km/hr?

Approximately one-half of this test is based on Chapter 6 material. The remainder is based on material covered in Chapters 1–5.

Do each problem. Simplify your answer.

1. Subtract: 9,824
                  − 3,796

2. Multiply: 608
                 × 305

3. Divide: $28 \overline{)\ 1932}$.

4. Add: $\dfrac{1}{7} + \dfrac{3}{14} + \dfrac{2}{21}$.

5. Subtract: $3\dfrac{1}{8} - 1\dfrac{3}{4}$.

6. Is this proportion true or false?

$$\frac{21}{35} = \frac{12}{20}$$

7. Solve the proportion: $\dfrac{0.4}{n} = \dfrac{2}{30}$.

8. A piece of wire 6.5-centimeters long weighs 68 grams. What will a 20-centimeter length of the same wire weigh?

9. What percent of 66 is 165?

10. Find 18% of 360.

11. 0.5% of what number is 100?

Convert. When necessary express your answer as a decimal rounded to the nearest hundredth.

12. 38 qt = _____ gal

13. 2.5 tons = _____ lb

14. 7 pt = _____ qt

15. 25 feet = _____ in.

1. _____

2. _____

3. _____

4. _____

5. _____

6. _____

7. _____

8. _____

9. _____

10. _____

11. _____

12. _____

13. _____

14. _____

15. _____

16.
17.
18.
19.
20.
21.
22.
23.
24.
25.
26.
27.
28.
29.
30.

Perform each conversion. Do not round your answer.

16. 3.7 km = _____ m   **17.** 62.8 g = _____ kg   **18.** 0.79 L = _____ mL

19. 5 cm = _____ m   **20.** 42 lb = _____ oz

Perform each conversion. Round to the nearest hundredth when necessary.

21. 28 gal = _____ L   **22.** 96 lb = _____ kg

23. 7.87 m = _____ ft   **24.** 9 mi = _____ km

Write in scientific notation.

25. 579,863   **26.** 0.00078

27. Find the perimeter in **meters** of this triangle.

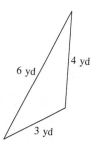

6 yd
4 yd
3 yd

28. Change 15°C to Fahrenheit temperature. Now find the difference between 15°C and 15°F. Which figure represents the higher temperature?

29. Ricardo traveled on a Mexican highway at 100 km/hr for $1\frac{1}{2}$ hours. He needs to travel a total distance of 100 miles. How far does he still need to travel? (Express your answer in miles.)

30. Two metal sheets are 0.72 centimeter thick and 0.98 centimeter thick, respectively. An insulating foil is placed between them that is 0.38 millimeter thick. When the three layers are placed together, what is the total thickness? Write the exact answer.

Geometry

Pilots learn to make a variety of mathematical calculations in planning for safe flying, including wind speed and flying time.

7

PRETEST CHAPTER 7

If you are familiar with the topics in this chapter, take this test now. Check your answers with those in the back of the book. If an answer was wrong or you couldn't do a problem, study the appropriate section of the chapter.

If you are not familiar with the topics in this chapter, don't take this test now. Instead, study the examples, work the practice problems, and then take the test.

This test will help you identify those concepts that you have mastered and those that need more study. In each problem round your answer to the nearest tenth when necessary. Use $\pi = 3.14$ in your calculation whenever a value of π is necessary.

Section 7.1   Find the perimeter of each rectangle or square.

1. Length = 6.5 m, width = 2.5 m

2. Length = width = 3.5 m

Find the area of each square or rectangle.

3. Length = width = 4.8 cm

4. Length = 2.7 cm, width = 0.9 cm

Section 7.2   Find the perimeter of each parallelogram or trapezoid.

5. Parallelogram with one side measuring 9.2 yd and another side measuring 3.6 yd.

6. Trapezoid with sides measuring 17 ft, 5 ft, 25 ft and 5 ft.

Find the area of parallelogram, trapezoid, or rectangle.

7. A parallelogram with a base of 27 in. and a height of 5 in.

8. A trapezoid with a height of 9 in. and bases of 16 in. and 22 in.

9.

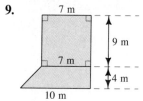

Section 7.3

10. Find the third angle in the triangle if two angles are 42° and 79°.

11. Find the perimeter of the triangle whose sides measure 7.2 m, 4.3 m, and 3.8 m.

Find the area of each triangle.

12. Base = 12 m, height = 5 m

13. Base = 5 km, height = 9 km

14. Base = 13 cm, height = 8 cm

Section 7.4

15. Find the diameter of a circle if the radius is 14 in.

16. Find the circumference of the circle whose diameter is 30 cm.

17. Find the area of the circle whose radius is 7 m.

18. Find the area of the shaded region.

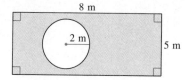

Section 7.5 Find the volume of each object.

19. A rectangular solid with the dimensions of 6 yd by 5 yd by 8 yd.

20. A sphere of radius 3 ft.

21. A cylinder of height 12 in. and radius 7 in.

22. A pyramid of height 21 meters with a square base measuring 25 meters on a side.

23. A cone of height 30 m and a radius of 6 m.

Section 7.6 Find n in each set of similar triangles.

24.

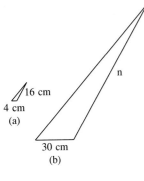

16 cm
4 cm
(a)
30 cm
(b)
n

25.

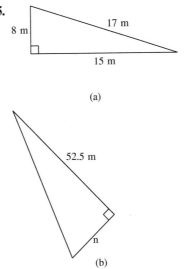

8 m
17 m
15 m
(a)

52.5 m
n
(b)

Section 7.7

26. A track field consists of two semicircles and a rectangle.
   (a) Find the area of the field.
   (b) It costs $0.15 per square yard to fertilize the field. What would it cost to complete this task?

30 yd
100 yd
15 yd

27. The side of a barn is shaped as shown. A painter will paint the barn for $1.10 per square foot. How much would the painter charge to paint this side?

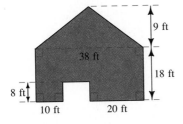

9 ft
38 ft
18 ft
8 ft
10 ft
20 ft

7.1 RECTANGLES AND SQUARES

Geometry has a visual aspect that many students find helps their learning. Numbers and abstract quantities may be hard to visualize, but we can take pen in hand and actually draw a picture of a rectangle that represents a room with certain dimensions. We can easily visualize problems such as "How many feet around the outside edges of the room (perimeter)?" or "How much carpeting will be needed for the room (area)?"

By learning about simple geometric shapes such as the rectangle we can build up to complex shapes that are made up from the simpler ones.

☐ After studying this section, you will be able to:

1 Find the perimeter of rectangles and squares

2 Find the perimeter of shapes made up of rectangles and squares

3 Find the area of rectangles and squares

4 Find the area of shapes made up of rectangles and squares

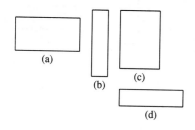

(a)

(b)

(c)

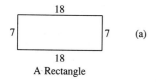

(d)

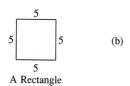

18

7   7   (a)

18

A Rectangle

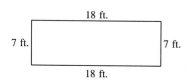

5

5   5   (b)

5

A Rectangle

Also known as a square

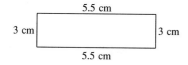

18 ft.

7 ft.   7 ft.

18 ft.

5.5 cm

3 cm   3 cm

5.5 cm

① Find the Perimeter of Rectangles and Squares

A **rectangle** is a four-sided figure like the ones shown here. A rectangle has two interesting properties: (1) that any two adjoining sides are perpendicular and (2) that opposite sides of a rectangle are equal.

By "any two adjoining sides are perpendicular" we mean that any two sides that are next to each other form an angle (called a right angle) that measures 90° and forms one of these shapes: ⌐ ⌐ ∟ ∟. When we say that "opposite sides of a rectangle are equal" we mean that the measure of a side is equal to the measure of the side opposite to it. Note that all four sides *may or may not* all be the same length. The **perimeter** of a rectangle is the sum of the lengths of all its sides. To find the perimeter we add up the lengths of all the sides of the figure. The perimeter of the rectangle above with sides of 18 ft, 7 ft, 18 ft and 7 ft is 50 ft. The perimeter of a rectangle can also be found if we refer to its length and width. Each of the longer sides is called the **length** of the rectangle. Each of the shorter sides is called the **width**. In the rectangle above the length is 18 feet and the width is 7 feet. We use letters to represent the measurements of the length and width. We let l represent the length and w represent the width. We also let p represent the perimeter. Since the perimeter is found by adding up the measurements all around the rectangle, we know that

$$P = w + l + w + l$$

$$P = 2l + 2w$$

When we write $2l$ and $2w$ we mean 2 *times* l and 2 *times* w. The perimeter can be found quickly by using the following formula.

> The **perimeter (P) of a rectangle** is twice the length plus twice the width:
>
> $$P = 2l + 2w$$

Example 1 Find the perimeter of the rectangle.

$$\text{Length} = l = 5.5 \text{ cm}$$

$$\text{Width} = w = 3 \text{ cm}$$

This is the formula for perimeter of a rectangle. We substitute 5.5 for l and 3 for w. $2l$ means 2 times l. $2w$ means 2 times w. Thus

$$P = 2l + 2w$$
$$= (2)(5.5) + (2)(3)$$
$$= 11 + 6 = 17 \text{ cm} \quad ■$$

Practice Problem 1 Find the perimeter of the rectangle.

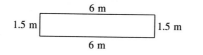

6 m

1.5 m   1.5 m

6 m   ■

Example 2 Find the perimeter of a rectangular field with a width of 700 yd and a length of 1300 yd.

$$\text{Width} = w = 700 \text{ yd}$$

$$\text{Length} = l = 1300 \text{ yd}$$

$$P = 2l + 2w$$
$$= (2)(1300) + (2)(700)$$
$$= 2600 + 1400 = 4000 \text{ yd} \quad ■$$

Practice Problem 2 Find the perimeter of a rectangular field with a width of 860 ft and length of 1200 ft. ■

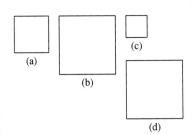

(a)

(b)

(c)

(d)

A **square** is a rectangle with all four sides equal. Here are some examples of squares.

A square is only a special type of rectangle. We can find the perimeter of a square just as we have found the perimeter of a rectangle—by adding the measurements of all the sides all around the figure. But in a square, the lengths of all sides are the same.

We let s represent the length of one side and P represent the perimeter. So to find the perimeter, we can write

> The **perimeter of a square** is four times the length of a side:
>
> $$P = 4s$$

Example 3 Find the perimeter of the square.

Side $= s = 8.6$ yd

$$P = 4s$$
$$= (4)(8.6)$$
$$= 34.4 \text{ yd} \quad ■$$

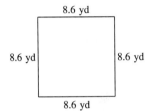

8.6 yd

8.6 yd   8.6 yd

8.6 yd

Practice Problem 3 Find the perimeter of the square.

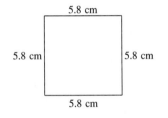

5.8 cm

5.8 cm   5.8 cm

5.8 cm

■

❷ Find the Perimeter of Shapes Made Up of Rectangles and Squares

Example 4 Find the perimeter of the shape above consisting of a rectangle and a square.

We want to find the distance around the object. Therefore, we look only at the outside edges. There are six sides to add together.

$$5 + 3 + 3 + 2 + 2 + 1 = 16$$

The perimeter of the shape is 16 m. ■

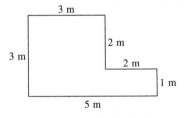

3 m

2 m

3 m

2 m

1 m

5 m

Practice Problem 4 Find the perimeter of the following shape which consists of a rectangle and a square.

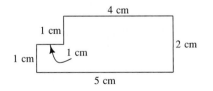

4 cm

1 cm

2 cm

1 cm   1 cm

5 cm

■

In practical situations, by determining the outside perimeter of geometric figures, we can find how many feet of picture framing will be needed or how many feet of weather sealing will be used. Consider the following problem.

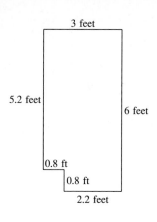

3 feet

5.2 feet

6 feet

0.8 ft

0.8 ft

2.2 feet

Example 5   Find the amount of weather-sealing materials at $0.12 per foot required to seal the outside edge of a window and these dimensions.

The actual perimeter is the sum of all the edges.

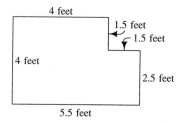

$$\begin{array}{r} 3.0 \\ 6.0 \\ 2.2 \\ 0.8 \\ 0.8 \\ + \ 5.2 \\ \hline 18.0 \text{ ft} \end{array}$$

Complete the cost.

$$18.0 \ \cancel{ft} \times \frac{0.12 \text{ dollar}}{\cancel{ft}} = \$2.16 \text{ for weather-sealing materials} \quad \blacksquare$$

Practice Problem 5   Find the amount of weather-sealing materials at $0.16 per foot required to seal the outside edges of a window with this shape.

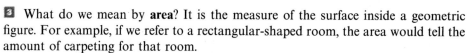

4 feet

1.5 feet

1.5 feet

4 feet

2.5 feet

5.5 feet

3   What do we mean by **area**? It is the measure of the surface inside a geometric figure. For example, if we refer to a rectangular-shaped room, the area would tell the amount of carpeting for that room.

One "square meter" is the measure of a square that is 1 m long and 1 m wide. We can abbreviate square meter as m^2. In fact, all areas are measured in square meters, square feet, square inches, and so on (written as m^2, ft^2, $in.^2$, and so on).

In this section we'll learn to calculate the area of a rectangular region if we know its length and its width. To find the area, *multiply* the length by the width.

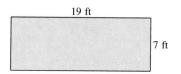

1 "square meter"

1 meter

1 meter

> The **area** (A) **of a rectangle** is the length times the width.
>
> $$A = lw$$

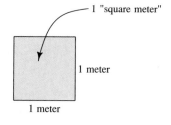

19 ft

7 ft

Example 6   Find the area of this rectangle.

We know that our answer will be measured in square feet.

$$w = 7 \text{ ft} \qquad l = 19 \text{ ft}$$

$$A = (l)(w) = (19)(7) = 133 \ ft^2 \quad \blacksquare$$

Practice Problem 6   Find the area of each rectangle.

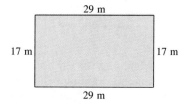

29 m

17 m

17 m

29 m

To find the area of a square, we multiply the length of a side by itself.

The **area of a square** is the square of the length of one side.

$$A = s^2$$

Example 7 A square measures 9.6 in on each side. Find its area.

We know our answer will be measured in square inches. We will write this as in.^{2}

$$\begin{aligned} A &= s^2 \\ &= (9.6)^2 \\ &= 92.16 \text{ in.}^2 \quad \blacksquare \end{aligned}$$

Practice Problem 7 Find the area of a square computer chip that measures 11.8 mm on each side. ■

4 Find the Area of Shapes Made Up of Rectangles and Squares

Example 8 Consider the following shape made up of a rectangle and a square. Find the area of the shaded region.

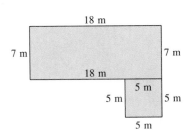

The shaded region is made up of two separate regions. You can think of each separately, and calculate the area of each one. The total area is just the sum of the two separate areas.

Area of rectangle $= (7)(18) = 126 \text{ m}^2$ Area of square $= 5^2 = 25 \text{ m}^2$

$$\begin{aligned} \text{The area of the rectangle} &= 126 \text{ m}^2 \\ + \text{ the area of the square} &= 25 \text{ m}^2 \\ \hline \text{the total area is} &= 151 \text{ m}^2 \quad \blacksquare \end{aligned}$$

Practice Problem 8 Find the area of the shaded region.

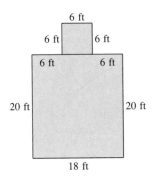

■

Find the perimeter of each rectangle or square.

1.
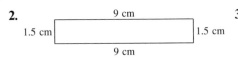
5.5 mi, 2 mi, 2 mi, 5.5 mi

2.
9 cm, 1.5 cm, 1.5 cm, 9 cm

3.
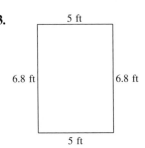
5 ft, 6.8 ft, 6.8 ft, 5 ft

4.

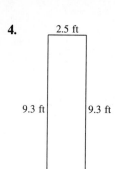

2.5 ft

9.3 ft   9.3 ft

2.5 ft

5.

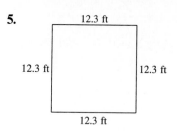

12.3 ft

12.3 ft   12.3 ft

12.3 ft

6.

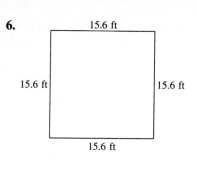

15.6 ft

15.6 ft   15.6 ft

15.6 ft

7. Length = 0.92 mm, width = 0.16 mm

8. Length = 9.3 in., width = 1.8 in.

9. Length = 7 in., width = 5.73 in.

10. Length = 0.54 km, width = 0.36 km

11. Length = width = 4.28 km

12. Length = width = 9.63 cm

Find the perimeter of each rectangle.

13. Length = 18.4 m, width = 5.35 m

14. Length = 15.2 m, width = 6.65 m

15. Length = 0.0093 cm, width = 0.0076 cm

16. Length = 0.0089 cm, width = 0.0034 cm

Find the perimeter of each square of given length of each side.

17. 9 m   **18.** 6 cm   **19.** 1.2 mi   **20.** 1.8 m

21. 6.32 cm   **22.** 7.96 cm   **23.** 0.0043 mm   **24.** 0.0052 mm

Find the perimeter of the objects made up of rectangles and squares.

25.

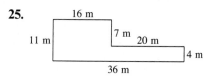

16 m

11 m   7 m   20 m

4 m

36 m

26.

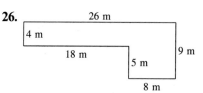

26 m

4 m

18 m   9 m

5 m

8 m

27.

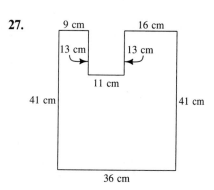

9 cm   16 cm

13 cm   13 cm

11 cm

41 cm   41 cm

36 cm

28.

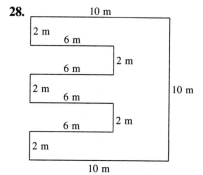

10 m

2 m

6 m

6 m   2 m

2 m

6 m

6 m   2 m

2 m

10 m

10 m

Find the area of each rectangle or square.

29. Length = 24 in., width = 7 in.

30. Length = 18 in., width = 4 in.

31. Length = width = 7.6 ft

32. Length = width = 9.8 ft

33. Length = 0.96 m, width = 0.3 m

34. Length = 0.132 m, width = 0.02 m

35. Length = 156 yd, width = 96 yd

36. Length = 183 yd, width = 81 yd

Find the shaded area.

37.

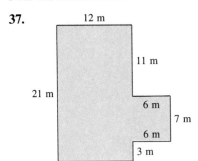

38.

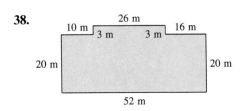

Some of the following problems will require that you find a perimeter. Others will require that you find an area. Read through each problem carefully to determine which you are to find.

39. A rectangular field measures 150 ft × 35 ft. Fertilizing the field will cost $0.03 per square root. How much does it cost to fertilize the field?

40. A rectangular field measures 196 ft × 40 ft. To lime the field costs $0.02 per square foot. How much does it cost to lime the field?

41. A rectangular picture requires a frame 6.5 ft × 4 ft. The framing material is $1.30 per foot. How much will the framing cost?

42. A rectangular picture requires a frame 3.6 ft × 7.5 ft. The framing material is $1.50 per foot. How much will the framing cost?

 To Think About

A family decides to have custom carpeting installed. It will cost $14.50 per square yard. The binding, which runs along the outside edges of the carpet, will cost $1.50 per yard. Find the cost of carpeting and binding for each room. Note that dimensions are given in feet.

43.

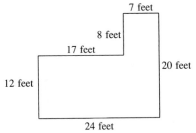

44.

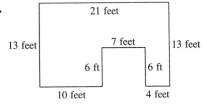

45. Add: $\begin{array}{r} 156.8 \\ 27.2 \\ + \ 39.3 \\ \hline \end{array}$

46. Subtract: $\begin{array}{r} 200.57 \\ -193.39 \\ \hline \end{array}$

47. Multiply: $\begin{array}{r} 1076 \\ \times \ 20.3 \\ \hline \end{array}$

48. Divide: $12.3\overline{)19.384}$

Calculator Problems

49. Find the area of a rectangle with length 256.00 ft and width 35.26 ft.

50. A rectangular field measures 125.52 ft × 72.6 ft. It will cost $0.05 per square foot to fertilize the field. How much does it cost to fertilize the field?

For Extra Practice Examples and Exercises, turn to page 393.

Solutions to Odd-Numbered Practice Problems

1. $P = 2l + 2w$
$= (2)(6) + (2)(1.5)$
$= 12 + 3 = 15$ m

3. $P = 4s$
$= (4)(5.8) = 23.2$ cm

5. $P = 4 + 4 + 5.5 + 2.5 + 1.5 + 1.5$
$= 19$ ft
$$\text{Cost} = 19\,\cancel{\text{ft}} \times \frac{0.16 \text{ dollar}}{1\,\cancel{\text{ft}}} = \$3.04$$

7. 139.24 m²

Answers to Even-Numbered Practice Problems

2. 4120 ft **4.** 14 cm **6.** 493 m² **8.** 396 ft²

7.2 PARALLELOGRAMS AND TRAPEZOIDS

After studying this section, you will be able to:

1 *Find the perimeter and area of a parallelogram*

2 *Find the perimeter and area of a trapezoid*

Parallelograms and trapezoids are figures related to the rectangles. Actually, they are in the same "family," the **quadrilaterals** (four-sided figures). For all these figures, the perimeter is the distance around the figure. But each has a different formula for area that we develop and use in this section.

1 **A parallelogram** is a four-sided figure with both pairs of opposite sides parallel. Parallel lines are two straight lines that are always the same distance apart. The opposite sides of a parallelogram are equal in length. These figures are parallelograms.

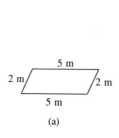
(a)

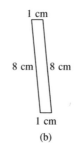
(b)

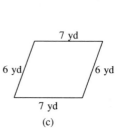
(c)

The **perimeter** of a parallelogram is the distance around it. It is found by adding the lengths of all the sides of the figure.

Example 1 Find the perimeter of the parallelogram.

$$P = (2)(1.2) + (2)(2.6)$$
$$= 2.4 + 5.2 = 7.6 \text{ m} \quad \blacksquare$$

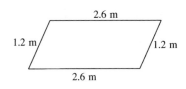

Practice Problem 1 Find the perimeter of the parallelogram.

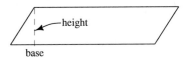

7.6 cm

3.5 cm

3.5 cm

7.6 cm ■

To find the *area* of a parallelogram we multiply the base times the height. Any side of a parallelogram can be considered the base. The height is the shortest distance between the base and the side opposite the base. The height is measured by a line segment that is perpendicular to the base. When we write the formula for area we use the length of the base (*b*) and the height (*h*).

height

base

The **area of a parallelogram** is the base (*b*) times the height (*h*).

$$A = bh$$

Example 2 Find the area of the parallelogram with base = 7.5 m and height = 3.2 m.

$$A = bh$$
$$= (7.5)(3.2)$$
$$= 24 \text{ m}^2 \quad \blacksquare$$

Practice Problem 2 Find the area of a parallelogram with base = 10.3 km and height = 1.5 km. ■

Example 3 Find the perimeter and the area of the parallelogram.

The perimeter is the sum of the lengths of its sides.

$$P = (2)(7.8) + (2)(12.3)$$
$$= 15.6 + 24.6 = 40.2 \text{ m}$$

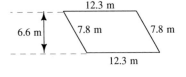

The area formula will use the height, which is labeled in the sketch as 6.6 m.

$$A = bh$$
$$= (12.3)(6.6) = 81.18 \text{ m}^2 \quad \blacksquare$$

Practice Problem 3 Find the perimeter and the area of the parallelogram.

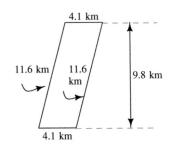

4.1 km

11.6 km

11.6 km

9.8 km

4.1 km ■

Why is the area of a parallelogram equal to the base times the height? What reasoning leads us to that formula? Suppose that we consider a parallelogram and "cut off" the triangular region on the left side and move it to the right side.

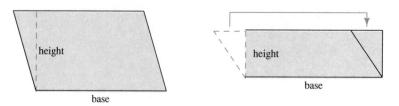

Now we have a rectangle.

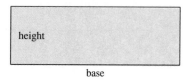

To find the area we multiply the width times the length. In this case $A = bh$. Thus we have seen that finding the area of a parallelogram is like finding the area of a rectangle of length b and width h. $A = bh$. We do some more exploring of this concept in problems 33 and 34. ■

2 In mathematics we often use parentheses as a way to group numbers together. The numbers inside parentheses should be combined first. To evaluate

$$(5)(7 + 2) = (5)(9) \qquad \text{First we add numbers inside the parentheses.}$$
$$= 45 \qquad \text{Then we multiply.}$$

The formula for the area of a trapezoid uses parentheses in this way.

A **trapezoid** is a four-sided figure with at least two parallel sides. The *perimeter* of a trapezoid is the sum of the lengths of all of its sides.

Example 4 Find the perimeter of the trapezoid with sides of 18 m, 5 m, 12 m and 5 m, respectively.

$$\text{Perimeter} = 18 + 5 + 12 + 5 = 40 \text{ m} \qquad ■$$

Practice Problem 4 Find the perimeter of the trapezoid with sides of 7 yd, 15 yd, 21 yd and 13 yd. ■

The area of a trapezoid is one-half of the product of the height times the sum of the bases.

The **area of a trapezoid** with a shorter base b and a longer base B and a height h is

$$A = \frac{h(b + B)}{2}$$

b

height = h

B

Example 5 Find the area of the trapezoid.

$$A = \frac{h(b + B)}{2} \qquad \text{when } h = 18, b = 24, \text{ and } B = 30$$

$$A = \frac{(18)(24 + 30)}{2}$$

$$= \frac{(18)(54)}{2} = \frac{972}{2} = 486 \text{ cm}^2 \quad \blacksquare$$

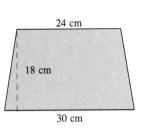

Practice Problem 5 Find the area of the trapezoid.

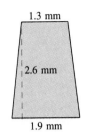

Example 6 A roadside sign is in the shape of a trapezoid. It has a height of 30 ft and the bases are 60 ft and 75 ft.
(a) What is the area of the sign?
(b) If 1 gallon of paint covers 200 ft², how many gallons of paint will be needed to paint the sign?

(a) $A = \frac{h(b + B)}{2} \qquad \text{when } h = 30, b = 60, \text{ and } B = 75$

$$= \frac{(30)(60 + 75)}{2}$$

$$= \frac{(30)(135)}{2} = \frac{4050}{2} = 2025 \text{ ft}^2$$

(b) Each gallon covers 200 ft², so we multiply by a unit fraction.

$$2025 \cancel{\text{ft}^2} \times \frac{1 \text{ gal}}{200 \cancel{\text{ft}^2}} = \frac{2025}{200} \text{ gal}$$

$$= 10.125 \text{ gal}$$

Thus 10.125 gallons of paint would be needed. (In this practical situation, we would probably buy 11 gallons of paint to be sure that we had enough.) $\quad \blacksquare$

Practice Problem 6 A corner parking lot is in the shape of a trapezoid. It has a height of 140 yd. The bases measure 180 yd and 130 yd.
(a) Find the area of the parking lot.
(b) If 1 gallon of sealer will cover 100 square yards of the parking lot, how many gallons are needed to cover the entire parking lot? $\quad \blacksquare$

Some area problems involve using two or more separate regions. Their areas can be added or subtracted.

Example 7 Find the area of the following manufactured part. The shape consists of one trapezoid and one rectangle.

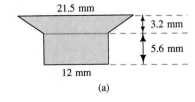

(a)

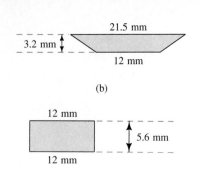

(b)

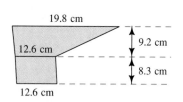

(c)

We need to separate the area into two portions and find each one separately.

The area of the trapezoid is

$$A = \frac{h(b + B)}{2}$$

$$= (3.2)\left(\frac{12 + 21.5}{2}\right)$$

$$= \frac{(3.2)(33.5)}{2}$$

$$= \frac{107.2}{2}$$

$$= 53.6 \text{ mm}^2$$

The area of the rectangle is

$$A = lw$$

$$= (12)(5.6)$$

$$= 67.2 \text{ mm}^2$$

We now add each area:

$$
\begin{array}{r}
67.2 \text{ mm}^2 \\
+ \quad 53.6 \text{ mm}^2 \\
\hline
120.8 \text{ mm}^2
\end{array}
$$

The total area of the part is 120.8 mm². ■

Practice Problem 7 Find the area of the manufactured part shown. The shape consists of one trapezoid and one parallelogram. ■

EXERCISES 7.2

Find the perimeter of each parallelogram.

1.
12.3 in
2.6 in
2.6 in
12.3 in

2.
15.6 in
9.2 in
9.2 in
15.6 in

3. One side measures 2.8 m, and a second side measures 17.3 m.

4. One side measures 4.6 m, and a second side measures 20.5 m.

Find the area of each parallelogram.

5. The base is 20.6 cm, and the height is 5.5 cm.

6. The base is 13.2 cm, and the height is 8.5 cm.

7. The base is 36 m, and the height is 38.5 m.

8. The base is 20.5 m, and the height is 21.5 m.

9. A field in the shape of a parallelogram has a base of 126 yd and a height of 28 yd. Find its area.

10. A field in the shape of a parallelogram has a base of 135 yd and a height of 32 yd. Find its area.

Find the perimeter of each trapezoid.

11.

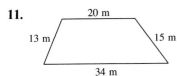

12.

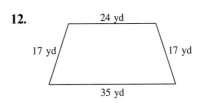

13. The four sides are 160 cm, 65 cm, 185 cm, and 60 cm.

14. The four sides are 130 cm, 100 cm, 70 cm, and 80 cm.

Find the area of each trapezoid.

15. The height is 9 m, and the bases are 5 m and 7 m.

16. The height is 12 m, and the bases are 8 m and 14 m.

17. The height is 2 cm and the bases are 26 cm and 31 cm.

18. The height is 4 cm, and the bases are 14 cm and 29 cm.

19. The height is 16 yd, and the bases are 15 yd and 28 yd.

20. The height is 22 yd, and the bases are 13 yd and 29 yd.

21. A provincial park in Canada in the shape of a trapezoid has a height of 20 km and has bases of 24 km and 31 km. Find the area of the park.

22. A park in Australia is in the shape of a trapezoid. It has a height of 14 km. The bases are 26 km and 38 km. Find the area of the park.

Find the following areas for shapes consisting of trapezoids, parallelograms, squares, and rectangles.

23.

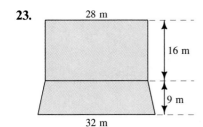

24.

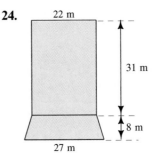

25.

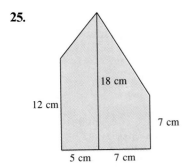

26.

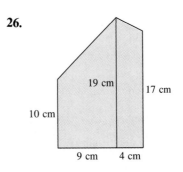

27.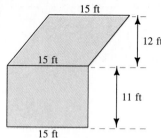
15 ft
12 ft
15 ft
11 ft
15 ft

28.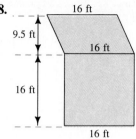
16 ft
9.5 ft
16 ft
16 ft
16 ft
16 ft

The following shapes represent the lobby of a conference center. It will be carpeted at a cost of $22.00 per square yard. How much will the carpeting cost?

29.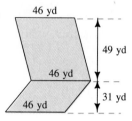
46 yd
49 yd
46 yd
46 yd
31 yd

30.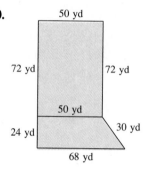
50 yd
72 yd
72 yd
50 yd
24 yd
30 yd
68 yd

 To Think About

31. See if you can find a formula that would find the area of a regular octagon. (A regular octagon is an eight-sided figure with all sides equal.) The dimensions of the rectangles and trapezoids are labeled on the sketch.

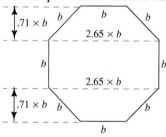

$.71 \times b$ b b b
$2.65 \times b$
b b
$2.65 \times b$
$.71 \times b$ b b
b

32. See if you can find a formula for the area of a regular hexagon with each side equal to *b* units. (A regular hexagon is a six-sided figure with all sides equal.)

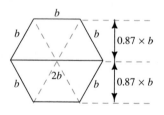

b
b b
$0.87 \times b$
b $2b$ b
$0.87 \times b$

Cumulative Review Problems

Complete each conversion.

33. 10 yd = _____ ft

34. 15,840 ft = _____ mi

35. 18 m = _____ cm

36. 26 mm = _____ cm

 Calculator Problems

37. Find the perimeter of a trapezoid. The sides measure 162.3 ft, 212.56 ft, 372.55 ft, and 192.34 ft.

38. Find the area of a trapezoid. The height is 12.26 cm, and the bases are 10.12 cm, and 21.34 cm.

For Extra Practice Examples and Exercises, turn to page 394.

1. Perimeter = $(2)(7.6) + (2)(3.5)$
$= 15.2 + 7.0 = 22.2$ cm

3. Perimeter = $(2)(4.1) + (2)(11.6)$
$= 8.2 + 23.2 = 31.4$ km

Area = bh
$= (4.1)(9.8) = 40.18$ km^2

5. $A = \dfrac{h(b + B)}{2} = \dfrac{(2.6)(1.3 + 1.9)}{2} = \dfrac{(2.6)(3.2)}{2} = \dfrac{8.32}{2} = 4.16$ mm^2

7. The area of the trapezoid

$A = \dfrac{(9.2)(12.6 + 19.8)}{2}$

$= \dfrac{(9.2)(32.4)}{2} = \dfrac{298.08}{2}$

$= 149.04$ cm^2

The area of the parallelogram is
$A = (8.3)(12.6) = 104.58$ cm^2
Total area $= 149.04$ cm^2 + 104.58 cm^2 = 253.62 cm^2

2. 15.45 km^2 **4.** 56 yd   **6. (a)** 21,700 yd^2 **(b)** 217 gal

---

7.3 ANGLES AND TRIANGLES

After studying this section, you will be able to:

1 *Understand angles with triangles and four-sided figures*

2 *Find the perimeter and area of a triangle*

1 Angles

A **line** $\longleftrightarrow$ extends indefinitely, but a portion of a line, called a **line segment**, has a beginning and an end. An **angle** is formed whenever two line segments meet. The two line segments are called the **sides** of the angle. The point at which they meet is called the **vertex** of the angle.

The "amount of opening" of an angle can be measured and a number written. Angles are commonly measured in degrees. In the above sketch the angle measures 30 degrees. The symbol ° indicates degrees. If you fix one side of an angle and keep moving the other side, the angle measure will get larger and larger until eventually you have gone around in one complete revolution.

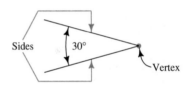

One complete revolution is 360°.

One half of a revolution is 180°.

One quarter of a revolution is 90°

We call two lines **perpendicular** when they meet at an angle of 90°. A 90° angle is called a **right angle**. A 90° angle is often indicated by a small □ at the vertex. Thus when you see ⌐ you know that the angle is 90° and also that the sides are perpendicular to each other.

A **triangle** is a three-sided figure with three angles.

> The sum of the measures of the angles in a triangle is 180°.

We can use this fact to find an unknown angle measurement when we know the other two.

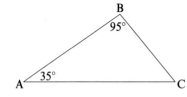

Example 1 If angle *B* measures 95° and angle *A* measures 35° in a triangle, what is the measure of angle *C*?

We will use the fact that the sum of the measures of the angles of a triangle is 180°.

$$35 + 95 + x = 180$$
$$130 + x = 180$$

What number, *x*, when added to 130 equals 180? *x* must equal 50.

Angle *C* must equal 50°. ■

Practice Problem 1 If angle *B* measures 125° and angle *C* measures 15° in a triangle, what is angle *A*? ■

We often label each vertex of a rectangle and a square with the symbol □ to indicate that this angle measures 90°. When you see these figures, you know that

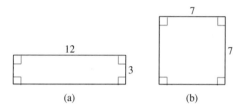

(a)   (b)

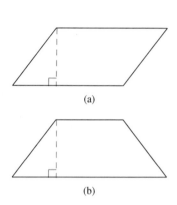

(a)

(b)

they are both rectangles, and, in addition, the one with the sides of length 7 is a square. Any four-sided figure with four right angles is a rectangle. If two adjacent sides are equal and the figure has four right angles, we know that the figure is a square.

The height of a parallelogram or a trapezoid is perpendicular to the base. We usually indicate this by a small □.

2 The perimeter of a triangle is the sum of the three sides.

Example 2 Find the perimeter of a triangle whose sides are 6 in., 5 in., and 8 in.

$$P = 6 + 5 + 8 = 19 \text{ in.} \blacksquare$$

Practice Problem 2 Find the perimeter of a triangle whose sides are 10.5 m, 10.5 m and 8.5 m. ■

A triangle with two sides equal is called an *isosceles triangle*.

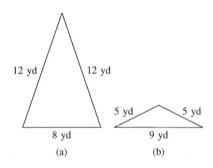

(a)   (b)

Isosceles triangles

A special kind of isosceles triangle with all three sides equal is called an *equilateral triangle*. All angles in an equilateral triangle are exactly 60°.

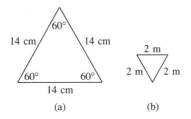

(a)   (b)

Equilateral triangles

A triangle with one 90° angle is called a *right triangle*.

The height of any triangle is the distance of a line drawn from a vertex perpendicular to the other side or an extension of the other side. The height may be one of the sides in a right triangle. The height may reach to an extension of one side if one angle of the triangle is greater than 90°.

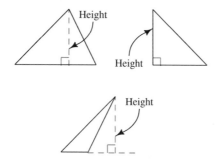

The area of any triangle is one half of the product of the base times the height of the triangle. The height is measured from the vertex above the base, to that base.

> The **area of a triangle** is the product of the base and the height divided by 2.
> $$A = \frac{bh}{2}$$

Example 3   Find the area of the triangle.

$$A = \frac{bh}{2} = \frac{(23)(16)}{2} = \frac{368}{2} = 184 \text{ m}^2 \quad \blacksquare$$

Practice Problem 3   Find the area of the triangle.

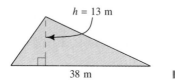

Example 4   Find the area of a triangle whose base is 18 mm and whose height is 15 mm.

$$A = \frac{bh}{2} = \frac{(18)(15)}{2} = \frac{270}{2} = 135 \text{ mm}^2 \quad \blacksquare$$

Practice Problem 4   Find the area of a triangle whose base is 5 cm and height is 16 cm.   ■

Some geometric shapes are a combination of a triangle and one or more of rectangles, squares, parallelograms, and trapezoids.

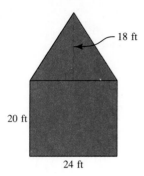

18 ft

20 ft

24 ft

Example 5 Find the area of the side of a house.

The triangle has a base of 24 feet since opposite sides of a rectangle are equal. Thus

$$A = \frac{bh}{2} = \frac{(24)(18)}{2} = \frac{432}{2} = 216 \text{ ft}^2$$

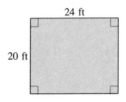

18 ft

24 ft

The area of the rectangle is $A = lw = (24)(20) = 480 \text{ ft}^2$

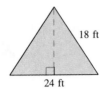

24 ft

20 ft

The sum of the two areas is

$$\begin{array}{r} 216 \text{ ft}^2 \\ + \ 480 \text{ ft}^2 \\ \hline 696 \text{ ft}^2 \end{array}$$ ■

Practice Problem 5 Find the area.

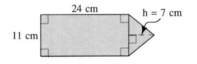

24 cm

$h = 7$ cm

11 cm

■

To Think About

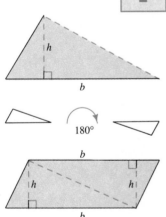

Where does the 2 come from in the formula $A = \frac{bh}{2}$? Why does this formula for the area of a triangle work? Suppose that we construct a triangle with base b and height h. Assume that we do *not* yet know how to find the area.

Now let us make an exact copy of the triangle and turn the copy around to the right exactly 180°. Carefully place the two triangles together. We now have a parallelogram of base b and height h. The height h could be measured at any of several places. The area of a parallelogram is $A = bh$.

Because the parallelogram has area $A = bh$ and is made up of two triangles of identical shape and area, the area of one of the triangles would be the area of the parallelogram divided by 2. Thus the area of a triangle is $A = \frac{bh}{2}$. ■

EXERCISES 7.3

Indicate if each of the following are true or false.

1. Two lines that meet at a 90° angle are perpendicular.

2. A right triangle has two angles of 90°.

3. The sum of the angles of a triangle is 180°.

4. An isosceles triangle has two equal sides.

5. An equilateral triangle has one angle greater than 90°.

6. All equilateral triangles are the same size.

Find the missing angle in each triangle.

7. Two known angles are 30° and 90°.

8. Two known angles are 90° and 45°.

9. Two known angles are 130° and 20°.

10. Two known angles are 16° and 18°.

11. Two known angles are both 45°.

12. Two known angles are both 60°.

13. Two known angles are 36° and 9°.

14. Two known angles are 140° and 24°.

Find the shaded area of each triangle.

15.

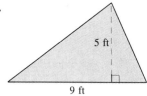

5 ft

9 ft

16.

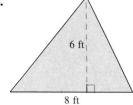

6 ft

8 ft

17. The base is 17.5 cm and the height is 9.5 cm.

18. The base is 3.6 cm and the height is 11.2 cm.

19. The base is 5.9 m and the height is 3.6 m.

20. The base is 7.5 m and the height is 2.4 m.

21. The base is 3.5 yd and the height is 7 yd.

22. The base is 5.6 yd and the height is 4.8 yd.

Find the perimeter *of each figure. Use* outside edges only.

23. A triangle whose sides are 8.5 in., 7.5 in., and 9.5 in.

24. A triangle whose sides are 6.5 in., 10.0 in., and 12.3 in.

25.

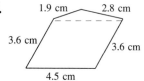

1.9 cm   2.8 cm

3.6 cm

3.6 cm

4.5 cm

26.

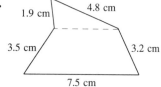

1.9 cm   4.8 cm

3.5 cm   3.2 cm

7.5 cm

Find the area of the shaded regions.

27.

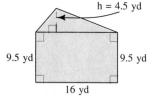

h = 4.5 yd

9.5 yd   9.5 yd

16 yd

28.

h = 8 yd

20 yd   20 yd

25 yd

? To Think About

The top surface of the wings of a test plane must be coated with a special lacquer that costs $90.00 per square yard. Find the cost to coat the shaded wing surface of each plane.

29.

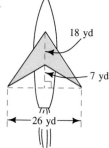

18 yd
7 yd
26 yd

30.

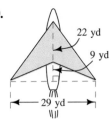

22 yd
9 yd
29 yd

Cumulative Review Problems

Solve for n. Round to the nearest hundredth.

31. $\dfrac{5}{n} = \dfrac{7.5}{18}$

32. $\dfrac{n}{29} = \dfrac{7}{3}$

33. If the ratio of faculty to students is 3 to 29, how many faculty should there be for 3799 students?

34. The manufacturing department found 176 faulty parts out of 2000 manufactured parts last month. If they examine 150 parts, how many defective ones would they expect to find? (Round to the nearest whole number.)

For Extra Practice Examples and Exercises, turn to page 394.

Solutions to Odd-Numbered Practice Problems

1. The sum of the angles in a triangle is 180°. The two given angles total $125 + 15 = 140°$. Thus $180° - 140° = 40°$. Angle A must be 40°.

3. $A = \dfrac{bh}{2} = \dfrac{(38)(13)}{2} = \dfrac{494}{2} = 247 \text{ m}^2$

5. Area of rectangle $= (11)(24) = 264 \text{ cm}^2$
Area of triangle $= \dfrac{(11)(7)}{2} = \dfrac{77}{2} = 38.5 \text{ cm}^2$
Total area $= 302.5 \text{ cm}^2$

Answers to Even-Numbered Practice Problems

2. 29.5 m   **4.** 40 cm^2

After studying this section, you will be able to:

1 Find the area and circumference of a circle

2 Solve area problems containing circles and other geometric shapes

7.4 CIRCLES

Every point on the rim of a circle is the same distance from the center, so a circle looks the same whether rotated or flipped over. No one position is "more important" than another. Diplomats have been known to choose a circular table for the site of their more difficult negotiations because every person's position at a circle has equal importance. In geometry we study the relationship between parts of a circle and we learn how to calculate "how far around" and "how large a region inside." We learn how to find these in this section.

1 A **circle** is a figure for which all points are at an equal distance from a given point. This given point is called the **center** of the circle.

(a)

C
Diameter

(b)

The **radius** is a line segment from the center C to a point on the circle.

The **diameter** is a line segment across the circle that passes through the center C.

We often use the words **radius** and **diameter** to mean the length of those segments. Note that the plural of radius is radii. Clearly, then,

$$\text{diameter} = 2 \times \text{radius} \qquad \text{or} \qquad d = 2r$$

We could also say that

$$\text{radius} = \text{diameter} \div 2 \qquad \text{or} \qquad r = \frac{d}{2}$$

The distance around the rim of the circle is called the **circumference**.

There is a special number called pi, which we denote by the symbol π. π is the number we get when we divide the circumference of a circle by the diameter $\frac{C}{d} = \pi$. π is approximately 3.14159265359. We can approximate π to any number of digits. For all work in this book we will use the following:

> π is approximately 3.14, rounded to the nearest hundredth.

> We find the circumference C of a circle by multiplying the length of the diameter d times π.
>
> $$C = \pi d$$

Example 1 Find the circumference of a circle when the diameter is 7 m. Use $\pi = 3.14$. Round your answer to the nearest tenth.

$C = \pi d = (3.14)(7) = 21.98 = 22.0$ m rounded to the nearest tenth   ■

Practice Problem 1 Find the circumference of a circle when the diameter is 9 meters. Use $\pi = 3.14$. Round your answer to the nearest tenth.   ■

An alternative formula is $C = 2\pi r$

when we find the circumference by using a formula involving the length of the radius.

Example 2 Find the circumference of a circle with radius 12 in. Use $\pi = 3.14$. Round to the nearest tenth.

$C = 2\pi r$

$= (2)(3.14)(12)$

$= (6.28)(12)$

$= 75.36$

$= 75.4$ in rounded to the nearest tenth   ■

Practice Problem 2 Find the circumference of a circle with radius 14 in. Use $\pi = 3.14$. Round to the nearest tenth. ■

> The **area of a circle** is the product of π times the radius squared.
>
> $$A = \pi r^2$$

Example 3

(a) Find an estimate of the area of a circle whose radius is 6 cm.

(b) Find the area of a circle whose radius is 6 cm. Use $\pi = 3.14$. Round your answer to the nearest tenth.

$$A = \pi r^2$$

(a) We can estimate the area. Since π is a number between 3 and 4 $\left(\text{it's close to } 3\frac{1}{4}\right)$ we know that the area we are looking for is between $3r^2$ and $4r^2$. So the area is between (3)(36), or 108, and (4)(36), or 144 (cm²). Because π is close to 3 we will select 3. Thus our *estimated* area is 108 cm².

(b) Let's compute the exact area.

$\quad A = \pi r^2$

$\qquad = (3.14)(6^2)$

$\qquad = (3.14)(36)$ We *must* square the radius first before multiplying by 3.14.

$\qquad = 113.04 \text{ cm}^2$

$\qquad = 113.0 \text{ cm}^2$ rounded to the nearest tenth

Our exact answer is close to the value 108 that we found in part (a). ■

Practice Problem 3 Find the area of a circle whose radius is 5 km. Use $\pi = 3.14$. Round your answer to the nearest tenth. ■

The formula for finding the area uses the length of the radius. If we are given a diameter, we can use the property that $r = \dfrac{d}{2}$.

Example 4 Find the area of a circle that has a diameter of 18 m. Use $\pi = 3.14$. Round your answer to the nearest tenth.

$$r = \frac{d}{2} = \frac{18}{2} = 9 \qquad \text{The radius is 9 m.}$$

$\quad A = \pi r^2$

$\qquad = (3.14)(9^2)$

$\qquad = (3.14)(81)$ We must square the radius first. Then multiply by 3.14.

$\qquad = 254.34 \text{ m}^2$

$\qquad = 254.3 \text{ m}^2$ rounded to the nearest tenth ■

Practice Problem 4 Find the area of a circle whose diameter is 22 meters. Use $\pi = 3.14$. Round your answer to the nearest tenth. ■

2 Several applied area problems have a circular region combined with another region.

Example 5 Find the area of the shaded region. Use $\pi = 3.14$. Round your answer to the nearest tenth.

We will subtract two areas to find the shaded region.

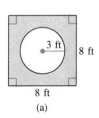

3 ft

8 ft

8 ft

(a)

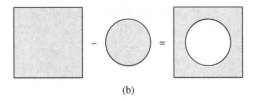

(b)

Area of the square − area of the circle = area of the shaded region

$$A = s^2 \qquad\qquad A = \pi r^2$$
$$= 8^2 \qquad\qquad\quad = (3.14)(3^2)$$
$$= 64 \text{ ft}^2 \qquad\quad\; = (3.14)(9)$$
$$\qquad\qquad\qquad\qquad = 28.26 \text{ ft}^2$$

Area of the square   area of the circle   area of the shaded region

$$64 \text{ ft}^2 \qquad - \qquad 28.26 \text{ ft}^2 \qquad = \qquad 35.74 \text{ ft}^2$$
$$= 35.7 \text{ ft}^2 \quad \text{rounded to}$$
$$\text{nearest tenth} \quad \blacksquare$$

Practice Problem 5   Find the area of the shaded region. Use $\pi = 3.14$. Round your answer to the nearest tenth.

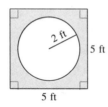

5 ft

5 ft

■

Many geometric shapes involve the semicircle. **A semicircle** is one-half of a circle. The area of a semicircle is therefore one-half of the area of a circle.

Example 6   Find the area of the shaded region. Use $\pi = 3.14$. Round to the nearest tenth.

First we will find the area of the semicircle with the diameter of 6 ft.

$$r = \frac{d}{2} = \frac{6}{2} = 3$$

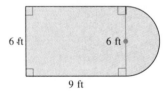

6 ft   6 ft

9 ft

The radius is 3 ft. The area of a semicircle with radius 3 ft is

$$A_{\text{semicircle}} = \frac{\pi r^2}{2} = \frac{(3.14)(3^2)}{2} = \frac{(3.14)(9)}{2}$$
$$= \frac{28.26}{2} = 14.13 \text{ ft}^2$$

Now we add the area of the rectangle.

$$A = lw = (9)(6) = 54 \text{ ft}^2$$

$$\begin{array}{ll} 54.00 \text{ ft}^2 & \text{area of rectangle} \\ + \; 14.13 \text{ ft}^2 & \text{area of semicircle} \\ \hline 68.13 \text{ ft}^2 & \text{total area} \end{array}$$

Rounded to the nearest tenth area $= 68.1 \text{ ft}^2$ ■

Practice Problem 6   Find the area of the shaded region. Use $\pi = 3.14$. Round to the nearest tenth.   ■

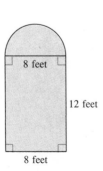

8 feet

12 feet

8 feet

In all problems use π = 3.14. Round each answer to the nearest tenth.

Find the length of the diameter of a circle if the radius has the value given.

1. $r = 29$ in.   **2.** $r = 33$ in.   **3.** $r = 7.5$ mm   **4.** $r = 6.6$ m

Find the length of the radius of a circle if the diameter has the value given.

5. $d = 45$ yd   **6.** $d = 65$ yd   **7.** $d = 3.8$ cm   **8.** $d = 5.2$ cm

Find the circumference of each circle.

9. Diameter = 24 cm   **10.** Diameter = 36 cm   **11.** Radius = 11 in.   **12.** Radius = 15 in.

Find the area of each circle.

13. Radius = 5 yd   **14.** Radius = 7 yd   **15.** Radius = 17 m   **16.** Radius = 14 m

17. Diameter = 32 cm   **18.** Diameter = 44 cm

Find the area of each semicircle.

19.
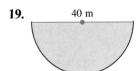
40 m

20.
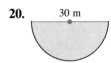
30 m

A water sprinkler sends water out in a circular pattern. Determine how large an area is watered.

21. The radius of watering is 7 ft.   **22.** The radius of watering is 10 ft.

A radio station sends out radio waves in all directions from a tower at the center of the circle of broadcast range. Determine how large an area is reached.

23. The diameter is 120 mi.   **24.** The diameter is 90 mi.

Find the area of the shaded regions.

25.
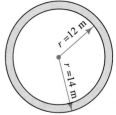
$r = 12$ m
$r = 14$ m

26.
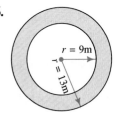
$r = 9$ m
$r = 13$ m

27.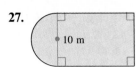

28.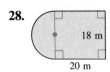

29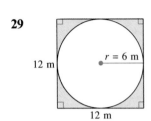

30.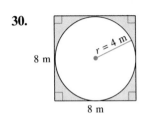

Find the cost of fertilizing a playing field at $0.20 per square yard for the conditions stated.

31. The rectangular part of the field is 120 yd long and the diameter of each semicircle is 40 yd.

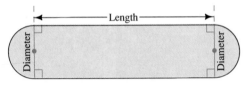

32. The rectangular part of the field is 110 yd long and the diameter of each semicircle is 50 yd.

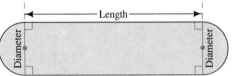

 To Think About

33. A 15-in.-diameter pizza costs $6.00. A 12-in.-diameter pizza costs $4.00. The 12-in.-diameter pizza is cut into six pieces. The 15-in.-diameter pizza is cut into eight pieces.
  (a) What is the cost per slice of the 15-in.-diameter pizza? How many square inches of pizza is in one slice?
  (b) What is the cost per slice of the 12-in.-diameter pizza? How many square inches of pizza is in one slice?
  (c) If you want more for your money, which slice of pizza should you buy?

34. A 14-in.-diameter pizza costs $5.50. It is cut into eight pieces. A 12.5 in. × 12.5 in. square pizza costs $6.00. It is cut into nine pieces.
  (a) What is the cost of one slice of the 14-in.-diameter pizza? How many square inches of pizza is in one slice?
  (b) What is the cost of one small square of the 12.5 in. × 12.5 in. square pizza? How many square inches of pizza is in one slice?
  (c) If you want more for your money, which slice of pizza should you buy?

35. Measure the circumference and diameter of a plate with a metric ruler and a piece of string. Divide the length of the diameter into the measure of the circumference. How close is the obtained value to $\pi \approx 3.14$?

36. Measure the circumference and diameter of a can with a metric ruler and a piece of string. Divide the length of the diameter into the measure of the circumference. How close is the obtained value to $\pi \doteq 3.14$?

37. Tony's car has tires with a radius of 15 in. Tony moved the car ahead 942 in. How many complete revolutions did the wheels turn?

38. Tanya's car has tires with a radius of 14 in. Tanya backed the car up a distance of 8792 in. How many complete revolutions did the wheels turn?

Cumulative Review Problems

39. Find 16% of 87.

40. What is 0.5% of 60?

41. 12% of what number is 720?

42. 19% of what number is 570?

Calculator Problem

Use $\pi \approx 3.14159$ in all calculations for problems 43 and 44.

43. Find the circumference of the circle with a 0.223 m diameter.

44. Find the area of the circle whose radius is 1.39 cm.

For Extra Practice Examples and Exercises, turn to page 395.

Solutions to Odd-Numbered Practice Problems

1. $C = \pi d$
= (3.14)(9)
= 28.26
= 28.3 m to nearest tenth

3. $A = \pi r^2$
= (3.14)(5²)
= (3.14)(25)
= 78.5 km²

5. Area of square − area of circle = shaded area

s^2	−	πr^2	= shaded area
5²	−	(3.14)(2²)	
25	−	(3.14)(4)	
25.00	−	12.56	= 12.4 ft² shaded area (rounded to nearest tenth)

Answers to Even-Numbered Practice Problems

2. 87.9 in   **4.** 379.9 m²   **6.** 121.1 ft²

After studying this section, you will be able to:

1 Find the volume of a box (rectangular solid)

2 Find the volume of a cylinder

3 Find the volume of a sphere

4 Find the volume of a pyramid

5 Find the volume of a cone

7.5 VOLUME

How much grain can that shed hold? How much water is in the polluted lake? How much air is inside a basketball? These are questions of **volume**. We can start with a box 1 in. × 1 in. × 1 in.

This box has volume of 1 cubic inch (written 1 in.³). We can use this as a **unit of volume**.

In this section we compute the volume of several of the three-dimensional geometric figures: the box, cylinder, sphere, pyramid, and cone.

1 Volume is measured in cubic units like cubic meters (abbreviated m³) or cubic feet (abbreviated ft³). When we measure volume we are measuring the space inside an object.

> The **volume of a box** (rectangular solid) is a product of the length times the width times the height:
>
> $$V = lwh$$

Example 1 Find the volume of a box of width 2 ft, length 3 ft, and height 4 ft.
$$V = lwh = (3)(2)(4) = (6)(4) = 24 \text{ ft}^3 \quad \blacksquare$$

Practice Problem 1 Find the volume of a box of width 5 m, length 6 m, height 2 m. ■

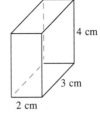

4 cm

3 cm

2 cm

2 Cylinders are the shape we observe when we see a tin can or a tube.

> The **volume of a cylinder** is the product of the area of a circle πr^2 and the height h.
> $$V = \pi r^2 h$$

We will continue to use π as 3.14 as we did in Section 7.4 on all volume problems requiring the use of π.

Example 2 Find the volume of a cylinder of radius 3 in and height 7 in. Round answer to nearest tenth.
$$V = \pi r^2 h = (3.14)(3^2)(7)$$

Be sure to square the radius before doing any other multiplication.
$$V = (3.14)(9)(7) = (28.26)(7)$$
$$= 197.82 = 197.8 \text{ in.}^3 \text{ rounded to nearest tenth} \quad \blacksquare$$

Practice Problem 2 Find the volume of a cylinder of radius 2 in and height 5 in. Round to nearest tenth. ■

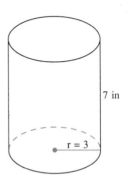

7 in

r = 3

3 Have you ever considered how you would find the volume of the inside of a ball? How many cubic inches of air are inside a basketball? To answer these questions we need a volume formula for a *sphere*.

> The **volume of a sphere** is the product of 4 times π times the radius cubed divided by 3.
> $$V = \frac{4\pi r^3}{3}$$

Example 3 Find the volume of a sphere with radius 3 m. Round your answer to the nearest tenth.
$$V = \frac{4\pi r^3}{3} = \frac{(4)(3.14)(3^3)}{3}$$
$$= \frac{(4)(3.14)(3)(3)(\cancel{3})}{\cancel{3}}$$

Notice here we can cancel the common factor of 3.
$$V = (4)(3.14)(9) = (12.56)(9) = 113.04 \text{ m}^3$$
$$= 113.0 \text{ m}^3 \quad \text{rounded to nearest tenth} \quad \blacksquare$$

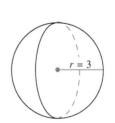

r = 3

Practice Problem 3 Find the volume of a sphere with radius 6 m. Round your answer to the nearest tenth. ■

4 We see the shape of a cone when we look at the sharpened end of a pencil or at an ice cream cone. To find the volume of a cone we use the following formula.

> The **volume of a cone** is the product of π times the radius squared times the height divided by 3.
>
> $$V = \frac{\pi r^2 h}{3}$$

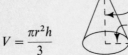

h

r

Example 4   Find the volume of a cone of radius 7 m and height 9 m.

$$V = \frac{\pi r^2 h}{3}$$

$$= \frac{(3.14)(7^2)(9)}{3}$$

$$= \frac{(3.14)(7^2)(\overset{3}{\cancel{9}})}{\underset{1}{\cancel{3}}}$$

$$= (3.14)(49)(3) = (153.86)(3) = 461.58 \text{ m}^3$$

$$= 461.6 \text{ m}^3 \quad \text{rounded to nearest tenth} \quad \blacksquare$$

Practice Problem 4   Find the volume of a cone of radius 5 m and height 12 m. Round to nearest tenth.   ■

5   You have seen pictures of the great pyramids of Egypt. These amazing stone structures are over 4000 years old.

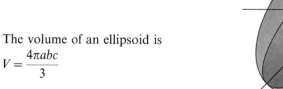

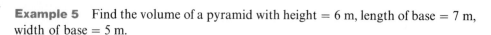

The volume of a pyramid is obtained by multiplying the area of the base *B* by the height *h* and dividing by 3.

$$V = \frac{Bh}{3}$$

Example 5   Find the volume of a pyramid with height = 6 m, length of base = 7 m, width of base = 5 m.

The base is a rectangle.

$$\text{Area of base} = (7)(5) = 35$$

Substituting the area of the base 35 and the height of 6, we have

$$V = \frac{Bh}{3} = \frac{(35)(\overset{2}{\cancel{6}})}{\underset{1}{\cancel{3}}} = 70 \text{ m}^3 \quad \blacksquare$$

Practice Problem 5   Find the volume of a pyramid having the dimensions given.
(a) Height 10 m, width 6 m, length 6 m
(b) Height 15 m, width 7 m, length 8 m   ■

?

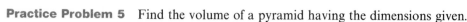
To Think About

What would happen if we try to find the volume of an object that is "almost" spherical but whose dimensions are longer in one direction than the other? This type of shape is called an *ellipsoid*. Shapes like cold capsules, footballs, and eggs are either ellipsoids or very close to ellipsoids.

The volume of an ellipsoid is
$$V = \frac{4\pi abc}{3}$$

The volume of an ellipsoid-shaped capsule with $a = 1.5$ cm, $b = 2.5$ cm, and $c = 1.0$ cm is given by

$$V = \frac{4\pi abc}{3} = \frac{(4)(3.14)(1.5)(2.5)(1)}{3} = \frac{47.1}{3} = 15.7 \text{ cm}^3$$

What would happen if $a = b = c$? Would this be similar to a sphere? Why? ■

EXERCISES 7.5

Find each volume. Use $\pi = 3.14$. Round each answer to the nearest tenth.

1. A rectangular solid with width = 2 m, length = 4 m, height = 3 m

2. A rectangular solid with width = 5 m, length = 4 m, height = 6 m

3. A rectangular solid with width = 26 mm, length = 30 mm, height = 1.5 mm

4. A rectangular solid with width = 15 mm, length = 20 mm, height = 2,5 mm

5. A cylinder with radius 2 m and height 7 m

6. A cylinder with radius 3 m and height 8 m

7. A cylinder with radius 12 m and height 5 m

8. A cylinder with radius 9 m and height 6 m

9. A sphere with radius 9 yd

10. A sphere with radius 12 yd

11. A sphere with radius 4 m

12. A sphere with radius 5 m

Problems 13 and 14 involve hemispheres. A hemisphere is exactly one-half of a sphere.

13. Find the volume of a hemisphere with radius = 7 m.

14. Find the volume of a hemisphere with radius = 6 m.

A collar of Styrofoam is made to insulate a pipe. Find the volume of the unshaded region (which represents the collar). The large radius R is to the outer rim. The small radius r is to the edge of the insulation.

15. $r = 3$ in.
     $R = 5$ in.
     $h = 20$ in.

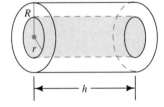

16. $r = 4$ in.
     $R = 6$ in.
     $h = 25$ in.

17. Desmond and Carolyn want to put down 3 in. of good loam on their front lawn, which measures 75 yd by 50 yd. How many cubic yards of loam do they need?

18. Mr. Bledsoe wants to put down a crushed stone driveway to his summer camp. The driveway is 7 yd wide and 120 yd long. The crushed stone is to be 4 in. thick. How many cubic yards of stone will he need?

19. The earth has a radius of about 4000 mi. Assuming that it is a sphere, what is its volume?

20. A tennis ball has a diameter of 2.5 in. A baseball has a diameter of 2.9 in. What is the difference in *volume* between the baseball and the tennis ball?

Find the volume of each pyramid or cone.

21. A pyramid with a height of 7 m and a square base of 3 m on a side

22. A pyramid with a height of 10 m and a square base of 7 m on a side

23. A pyramid with a height of 5 m and a rectangular base measuring 6 m by 12 m

24. A pyramid with a height of 10 m and a rectangular base measuring 8 m by 14 m

25. A cone with a height of 14 cm and a radius of 6 cm

26. A cone with a height of 12 cm and a radius of 9 cm

27. A cone with a height of 20 ft and a radius of 7 ft

28. A cone with a height of 15 ft and a radius of 8 ft

 To Think About

A ceramic cover is made in the shape of an ellipsoid.

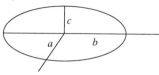

29. What is the volume if $a = 8$ cm, $b = 10$ cm, and $c = 12$ cm? How does it compare to the volume of a sphere with $r = 10$ cm?

30. What is the volume if $a = 6$ cm, $b = 8$ cm, and $c = 10$ cm? How does it compare to the volume of a sphere with $r = 8$ cm?

Suppose that a new pyramid has been found in South America. Find the volume of the pyramid for the conditions given.

31. The rectangular base measures 90 yd by 110 yd and the pyramid has a height of 65 yd.

32. The pyramid given in problem 31 is made of solid stone. It is not hollow like the pyramids of Egypt. It is comprised of layer after layer of cut stone. The stone weighs 422 lb per cubic yard. How many *pounds* will the pyramid weigh? How many tons will the pyramid weigh?

Cumulative Review Problems

33. Add: $7\frac{1}{3} + 2\frac{1}{4}$.

34. Subtract: $9\frac{1}{8} - 2\frac{3}{4}$.

35. Multiply: $2\frac{1}{4} \times 3\frac{3}{4}$.

36. Divide: $7\frac{1}{2} \div 4\frac{1}{5}$.

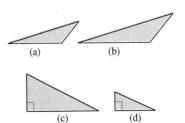

Calculator Problems

37. Find the volume of a **box** with dimensions 5.88 m by 4.26 m by 6.82 m.

38. Find the volume of a **cylinder** of radius 3.22 cm and height 8.43 cm.

39. Find the volume of a **sphere** with radius 5.21 m. Round your answer to the nearest tenth.

40. Find the volume of a **cone** of radius 21.12 mm and height 32 mm.

41. Find the volume of a **pyramid** with height = 9.212 ft, length of rectangular base = 6.22 ft, and length of width = 5.01 ft.

For Extra Practice Examples and Exercises, turn to page 396.

Solutions to Odd-Numbered Practice Problems

1. $V = lwh$
$= (6)(5)(2)$
$= (30)(2)$
$= 60 \text{ m}^3$

3. $V = \dfrac{4\pi r^3}{3} = \dfrac{(4)(3.14)(6^3)}{3}$

$= \dfrac{(4)(3.14)(6)(6)(\overset{2}{\cancel{6}})}{\underset{1}{\cancel{3}}}$

$= (12.56)(36)(2) = 904.32$
$= 904.3 \text{ m}^3$ Rounded to nearest tenth

5. (a) $V = \dfrac{Bh}{3}$

$= \dfrac{(6)(6)(10)}{3} = \dfrac{(36)(10)}{3}$

$= \dfrac{(\overset{12}{\cancel{36}})(10)}{\underset{1}{\cancel{3}}} = 120 \text{ m}^3$

(b) $V = \dfrac{Bh}{3}$

$= \dfrac{(7)(8)(15)}{3} = \dfrac{(7)(8)(\overset{5}{\cancel{15}})}{\underset{1}{\cancel{3}}}$

$= (56)(5) = 280 \text{ m}^3$

Answers to Even-Numbered Practice Problems

2. 62.8 in.^{3} **4.** 314 m^3

7.6   SIMILAR GEOMETRIC FIGURES

☐ After studying this section, you will be able to:

1 *Find corresponding parts of similar triangles*

2 *Find corresponding parts of similar geometric figures*

In English, "similar" means that two things are, in general, alike. But in mathematics, "similar" means that two things are alike in a certain specialized way—they are *alike in shape*, even though they may be different in size. So photographs that are enlarged produce images *similar* to the original; a floor plan of a building is *similar* to the actual building; a model car is *similar* to the actual vehicle.

1 Two triangles with the same shape but not necessarily the same size are called **similar triangles**. Here are two pairs of similar triangles.

The **corresponding angles** of similar triangles are equal.

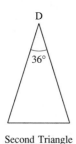

A

36°

First Triangle

D

36°

Second Triangle

These two triangles are similar. The smallest angle in the first triangle is angle *A*. The smallest angle in the second triangle is angle *D*. Both angles measure 36°. Similarly, we could say that the largest angles in each of these two triangles are equal.

> The **corresponding sides** of similar triangles have the same ratio.

These two triangles are similar.

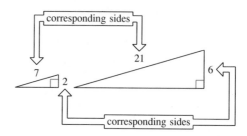

We see that the ratio of 7 to 21 is the same as the ratio of 2 to 6.

$$\frac{7}{21} = \frac{2}{6} \quad \text{is obviously true since} \quad \frac{1}{3} = \frac{1}{3}$$

Example 1 These two triangles are similar. Find the length of side *n*. Round to the nearest tenth.

12 m

5 m

(a)

19 m

n

(b)

The ratio of 12 to 19 is the same as the ratio of 5 to *n*.

$$\frac{12}{19} = \frac{5}{n}$$

$12n = (5)(19)$ Cross-multiplying.

$12n = 95$ Simplifying.

$$\frac{12n}{12} = \frac{95}{12}$$ Dividing each side by 12.

$n = 7.91\overline{6}$ Performing the division.

$= 7.9$ Rounding to the nearest tenth.

Side *n* is of length 7.9 meters. ■

Practice Problem 1 These two triangles are similar. Find the length of side *n*. Round to the nearest tenth. ■

11 m

15 m

27 m

n

Similar triangles are not always oriented the same way. You may find it helpful to rotate one of the triangles so that the similarity is more apparent.

Example 2 These two triangles are similar. Name the sides that correspond.

First we turn the second triangle so that the shortest side is on the top, the intermediate side is to the left, and the longest side is on the right.

a

b

c

d

f

e

Now the shortest sides of each triangle are on the top, the longest side of each triangle is on the right, and so on.

d

a

e

f

b

c

> *a* corresponds to *d*
> *b* corresponds to *e*
> *c* corresponds to *f* ■

Practice Problem 2 These two triangles are similar. Name the sides that correspond.

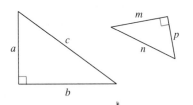

The perimeters of similar triangles have the same ratios as the corresponding sides.

Example 3 Two triangles are similar. The smaller triangle has sides 5 yd, 7 yd and 10 yd. The 7 yd side on the smaller triangle corresponds to a side of 21 yd on the larger triangle. What is the perimeter of the larger triangle?

The perimeter of the smaller triangle is 22 yards.

Let p = the unknown perimeter. Now the ratio of 7 to 21 is the same as the ratio of 22 to the unknown perimeter. Thus

$$\frac{7}{21} = \frac{22}{p}$$

$$7p = (21)(22)$$

$$7p = 462$$

$$\frac{7p}{7} = \frac{462}{7}$$

$$p = 66$$

The perimeter of the larger triangle is 66 yards.

Practice Problem 3 Two triangles are similar. The smaller triangle has sides of 5 yd, 12 yd and 13 yd. The 12 yd side of the smaller triangle corresponds to a side of 40 yards on the larger triangle. What is the perimeter of the larger triangle?

Example 4 A flagpole casts a shadow of 36 feet. At the same time a tree that is 3 feet tall has a shadow of 5 feet. How tall is the flagpole?

The sun shining on vertical objects and their shadows at the same time of day forms similar triangles.

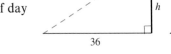

Let n = the height of the flagpole. Thus we can say n is to 3 as 36 is to 5.

$$\frac{n}{3} = \frac{36}{5}$$

$$5n = (3)(36)$$

$$5n = 108$$

$$\frac{5n}{5} = \frac{108}{5}$$

$$n = 21.6$$

The flagpole is about 21.6 feet tall.

Practice Problem 4 How tall (h) is the building if the two triangles are similar?

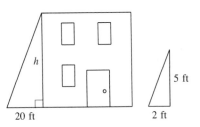

2 Geometric figures such as rectangles, trapezoids, and circles can be similar figures.

> The **corresponding sides of similar geometric figures** have the same ratio.

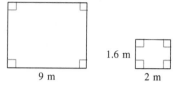

9 m

1.6 m

2 m

Example 5   The two rectangles shown here are similar because the corresponding sides of the two rectangles have the same ratio. Find the width of the larger rectangle.

Let w = the width of the larger rectangle.

$$\frac{w}{1.6} = \frac{9}{2}$$

$$2w = (1.6)(9)$$

$$2w = 14.4$$

$$\frac{2w}{2} = \frac{14.4}{2}$$

$$w = 7.2$$

The width of the larger rectangle is 7.2 meters.   ◼

Practice Problem 5   The two rectangles shown below are similar. Find the width of the larger rectangle.

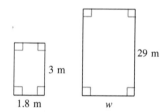

3 m

29 m

1.8 m

w

◼

> The perimeter of similar figures—whatever the figures—is proportional to their corresponding sides. Circles are special cases. All circles are similar. The ratio of circumferences of two circles is proportional to the ratio of their radii.

?

> *To Think About*

How would you find the relationship between the areas of two similar geometric figures? Consider the following two similar rectangles.

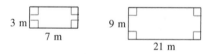

3 m

7 m

9 m

21 m

The area of the smaller rectangle is $(3)(7) = 21$ m². The area of the larger rectangle is $(9)(21) = 189$ m². How could one have predicted this amount?

The ratio of small width to large width is $\frac{3}{9} = \frac{1}{3}$. The small rectangle has sides that are $\frac{1}{3}$ as large as the large rectangle. The ratio of area of the small rectangle to the large rectangle is $\frac{21}{189} = \frac{1}{9}$. Note that $\left(\frac{1}{3}\right)^2 = \frac{1}{9}$.

Thus we can develop the following principle: The areas of two similar figures are in the same ratio as the square of the ratio of two corresponding sides.

What is the area of the small parallelogram if the two parallelograms are similar? Let A = the area of the small parallelogram. The ratio of the corresponding sides is 2 to 10. If we square that ratio, we will have the ratio of the areas.

$$\left(\frac{2}{10}\right)^2 = \frac{A}{80}$$

$$\left(\frac{1}{5}\right)^2 = \frac{A}{80}$$

$$\frac{1}{25} = \frac{A}{80}$$

$$80 = 25A$$

$$\frac{80}{25} = \frac{25A}{25}$$

$$3.2 = A$$

Thus the area of the smaller parallelogram is 3.2 m². ∎

EXERCISES 7.6

For each pair of similar triangles, find the missing side n. *Round your answer to the nearest tenth if necessary.*

1.

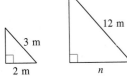

3 m
2 m
12 m
n

2.

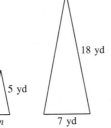

8 m
3 m
24 m
n

3.

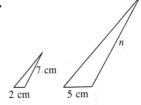

7 cm
2 cm
5 cm
n

4.

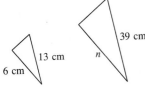

13 cm
6 cm
39 cm
n

5.

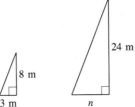

5 yd
n
18 yd
7 yd

6.

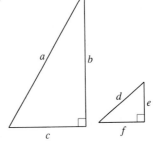

23 yd
42 yd
6 yd
n

Each pair of triangles is similar. Determine which pair of sides correspond in each case.

7.

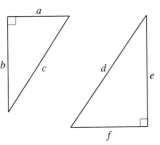

a
b
c
d
e
f

8.

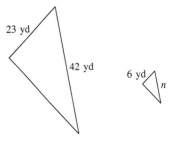

a
b
c
d
e
f

For each pair of similar triangles, find the missing side n. Round your answer to the nearest tenth if necessary.

9.

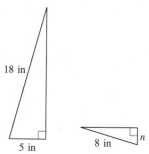

18 in

5 in

8 in n

10.

3 in

15 in 20 in

n

A flagpole casts a shadow. At the same time a small tree casts a shadow. Use the sketch to find the length n of each flagpole.

11.

n

24 ft

6 ft

4 ft

12.

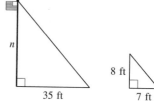

n

35 ft

8 ft

7 ft

Each pair of figures is similar. Find the missing side. Round to the nearest tenth if necessary.

13.

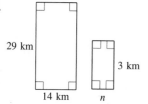

29 km

14 km

3 km

n

14.

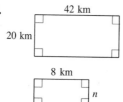

42 km

20 km

8 km

n

15.

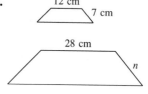

12 cm

7 cm

28 cm

n

16.

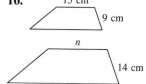

15 cm

9 cm

n

14 cm

Each pair of figures is similar. Find the missing perimeter.

17. Perimeter = 100m

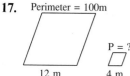

P = ?

12 m 4 m

18. Perimeter = 160m

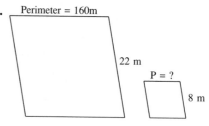

22 m

P = ?

8 m

 To Think About

Each pair of geometric figures is similar. Find the unknown area.

19.

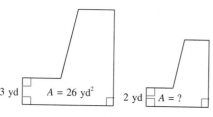

3 yd   A = 26 yd²

2 yd A = ?

20.

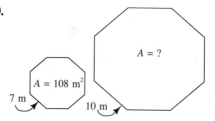

A = ?

A = 108 m²

7 m

10 m

Perform the calculation. Use the correct order of operations.

21. $2 \times 3^2 + 4 - 2 \times 5$

22. $2^2 \times 3^2 + 4^2 - 20 \div 2$

For Extra Practice Examples and Exercises, turn to page 396.

Solutions to Odd-Numbered Practice Problems

1.
$$\frac{11}{27} = \frac{15}{n}$$
$$11n = (27)(15)$$
$$11n = 405$$
$$\frac{11n}{11} = \frac{405}{11}$$
$$n = 36.\overline{81}$$
$$= 36.8 \text{ meters measured to the nearest tenth.}$$

3. Small triangle has a perimeter of 30 yards. Let p = the unknown perimeter.
$$\frac{12}{40} = \frac{30}{p}$$
$$12p = (40)(30)$$
$$12p = 1200$$
$$\frac{12p}{12} = \frac{1200}{12}$$
$$p = 100$$

The larger triangle has a perimeter of 100 yards.

5.
$$\frac{3}{29} = \frac{1.8}{n}$$
$$3n = (1.8)(29)$$
$$3n = 52.2$$
$$\frac{3n}{3} = \frac{52.2}{3}$$
$$n = 17.4 \qquad \text{The width is 17.4 meters.}$$

Answers to Even-Numbered Practice Problems

2. a corresponds to p, b corresponds to m, c corresponds to n

4. 50 ft

7.7 APPLICATIONS OF GEOMETRY

☐ After studying this section you will be able to

1 Solve application problems involving geometric shapes

We can solve many real-life problems with the geometric knowledge we now have. Our everyday world is filled with objects that are geometric in shape, so we can use our knowledge of geometry to find length, area, or volume. How far is the automobile trip? How much framing, edging, or fencing is required? How much paint, siding, or roofing is required? How much can we store? We can also find how much certain activities cost if we know the unit cost. It helps, when confronted with a real-life problem, to write it down, visualize it, and if possible, to draw a picture. Try to start each of the problems in this section that way.

Example 1 A professional painter can paint 90 ft^2 of wall space in 20 minutes. How long will it take the painter to paint these four walls; 14 ft $\times$ 8 ft, 12 ft $\times$ 8 ft, 10 ft $\times$ 7 ft, and 8 ft $\times$ 7 ft?

Each wall is a rectangle. The first one is 8 ft wide and 14 ft long.

$$A = lw$$
$$= (14)(8) = 112 \text{ ft}^2$$

We continue with the other three walls.

$$(12)(8) = 96 \text{ ft}^2 \qquad (10)(7) = 70 \text{ ft}^2 \qquad (8)(7) = 56 \text{ ft}^2$$

Total area is obtained by adding.

$$\begin{array}{r} 112 \text{ ft}^2 \\ 96 \text{ ft}^2 \\ 70 \text{ ft}^2 \\ + \ \ 56 \text{ ft}^2 \\ \hline 334 \text{ ft}^2 \end{array}$$

Now we multiply the total area by a unit fraction that tells us 90 ft² can be done in 20 minutes.

$$334 \text{ ft}^2 \times 1 = 334 \text{ ft}^2 \times \frac{20 \text{ min}}{90 \text{ ft}^2}$$

$$= \frac{(334)(20)}{90} \text{ min}$$

$$= \frac{6680}{90} = 74 \text{ minutes} \quad \text{(rounded to the nearest minute)} \quad \blacksquare$$

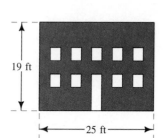

19 ft

25 ft

Practice Problem 1   Mike rented an electric floor sander. It will sand 80 ft² of hardwood floor in 15 minutes. He needs to sand the floors in three rooms. The floor dimensions are 24 ft × 13 ft, 12 ft × 9 ft, and 16 ft × 3 ft. How long will it take him to sand the floors in all three rooms?   ■

Example 2   Joan and Rosetta want to put vinyl siding on the front of their home. The house dimensions are shown. The door dimensions are 6 ft × 3 ft. The windows measure 2 ft × 4 ft.

(a) Excluding windows and doors how many square feet of siding will be needed?
(b) If the siding costs $2.25 per square foot, how much will it cost to side the front of the house?

(a) We will find the area of the large rectangle representing the front of the house. Then we will subtract the area of windows and the door.

$$\text{Area each window} = (2)(4) = 8 \text{ ft}^2$$

$$\text{Area of 9 windows} = (9)(8) = 72 \text{ ft}^2$$

$$\text{Area of 1 door} = (6)(3) = 18 \text{ ft}^2$$

$$\text{Area of 9 windows} + 1 \text{ door} = 90 \text{ ft}^2$$

$$\text{Area of large rectangle} = (19 \text{ ft})(25 \text{ ft}) = 475 \text{ ft}^2$$

Total area of front of house	475 ft²
− Area of 9 windows and 1 door	− 90 ft²
= Total area to be covered	385 ft²

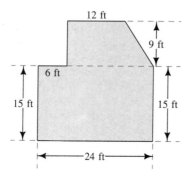

12 ft

9 ft

6 ft

15 ft

15 ft

24 ft

(b) Cost $= 385 \text{ ft}^2 \times \dfrac{\$2.25}{1 \text{ ft}^2} = \$866.25$

The cost to put up siding on the front of the house is $866.25.   ■

Practice Problem 2   At left is a sketch of a roof.
(a) What is the area of the roof?
(b) How much would it cost to install new roofing on the roof area shown if the roofing costs $2.75 per square yard?   ■

EXERCISES 7.7

Round all answers to the nearest tenth unless otherwise directed.

1. A rectangular television screen measures 44 cm in length and 36 cm in width. It is surrounded by a chrome border that costs $0.06 per centimeter. What is the cost of the chrome border?

2. The Kim family started a flower garden. The rectangular garden measures 12 ft long and 7 ft wide. To keep out small animals they had to surround it with fencing which costs $0.85 per foot. How much did the fencing cost?

3. Monica drives to work each day from Bethel to Bridgeton. The sketch shows the two possible routes.

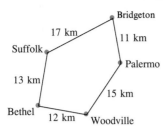

(a) How many kilometers is the trip if she drives through Suffolk? What is her average speed if this trip takes her 0.4 hour?
(b) How many kilometers is the trip if she drives through Woodville and Palermo? What is her average speed if this trip takes her 0.5 hour?
(c) Over which route does she travel at a more rapid rate?

4. Robert drives from work to either a convenience food store and then home or to a food supermarket and then home. The sketch shows the distances.

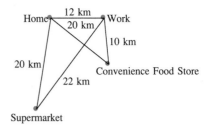

(a) How far does he travel if he goes from work to the supermarket and then home? How fast does he travel if the trip takes 0.6 hour?
(b) How far does he travel if he goes from work to the convenience food store and then home? How fast does he travel if the trip takes 0.5 hour?
(c) Over which route does he travel at a more rapid rate?

5. Alyce is using a new power painter that can spray paint on a wall at the rate of 150 ft² in 12 minutes. She will be painting four walls in her house. They measure 8 ft × 10 ft, 8 ft × 14 ft, 7 ft × 6 ft, and 7 ft × 10 ft. How many minutes will it take her to paint these walls?

6. A professional wallpaper hanger can wallpaper 120 ft² in 15 minutes. She will be papering four walls in a house. They measure 7 ft × 10 ft, 7 ft × 14 ft, 6 ft × 10 ft, and 6 ft × 8 ft. How many minutes will it take her to paper all four walls?

7. The floor area of the recreation room at Yvonne's house is shown in the drawing. How much will it cost to carpet the room if the carpet costs $15 per *square yard*?

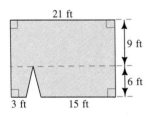

8. The side view of a barn is shown in the diagram. The cost of aluminum siding is $18 per square yard. How much will it cost to put siding on this side of the barn?

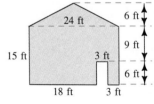

9. A cylindrical gasoline tank is 40 ft high and has a diameter of 80 ft. How many gallons of gasoline will the tank hold if there are 7.5 gal in 1 ft³?

10. A conical pile of grain is 18 ft in diameter and 12 ft high. How many bushels of grain are in the pile if 1 cubic foot = 0.8 bushel?

11. Find the volume of a concrete connector for the city sewer system. A diagram of the connector is shown. It is shaped like a box with a hole of diameter 2 m. If it is formed using concrete that costs $1.20 per cubic meter, how much will the necessary concrete cost?

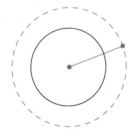

4 m   7 m   3 m

12. A dentist places a gold filling in a tooth in the shape of a cylinder with a hemispherical top. The radius r of the filling is 1 mm. The height is 2 mm. Find the volume of the filling. If dental gold costs $95.00 per cubic millimeter, how much did the gold cost for the filling?

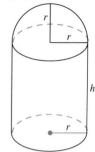

r r h r

 To Think About

13. The *Landstat* satellite orbits the earth in an almost circular pattern. Assume that the radius of orbit (distance from center of earth to the satellite) is 6500 km.

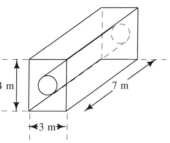

(a) How many kilometers long is one orbit of the satellite? (That is, what is the circumference of the orbit path?)
(b) If the satellite goes through one orbit around the earth in 2 hours, what is its speed in kilometers per hour?

14. The North City Park is constructed in a shape that includes one-fourth of a circle. It is shaded yellow in this sketch.

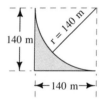

140 m $r = 140$ m   140 m

(a) What is the area of the park?
(b) How much will it cost to re-sod the park with new grass at $4.00 per square meter?

Cumulative Review Problems

Divide.

15. $16 \overline{)2048}$

16. $27 \overline{)24705}$

17. $1.3 \overline{)0.325}$

18. $0.52 \overline{)2.5324}$

 Calculator Problems

19. A rectangular field measures 121.2 yd in length and 96.35 yd in width. To fence the field it will cost $3.98 per yard. What will it cost to fence the field?

20. A cylindrical water tank is 48.8 ft high and has a diameter of 72.32 ft. How many gallons of water will the tank hold if there are 7.5 gal in 1 ft³? Round to the nearest hundredth.

For Extra Practice Examples and Exercises, turn to page 397.

Solution to Odd-Numbered Practice Problem

1. The area of the three rooms

$A = 13 \times 24 = 312 \text{ ft}^2$

$A = 9 \times 12 = 108 \text{ ft}^2$

$A = 3 \times 16 = 48 \text{ ft}^2$ Total area: $312 \text{ ft}^2 + 108 \text{ ft}^2 + 48 \text{ ft}^2 = 468 \text{ ft}^2$

Now we multiply by a unit fraction.

$$468 \text{ ft}^2 \times \boxed{1} = 468 \text{ ft}^2 \times \boxed{\frac{15 \text{ min}}{80 \cancel{\text{ft}^2}}}$$

$$= \frac{(468)(15)}{80} \text{ min} = \frac{7020}{80} = 88 \text{ minutes (Rounded to the nearest minute.)}$$

Answer to Even-Numbered Practice Problem

2. (a) 55 yd^2 **(b)** $\$151.25$

EXTRA PRACTICE: EXAMPLES AND EXERCISES

Section 7.1

Find the perimeter of a rectangle with the following sides.

Example   Length = 12.4 m, width = 3.55 m

$$P = 2l + 2w$$
$$= (2)(12.4) + (2)(3.55)$$
$$24.8 + 7.1 = 31.9 \text{ m}$$

1. Length = 11.7 m, width = 6.9 m

2. Length = 16 m, width = 12.8 m

3. Length = 2.55 cm, width = 1.09 cm

4. Length = 7.94 cm, width = 3.61 cm

5. Length = 21 km, width = 13 km

Find the area of each rectangle or square.

Example   Length = 16 in., width = 5 in. This is a rectangle so the area is $A = lw$. Now $l = 16$ in., $w = 5$ in. so $A = (16)(5) = 80 \text{ in.}^2$

6. Length = 13 in., width = 7 in.

7. Length = 25 yds, width = 11 yds

8. All sides = 5 m

9. All sides = 12 in.

10. Length = 8.4 ft, width = 3.1 ft

Example   A rectangular field measures 130 ft × 25 ft. To fertilize the field will cost $\$0.05$ per square foot. **(a)** How much does it cost to fertilize the field? **(b)** How many feet of fencing would be needed to put a fence around all sides of the field?

(a) $A = lw$, $l = 130$ ft, $w = 25$ ft

$A = (130)(25) = 3250 \text{ ft}^2$ the number of square feet to fertilize

Number of square feet × cost for each square foot = total cost

     3250   ×   0.05   = $\$162.50$

(b) $P = 2l + 2w$

$= (2)(130) + (2)(25) = 260 + 50 = 310$ ft of fencing

11. A rectangular field measures 120 ft by 30 ft. To fertilize the field will cost $\$0.07$ per square foot. How much does it cost to fertilize the field?

12. A rectangular field measures 188 ft × 30 ft. To lime the field costs $\$0.04$ per square foot. How much does it cost to lime the field?

13. A farmer wants to fence his field. The field measures 175 ft × 65 ft. How much fencing will he need?

14. A rectangular picture requires a frame 5.5 ft × 3 ft. The framing material is $\$1.25$ per foot. How much will the framing cost?

15. A rectangular picture requires a frame 4.5 ft × 8.6 ft. The framing material is $\$1.75$ per foot. How much will the framing cost?

Find the area of each shaded object.

Example

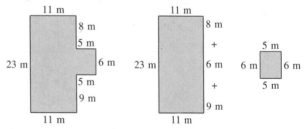

There are two rectangles: one 11 m × 23 m and a smaller one 5 m × 6 m.

$$\underset{\text{large rectangle}}{\text{Area of the}} + \underset{\text{small rectangle}}{\text{area of the}} = \underset{\text{area}}{\text{Total shaded}}$$

    (11)(23)   +   (5)(6)

     253   +   30   = 283 m^2

16.

7 mi, 10 mi, 5 mi, 18 mi, 8 mi, 2 mi

17.

22 m, 10 m, 11 m, 9 m, 9 m, 16 m, 16 m, 43 m

Section 7.2

Find the area and perimeter of each parallelogram.

Example

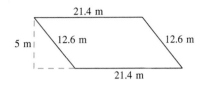

The perimeter is the sum of the length of its sides.

$$P = (2)(12.6) + (2)(21.4)$$
$$= 25.2 + 42.8 = 68 \text{ m}$$

The area is the length of the base times the height:

$$A = (21.4)(5) = 107 \text{ in.}^2$$

1.

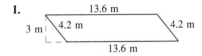

2.

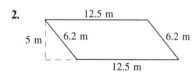

Find the perimeter of a trapezoid with four sides of the given length.

Example Sides: 12 m, 22 m, 14 m, 35 m

The perimeter is the sum of the sides

$$p = 12 + 22 + 14 + 35 = 83 \text{ m}$$

3. Sides: 17 m, 33 m, 15 m, 42 m
4. Sides: 6 m, 12 m, 8 m, 21 m
5. Sides: 15 yd, 32 yd, 15 yd, 35 yd
6. Sides: 6 cm, 13 cm, 6 cm, 15 cm
7. Sides: 21 cm, 60 cm, 21 cm, 75 cm

Find the area of a trapezoid with bases b and B and height h.

Example $b = 4$ m, $B = 12$ m, $h = 7$ m

$$A = \frac{h(B + b)}{2} = \frac{7(12 + 4)}{2} = \frac{7(16)}{2} = \frac{112}{2} = 56 \text{ m}^2$$

8. $b = 5$ cm, $B = 13$ cm, $h = 3$ cm
9. $b = 15$ m, $B = 25$ m, $h = 6$ m
10. $b = 125$ m, $B = 145$ m, $h = 35$ m
11. $b = 46$ cm, $B = 84$ cm, $h = 12$ cm
12. $b = 4.5$ m, $B = 7$ m, $h = 3.2$ m

Find the area for each trapezoid, parallelogram, square, or rectangle used to form each figure.

Example

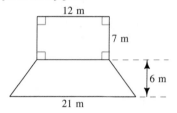

Area of rectangle + Area of trapezoid

$$A = lw \qquad A = \frac{h(B + b)}{2}$$

$$\text{Total area} = (12)(7) \qquad + \qquad \frac{(6)(21 + 12)}{2}$$

$$84 \qquad + \qquad 99 = 183 \text{ m}^2$$

13.

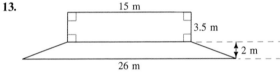

14.

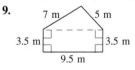

Section 7.3

In each of the following problems find the third angle of a triangle if the first two angles are listed.

Example 30°, 105°, ?

The sum of the angles of a triangle is 180°. Let $x =$ the third angle.

$$30 + 105 + x = 180$$
$$135 + x = 180$$
$$x = 45° \quad \text{because } 135 + 45 = 180$$

1. 125°, 25°, ?   **2.** 45°, 50°, ?
3. 85°, 30°, ?   **4.** 30°, 45°, ?
5. 110°, 20°, ?

Find the perimeter of each figure.

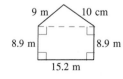

Example Find the perimeter of this figure (solid lines only).

The perimeter is the sum of the outer edges.

$$P = 9 + 10 + 8.9 + 15.2 + 8.9 = 52 \text{ m}$$

Find the perimeter of a triangle whose sides are:
6. 6.3 cm, 8 cm, 13.2 cm
7. 4 in., 4 in., 7 in.
8. 3 yd, 4 yd, 5 yd

Find the perimeter of these figures.

9.

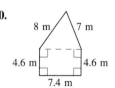

10.

Find the area of a triangle with given dimensions.

Example Find the area of a triangle with base = 5 feet and height = 8 feet.

$$A = \frac{bh}{2} = \frac{(5)(8)}{2} = \frac{40}{2} = 20 \text{ ft}^2$$

11. base = 7 ft, height = 4 ft
12. base = 5 m, height = 2 m
13. base = 7.2 cm, height = 5.5 cm
14. base = 6 yd, height = 2.7 yd
15. base = 3.2 in., height = 8.1 in.

Find the area of the shaded region.

Example

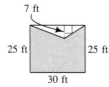

The area of the shaded region is

Area of rectangle − area of triangle

$$lw \qquad - \qquad \frac{bh}{2} \qquad\qquad 750 - 105 = 645 \text{ ft}^2$$

16.

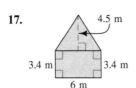

17.

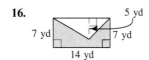

Section 7.4

Find the circumference of a circle with a given radius or diameter.

Example Find the circumference of a circle with *radius* 31 cm.

$$C = 2\pi r = 2(3.14)(31) = (6.28)(31) = 194.68 \text{ cm}$$

Find the circumference of a circle with a

1. *radius* of 14 cm
2. *radius* of 8 cm
3. *diameter* of 11 m
4. *diameter* of 5 yd
5. *radius* of 2 ft

Find the area of a circle with a given radius or diameter.

Example Find the area of a circle with diameter 4 feet.

$$A = \pi r^2$$

If diameter = 4 feet then radius = 2 feet

$$A = (3.14)(2)^2 = (3.14)(4) = 12.56 \text{ ft}^2$$

Find the area of a circle with a

6. *diameter* of 3 ft
7. *diameter* of 12 m
8. *diameter* of 16 mm
9. *radius* of 2 m
10. *radius* of 5 m

Example

$A = \pi r^2$. The area of the shaded region is:

Area of the − area of the
large circle   small circle

$$(3.14)(12^2) \; - \; (3.14)(9^2)$$

$$(3.14)(144) \; - \; (3.14)(81)$$

$$452.16 \quad - \quad 254.34 \quad = 197.82 \text{ m}^2$$

Find the area of the shaded region.

11.

12.

Find the area of a semicircle.

Example Find the area of a semicircle with diameter = 34 m

$$A = \frac{\pi r^2}{2}$$

Since the area of a semicircle is one-half the area of a whole circle. If $d = 34$, then $r = 17$.

$$A = \frac{(3.14)(17)^2}{2} = \frac{(3.14)(289)}{2} = \frac{907.46}{2} = 453.73 \text{ m}^2$$

Round each answer to the nearest hundredth.

Find the area of a semicircle with

13. diameter = 14 m
14. diameter = 40 ft
15. diameter = 7 cm
16. diameter = 12 yd
17. diameter = 18 m

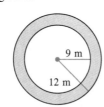

Find the volume of a box having the dimensions given.

Example   Width = 3 ft, length = 4 ft, height = 6 ft

$$V = lwh$$
$$= (4)(3)(6) = 72 \text{ ft}^3$$

1. Length = 5 ft, width = 4 ft, height = 7 ft
2. Length = 16 cm, width = 6 cm, height = 12 cm
3. Length = 6 yd, width = 2 yd, height = 9 yd
4. Length = 10 ft, width = 8 ft, height = 13 ft
5. Length = 2.9 ft, width = 1.6 ft, height = 3.3 ft

Find the volume of a cylinder having the dimensions given. Round to the nearest hundredth.

Example   radius 3 cm, height 5 cm

$$V = \pi r^2 h = (3.14)(3)^2(5) = (3.14)(9)(5) = 141.3 \text{ cm}^3$$

6. radius = 2 cm, height = 6 cm
7. radius = 4 cm, height = 4.4 cm
8. radius = 5 m, height = 8 m
9. diameter = 6 feet, height = 8.1 feet
10. diameter = 12 meters, height = 21 m

Find the volume of a sphere with the given radius. Round to the nearest hundredth.

Example   r = 5 m

$$V = \frac{4\pi r^3}{3} = \frac{4(3.14)(5)^3}{3} = \frac{4(3.14)(125)}{3} = \frac{1570}{3} = 523.33 \text{ m}^3$$

11. r = 7 cm   12. r = 4 m
13. r = 3 cm   14. r = 8 cm
15. r = 6 yd

Find the volume of a cone with given dimensions. Round to nearest hundredth.

Example   radius = 4 m, height = 7 m

$$V = \frac{\pi r^2 h}{3} = \frac{(3.14)(4)^2(7)}{3} = \frac{(3.14)(16)(7)}{3} = \frac{351.68}{3} = 117.23 \text{ m}^3$$

16. radius = 2 m, height = 3 m
17. radius = 3.1 m, height = 4 m
18. radius = 1.1 cm, height = 2 cm
19. radius = 2.5 cm, height = 3.6 cm
20. radius = 2 feet, height = 5 ft

Each pair of triangles is similar. In each case find the missing side n. Round your answer to the nearest tenth if necessary.

Example

$$\frac{20}{9} = \frac{6}{n} \qquad 20n = (9)(6)$$
$$20n = 54$$
$$n = 2.7 \text{ in.}$$

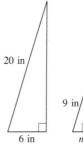

20 in

9 in

6 in   n

Rotate the triangles so that the smallest sides are at the bottom and the right angles are on the left.

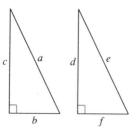

c   a   d   e

b   f

a corresponds to e, c corresponds to d, b corresponds to f.

1.

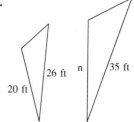

26 ft   n

35 ft

20 ft

2.

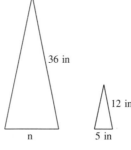

36 in

n   5 in

12 in

3.   **4.**

Each pair of triangles is similar. Determine which pair of sides corresponds in each case.

Example

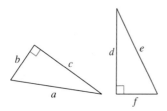

b
c
a

d   e

f

Each pair of rectangles is similar. Find the missing side. Round to the nearest tenth.

Example

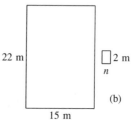

22 m

2 m
n

(b)

15 m

$$\frac{n}{15} = \frac{2}{22} \qquad (n)(22) = (15)(2)$$
$$22n = 30$$
$$n = 1.36 \quad \text{which rounds to 1.4 meters}$$

5.

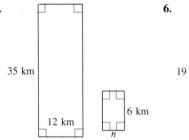

35 km

12 km

6 km

n

6.

19 km

5 km

3 km

n

7.

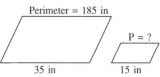

Perimeter = 185 in

35 in

P = ?

15 in

8.

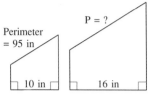

Perimeter
= 95 in

P = ?

10 in

16 in

Each pair of figures is similar. Find the missing perimeter or circumference.

Example Perimeter = 75 in. P = ?

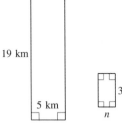

$$\frac{15}{6} = \frac{75}{P}$$

$15P = (6)(75)$

$15P = 450$

$P = 30$ in.

Perimeter = 75 in

15 in

P = ?

6 in

Section 7.7

Example A rectangular picture measures 39 in. in length and 25 in. in width. To frame the picture it will cost $0.08 per inch. What is the cost of the picture frame?

First, find the perimeter of a rectangle with $l = 39$ in. and $w = 25$ in.

$$P = 2l + 2w$$
$$= (2)(39) + (2)(25)$$
$$= 78 + 50 = 128 \text{ in.}$$

Next, find the cost of 128 inches of framing.

$$\frac{128 \text{ in.}}{1} \times \frac{\$0.08}{1 \text{ in.}} = \$10.24$$

1. A rectangular picture measures 35 in. in length and 30 in. in width. To frame the picture will cost $0.09 per inch. What is the cost of the picture frame?

2. A farmer wants to fence his field. A sketch of the field is shown. The fencing cost $3.95 per yard. How much will the fencing cost?

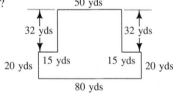

50 yds

32 yds

32 yds

20 yds

15 yds

15 yds

20 yds

80 yds

3. The Johnson family started a vegetable garden. The rectangular garden measures 15 ft long and 11 ft wide. They surrounded it with fencing that costs $0.95 per foot. How much did the fencing cost?

Example The Matthews want to put carpet in the family room. The shape of the room is rectangular, of length 21 ft and width 15 ft. If carpet cost $15.95 per square yard, how much will it cost to carpet their family room?

First change the length and width from feet to yards:

$$\frac{1 \text{ yd}}{3 \text{ ft}} \times 21 \text{ ft} = 7 \text{ yd} \qquad \frac{1 \text{ yd}}{3 \text{ ft}} \times 15 \text{ ft} = 5 \text{ yd}$$

Next find the area: $A = lw = (7)(5) = 35$ yd². Then find the cost:

$$35 \text{ yd}^2 \times \frac{\$15.95}{1 \text{ yd}^2} = (35)(15.95) = \$558.25$$

4. Jack is using a new power painter that can spray paint on a wall at the rate of 175 ft² in 10 minutes. He will be painting three walls in his house. They measure 8 ft × 9 ft, 8 ft × 13 ft, and 7 ft × 5 ft. How many minutes will it take him to paint these walls?

5. A professional wallpaper hanger can wallpaper 115 ft² in 20 minutes. He will be papering four walls in the house. They measure 8 ft × 10 ft, 8 ft × 12 ft, 7 ft × 11 ft, and 6 ft × 9 ft. How many minutes will it take him to paper all four walls?

6. The floor area of a department store is rectangular with a length of 35 yd and a width of 22 yd. How much will it cost to tile the floor if the tile costs $12 per square yard?

7. Jack is putting carpet in his living room of length 15 ft and width 12 ft. If carpet cost $11.95 a square yard, how much will it cost Jack to carpet his living room?

8. Mike has made a scale drawing of a rectangular garden as shown. He wants to plant a garden with a width (shortest side) of 20 ft.

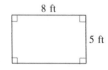

8 ft

5 ft

  (a) If the rectangles are similar, what will be the length of the garden?

  (b) Around the edge he wants to put a fence that will cost $1.75 per foot. What will be the cost of the fence?

9. Pam made a scale drawing of a rectangular picture frame she wants to make with length 11 in. and width 7 in. The frame will be 66 in. long (longest dimension) and similar to the scale drawing.

  (a) How wide will the frame be? (Round to the nearest tenth.)

  (b) If she uses gold-edged framing that costs $0.85 per inch, how much will the framing cost? (Round to the nearest cent.)

11 in

7 in

CHAPTER ORGANIZER

Topic	Procedure	Examples
Perimeter of a rectangle, p. 354	$P = 2l + 2w$	Find the perimeter of a rectangle with width = 3 m, length = 8 m $p = (2)(8) + (2)(3) = 16 + 6 = 22$ m
Perimeter of a square, p. 355	$P = 4s$	Find the perimeter of a square with side length $s = 6$ m $p = (4)(6) = 24$ m
Area of a rectangle, p. 356	$A = lw$	Find the area of a rectangle with width = 2 m, length = 7 m $A = (7)(2) = 14$ m^2
Area of a square, p. 357	$A = s^2$	Find the area of a square with a side of 4 m $A = s^2 = 4^2 = 16$ m^2
Perimeter of parallelograms, trapezoids, and triangles, p. 360	Add up the lengths of all sides.	Find the perimeter of a triangle with sides of 3 m, 6 m, and 4 m $3 + 6 + 4 = 13$ m
Area of a parallelogram, p. 361	$A = bh$ b = length of base $\quad h$ = height	Find the area of a parallelogram with a base of 12 m and a height of 7 m $A = bh = (12)(7) = 84$ m^2
Area of a trapezoid, p. 362	$A = \dfrac{h(b + B)}{2}$ b = length of shortest base B = length of longest base h = height	Find the area of a trapezoid whose height is 12 m and whose bases are 17 m and 25 m. $A = \dfrac{(12)(17 + 25)}{2} = \dfrac{(12)(42)}{2}$ $= \dfrac{504}{2} = 252$ m^2
The sum of the measures of the three interior angles of a triangle is 180°, p. 367	In a triangle to find one missing angle if two are given: **1.** Add up the two known angles. **2.** Subtract the sum from 180°.	Find the missing angle if two known angles in a triangle are 60° and 70°. **1.** $60° + 70° = 130°$ **2.** $\begin{array}{r} 180° \\ -\ 130° \\ \hline 50° \end{array}$ The missing angle is 50°.
Area of a triangle, p. 369	$A = \dfrac{bh}{2}$ b = base h = height	Find the area of the triangle whose base is 1.5 m and whose height is 3 m. $A = \dfrac{bh}{2} = \dfrac{(1.5)(3)}{2} = \dfrac{4.5}{2} = 2.25$ m^2
Radius and diameter of a circle, p. 373	r = radius d = diameter $r = \dfrac{d}{2}$ $d = 2r$	What is the radius of a circle with diameter 50 in.? $r = \dfrac{50}{2} = 25$ in. What is the diameter of a circle with radius 16 in.? $d = (2)(16) = 32$ in.
Pi, p. 373	Pi is a decimal that goes on forever. It can be approximated by as many decimal places as needed. $\pi = \dfrac{\text{circumference of a circle}}{\text{diameter of same circle}}$	Use $\pi = 3.14$ for all calculations. Unless otherwise directed, round your final answer to the nearest tenth when any calculation involves π.

Topic	Procedure	Examples
Circumference of a circle, p. 373	$C = \pi d$	Find the circumference of a circle with a diameter of 12 feet. $C = \pi d = (3.14)(12) = 37.68$ $= 37.7 \text{ ft}$ (to nearest tenth)
Area of a circle, p. 374	$A = \pi r^2$ **1.** Square the radius first. **2.** Then multiply the result by 3.14.	Find the area of a circle with radius 7 feet. $A = \pi r^2 = (3.14)(7^2) = (3.14)(49) = 153.86$ $= 153.9 \text{ ft}^2$ (to nearest tenth)
Volume of a box (rectangular solid), p. 378.	$V = lwh$	Find the volume of a rectangular box whose dimensions are 5 m by 8 m by 2 m. $V = (5)(8)(2) = (40)(2) = 80 \text{ m}^3$
Volume of a cylinder, p. 379	$r = \text{radius} \quad h = \text{height} \quad V = \pi r^2 h$ **1.** Square the radius first. **2.** Then multiply the result by 3.14 and by the height. 	Find the volume of a cylinder with a radius of 7 m and a height of 3 m. $V = \pi r^2 h = (3.14)(7^2)(3) = (3.14)(49)(3)$ $= (153.86)(3) = 461.58$ $= 461.6 \text{ m}^3$ (rounded to nearest tenth)
Volume of a sphere, p. 379	$V = \dfrac{4\pi r^3}{3}$ $r = \text{radius}$	Find the volume of a sphere of radius 3 m. $V = \dfrac{4\pi r^3}{3} = \dfrac{(4)(3.14)(3^3)}{3}$ $= \dfrac{(4)(3.14)(\overset{9}{\cancel{27}})}{\underset{1}{\cancel{3}}} = (4)(3.14)(9)$ $= (12.56)(9) = 113.04$ $= 113.0 \text{ m}^3$ (rounded to nearest tenth)
Volume of a cone, p. 379	$V = \dfrac{\pi r^2 h}{3}$ $r = \text{radius}$ $h = \text{height}$ 	Find the volume of a cone of height 9 m and radius 7 m. $V = \pi r^2 h = \dfrac{(3.14)(7^2)(9)}{3}$ $= \dfrac{(3.14)(7^2)(\overset{3}{\cancel{9}})}{\underset{1}{\cancel{3}}} = (3.14)(49)(3)$ $= (153.86)(3) = 461.58$ $= 461.6 \text{ m}^3$ (rounded to the nearest tenth)
Volume of a pyramid, p. 380	$V = \dfrac{Bh}{3}$ $B = \text{area of the base}$ $h = \text{height}$ **1.** Find the area of the base. **2.** Multiply this area by the height and divide the result by 3. 	Find the volume of a pyramid whose height is 6 meters and whose rectangular base is 10 meters by 12 meters. **1.** $B = (12)(10) = 120$ **2.** $V = \dfrac{(120)(\overset{2}{\cancel{6}})}{\underset{1}{\cancel{3}}} = (120)(2) = 240 \text{ m}^3$
Similar figures, corresponding sides p. 386	The corresponding sides of similar figures have the same ratio.	Find n in the following similar figures. $\dfrac{n}{4} = \dfrac{9}{3}$ $3n = 36$ $n = 12 \text{ m}$
Similar figures, corresponding perimeters p. 386	The perimeters of similar figures have the same ratio as the corresponding sides. For reasons of space, the procedure for the areas of similar figures is not given here but may be found in the text (see p. 386).	These two figures are similar. Find the perimeter of the larger figure. $\dfrac{6}{12} = \dfrac{29}{p}$ $6p = (12)(29)$ $6p = 348$ $\dfrac{6p}{6} = \dfrac{348}{6}$ $p = 58$ The perimeter of the larger figure is 58 m.

In each problem in the chapter review, round your answer to the nearest tenth when necessary. Use $\pi = 3.14$ in all calculations requiring the use of π.

7.1 *Find the perimeter of each square and rectangle.*

1. Length = 8.3 m, width = 1.6 m

2. Length = 9.6 cm, width = 2.8 cm

3. Length = width = 5.8 yd

4. Length = width = 2.4 yd

Find the area of each square and rectangle.

5. Length = 5.9 cm, width = 2.8 cm

6. Length = 9.3 m, width = 7.9 m

7. Length = width = 4.3 in.

8. Length = width = 7.2 in.

Find the perimeter of each object, made up of rectangles and squares.

9.

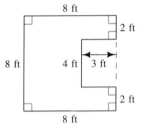

10.

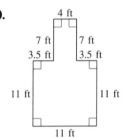

Find the shaded area of each object, made up of rectangles and squares.

11.

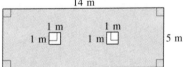

12.

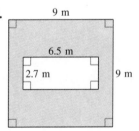

7.2 *Find the perimeter of each parallelogram or trapezoid.*

13. Two sides of the parallelogram are 43 m and 7.2 m.

14. Two sides of the parallelogram are 26 cm and 12.2 cm.

15. The sides of the trapezoid are 22 in, 13 in, 32 in, and 13 in.

16. The sides of the trapezoid are 5 mi, 22 mi, 5 mi, and 30 mi.

Find the area of each parallelogram or trapezoid.

17. The parallelogram has a base of 90 m and a height of 30 m.

18. The parallelogram has a base of 82 m and a height of 25 m.

19. The trapezoid has a height of 14 yd and bases of 20 yd and 28 yd.

20. The trapezoid has a height of 36 yd and bases of 17 yd and 23 yd.

Find the shaded area of each region, made up of parallelograms, trapezoids, and rectangles.

21.

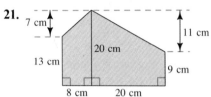

22.

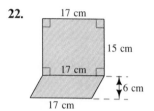

7.3 *Find the perimeter of each triangle.*

23. The sides of the triangle are 10 ft, 5 ft, and 7 ft.

24. The sides of the triangle are 5.5 ft, 3 ft, and 5.5 ft.

Find the measure of the third angle in each triangle.

25. Two known angles are 15° and 12°.

26. Two known angles are 62° and 78°.

Find the area of each triangle.

27. Base = 9.6 m, height = 5.1 m

28. Base = 12.5 m, height = 9.5 m

29. Base = 12 cm, height = 7.6 cm

30. Base = 18.2 m, height = 4.4 m

Find the area of each region, made up of triangles and rectangles.

31.

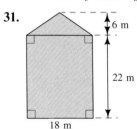

32.

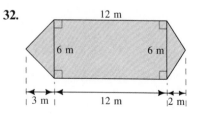

33. What is the *diameter* of a circle whose radius is 53 cm?

34. What is the *radius* of a circle whose diameter is 48 cm?

35. Find the *circumference* of a circle with diameter 12 in.

36. Find the *circumference* of a circle with radius 7 in.

Find the area of each circle.

37. Radius = 6 m

38. Radius = 9 m

39. Diameter = 16 ft

40. Diameter = 14 ft

Find the area of each shaded region, made up of circles, semicircles, rectangles, trapezoids, and parallelograms.

41.

42.
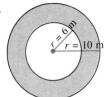

43.
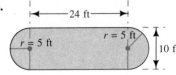

44.
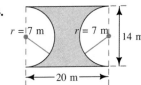

45.
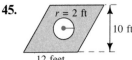

46.
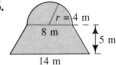

7.5 *Find the volume of each object.*

47. A rectangular box measuring 3 ft by 6 ft by 2.5 ft.

48. A rectangular box measuring 20 m by 15 m by 1.5 m.

49. A sphere with radius 15 feet.

50. A sphere with radius 1.2 feet.

51. Find the volume of a storage can that is 2 m tall and has a radius of 7 m.

52. Find the volume of a soup can 18 cm tall and having a radius of 9 cm.

53. A pyramid that is 18 m high and whose rectangular base measures 16 m by 18 m.

54. A pyramid that is 12 m high and whose square base measures 5 m by 5 m.

55. Find the volume of a cone of sand 9 ft tall with a radius of 20 ft.

56. A chemical has polluted a volume of ground in a cone shape. The depth of the cone is 30 yd. The radius of the cone is 17 yd. Find the volume of polluted ground.

7.6 *Find n in each set of similar triangles.*

57.

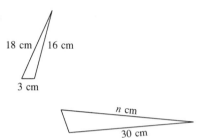

3 m
2 m
45 m
n

58.

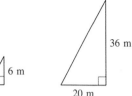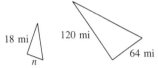

6 m
n
36 m
20 m

59.

18 cm / 16 cm
3 cm

n cm
30 cm

60.

18 mi
n
120 mi
64 mi

Find the perimeter of the larger of the two similar figures.

61.

18 cm
5 cm / \ 5 cm
26 cm

108 cm

62.

20 ft
12 ft / 13 ft
25 ft

32.5 ft

7.7

63. The fencing to surround a flower garden 14 ft by 8 ft costs $2.10 per foot. What is the total cost to fence in the garden?

64. (a) How many kilometers is it from Homeville to Seaview if you drive through Ipswich? How fast do you travel if it takes 0.5 hour to travel that way?

(b) How many kilometers is it from Homeville to Seaview if you drive through Acton and West-ville? How fast do you travel if it takes 0.8 hour to travel that way?

(c) Over which route do you travel at a more rapid rate?

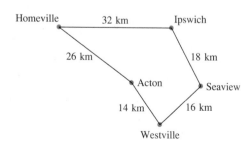

Homeville 32 km Ipswich
26 km 18 km
Acton Seaview
14 km 16 km
Westville

65. The Wilsons are carpeting a recreation room with the dimensions shown. Carpeting costs $8.00 per square yard. How much will the carpeting cost?

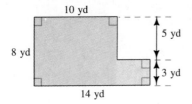

66. A conical tank holds acid in a chemistry lab. The tank has a radius of 9 in. and a height of 24 in. How many cubic inches does the tank hold? The acid weighs 16 g per cubic inch. What is the weight of the acid if the tank is full?

67. A silo has a cylindrical shape with a hemisphere dome. It has dimensions as shown.
   (a) What is its volume in cubic feet?
   (b) If 1 cubic foot = 0.8 bushel, how many bushels of grain will it hold?

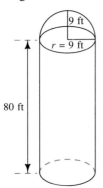

68. A picture measures 3 cm by 5 cm. A photographer wishes to enlarge it to a width (shorter distance) of 126 cm.
   (a) What will be the length of the new picture?
   (b) How much larger is the *area* of the new picture compared to the old picture?

In each problem round to the nearest tenth when necessary. Use $\pi = 3.14$ in your calculations.

Find the perimeter of each square or rectangle.

1. Length = width = 0.8 m

2. Length = 6.7 cm, width = 1.2 cm

Find the area of each square or rectangle.

3. Length = 12 yd, width = 8 yd

4. Length = width = 15 yd

Find the perimeter of each parallelogram or trapezoid.

5. A parallelogram with one side measuring 3.6 m and another side measuring 9.2 m.

6. A trapezoid with sides of 21 cm, 13 cm, 31 cm, and 13 cm.

Find the area of each parallelogram or trapezoid.

7. A parallelogram with a base of 22 km and a height of 18 km.

8. A trapezoid with a height of 20 yd and bases of 18 yd and 37 yd.

9. Find the diameter of a circle with radius 3.70 in.

10. Find the circumference of the circle with diameter 12 in.

11. Find the area of the circle with radius of 7 mi.

1. _____

2. _____

3. _____

4. _____

5. _____

6. _____

7. _____

8. _____

9. _____

10. _____

11. _____

12. _____

Find the shaded areas of each region, make up of squares, rectangles, trapezoids, parallelograms, semicircles, or circles.

12.

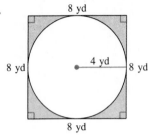

13. _____

13.

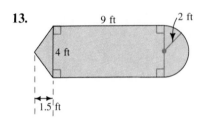

14.

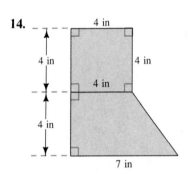

14. _____

15.

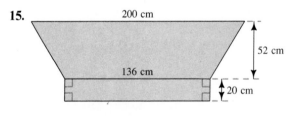

15. _____

16.

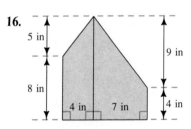

16. _____

Find the volume of each object.

1. A rectangular solid measuring 8 yd by 7 yd by 2 yd.

2. A sphere with radius 6 yd.

3. A cylinder with radius 9 in. and a height 7 in.

4. A pyramid with a height of 6 ft and a rectangular base measuring 11 ft by 10 ft.

5. A cone with height 14 in. and radius 9 in.

6.

9 cm $r = 2$ cm

Find n in each similar triangle.

7.

8   7 n 4

8. Find *n* in this pair of similar rectangles.

24 m   8 m
n 1.6 m

9. Here are a pair of similar pentagons (five-sided figures). Find the perimeter of the larger pentagon.

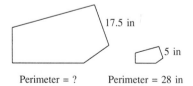
17.5 in   5 in
Perimeter = ?   Perimeter = 28 in

10. Find the height *h* of the building if each triangle is similar.

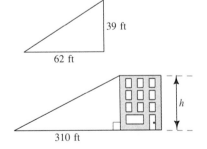
39 ft
62 ft
h
310 ft

1. _____

2. _____

3. _____

4. _____

5. _____

6. _____

7. _____

8. _____

9. _____

10. _____

11. _____

12. _____

13. _____

14. _____

15. _____

16. _____

11. Jimmy and Lou-Ann are having a concrete driveway installed. The driveway is 17 yd long, 6 yd wide, and 6 in. thick.
  (a) What is the volume of the driveway in *cubic yards*?
  (b) How much will it cost if concrete is $14.00 per *cubic yard*?

12. John can paint 100 ft^2 in 3 hours. How long will it take him to paint four walls that measure 6 ft × 12 ft, 6 ft × 10 ft, 7 ft × 14 ft, and 7 ft × 12 ft?

13. (a) Find the volume of a conical shaped pile of wheat with a diameter of 30 ft and a height of 12 ft.
  (b) If each cubic foot of wheat is worth $3.50, what is the value of the pile of wheat?

14. Lexie and John are putting down new carpet in the living room and hallway. The dimensions are shown.
  (a) What is the area of the living room and hallway in *square yards*?
  (b) What is the cost of carpeting this area if the carpet costs $9.00 per square yard?

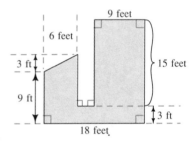

15. Find the *volume* of the iron washer shown in the sketch.

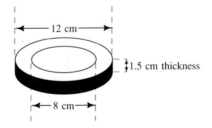

16. A rectangular fish tank measures 50 in. × 30 in. × 45 in.
  (a) How many cubic inches does the tank hold?
  (b) How many gallons of water are in the tank when it is filled? (1 gallon = 231 cubic inches.)

Find the perimeter for each.

1. A rectangle that measures 9 yd × 11 yd

2. A square with side 8.5 ft

3. A parallelogram with sides measuring 6.5 m and 3.5 m

4. A trapezoid with sides measuring 22 m, 13 m, 32 m, and 13 m

5. A triangle with sides measuring 4.4 m, 10.8 m, and 9.6 m

Find the area for each. Round each area to the nearest tenth.

6. A rectangle that measures 10 yd × 18 yd

7. A square 10.2 m on a side

8. A parallelogram with a height of 6 m and a base of 13 m

9. A trapezoid with a height of 9 m and bases of 7 m and 25 m

10. A triangle with a base of 4 cm and a height of 6 cm

11. A triangle with a base of 15 m and a height of 7 m

12. Find the circumference of a circle with *radius* 6 in.

13. Find the area of a circle with *diameter* 18 ft.

Find the shaded area of each region, made up of circles, semicircles, rectangles, squares, trapezoids, or parallelograms.

14.

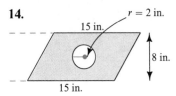

15.

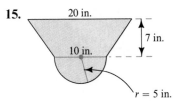

16.

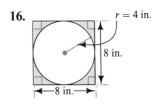

1.	_____
2.	_____
3.	_____
4.	_____
5.	_____
6.	_____
7.	_____
8.	_____
9.	_____
10.	_____
11.	_____
12.	_____
13.	_____
14.	_____
15.	_____
16.	_____

17. _____

18. _____

19. _____

20. _____

21. _____

22. _____

23. _____

24. _____

25. _____

Find the volume of each.

17. A rectangle box measuring 7 m by 12 m by 10 m

18. A cone with a height of 12 m and a radius of 8 m

19. A sphere of radius 2 m

20. A cylinder of height 2 feet and radius 9 feet

21. A pyramid of height 14 m and whose rectangular base measures 4 m by 3 m.

Each pair of triangles is similar. Find the missing side n.

22.

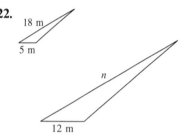

23.

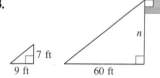

Solve each applied problem.

24. An athletic field has the dimensions shown.

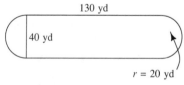

(a) What is the area of the athletic field?

(b) How much will it cost to fertilize it at $0.40 per square yard?

25. A metal sphere is filled with a mixture of acid. The radius of the sphere is 3 m.

(a) What is the volume of the sphere?

(b) The empty metal sphere weighs 16 kg. How much does the sphere weigh when it is filled with acid that weighs 0.5 kg per cubic meter?

Approximately one-half of this test is based on Chapter 7 material. The remainder is based on material covered in Chapters 1–6.

Do each problem. Simplify your answer.

1. Add: 126,350
278,120
+ 531,290

2. Multiply: 163
× 205

3. Subtract: $\dfrac{17}{18} - \dfrac{11}{30}$.

4. Divide: $\dfrac{3}{7} \div 2\dfrac{1}{4}$.

5. Round to the nearest hundredth: 56.1279.

6. Multiply: 9.034
× 0.8

7. Divide: $0.021\,\overline{)\,1.743}$.

8. Find *n:* $\dfrac{3}{n} = \dfrac{2}{18}$.

9. There are seven teachers for every 100 students. If there are 56 teachers at the university, how many students would you expect?

10. What is 18.5% of 220?

11. 0.8% of what number is 16?

12. Michael scored 18 baskets out of 24 shots on the court. What percent of his shots went into the basket?

13. Convert 586 cm to m.

14. Convert 42 yd to in.

15. Ben traveled 88 km. How many miles did he travel? (1 kilometer = 0.62 mile; 1 mile = 1.61 kilometers.)

In problems 16–30, round your answer to the nearest tenth when necessary. Use π = 3.14 in calculations involving π.

Find the perimeter of each object.

16. A rectangle of length 17 m and width 8 m

17. A trapezoid with sides of 86 cm, 13 cm, 96 cm, and 13 cm

18. Find the circumference of a circle with diameter 18 yd.

1. _____

2. _____

3. _____

4. _____

5. _____

6. _____

7. _____

8. _____

9. _____

10. _____

11. _____

12. _____

13. _____

14. _____

15. _____

16. _____

17. _____

18. _____

19. _____

20. _____

21. _____

22. _____

23. _____

24. _____

25. _____

26. _____

27. _____

28. _____

29. _____

30. _____

31. _____

Find the area of each object.

19. A triangle with base 1.2 cm and height 2.4 cm

20. A trapezoid with height 18 m and bases of 26 m and 34 m

21.

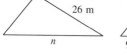

12 m

12 m

12 m

4 m

12 m

22.

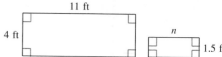

35 yd

20 yd 6 yd   20 yd

6 yd

5 yd   24 yd

23. A circle with radius 4 m.

Find the volume of each object.

24. A cylinder with a height of 12 m and a radius of 8 m

25. A sphere with a radius of 9 cm

26. A pyramid with height 32 cm and a rectangular base 14 cm by 21 cm

27. A cone of height 18 m and a radius of 12 m

Find the value of n in each pair of similar figures.

28.

26 m

n

7 m

9 m

29.

11 ft

4 ft

n

1.5 ft

Solve each problem.

30. Mary Ann and Wong Twan have a recreation room with the dimensions shown in the figure. They wish to carpet it at a cost of $8.00 per square yard.

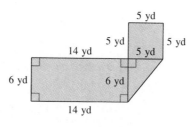

5 yd

5 yd   5 yd

14 yd   5 yd

6 yd   6 yd

14 yd

(a) How many square yards of carpet are needed?

(b) How much will it cost?

31. A cylindrical silo has a diameter of 12 ft and a height of 30 ft. It is topped by a hemispherical dome.

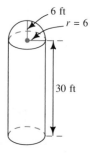

6 ft

$r = 6$

30 ft

(a) How many cubic feet are in the entire silo (including cylinder and the dome)?

(b) How many bushels of grain will it hold if 1 cubic foot = 0.8 bushel?

Statistics, Square Roots and the Pythagorean Theorem

8

Advances in modern medicine are presented to doctors as statistical studies. Doctors need to know mathematics to evaluate the effectiveness of a new medical procedure.

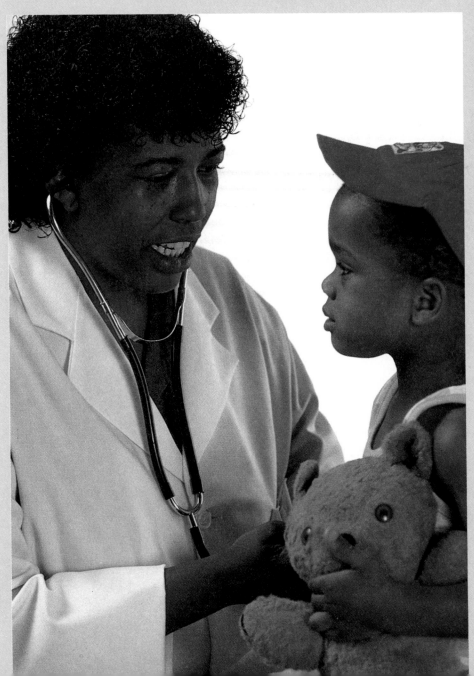

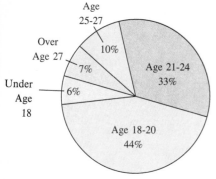

PRETEST CHAPTER 8

If you are familiar with the topics in this chapter, take this test now. Check your answers with those in the back of the book. If an answer was wrong or you couldn't do a problem, study the appropriate section of the chapter.

If you are not familiar with the topics in this chapter, don't take this test now. Instead, study the examples, work the practice problems, and then take the test.

This test will help you identify those concepts that you have mastered and those that need more study.

Section 8.1 The 5000 students on campus were recorded by age. The following circle graph depicts their distribution by age.

1. What age group has the smallest percent of the student body?
2. What percent of the students are between 21 and 27?
3. What percent of the students are 25 or older?
4. If 5000 students are at the university, *how many students* are between 18 and 20 years of age?
5. How *many students* are under age 18?

Section 8.2 The following double-bar graph indicates the number of people unemployed in Pacerville during each quarter of 1989 and 1990. Use this graph to answer questions 6–11.

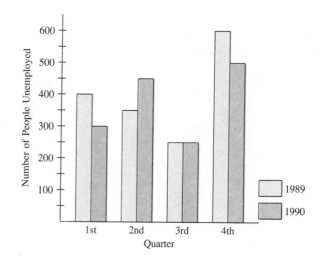

6. How many people were unemployed in Pacerville during the second quarter of 1990?
7. How many people were unemployed in Pacerville during the fourth quarter of 1989?
8. When were the fewest number of people unemployed in Pacerville?
9. When were the greatest number of people unemployed in Pacerville?
10. How many more people were unemployed in the fourth quarter of 1989 than the fourth quarter of 1990?
11. How many more people were unemployed during the second quarter of 1990 than the second quarter of 1989?

The following comparison line graph indicates sales and production of color television sets by a major manufacturer during the specified months. Use this graph to answer questions 12–16.

12. During what months was the production of television sets the lowest?

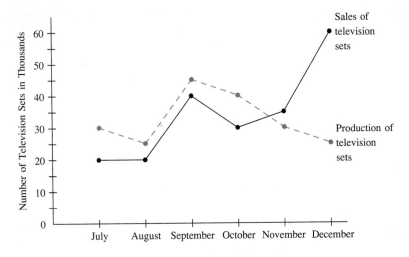

13. During what month was the sales of television sets the highest?

14. What was the first month in which the production of television sets was lower than the sales of television sets?

15. How many television sets were sold in September?

16. How many television sets were produced in October?

Section 8.3 The following histogram prepared by a state traffic and highway department tells us about the number of miles a car was driven before the car was discarded or sold to a junk dealer. Use this histogram to answer questions 17–20.

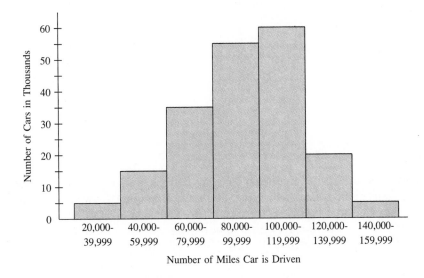

17. How many discarded or junked cars had been driven 80,000 to 99,999 mi?

18. How many discarded or junked cars had been driven 100,000 to 119,999 mi?

19. How many discarded or junked cars had been driven 120,000 mi or more?

20. How many discarded or junked cars had been driven less than 60,000 mi?

Section 8.4 A typist produced the following number of pages.

Mon.	Tues.	Wed.	Thurs.	Fri.
39	41	43	26	31

21. Find the *mean* number of pages typed per day.

22. Find the *median* number of pages typed per day.

Six students at the university reported earning the following hourly salaries at their summer jobs: $6.50, $4.00, $4.50, $3.90, $5.20, $8.00.

23. Find the *median* hourly salary for their summer jobs.

24. Find the *mean* hourly salary for their summer jobs.

Section 8.5 Evaluate *exactly.*

25. $\sqrt{64}$

26. $\sqrt{49}$

27. $\sqrt{4} + \sqrt{100}$

28. $\sqrt{9} + \sqrt{81}$

Approximate using the square root table on page 578 or by using a calculator with a square root key. Round your answer to the nearest thousandth when necessary.

29. $\sqrt{46}$

30. $\sqrt{89}$

31. $\sqrt{131}$

Section 8.6 Find the unknown side of the right triangle. Use a calculator or the square root table on page 578 when necessary and round your answer to the nearest thousandth.

32.

10 ft

6 ft

33.

8 ft

11 ft

34. A boat travels 2 mi east and 3 mi south. How far is the boat from its starting point?

2

3

35. How far up on the antenna does a 17-ft support wire reach if the wire is anchored 15 ft from the base?

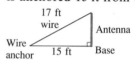

17 ft wire

Antenna

Wire anchor

15 ft

Base

8.1 CIRCLE GRAPHS

Statistics is a branch of mathematics that collects and studies data. The data (information) can be collected by observation and counting ("How many people who took a new medicine had an improvement in their health?") or by question-and-answer techniques ("If you favor raising the voting age, call your opinion in to phone number A; if you oppose raising the voting age, call your opinion in to phone number B.")

After the data are collected, they must be organized. We can use graphs to give a visual representation of the facts. Sections 8.1, 8.2, and 8.3 present circle graphs, bar graphs, line graphs, and histograms as ways of organizing and displaying data.

But what do all the data *mean*? We need to analyze the information to be able to understand it. We use graphs to tell general patterns or ratios. Then Section 8.4 presents one of the most common statistical ideas, that of the "center" of a group of numbers. Since all studies of statistics are based on data collection, organization, and analysis, we'll start there.

1 Graphs appeal to our eye. By their visual nature they can communicate directly, and complicated relationships among statistical data become clearer than if the same relationships were presented in a long list of numbers. For this reason, newspapers often use graphs to help their readers quickly grasp information.

Circle graphs are especially helpful in showing the relationship of parts to a whole. The entire circle represents 100%; the pie-shaped pieces represent the subcategories.

The following circle graph divides the 10,000 students at Westline College into categories.

Example 1 What is the largest category of students?

The largest pie-shaped section of the circle corresponds to the freshmen. The largest category is freshmen students. ∎

Practice Problem 1 What is the smallest category of students? ∎

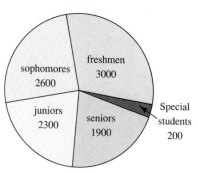

Distribution of Students at Westline College

Example 2 How many students are either sophomores or juniors?

There are 2600 sophomores and 2300 juniors. If we add these two numbers, we have $2600 + 2300 = 4900$. Thus we see that there are 4900 students who are either sophomores or juniors. ∎

Practice Problem 2 How many students are either freshmen or special students? ∎

Example 3 What is the ratio of freshmen to seniors?

Number of freshmen ⟶ 3000
Number of seniors ⟶ 1900

Thus $\dfrac{3000}{1900} = \dfrac{30}{19}$

The ratio of freshmen to seniors is $\dfrac{30}{19}$ ∎

Practice Problem 3 What is the ratio of freshmen to sophomores? ∎

Example 4 What is the ratio of seniors to the total number of students?

There are 1900 seniors out of 10,000 students. The ratio of seniors to the total number of students is

$$\frac{1900}{10,000} = \frac{19}{100}$$ ∎

Practice Problem 4 What is the ratio of freshmen to the total number of students? ∎

2 Together, the Great Lakes form the largest body of fresh water in the world. The total area of these five lakes is about 290,000 mi². The percentage of this total area taken up by each of the Great Lakes is shown in the following circle graph.

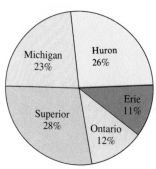

Percentage of Area Occupied by Each of the Great Lakes

Example 5 Which of the Great Lakes occupies the second largest area in square miles?

The largest percent corresponds to the biggest area, which is occupied by Lake Superior. The second largest percent corresponds to Lake Huron's area. Therefore, Lake Huron has the second largest area in square miles. ∎

Practice Problem 5 Which of the Great Lakes occupies the third largest area in square miles? ∎

Example 6 What percent of the total area is occupied either by Lake Erie or Lake Ontario?

If we add 11% for Lake Erie and 12% for Lake Ontario, we get

$$11\% + 12\% = 23\%$$

Thus 23% of the area is occupied by either Lake Erie or Lake Ontario. ∎

Practice Problem 6 What percent of the total area is occupied by either Lake Superior or Lake Michigan? ∎

Example 7 How many of the total 290,000 mi^2 are occupied by Lake Michigan?

Remember that we multiply percent times base to obtain amount. Here 23% of 290,000 is what is occupied by Lake Michigan.

$$(0.23)(290,000) = n$$
$$= 66,700$$

Thus about 66,700 mi^2 are occupied by Lake Michigan. ■

Practice Problem 7 How many of the total 290,000 mi^2 are occupied by Lake Superior? ■

Example 8 How many square miles are occupied by Lake Ontario?

Now 12% of 290,000 mi^2 is occupied by Lake Ontario.

$$(0.12)(290,000) = n$$
$$= 34,800$$

Thus about 34,800 mi^2 are occupied by Lake Ontario. ■

Practice Problem 8 How many of the total square miles are occupied by Lake Erie? ■

EXERCISES 8.1

The following circle graph displays Bob and Linda McDonald's monthly $2000 family budget. Use the circle graph to answer questions 1–10.

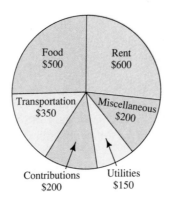

1. What category takes the largest amount of the budget?

2. What category takes the least amount of the budget?

3. How much money is alloted each month for transportation?

4. How much money is alloted each month for contributions?

5. How much money in total is alloted for either utilities or transportation?

6. How much money in total is alloted for either food or rent?

7. What is the ratio of money spent for rent to the amount of money spent for utilities?

8. What is the ratio of money spent for food to the amount of money spent for transportation?

9. What is the ratio of money spent for rent to the total amount in the monthly budget?

10. What is the ratio of money spent for food to the total amount in the monthly budget?

A major league pitcher has thrown 650 pitches during the first part of the baseball season. The following circle graph shows the results of his pitches. Use the circle graph to answer questions 11–20.

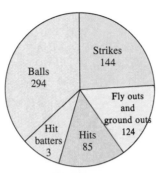

Results of 650 Pitches

11. What category had the least number of pitches?

12. What category had the second-highest number of pitches?

13. How many pitches were strikes?

14. How many pitches were balls?

15. How many pitches were either hits or balls?

16. How many pitches were either strikes or fly outs and ground outs?

17. What is the ratio of the number of balls to the total number of pitches?

18. What is the ratio of the number of strikes to the total number of pitches?

19. What is the ratio of the number of balls to the number of strikes?

20. What is the ratio of the number of fly outs and ground outs to the number of hits?

In 1986 there were approximately 87,000 women physicians in the United States. The following circle graph divides them by age. Use the circle graph to answer questions 21–26.

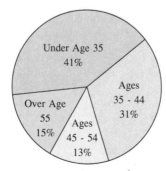

Ages of the 87,000 Women Physicians in the United States in 1986

21. What percent of the women physicians are between the ages of 45 and 54?

22. What percent of the women physicians are between the ages of 35 and 44?

23. What percent of the women physicians are over the age of 45?

24. What percent of the women physicians are under the age of 45?

25. How many of the 87,000 women physicians are between the ages of 35 and 44?

26. How many of the 87,000 women physicians are over age 55?

Researchers estimate that the religious faith distribution of the 5,000,000,000 people in the world in 1988 was approximately that displayed in the following circle graph. Use the graph to answer questions 27–32.

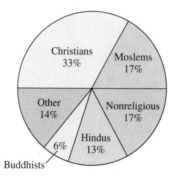

Distribution of Religious Faith in
the World Population of 1988

27. Approximately how many of the 5,000,000,000 people are Christians?

28. Approximately how many of the 5,000,000,000 people are Moslems?

29. What percent of the world's population is either Hindu or Buddhist?

30. What percent of the world's population is either Moslem or nonreligious?

31. What percent of the world's population is *not* Moslem?

32. What percent of the world's population is *not* Christian?

 To Think About

The researchers who presented this information said that their percentages could be in error by 2%. This means that the actual percent for each faith could be 2% larger or 2% smaller than that given.

33. Of the 5,000,000,000 people, what is the *largest* number of Buddhists possible according to this estimation of error?

34. Of the 5,000,000,000 people, what is the *smallest* number of Buddhists possible according to this estimation of error?

Cumulative Review Problems

35. Find the area of a triangle with base 6 in. and height 14 in.

36. Find the area of a parallelogram with base of 17 in. and height 12 in.

For Extra Practice Examples and Exercises, turn to page 446.

1. The smallest category of students is *special students*.

3. There are 3000 freshmen but only 2600 sophomores. The ratio of freshmen to sophomores is $\dfrac{3000}{2600} = \dfrac{15}{13}$.

5. Lake Michigan occupies the third largest area with 23%.

7. Lake Superior has 28% of the area. Now 28% of 290,000 mi^2 is

$$(0.28)(290,000) = 81,200 \text{ mi}^2$$

Answers to Even-Numbered Practice Problems

2. 3200   **4.** $\dfrac{3}{10}$ **6.** 51%   **8.** About 31,900 mi^2

8.2 BAR GRAPHS

☐ After studying this section, you will be able to:

1 Read and interpret a bar graph

2 Read and interpret a double-bar graph

3 Read and interpret a line graph

4 Read and interpret a comparison line graph

1 Bar graphs are helpful in seeing changes over a period of time. Bar graphs or line graphs are especially helpful when the same type of data is repeatedly studied.

Since graphs are so frequently encountered, one of the main tasks for a student of statistics is to interpret and draw proper conclusions from various kinds of graphs. This we do in the first few sections of this chapter. The population of California for various years is displayed in the following bar graph.

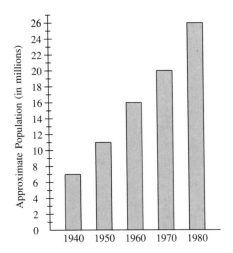

Example 1 What was the approximate population of California in 1970?

The bar for 1970 rises to 20. This represents 20 million; thus the approximate population is 20,000,000 ■

Practice Problem 1 What was the approximate population of California in 1980?
■

Example 2 What was the approximate population of California in 1950?

The bar for 1950 rises to 11. Thus the approximate population is 11,000,000.
■

Practice Problem 2 What was the approximate population of California in 1940?
■

2 Double-bar graphs are useful in making comparisons. For example, when a company is analyzing its sales, it may want to compare different years or different quarters. The following double-bar graph illustrates the sales of new cars at a local car dealership for two different years, 1989 and 1990. The sales are recorded for each quarter of the year.

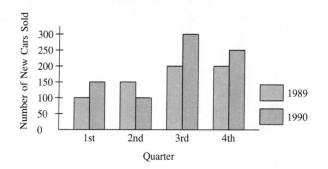

Example 3 How many cars were sold in the second quarter of 1989?

The bar rises to 150 for the second quarter of 1989. Therefore, 150 cars were sold. ■

Practice Problem 3 How many cars were sold in the fourth quarter of 1990? ■

Example 4 How many more cars were sold in the third quarter of 1990 than the third quarter of 1989?

From the double-bar graph, we see that 300 cars were sold in the third quarter of 1990 and that 200 cars were sold in the third quarter of 1989.

$$\begin{array}{r} 300 \\ -\ 200 \\ \hline 100 \end{array}$$

Thus 100 more cars were sold. ■

Practice Problem 4 How many fewer cars were sold in the second quarter of 1990 than in the second quarter of 1989? ■

3 A line graph is useful for showing trends over a period of time. In a line graph only a few points are actually plotted from measured values. The points are then connected by straight lines to show a trend. The intervening values between points may not lie exactly on the line. The following line graph shows the number of customers per month coming to a restaurant in a tourist vacation community.

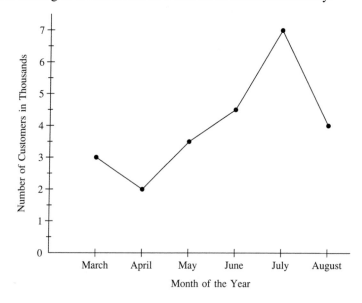

Example 5 Which month had the fewest number of customers come to the restaurant?

The lowest point on the graph occurs in the month of April. Thus the fewest number of customers came in April. ■

Practice Problem 5 Which month had the greatest number of customers come to the restaurant? ▨

Example 6

(a) Approximately how many customers per month came during the month of June?
(b) From May to June did the number of customers increase or decrease?

(a) Notice that the dot is halfway between 4 and 5. This represents a value halfway between 4000 and 5000 customers. Thus we would estimate 4500 customers came during the month of June.
(b) Between May and June the line goes up, so the number of customers increased. ▨

Practice Problem 6

(a) Approximately how many customers per month came during the month of May?
(b) From March to April did the number of customers increase or decrease? ▨

Example 7 Between what two months is the *increase* in attendance the largest?

The line from June to July goes upward at the steepest angle. This represents the largest increase. (You can check this by reading the numbers from the left axis.) Thus the greatest increase in attendance is between June and July. ▨

Practice Problem 7 Between what two months does the biggest *decrease* occur? ▨

◢ Two or more sets of data can be compared by using a *comparison line graph*. A comparison line graph shows two or more line graphs together. A different style for each line distinguishes them. Note that using a solid line and a dashed line in the following graph makes it easy to read.

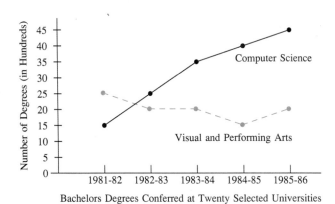

Bachelors Degrees Conferred at Twenty Selected Universities

Example 8 How many bachelor's degrees in computer science were awarded in 1983–84?

Because the dot corresponding to 1983–84 is opposite 35 and the scale is in hundreds, we have 35 × 100 = 3500. Thus 3500 degrees were awarded in computer science in 1983–84. ▨

Practice Problem 8 How many bachelor's degrees in visual and performing arts were awarded in the academic year 1984–85? ▨

Example 9 In what academic year were more degrees awarded in the visual and performing arts than in computer science?

The only year when the visual and performing arts had more bachelor's degrees conferred was in the academic year 1981–82. ▨

Practice Problem 9 What was the first academic year in which more degrees were conferred in computer science than in the visual and performing arts? ▨

The following bar graph shows the approximate population of Texas for various years. Use the graph to answer questions 1–6.

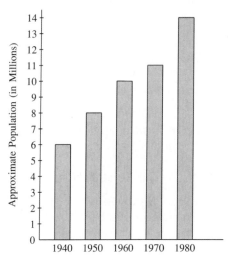

1. What was the approximate population of Texas in 1950?

2. What was the approximate population of Texas in 1960?

3. What was the approximate population of Texas in 1970?

4. What was the approximate population of Texas in 1980?

5. According to this bar graph, between what years did the population of Texas increase by the smallest amount?

6. According to this bar graph, between what years did the population of Texas increase by the largest amount?

The following double-bar graph illustrates the number of students at 10 selected universities at 20-year intervals. The student population is divided into men and women. Use the graph to answer questions 7–14.

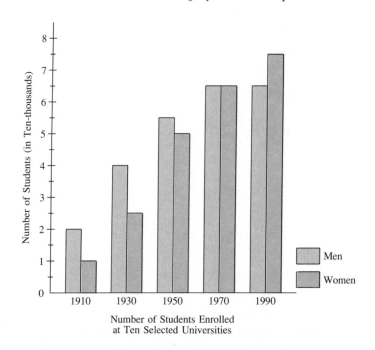

Number of Students Enrolled
at Ten Selected Universities

7. How many men students were enrolled in 1930?

8. How many women students were enrolled in 1950?

9. In what year was the number of women students enrolled greater than the number of men students enrolled?

10. In what year was the number of women students enrolled equal to the number of men students enrolled?

11. How many more women students were enrolled in 1950 than in 1930?

12. How many more men students were enrolled in 1970 than in 1950?

13. In what 20-year period was the greatest increase in the enrollment of men?

14. In what 20-year period was the greatest increase in the enrollment of women?

The following line graph shows Wentworth Construction Company's profits during the last six years. Use the graph to answer questions 15–20.

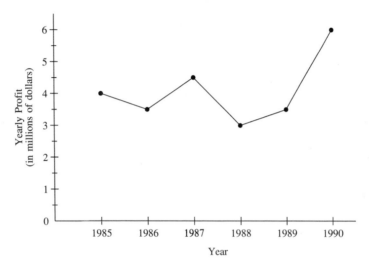

15. What was the profit in 1985?

16. What was the profit in 1989?

17. What year had the lowest profit?

18. What year had the highest profit?

19. How much greater was the profit in 1987 than in 1986?

20. How much greater was the profit in 1990 than in 1989?

The following comparison line graph indicates the rainfall for the last six months of two different years in Springfield. Use the graph to answer questions 21–26.

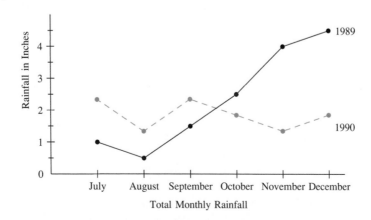

21. In September 1990, how many inches of rain were recorded?

22. In October of 1989, how many inches of rain were recorded?

23. During what months was the rainfall of 1990 greater than the rainfall of 1989?

24. During what months was the rainfall of 1990 less than the rainfall of 1989?

25. How many more inches of rain fell in November of 1989 than in October 1989?

26. How many more inches of rain fell in September of 1989 than in August 1989?

Cumulative Review Problems

Do each calculation in the proper order.

27. $7 \times 6 + 3 - 5 \times 2$

28. $6^2 + 2^3 + 20 \div 5$

For Extra Practice Examples and Exercises, turn to page 447.

Solutions to Odd-Numbered Practice Problems

1. The bar rises to 24. The approximate population was 24,000,000.

3. The bar rises to 250. The number of new cars sold in the fourth quarter of 1990 was 250.

5. The greatest number of customers came in July since the highest peak of the graph occurs in July.

7. The sharpest line of decrease is from July to August. The line slopes downward most steeply between those two months. Thus the biggest decrease in customers is between July and August.

9. The computer science line goes above the visual and performing arts line first in 1982–83. Thus the first academic year for more degrees in computer science was 1982–83.

Answers to Even-Numbered Practice Problems

2. 7,000,000   **4.** 50   **6. (a)** 3500   **(b)** The number decreased   **8.** 1500

☐ After studying this section, you will be able to:

1 *Understand and interpret a histogram*

2 *Construct a histogram from raw data*

8.3   HISTOGRAMS

In business or in higher education you are often asked to take data and organize it in some way. This section shows you the technique for making a *histogram*—a type of bar graph.

1 Suppose that a mathematics professor announced the results of a class test. The 40 students in the class scored between 50 and 99 on the test. The results are displayed on this chart.

Scores on the Test	Class Frequency
50–59	4
60–69	6
70–79	16
80–89	8
90–99	6

The results in the table can be organized in a special type of bar graph known as a *histogram*. In a histogram the width of each bar is the same. The width represents the range of scores on the test. This is called a *class interval*. The height of each bar gives the class frequency of each class interval. The *class frequency* is the number of times a score occurs in a particular class interval.

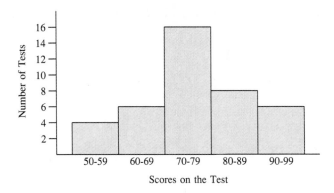

Scores on the Test

Example 1 How many students scored a B on the test if the professor considers a test score of 80–89 as a B?

Since the 80–89 bar rises to a height of 8, there were eight students who scored a B on the test. ■

Practice Problem 1 How many students scored a D on the test if the professor considers a test score of 60–69 as a D? ■

Example 2 How many students scored less than 80 on the test?

From the histogram, we see that there are three different bar heights to be included. Four tests were 50–59, six tests were 60–69, and 16 tests were 70–79. When we combine $4 + 6 + 16 = 26$ we can see that 26 students scored less than 80 on the test. ■

Practice Problem 2 How many students scored greater than 69 on the test? ■

The following histogram tells us about the length of life of 110 new light bulbs tested at a research center. The number of hours the bulbs lasted is indicated on the horizontal scale. The frequency of bulbs lasting that long is indicated on the vertical scale.

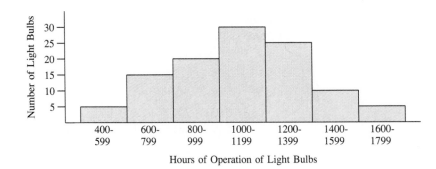

Hours of Operation of Light Bulbs

Example 3 How many light bulbs lasted between 1400 and 1599 hours?

The bar with a range of 1400–1599 hours rises to 10. Thus 10 light bulbs lasted that long. ■

Practice Problem 3 How many light bulbs lasted between 800 and 999 hours? ■

Example 4 How many light bulbs lasted fewer than 1000 hours?

We see that there are three different bar heights to be included. Five bulbs lasted 400–599 hours, 15 bulbs lasted 600–799 hours, and 20 bulbs lasted 800–999 hours. We add $5 + 15 + 20 = 40$. Thus 40 light bulbs lasted less than 1000 hours. ■

Practice Problem 4 How many light bulbs lasted more than 1199 hours? ■

☑ Constructing a Histogram from Raw Data

To construct a histogram we start with *raw data*, data that has not yet been organized or interpreted. We perform the following steps:

> **1.** Select data class intervals of equal width for the data.
> **2.** Make a table with class intervals and a *tally* (count) of how many numbers occur in each interval. Add up the tally to find the class frequency for each class interval.
> **3.** Draw the histogram.

First we will practice making the table. Later we will use the table to draw the histogram.

Example 5 Complete a table to determine the frequency of each class interval for the following data. Each number represents the number of kilowatt-hours of electricity used in a home during a one-month period.

770	520	850	900	1100
1200	1150	730	680	900
1160	590	670	1230	980

1. We select class intervals of equal width for the data. We choose intervals of 200. We might have chosen smaller or larger intervals, but we choose 200 because it gives us a convenient number of intervals to work with, as we will see.
2. We make a table. We write down the class intervals, then count (tally) how many numbers occur within each interval. Then we write the total. This is the class frequency.

Kilowatt-Hours Used Class Interval	Tally	Frequency
500–699	||||	4
700–899	|||	3
900–1099	|||	3
1100–1299	|||||	5

Practice Problem 5 Complete the following table to determine the frequency of each class interval for the following data. Each number represents the weight in pounds of a new car.

2250	1760	2000	2100	1900
1640	1820	2300	2210	2390
2150	1930	2060	2350	1890

Weight in Pounds Class Interval	Tally	Frequency
1600–1799		
1800–1999		
2000–2199		
2200–2399		

Example 6 Draw a histogram from the table in Example 5. ■

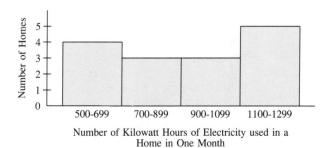

Number of Kilowatt Hours of Electricity used in a Home in One Month

Practice Problem 6 Draw a histogram using the margin from the table in Example 5. ■

A total of 190 rental cars were studied. The following histogram shows the number of rental cars fitting a certain rating classification in miles per gallon. Use the histogram to answer questions 1–8.

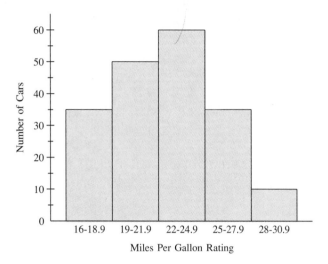

1. How many cars achieve between 19 and 21.9 mi per gallon?

2. How many cars achieve between 28 and 30.9 mi per gallon?

3. How many cars achieve between 25 and 27.9 mi per gallon?

4. How many cars achieve between 16 and 18.9 mi per gallon?

5. How many cars achieve less than 22 mi per gallon?

6. How many cars achieve more than 24.9 mi per gallon?

7. How many cars achieve between 19 and 27.9 mi per gallon?

8. How many cars achieve between 16 and 24.9 mi per gallon?

In an air-pollution study a large factory was studied for 47 operating days. Each day the number of tons of sulfur oxides emitted was measured. The following histogram shows the number of days that emissions of sulfur oxide reached a certain class interval. Use the histogram to answer questions 9–16.

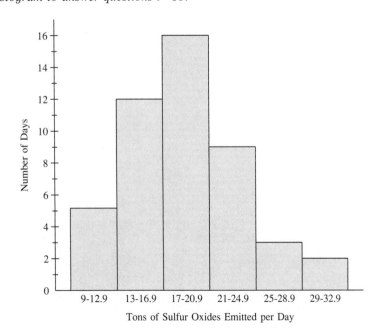

9. For how many days did the factory emit between 17 and 20.9 tons of sulfur oxides?

10. For how many days did the factory emit between 13 and 16.9 tons of sulfur oxides?

11. For how many days did the factory emit between 25 and 28.9 tons of sulfur oxides?

12. For how many days did the factory emit between 21 and 24.9 tons of sulfur oxides?

13. For how many days did the factory emit more than 16.9 tons of sulfur oxides?

14. For how many days did the factory emit less than 25 tons of sulfur oxides?

15. For how many days did the factory emit between 13 and 24.9 tons of sulfur oxides?

16. For how many days did the factory emit between 17 and 28.9 tons of sulfur oxides?

 To Think About

17. On what percent of the 47 operating days were 17 or more tons of sulfur oxides emitted? (Round to the nearest tenth of a percent.)

18. On what percent of the 47 operating days were less than 21 tons of sulfur oxides emitted? (Round to the nearest tenth of a percent.)

Complete a table to determine the frequency of each class interval of the following data. Each number represents the highest temperature in degrees Fahrenheit for the day in Boston for February.

23°	26°	30°	18°	42°	17°	19°
51°	42°	38°	36°	12°	18°	14°
20°	24°	26°	30°	18°	17°	16°
35°	38°	40°	33°	19°	22°	26°

	Temperature Class Interval	*Tally*	*Frequency*			*Temperature Class Interval*	*Tally*	*Frequency*
19.	12°–16°	‒‒‒‒	‒‒‒‒		**20.**	17°–21°	‒‒‒‒	‒‒‒‒
21.	22°–26°	‒‒‒‒	‒‒‒‒		**22.**	27°–31°	‒‒‒‒	‒‒‒‒
23.	32°–36°	‒‒‒‒	‒‒‒‒		**24.**	37°–41°	‒‒‒‒	‒‒‒‒
25.	42°–46°	‒‒‒‒	‒‒‒‒		**26.**	47°–51°	‒‒‒‒	‒‒‒‒

27. Construct a histogram using the table prepared in problems 19–26.

Complete a table to determine the frequency of each class interval of the following data. Each number represents the cost of a prescription purchased by the Lin family this year.

$28.50	$16.00	$32.90	$46.20	$ 9.85
$27.30	$16.00	$41.95	$36.00	$24.20
$ 7.65	$ 8.95	$ 4.50	$11.35	$ 7.75
$12.30	$21.85	$46.20	$15.50	$ 4.50

	Purchase Price Class Interval	Tally	Frequency		Purchase Price Class Interval	Tally	Frequency
28.	$ 4.00–$ 9.99	———	———	29.	$10.00–$15.99	———	———
30.	$16.00–$21.99	———	———	31.	$22.00–$27.99	———	———
32.	$28.00–$33.99	———	———	33.	$34.00–$39.99	———	———
34.	$40.00–$45.99	———	———	35.	$46.00–$51.99	———	———

36. Construct a histogram using the table constructed in problems 28–35.

Cumulative Review Problems

37. Solve for n: $\dfrac{n}{18} = \dfrac{3.5}{9}$.

38. Solve for n: $\dfrac{126}{n} = \dfrac{36}{17}$.

For Extra Practice Examples and Exercises, turn to page 448.

Solutions to Odd-Numbered Practice Problems

1. The 60–69 bar rises to a height of 6. Thus six tests would have a D grade.
3. The 800–999 bar rises to a height of 20. Thus 20 light bulbs operated between 800 and 999 hours.
5. *Weight in Pounds*

Class Interval	Tally	Frequency
1600–1799	||	2
1800–1999	||||	4
2000–2199	||||	4
2200–2399	|||||	5

Answers to Even-Numbered Practice Problems

2. 30 students   **4.** 40 light bulbs   **6.**

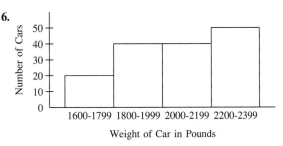

Weight of Car in Pounds

8.4 MEAN AND MEDIAN

We often want to know the "middle value" of a group of numbers. In this section we learn that, in statistics, there is more than one way of describing this middle value: there is the *mean* of the group of numbers and there is the *median* of the group of numbers. In some situations it's more helpful to look at the mean, and in others it's more helpful to look at the median. We'll learn to tell which situations lend themselves to one or the other.

☐ After studying this section, you will be able to:

1 *Find the mean of a set of numbers*

2 *Find the median value of a set of numbers*

◻ The Mean of a Set of Numbers

The **mean** of a set of values is the sum of the values divided by the number of values. The mean is often called the **average**.

Example 1   Find the average or mean test score of a student who has test grades of 71, 83, 87, 99, 80, and 90.

We take the sum of the six tests and divide the sum by 6.

$$\begin{array}{c} \text{Sum of tests} \longrightarrow \\ \text{Number of tests} \longrightarrow \end{array} \frac{71 + 83 + 87 + 99 + 80 + 90}{6} = \frac{510}{6} = 85$$

The mean is 85.   ■

Practice Problem 1   Find the average or mean of the following test scores: 88, 77, 84, 97, 89.   ■

The mean value is often rounded to a certain decimal-place accuracy.

Example 2   Carl recorded the miles per gallon achieved by his car for the last two months. His results were:

Week	1	2	3	4	5	6	7	8
Miles per Gallon	26	24	28	29	27	25	24	23

What is the mean miles per gallon figure for the last 8 weeks? Round your answer to the nearest mile per gallon.

$$\begin{array}{c} \text{Sum of values} \longrightarrow \\ \text{Number of values} \longrightarrow \end{array} \frac{26 + 24 + 28 + 29 + 27 + 25 + 24 + 23}{8}$$

$$= \frac{206}{8} = 26 \quad \text{rounded to the nearest whole number}$$

Mean miles per gallon rating is 26.   ■

Practice Problem 2   Bob and Wally kept records of their phone bill for the last six months. Their bills were $39.20, $43.50, $81.90, $34.20, $51.70, and $48.10. Find the mean monthly phone bill. Round your answer to the nearest cent.   ■

◻ The Median Value

If a set of numbers is arranged in order from smallest to largest, the **median** is that value that has the same number of values above it as below it.

Example 3   Find the median value of the following package weights: 6 lb, 7 lb, 9 lb, 12 lb, 13 lb, 16 lb, 17 lb.

The numbers are arranged in order.

$$\underbrace{6, 7, 9}_{\text{three numbers}} \qquad \underset{\underset{\substack{\text{middle} \\ \text{number}}}{\uparrow}}{12} \qquad \underbrace{13, 16, 17}_{\text{three numbers}}$$

There are three numbers smaller than 12 and three numbers larger than 12. Thus 12 is the median.   ■

Practice Problem 3 Find the median value of the following lengths of telephone calls: 2 minutes, 3 minutes, 7 minutes, 12 minutes, 13 minutes, 15 minutes, 18 minutes, 20 minutes, 21 minutes. ■

If the numbers are not arranged in order, then the first step is to do so.

Example 4 Find the median cost of the following costs for a microwave oven: $100, $60, $120, $200, $190, $130, $320, $290, $180.

We must arrange the costs in order from smallest to largest (or largest to smallest).

$60, $100, $120, $130 $180 $190, $200, $290, $320

      four numbers   middle   four numbers
                           number

Thus $180 is the median cost. ■

Practice Problem 4 Find the median weekly salary of the following salaries: $320, $150, $400, $600, $290, $180, $450. ■

If a list of numbers contains an even number of items, then of course there is no one middle number In this situation we obtain the median by taking the average of the two middle numbers.

Example 5 Find the median of the following numbers: 13, 16, 18, 26, 31, 33, 38, 39.

      13, 16, 18   26, 31   33, 38, 39

    three numbers   two middle   three numbers
                       numbers

The average (mean) of 26 and 31 is

$$\frac{26 + 31}{2} = \frac{57}{2} = 28.5$$

Thus, the median value is 28.5. ■

Practice Problem 5 Find the median value of the following numbers: 88, 90, 100, 105, 118, 126. ■

To Think About

When would someone want to use the mean, and when would someone want to use the median? Which is more helpful? The mean, or average, is used more frequently. It is most helpful when the data are distributed fairly evenly, that is, when no one value is "much larger" or "much smaller" than the rest. For example, a company had employees with annual salaries of $9000, $11,000, $14,000, $15,000, $17,000, and $20,000. All the salaries fall within a fairly limited range. The mean salary

$$\frac{9000 + 11,000 + 14,000 + 15,000 + 17,000 + 20,000}{6} = \$14,333.33$$

gives us a reasonable idea of the "average" salary. However, suppose the company had six employees with salaries of $9000, $11,000, $14,000, $15,000, $17,000, and $90,000. Talking about the mean salary, which is $26,000, is deceptive. No one earns a salary very close to the mean salary. The "average" worker does not earn $26,000. In this case the median value is more appropriate. Here the median is $14,500. Some problems of this type are included in problems 29 and 30. ■

In problems 1–14, find the mean. Round to the nearest tenth when necessary.

1. A student received grades of 89, 92, 83, 96, and 99 on math quizzes.

2. A student received grades of 77, 88, 90, 92, 83, and 84 on history quizzes.

3. A receptionist received the following number of phone calls: 38, 29, 43, 36, 40, and 18.

4. A taxicab company had the following number of requests for a ride: 67, 72, 80, 77, 93, 47 and 54.

5. The last five houses built in town sold for the following: $89,000, $93,000, $62,000, $102,000, $89,000.

6. Luiz priced a sofa at six local stores. The prices were $499, $359, $600, $450, $529, $629.

7. Sam watched television last week for the following number of hours:

Mon.	Tues.	Wed.	Thurs.	Fri.	Sat.	Sun.
8	3	2	5	6.5	7.5	3

8. Steve's car got the following miles per gallon results during the last 6 months:

Jan.	Feb.	Mar.	Apr.	May	June
23	22	25	28	29	30

9. Alicia has taken four three-credit courses at college each of four semesters. Her semester grade point averages are:

Semester 1	Semester 2	Semester 3	Semester 4
2.02	2.31	3.05	2.98

10. Desmond has taken four three-credit courses at college each of four semesters. His semester grade averages are:

Semester 1	Semester 2	Semester 3	Semester 4
2.90	2.66	2.52	2.48

11. The captain of the college baseball team achieved the following results:

	Game 1	Game 2	Game 3	Game 4	Game 5
Hits	0	2	3	2	2
Times at Bat	5	4	6	5	4

Find his batting average by dividing his total number of hits by the total times at bat.

12. The captain of the college bowling team had the following results after practice:

	Practice 1	Practice 2	Practice 3	Practice 4
Score (Pins)	541	561	840	422
Number of Games	3	3	4	2

Find her bowling average by dividing the total number of pins scored by the total number of games.

13. Frank and Wally traveled to the west coast during the summer. The number of miles they drove and the number of gallons of gas used are recorded below:

	Day 1	Day 2	Day 3	Day 4
Miles Driven	276	350	391	336
Gallons of Gas	12	14	17	14

14. Cindy and Andrea traveled to Boston this fall. The number of miles they drove and the number of gallons of gas used are recorded below:

	Day 1	Day 2	Day 3	Day 4
Miles Driven	260	375	408	416
Gallons of Gas	10	15	17	16

Find the average miles per gallon achieved by the car on the trip by dividing the total number of miles driven by the total number of gallons used.

Find the average miles per gallon achieved by the car on the trip by dividing the total number of miles driven by the total number of gallons used.

In Problems 15–30, find the median value.

15. 22, 36, 45, 47, 48, 50, 58

16. 37, 39, 46, 53, 57, 60, 63

17. 1052, 968, 1023, 999, 865, 1152

18. 1400, 1329, 1200, 1386, 1427, 1350

19. 0.52, 0.69, 0.71, 0.34, 0.58

20. 0.26, 0.12, 0.35, 0.43, 0.28

21. The annual salaries of the employees of the company are $6500, $8900, $10,300, $12,800, $19,650, $28,000.

22. The costs of six cars recently purchased by the Weston Company are $18,270, $11,300, $16,400, $9100, $12,450, $13,800.

23. The number of minutes spent on the phone per day by each of eight salespeople is 16 minutes, 43 minutes, 24 minutes, 62 minutes, 19 minutes, 38 minutes, 17 minutes, 57 minutes.

24. The age of 10 students in the senior psychology seminar is 28 years, 19 years, 24 years, 31 years, 22 years, 21 years, 26 years, 27 years, 32 years, 30 years.

25. The cost of a family vacation taken by the Stevens family for the last seven years is $2300, $562, $4100, $1955, $327, $3100, $1820.

26. The cost of a new stereo system priced at nine different stores in the city is $320, $299, $269, $400, $354, $289, $315, $349, $279.

27. The number of overtime hours worked last month by officers of a town police department was 18, 26, 13, 7, 42, 31, 27, 19, 22, 17.

28. The number of students who tried out for the college tennis team over the last eight years was 17, 19, 23, 16, 42, 33, 27, 38.

? **To Think About**

29. A company has 10 employees. Their monthly salaries are $1500, $1700, $1650, $1300, $1440, $1580, $1820, $1380, $2900, $6300.
(a) Find the mean.
(b) Find the median.
(c) Which of these numbers best represents "what the average person earns?" Why?

30. A college track star ran the 100-m event in eight track meets. Her times were 11.7 seconds, 11.6 seconds, 12.0 seconds, 12.1 seconds, 11.9 seconds, 18 seconds, 11.5 seconds, 12.4 seconds.
(a) Find the mean.
(b) Find the median.
(c) Which of these numbers best represents "her average running time"? Why?

Find the mean.

31. The number of bushels of oats produced in the United States for four recent years was:

1987	1986	1985	1984
373,765,000	386,356,000	520,800,000	473,661,000

32. The amount of money spent in the United States for four recent years for food (not including restaurant meals) was:

1987	1986	1985	1984
$352,000,000,000	$339,800,000,000	$322,700,000,000	$305,800,000,000

Cumulative Review Problems

33. Find the area of a circle with radius 3 in. (Use $\pi = 3.14$.)

34. Find the area of a triangle with altitude 22 m and a base of 31 m.

 Calculator Problems

Find the mean. Round to the nearest tenth when necessary.

35. The salaries of eight lawyers; $90,000, $105,750, $96,720, $121,200, $115,000, $129,900, $112,000, $109,750.

36. The price of nine automobiles: $9750, $10,200, $7999, $11,260, $8700, $6800, $9200, $12,700, $11,900.

Find the median value.

37. 2576, 8764, 3700, 5000, 7200, 4700

38. 15.276, 21.375, 18.90, 29.2, 14.77, 19.02

For Extra Practice Examples and Exercises, turn to page 449.

Solutions to Odd-Numbered Practice Problems

1. $\dfrac{88 + 77 + 84 + 97 + 89}{5} = \dfrac{435}{5} = 87$ mean value

3. 2, 3, 7, 12,   13,   15, 18, 20, 21

four values   middle value   four values

The median value is 13 minutes.

5. 88, 90,   100, 105,   118, 126

two values   two middle values   two values

$\dfrac{100 + 105}{2} = \dfrac{205}{2} = 102.5$ median value

Answers to Even-Numbered Practice Problems

2. $49.77   **4.** $320

After studying this section, you will be able to:

❶ *Evaluate the square root of a number that is a perfect square*

❷ *Approximate the square root of a number that is not a perfect square*

8.5 SQUARE ROOTS

In this section we focus on the word "square" in different ways. We use the physical notion of a square to learn about square roots.

❶ Evaluating the Square Root of a Number That Is a Perfect Square

We know that by using the formula $A = s^2$ we can quickly find the area of a square with a side of 3 in. We sometimes want to ask the question the other way. A square has an area of 64 in.^{2}, what is the length of its sides?

$A = s^2 = 3^2 = 9$

Area = 9 square inches

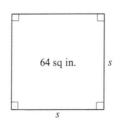

64 sq in. s

$s^2 = 64$

What is s?

The answer is 8 in. Do you see why? The skill we need is the ability to find a square root.

The **square root** of a number is one of two identical factors of that number.

The square root of 64 is 8 because $(8)(8) = 64$.
The square root of 9 is 3 because $(3)(3) = 9$.

The symbol for finding the square root of a number is $\sqrt{}$. To find the square root of 64, we write $\sqrt{64} = 8$. Sometimes we speak of finding the square root of a number as *taking* the square root of the number. Or we can say that we will *evaluate* the square root of the number. Thus to take the square root of 9 we write $\sqrt{9} = 3$; to evaluate the square root of 9 we write $\sqrt{9} = 3$.

Example 1   Find $\sqrt{25}$.

$$\sqrt{25} = 5 \text{ because } (5)(5) = 25 \quad \blacksquare$$

Practice Problem 1   Find $\sqrt{49}$.   ■

Example 2   Find $\sqrt{121}$.

$$\sqrt{121} = 11 \text{ because } (11)(11) = 121 \quad \blacksquare$$

Practice Problem 2   Find $\sqrt{169}$.   ■

If square roots are added or subtracted, they must be evaluated first, then added or subtracted.

Example 3   Find $\sqrt{25} + \sqrt{36}$.

$$\sqrt{25} = 5 \text{ because } (5)(5) = 25$$
$$\sqrt{36} = 6 \text{ because } (6)(6) = 36$$

Thus $\sqrt{25} + \sqrt{36} = 5 + 6 = 11.$ ■

Practice Problem 3   Find $\sqrt{49} - \sqrt{4}$.   ■

When a whole number is multiplied by itself, the number that is obtained is called a **perfect square**.

36 is a perfect square because $(6)(6) = 36$.

49 is a perfect square $(7)(7) = 49$.

The numbers 20 or 48 are *not* perfect squares. There is no whole number that when squared—multiplied by itself—yields 20 or 48. Consider 20. $4^2 = 16$, less than 20. $5^2 = 25$, more than 20. We realize, then, that no number squared equals 20. The square root of a perfect square is a whole number.

Example 4

(a) Is 81 a perfect square?   **(b)** If so, find $\sqrt{81}$.

(a) Yes. 81 is a perfect square because $(9)(9) = 81$.
(b) $\sqrt{81} = 9$ ■

Practice Problem 4
(a) Is 144 a perfect square?   **(b)** If so, find $\sqrt{144}$.   ■

It is helpful to make a list of the first 15 perfect squares. Take a minute to complete the following table.

Number, n	1	2	3	4	5	6	7	8	9	10	11	12	13	14	15
Number Squared, n^2	1	4	9	16										196	225

☑ Approximating the Square Root of a Number That Is Not a Perfect Square

If a number is not a perfect square, we can only approximate its square root. This can be done by using a square root table such as the one that follows. Except for exact values such as $\sqrt{4} = 2.000$, all values are rounded to the nearest thousandth.

Number, n	Square Root of the Number, $\sqrt{n}$	Number, n	Square Root of the Number, $\sqrt{n}$
1	1.000	8	2.828
2	1.414	9	3.000
3	1.732	10	3.162
4	2.000	11	3.317
5	2.236	12	3.464
6	2.449	13	3.606
7	2.646	14	3.742

A square root table is located on page 578. It will allow you to find the square root of whole numbers up to 200. Square roots can also be found with any calculator that has a square root key. Usually the key looks like this $\boxed{\sqrt{}}$ or this $\boxed{\sqrt{x}}$. To find the square root of 8 on a calculator enter the number 8 and press $\boxed{\sqrt{}}$ or $\boxed{\sqrt{x}}$. You will see displayed 2.8284271. (Your calculator may display fewer or more digits.) Remember, no matter how many digits your calculator displays when we find $\sqrt{8}$, we have only an **approximation**. It is not an exact answer. To emphasize this we use the $\approx$ notation to represent "is approximately equal." Thus $\sqrt{8} \approx 2.828$.

Example 5 Find an approximate value for $\sqrt{2}$ using the square root table or a calculator. Round your answer to the nearest thousandth.

$$\sqrt{2} \approx 1.414 \quad \blacksquare$$

Practice Problem 5 Approximate $\sqrt{3}$ to the nearest thousandth. ■

Example 6 Approximate $\sqrt{12}$ to the nearest thousandth.

$$\sqrt{12} \approx 3.464 \quad \blacksquare$$

Practice Problem 6 Approximate $\sqrt{13}$ to the nearest thousandth. ■

Example 7 Approximate $\sqrt{7}$ to the nearest thousandth.

$$\sqrt{7} \approx 2.646 \quad \blacksquare$$

Practice Problem 7 Approximate $\sqrt{5}$ to the nearest thousandth. ■

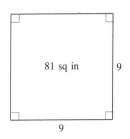

Example 8 Find the length of the side of a square that contains (that has an area of) 81 in^2.

$\sqrt{81} = 9$. The side measures 9 in. ■

Practice Problem 8 Find the length of the side of a square that contains (that has an area of) 100 in^2. ■

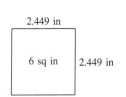

Example 9 Approximate to the nearest thousandth of an inch the length of the side of a square that contains (that has an area of) 6 in^2.

$\sqrt{6} \approx 2.449$. Thus to the nearest thousandth of an inch the side measures 2.449 in. ■

Practice Problem 9 Approximate the length to the nearest thousandth of a meter of a square that contains (that has an area of) 22 m^2. ■

Find each square root. Do not use a calculator. Do not refer to a table of square roots.

1. $\sqrt{1}$　　**2.** $\sqrt{4}$　　**3.** $\sqrt{16}$　　**4.** $\sqrt{9}$　　**5.** $\sqrt{25}$　　**6.** $\sqrt{36}$　　**7.** $\sqrt{49}$　　**8.** $\sqrt{64}$

9. $\sqrt{100}$　　**10.** $\sqrt{81}$　　**11.** $\sqrt{121}$　　**12.** $\sqrt{144}$　　**13.** $\sqrt{169}$　　**14.** $\sqrt{196}$　　**15.** $\sqrt{0}$　　**16.** $\sqrt{225}$

Evaluate the square roots first, then add or subtract the results. Do not use a calculator or a square root table.

17. $\sqrt{49} + \sqrt{9}$　　**18.** $\sqrt{25} + \sqrt{64}$　　**19.** $\sqrt{100} + \sqrt{1}$　　**20.** $\sqrt{0} + \sqrt{121}$　　**21.** $\sqrt{36} + \sqrt{64}$　　**22.** $\sqrt{1} + \sqrt{25}$

23. $\sqrt{0} + \sqrt{81}$　　**24.** $\sqrt{169} + \sqrt{4}$　　**25.** $\sqrt{144} - \sqrt{4}$　　**26.** $\sqrt{169} - \sqrt{64}$　　**27.** $\sqrt{121} - \sqrt{36}$　　**28.** $\sqrt{225} - \sqrt{100}$

29. (a) Is 49 a perfect square?
    (b) If so, find $\sqrt{49}$.

30. (a) Is 25 a perfect square?
    (b) If so, find $\sqrt{25}$.

31. (a) Is 256 a perfect square?
    (b) If so, find $\sqrt{256}$.

32. (a) Is 289 a perfect square?
    (b) If so, find $\sqrt{289}$.

Use a table of square roots or a calculator with a square root key to approximate the following to the nearest thousandth.

33. $\sqrt{15}$　　**34.** $\sqrt{17}$　　**35.** $\sqrt{31}$　　**36.** $\sqrt{42}$　　**37.** $\sqrt{83}$

38. $\sqrt{76}$　　**39.** $\sqrt{120}$　　**40.** $\sqrt{136}$　　**41.** $\sqrt{184}$　　**42.** $\sqrt{150}$

Find the length of the side of each square. If the area is not a perfect square, approximate by using a square root table or a calculator with a square root key. Round your answer to the nearest thousandth.

43. A square with area 121 m^2　　**44.** A square with area 169 m^2　　**45.** A square with area 26 m^2

46. A square with area 34 m^2　　**47.** A square with area 75 m^2　　**48.** A square with area 90 m^2

?　**To Think About**

49. Find each square root.
    (a) $\sqrt{4}$
    (b) $\sqrt{0.04}$
    (c) $\sqrt{0.0004}$
    (d) What pattern do you observe?
    (e) Can you find $\sqrt{0.004}$ exactly? Why?

50. Find each square root.
    (a) $\sqrt{25}$
    (b) $\sqrt{0.25}$
    (c) $\sqrt{0.0025}$
    (d) What pattern do you observe?
    (e) Can you find $\sqrt{0.025}$ exactly? Why?

51. Find an exact value for
$$\sqrt{1} + \sqrt{9} + \sqrt{25} + \sqrt{36} + \sqrt{49}$$

52. Find an exact value for
$$\sqrt{4} + \sqrt{9} + \sqrt{16} + \sqrt{81} + \sqrt{121}$$

Cumulative Review Problems

53. An aquarium is 12 in. high, 10 in. wide, and 20 in. long. How many gallons of water will fill the aquarium if 1 gal = 231 in.^{3}? (Round your answer to the nearest tenth.)

54. How many meters did Anita travel in a trip to Canada of 580 km?

Calculator Problems

Use a table of square roots or a calculator with a square root key to approximate each calculation to the nearest thousandth. Your answers may vary slightly, depending on the calculator used.

55. $\sqrt{192} + \sqrt{120} + \sqrt{167} + \sqrt{56}$

56. $\sqrt{155} + \sqrt{90}$

57. $\sqrt{17} + \sqrt{21} + \sqrt{6} + \sqrt{19} + \sqrt{12}$

For Extra Practice Examples and Exercises, turn to page 450.

Solutions to Odd-Numbered Practice Problems

1. $\sqrt{49} = 7$ because $(7)(7) = 49$ **3.** $\sqrt{49} - \sqrt{4} = 7 - 2 = 5$ **5.** $\sqrt{3} \approx 1.732$ rounded to the nearest thousandth
7. $\sqrt{5} \approx 2.236$ rounded to the nearest thousandth   **9.** $\sqrt{22} \approx 4.690$ m rounded to the nearest thousandth

Answers to Even-Numbered Practice Problems

2. 13   **4. (a)** Yes **(b)** 12   **6.** 3.606   **8.** Exactly 10 in

---

☐ **After studying this section, you will be able to:**

1 *Find the hypotenuse of a right triangle when given the length of each leg*

2 *Find the leg of a right triangle when given the length of the other leg and the hypotenuse*

3 *Solve applied problems involving the use of the Pythagorean Theorem*

4 *Solving special right triangles*

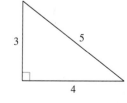

8.6 THE PYTHAGOREAN THEOREM

The Pythagorean Theorem is a mathematical idea formulated long ago. It is as useful today as ever. The Pythagorean theorem can help solve problems in building, distance, and other types of problems.

   The Pythagoreans lived in Italy about 2500 years ago. They studied various mathematical properties. They discovered that for any right triangle, the square of the hypotenuse equals the sum of the squares of the two legs of the triangle. This relationship is known as the *Pythagorean Theorem*. The side opposite the right angle is called the *hypotenuse*; the other two sides are called the *legs* of the right triangle.

$$(\text{hypotenuse})^2 = (\text{leg})^2 + (\text{leg})^2.$$

For example, in the right triangle (in margin)

$$5^2 = 3^2 + 4^2$$
$$25 = 9 + 16$$
$$25 = 25 \checkmark$$

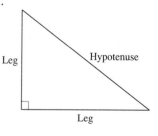

1 Finding the Length of the Hypotenuse When the Length of Each Leg Is Given

In a right triangle, the hypotenuse is the longest side. It is always opposite the largest angle, the right angle. The legs are the two shorter sides. When we know each leg of a right triangle we use the property

$$\text{hypotenuse} = \sqrt{(\text{leg})^2 + (\text{leg})^2}$$

Example 1 Find the hypotenuse of a triangle with legs of 5 in. and 12 in., respectively.

$$\begin{aligned}
\text{Hypotenuse} &= \sqrt{(5)^2 + (12)^2} \\
&= \sqrt{25 + 144} \qquad &&\text{Squaring each value first.} \\
&= \sqrt{169} \qquad &&\text{Adding together the two values.} \\
&= 13 \qquad &&\text{Taking the square root.} \quad \blacksquare
\end{aligned}$$

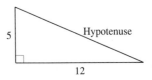

Practice Problem 1 Find the hypotenuse of a triangle with legs of 8 m and 6 m, respectively. ∎

Example 2 Find the hypotenuse of a triangle with legs of 8 m and 15 m, respectively.

$$\begin{aligned}
\text{Hypotenuse} &= \sqrt{(8)^2 + (15)^2} \\
&= \sqrt{64 + 225} \qquad &&\text{Squaring each value first.} \\
&= \sqrt{289} \qquad &&\text{Adding together the two values.} \\
&= 17 \qquad &&\text{Taking the square root.}
\end{aligned}$$

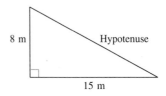

[Do you see why $\sqrt{289} = 17$? Verify that $(17)(17) = 289$. Finding square roots of large numbers like this sometimes takes a little work.] ∎

Practice Problem 2 Find the hypotenuse of a triangle with legs of 12 m and 16 m, respectively. ∎

Sometimes we cannot find the hypotenuse exactly. In those cases we often approximate the square root by using a calculator or a square root table.

Example 3 Find the hypotenuse of a right triangle with legs of 4 m and 5 m, respectively. Round your answer to the nearest thousandth.

$$\begin{aligned}
\text{Hypotenuse} &= \sqrt{(4)^2 + (5)^2} \\
&= \sqrt{16 + 25} \qquad &&\text{Squaring each value first.} \\
&= \sqrt{41} \qquad &&\text{Adding the two values together.}
\end{aligned}$$

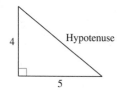

Using the square root table or a calculator, we have hypotenuse ≈ 6.403 m. ∎

Practice Problem 3 Find to the nearest thousandth the hypotenuse of a right triangle with legs of 3 cm and 7 cm, respectively. ∎

2 Finding the Length of a Leg When the Lengths of the Hypotenuse and the Other Leg Are Given

When we know the hypotenuse and one leg of a right triangle, we use the property

$$\text{leg} = \sqrt{(\text{hypotenuse})^2 - (\text{leg})^2}$$

to find the length of the other leg.

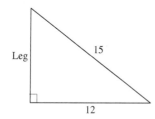

Example 4 A right triangle has a hypotenuse of 15 cm and a leg of 12 cm. Find the length of the other leg.

$$\text{Leg} = \sqrt{(15)^2 - (12)^2}$$
$$= \sqrt{225 - 144} \qquad \text{Squaring each value first.}$$
$$= \sqrt{81} \qquad \text{Subtracting.}$$
$$= 9 \text{ cm} \qquad \text{Finding the square root.} \quad \blacksquare$$

Practice Problem 4 A right triangle has a hypotenuse of 17 m and a leg of 15 m. Find the length of the other leg. ■

Example 5 A right triangle has a hypotenuse of 9 m and a leg of 7 m. Find the length of the other leg correct to the nearest thousandth.

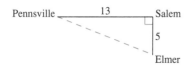

$$\text{Leg} = \sqrt{(9)^2 - (7)^2}$$
$$= \sqrt{81 - 49} \qquad \text{Squaring each value first.}$$
$$= \sqrt{32} \qquad \text{Subtracting the two numbers.}$$

Using a calculator or the square root table, leg ≈ 5.657 m. ■

Practice Problem 5 A right triangle has a hypotenuse of 10 m and a leg of 5 m. Find the length of the other leg correct to the nearest thousandth. ■

3 Certain applied problems call for the use of the Pythagorean Theorem in the solution.

Example 6 A pilot flies 13 mi east from Pennsville to Salem. She then flies 5 mi south from Salem to Elmer. What is the straight-line distance from Pennsville to Elmer? Round your answer to the nearest tenth of a mile.

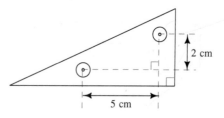

The distance from Pennsville to Elmer is the hypotenuse of the triangle.

$$\text{Hypotenuse} = \sqrt{(\text{leg})^2 + (\text{leg})^2}$$
$$= \sqrt{(13)^2 + (5)^2}$$
$$= \sqrt{169 + 25}$$
$$= \sqrt{194}$$

$$\sqrt{194} = 13.928$$

Rounded to the nearest tenth, the distance is 13.9 mi. ■

Practice Problem 6 Find the distance to the nearest thousandth between the centers of the holes in the triangular metal plate.

Example 7 A 25-ft ladder is placed against a building at a point 22 ft from the ground. What is the distance of the base of the ladder from the building? Round your answer to the nearest tenth.

$$\text{Leg} = \sqrt{(\text{hypotenuse})^2 - (\text{leg})^2}$$
$$= \sqrt{(25)^2 - (22)^2}$$
$$= \sqrt{625 - 484}$$
$$= \sqrt{141}$$

$$\sqrt{141} \approx 11.874$$

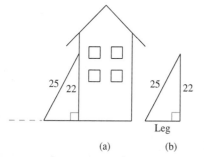

(a)   (b)

If we round to the nearest tenth, the ladder is 11.9 ft from the base of the house. ■

Practice Problem 7 A kite is out on 30 yd of string. The kite is directly above a rock. The rock is 27 yd from the boy flying the kite. How far above the rock is the kite? Round your answer to the nearest tenth. ■

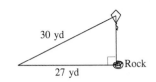

◢ Solving Special Right Triangles

If we use the Pythagorean Theorem and some other facts from geometry, we can find a relationship among the sides of two special right triangles. The first special right triangle is one that contains an angle that measures 30° and one that measures 60°. We call this the 30°–60°–90° right triangle.

In a 30°–60°–90° triangle the length of the leg opposite the 30° angle is $\frac{1}{2}$ the length of the hypotenuse.

Notice that the hypotenuse of the first triangle is 10 m and the side opposite the 30° angle is exactly $\frac{1}{2}$ of that, namely 5 m. The second triangle has a hypotenuse of 15 yd. The side opposite the 30° angle is exactly $\frac{1}{2}$ of that, namely 7.5 yd.

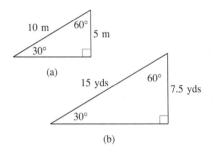

The second special right triangle is one that contains exactly two angles which each measure 45°. We call this the 45°–45°–90° right triangle.

In a 45°–45°–90° triangle the sides opposite the 45° angles are equal. The hypotenuse is equal to $\sqrt{2} \times$ the length of either leg.

We will usually use the decimal approximation that $\sqrt{2} = 1.414$ with this property.

$$\text{Hypotenuse} = \sqrt{2} \times 7$$
$$\approx 1.414 \times 7$$
$$\approx 9.898 \text{ cm}$$

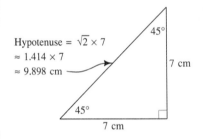

Example 8 Find the requested sides for each of the special triangles. Round to the nearest tenth.

(a) Find the length of sides y and x.   **(b)** Find the length of hypotenuse z.

(a) In a 30°–60°–90° triangle the side opposite the 30° angle is $\frac{1}{2}$ of the hypotenuse. $\frac{1}{2} \times 16 = 8$. Therefore, $y = 8$ yd.

When we know two sides of a right triangle we find a third by the Pythagorean Theorem.

$$\text{Leg} = \sqrt{(\text{hypotenuse})^2 - (\text{leg})^2}$$
$$= \sqrt{16^2 - 8^2} = \sqrt{256 - 64}$$
$$= \sqrt{192} \approx 13.856$$

$x = \text{leg} = 13.9$ yd   rounded to the nearest tenth.

(b) In a 45°–45°–90° triangle the

$$\text{Hypotenuse} = \sqrt{2} \times \text{leg}$$
$$= 1.414(6)$$
$$= 8.484$$

Rounding to the nearest tenth, the hypotenuse = 8.5 m. ■

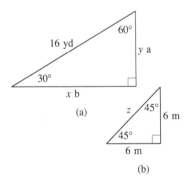

Practice Problem 8 Find the requested sides for each of the special triangles. Round your answer to the nearest tenth.
(a) Find the length of sides y and x.   **(b)** Find the length of hypotenuse z.

■

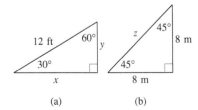

Find the unknown side of each triangle. Use a calculator or square root table when necessary and round your answer to the nearest thousandth.

1.
3 in.
4 in.

2.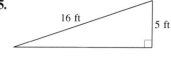
5 in.
12 in.

3.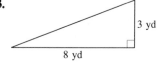
3 yd
8 yd

4.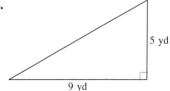
5 yd
9 yd

5.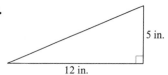
16 ft
5 ft

6.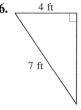
4 ft
7 ft

7.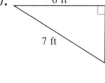
6 ft
7 ft

8.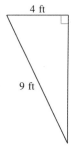
4 ft
9 ft

Find the unknown side of each right triangle to the nearest thousandth using the information given.

9. leg = 8 km, hypotenuse = 11 km

10. leg = 6 km, hypotenuse = 12 km

11. leg = 11 m, leg = 3 m

12. leg = 5 m, leg = 4 m

13. leg = 5 m, leg = 5 m

14. leg = 6 m, leg = 6 m

15. hypotenuse = 13 yd, leg = 11 yd

16. hypotenuse = 14 yd, leg = 10 yd

Solve each applied problem. Round your answer to the nearest tenth.

17. Find the length of the guy wire supporting the telephone pole.

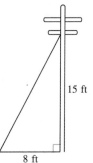
15 ft
8 ft

18. Find the length of this ramp to a loading dock.

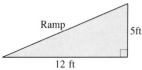
Ramp
5ft
12 ft

19. A boat goes 4 mi west and 3 mi south. How far is the boat from its starting point?

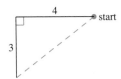

20. Find the distance between the centers of the holes in this triangular plate.

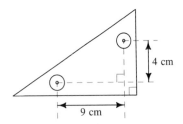

Using your knowledge of special right triangles, find the length of each leg. Round your answers to the nearest tenth.

21.

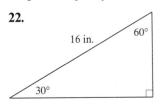

22.

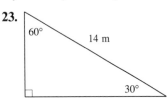

23.

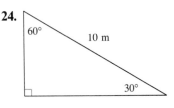

24.

25.

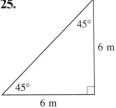

26.

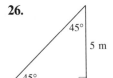

27.

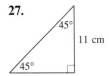

28.

? To Think About

29. A carpenter is going to use a wooden pole 10 in. in diameter. He wishes to shape a rectangular base. The base will be 7 in. wide. The carpenter wishes to make the base as tall as possible, minimizing any waste. How tall will the rectangular base be? (Round to the nearest tenth.)

30. A 4-m antenna is placed on a garage roof that is 2 m above the lower part of the roof. The base of the garage is 16 m wide. How long is an antenna support from point *A* to point *B*?

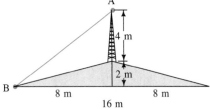

Find the exact value of the unknown side of each right triangle.

31. The two legs of the triangle are 48 yd and 20 yd.

32. The hypotenuse of the triangle is 65 yd and one leg is 33 yd.

Cumulative Review Problems

33. Add: $3.82 + 5.63 + 1.92$.

34. Subtract: $12.0078 - 10.1534$.

35. Multiply: 8.96×0.34.

36. Divide: $12.92 \div 7.6$.

 Calculator Problems

Find the unknown side of the triangle. Use a calculator or square root table when necessary and round your answer to the nearest thousandth.

37. leg = 120 in., hypotenuse = 200 in.

38. leg = 70 ft leg = 90 ft

For Extra Practice Examples and Exercises, turn to page 450.

Solutions to Odd-Numbered Practice Problems

1. Hypotenuse $= \sqrt{(6)^2 + (8)^2}$
$= \sqrt{36 + 64}$
$= \sqrt{100}$
$= 10$ m

3. Hypotenuse $= \sqrt{(3)^2 + (7)^2}$
$= \sqrt{9 + 49}$
$= \sqrt{58}$
Rounded to the nearest thousandth, hypotenuse ≈ 7.616 cm.

5. Leg $= \sqrt{(10)^2 - (5)^2}$
$= \sqrt{100 - 25}$
$= \sqrt{75}$
Rounded to the nearest thousandth, the leg ≈ 8.660 m.

7. Leg $= \sqrt{(30)^2 - (27)^2}$
$= \sqrt{900 - 729}$
$= \sqrt{171}$
The distance to the nearest tenth is 13.1 yd.

Answers to Even-Numbered Practice Problems

2. 20 m   **4.** 8 m   **6.** 5.385 cm   **8.** In problem 8 all answers are rounded to the nearest tenth.
(a) Side $y = 6$ ft   (b) Hypotenuse $z = 11.3$ m
side $x = 10.4$ ft

EXTRA PRACTICE: EXAMPLES AND EXERCISES

Section 8.1

Example J & J Cafe's monthly expenses are displayed in the circle graph. Use this circle graph to answer the following questions.

A. What category takes the least amount of the budget?
B. How much money is allotted each month for utilities?
C. How much money is allotted for either food or wages?

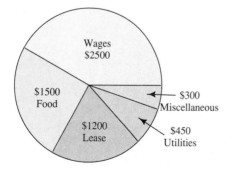

D. What is the ratio of money spent for wages to the total amount in the monthly budget?
A. Miscellaneous
B. $450
C. food $1500, wages $2500
food or wages: $1500 + 2500 = \$4000$
D. Ratio of wages to total amount:
Wages = 2500,
total amount $= 2500 + 1200 + 300 + 450 + 1500 = \$5950$

$$\frac{\text{Wages}}{\text{Total amount}} = \frac{2500}{5950} = \frac{50}{119}$$

Use the circle graph to answer each question.

1. What category takes the largest amount of the budget?
2. How much money is allotted each month for miscellaneous?
3. How much money is allotted for either the lease or utilities?
4. What is the ratio of money spent for utilities to amount of money spent for wages?

5. What is the ratio of money spent for the lease to the total amount in the monthly budget?

6. How much money is spent each month for categories that are not wages and are not food?

Example The circle graph displays the major field of study for students in the graduating class at a university. Use this graph to answer each question.

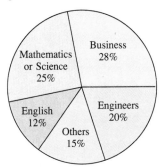

A. What percent of the graduating class were English or engineering majors?

B. If the total graduating class consisted of 3520 students, *how many students* were mathematics or science majors?

A. English = 12%; engineering = 20%

English or engineering = 12% + 20% = 32%

B. 25% of the graduating class were mathematics or science majors.

25% of 3520 = (0.25)(3520) = 880 students

Use the circle graph to answer each question.

7. What percent of the graduating class were English or business majors?

8. What percent of the graduating class were engineering or mathematics/science majors?

9. What percent of the graduating class are not engineering majors?

10. What percent of the graduating class are not English majors?

11. If the total graduating class consisted of 4200 students, how many students were mathematics or science majors?

Section 8.2

Example The following bar graph shows the approximate number of people who stayed at a ski resort over a period of years. Use the graph to answer each question.

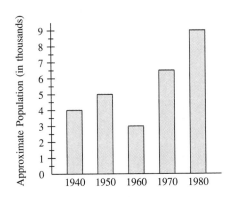

A. What was the approximate number of people at the ski resort in 1950?

B. According to the bar graph, between what years did the number of people staying at the ski resort increase by the smallest amount?

A. 5000

B. 1940 and 1950

1. What was the approximate number of people at the ski resort in 1960?

2. What was the approximate number of people at the ski resort in 1970?

3. What was the approximate number of people at the ski resort in 1980?

4. What was the approximate number of people at the ski resort in 1940?

5. According to this bar graph, between what years did the number of people staying at the ski resort increase by the largest amount?

Example The following double-bar graph illustrates the sales of men's shoes at Newmart Shoe Store for the years 1988 and 1989. Use the graph to answer each question.

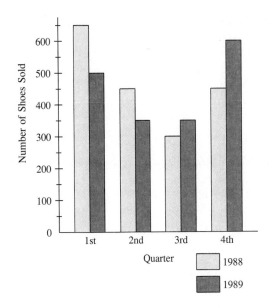

A. In what quarter of 1988 were the fewest shoes sold?

B. How many more shoes were sold in the second quarter of 1988 than in the second quarter of 1989?

A. Third quarter

B. 100 more shoes were sold in the second quarter of 1988.

6. In what quarter of 1989 were the most shoes sold?

7. How many shoes were sold in the second quarter of 1989?

8. How many shoes were sold in the first quarter of 1988?

9. How many more shoes were sold in the fourth quarter of 1989 than in the fourth quarter of 1988?

10. How many shoes were sold in the first quarter of 1989?

Example The following line graph shows the number of customers per month coming to a museum during the first half of the year. Use the graph to answer each question.

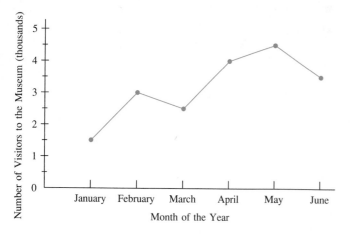

A. During what month did the greatest number of people come to the museum?
B. How many more people visited the museum in April than in January?
A. May
B. In April, 4000 visited the museum; in January, 1500 visited the museum.

$$4000 - 1500 = 2500 \text{ more people visited in April}$$

11. During what month did the fewest number of people come to the museum?
12. Approximately how many people visited the museum in March?
13. How many more people visited the museum in June than in March?
14. Between what two months is the increase the smallest?

15. Approximately how many people visited the museum in May?

Example The following comparison line graph indicates the snowfall for the last four months of two different years in a city in the midwest. Use the graph to answer each question.

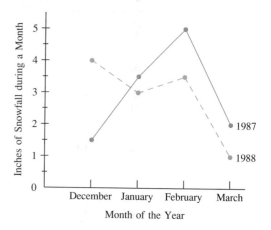

A. In March 1988, how many inches of snowfall were recorded?
B. During what months was the snowfall of 1987 greater than 1988?
A. 1 inch of snowfall
B. January, February, and March
16. In February 1987, how many inches of snowfall were recorded?
17. In March 1987, how many inches of snowfall were recorded?
18. During what month was the snowfall of 1988 greater than 1987?
19. How many more inches of snow fell in January 1987 than December 1987?
20. How many more inches of snow fell in December 1988 than January 1988?

Section 8.3

Example A car manufacturer kept track of the service required for new cars that were sold. The following histogram indicates the number of times these cars required repairs due to mechanical failure over a period of seven years. Use the histogram to answer each question.

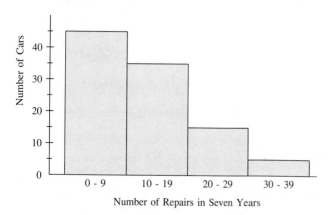

A. How many cars required between 10 and 19 repairs?
B. How many cars required fewer than 20 repairs?

C. How many cars required between 20 and 39 repairs?
A. 35   B. 80   C. 20

1. How many cars required between 0 and 9 repairs?
2. How many cars required between 20 and 29 repairs?
3. How many cars required between 30 and 39 repairs?
4. How many cars required more than 29 repairs?
5. How many cars required between 10 and 29 repairs?
6. How many cars required between 0 and 19 repairs?

Example Complete the following table to determine the frequency of each class interval for the following:

150	132	146	90	126	138
158	120	122	140	152	142
128	110	96	102	148	116

Each number represents the weight in pounds of high school freshman boys.

A. Complete the table for 90–99 lb.
B. Complete the table for 100–109 lb.

Weight in Pounds Class Interval	Tally	Frequency
90–99	||	2
100–109	|	1
110–119		
120–129		
130–139		
140–149		
150–159		

Complete the table for

7. 110–119 lb
8. 120–129 lb
9. 130–139 lb
10. 140–149 lb
11. 150–159 lb

Section 8.4

Find the mean (average) in each situation. Round to the nearest tenth when necessary.

Example A car salesman sold the following number of cars:

Sept.	Oct.	Nov.	Dec.	Jan.	Feb.
12	18	20	24	16	12

$$\frac{12 + 18 + 20 + 24 + 16 + 12}{6} = \frac{102}{6} = 17$$

The car salesman sold an average of 17 cars per month.

1. A student received grades of 79, 82, 83, 91, and 89.
2. A telephone operator had the following number of phone calls:

Mon.	Tues.	Wed.	Thurs.	Fri.	Sat.	Sun.
100	120	115	90	135	100	110

3. The last six houses sold in town were sold for $80,000, $105,000, $95,000, $82,000, $90,000, and $88,000.
4. Joe received the following miles per gallon rating on his car during the last six months: 21, 23, 26, 28, 27 and 24.
5. Wendy has taken four three-credit courses at college each of four semesters. Her semester grade point average for each of 4 semesters is 2.45, 2.61, 3.10 and 3.15.

Example Jerry and Sara traveled to the east coast during the summer. The number of miles they drove and the number of gallons of gas used are recorded below.

	Day 1	Day 2	Day 3	Day 4
Miles Driven	320	360	285	340
Gallons of Gas	13	15	11	14

Find the average miles per gallon achieved by the car on the trip by dividing the total number of miles driven by the total number of gallons used.

Total miles $320 + 360 + 285 + 340 = 1305$

Total gallons used $13 + 15 + 11 + 14 = 53$

$\dfrac{\text{Total miles}}{\text{total gas}} = \dfrac{1305}{53} = 24.6$ rounded to the nearest tenth

6. John had the following scores after practicing his bowling:

	Practice 1	Practice 2	Practice 3	Practice 4
Pins	562	540	760	621
Games	3	3	4	3

Find his bowling average by dividing the total number of pins scored by the total number of games.

7. The Sanson family traveled to Florida this fall. The number of miles they drove and the number of gallons of gas used are recorded below.

	Day 1	Day 2	Day 3	Day 4
Miles Driven	275	310	390	325
Gallons of Gas	11	13	17	14

Find the average miles per gallon achieved by the car on the trip by dividing the total number of miles driven by the total number of gallons used.

8. A player on a college baseball team had the following results:

	Game 1	Game 2	Game 3	Game 4	Game 5
Hits	3	0	4	1	2
Times at Bat	5	3	6	4	3

Find his batting average by dividing his total number of hits by his total times at bat. (Assume that he did not walk or sacrifice hit during any time at bat.)

Find the median value of each of the following problems.

Example 0.42, 0.58, 0.73, 0.31, 0.69

Arrange the numbers in order from smallest to largest.

0.31, 0.42,   0.58,   0.69, 0.73

two numbers   middle number   two numbers

9. 32, 26, 49, 57, 46, 52, 39
10. 16, 32, 21, 47, 59, 19, 27
11. 1234, 768, 1331, 897, 1051, 799
12. 0.35, 0.52, 0.21, 0.78, 0.69, 0.45, 0.66
13. 0.33, 0.10, 0.21, 0.19, 0.69

Example   The annual salaries of the employees of the company are $18,500, $19,700, $11,200, $10,300, $15,700, and $13,600.

Arrange in order:

$10,300, $11,200 $13,600, $15,700 $18,500, $19,700

two numbers   two middle   two numbers

Average mean of 13,600 and 15,700 is

$$\frac{13,600 + 15,700}{2} = \frac{29,300}{2} = \$14,650$$

The median value is $14,650.

Find the median value of each of the following:

14. The cost of a color television set priced at eight different

stores in the city is $315, $389, $420, $305, $375, $435, $359, and $299.

15. The number of students who tried out for the college softball team over the last six years was 16, 18, 22, 14, 29, and 25.

16. The cost of six cars recently purchased by the M & G Company are $17,250, $15,400, $18,970, $11,600, $9,750, and $13,600.

17. The number of overtime hours worked last month by employees of Jon's Auto Shop are 16, 21, 12, 9, 31, 26, 25, 19, and 13.

18. The number of minutes spent on the phone per day by each of seven receptionists is 13 minutes, 26 minutes, 41 minutes, 32 minutes, 19 minutes, 29 minutes and 21 minutes.

Section 8.5

Example   Is 64 a perfect square? If so, find $\sqrt{64}$.
Yes; the $\sqrt{64}$ is 8 because $(8)(8) = 64$.

1. Is 121 a perfect square? If so, find $\sqrt{121}$.
2. Is 169 perfect square? If so, find $\sqrt{169}$.
3. Is 1 a perfect square? If so, find $\sqrt{1}$.
4. Is 256 a perfect square? If so, find $\sqrt{256}$.
5. Is 144 a perfect square? If so, find $\sqrt{144}$.

Evaluate the square roots first, then add or subtract the results. Do not use a calculator or square root table.

Example $\sqrt{25} + \sqrt{0}$

$$\sqrt{25} = 5 \qquad \sqrt{0} = 0$$

$$5 + 0 = 5$$

6. $\sqrt{1} + \sqrt{64}$
7. $\sqrt{25} + \sqrt{169}$
8. $\sqrt{4} + \sqrt{49}$
9. $\sqrt{121} - \sqrt{9}$
10. $\sqrt{225} + \sqrt{100}$

Use a table of square roots or a calculator with a square root key to approximate the following to the nearest thousandth.

Example $\sqrt{18}$

Using a calculator:

$$\sqrt{18} = 4.24264 \quad \text{rounded to the nearest thousandth}$$

$$= 4.243$$

11. $\sqrt{32}$
12. $\sqrt{77}$
13. $\sqrt{105}$
14. $\sqrt{146}$
15. $\sqrt{79}$

Find the length of the side of the following squares. If the area is not a perfect square, approximate by using a square root table or a calculator with a square root key. Round your answer to the nearest thousandth.

Example   A square with area 96 m²

$$\sqrt{96} = 9.798$$

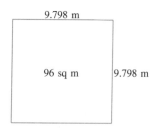

9.798 m

96 sq m   9.798 m

16. A square with area 16 m²
17. A square with area 100 m²
18. A square with area 31 m²
19. A square with area 76 m²
20. A square with area 109 yd²

Section 8.6

Find the missing side of each triangle to the nearest thousandth with the values given.

Example   Leg = 12 km, hypotenuse = 20 km

$$\text{Leg} = \sqrt{\text{hypotenuse}^2 - \text{leg}^2}$$
$$= \sqrt{20^2 - 12^2} = \sqrt{400 - 144} = \sqrt{256} = 16 \text{ km}$$

1. Leg = 5 km, hypotenuse = 13 km
2. Leg = 11 yd, hypotenuse = 17 yd
3. Leg = 12 m, leg = 4 m
4. Leg = 7 cm, leg = 9 cm
5. Leg = 8 m, leg = 8 m

Example   To strengthen a wall a 20-ft wire is tied from the top of the wall to a pin in the ground. The base of the wall is 16 ft from the pin. How far up on the wall is the wire?

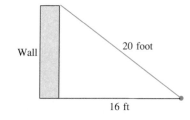

Wall   20 foot

16 ft

$$\text{Leg} = \sqrt{\text{hypotenuse}^2 - \text{leg}^2}$$
$$= \sqrt{20^2 - 16^2} = \sqrt{400 - 256} = \sqrt{144} = 12 \text{ ft}$$

6. A rectangle measures 8 cm × 6 cm. How long is its diagonal?

7. Find the length of the guy wire supporting the telephone pole.

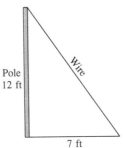

8. Find the length of this ramp.

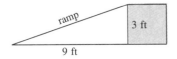

9. A car goes 6 mi west and 2 mi south. How far is the car from its starting point?

10. To strenghten a wall a 25-ft rope is tied from the top of the wall to a pin in the ground. The base of the wall is 20 ft from the pin. How far up on the wall is the rope?

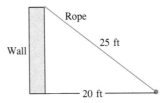

CHAPTER ORGANIZER

Topic	Procedure	Examples
Circle graphs, p. 416	The percentage of the 200 police officers within a given age range is illustrated. The following circle graph describes the age of the 200 men and women of the Glover City police force. *Under 23 10%* — *Over 50 years old 12%* — *Age 23-32 years 30%* — *Age 32-50 years 48%*	**1.** What percent of the police force is between 23 and 32 years old? 30% **2.** How many men and women in the police force are over 50 years old? 12% of 200 = (0.12)(200) = 24 people
Bar graphs and double-bar graphs, p. 421	The following double-bar graph illustrates the sales of color television sets by a major store chain for 1989 and 1990 in three regions of the country. *Number of Televisions Sold (in thousands)* — West Coast, Midwest, East Coast — 1989, 1990	**1.** How many color television sets were sold by the chain on the east coast in 1990? 6000 sets **2.** How many *more* color television sets were sold in 1990 than in 1989 on the west coast? 3000 sets were sold in 1990; 2000 sets were sold in 1989. 3000 − 2000 ———— 1000 sets more in 1990

Topic	Procedure	Examples
Line graphs and comparison line graphs, p. 422–23	The following line graph indicates the number of visitors to Wetlands State Park during a four-month period in 1989 and 1990. 	1. How many visitors came to the park in July of 1989? 3000 visitors 2. In what months were there more visitors in 1989 than in 1990? September and October 3. The sharpest *decrease* in attendance took place between what two months? Between August 1990 and September 1990
Histograms, p. 426	The following histogram indicates the number of quizzes in a math class that scored within each interval on a 15-point quiz. 	1. How many quizzes had a score between 8 and 11? 20 quizzes 2. How many quizzes had a score of less than 8? $12 + 6 = 18$ quizzes
Finding the mean, p. 432	The *mean* of a set of values is the sum of the values divided by the number of values. The *mean* is often called the average.	1. Find the mean of these numbers: 19, 13, 15, 25, and 18. $$\frac{19 + 13 + 15 + 25 + 18}{5} = \frac{90}{5} = 18$$ The mean is 18.

Finding the median, p. 432	1. Arrange the numbers in order from smallest to largest.
	2. If there are an odd number of values, the middle value is the median.
	3. If there are an even number of values, the average of the two middle values is the median.
		1. Find the median of 19, 29, 36, 15, and 20.
		  First we arrange in order from smallest to largest: 15, 19, 20, 29, 36.
		
		15, 19 — two numbers 20 — middle number 29, 36 — two numbers
		
		The median is 20.
		2. Find the median of 67, 28, 92, 37, 81, and 75.
		  First we arrange in order from smallest to largest: 28, 37, 67, 75, 81, 92. There are an even number of values.
		
		28, 37 67, 75 — two middle numbers 81, 92
		
		$$\frac{67 + 75}{2} = \frac{142}{2} = 71$$
		The median is 71.

Evaluating square roots of numbers that are perfect squares, p. 436	The square root of a number is one of two identical factors of that number.	$\sqrt{0} = 0$ because $(0)(0) = 0$ $\sqrt{4} = 2$ because $(2)(2) = 4$ $\sqrt{100} = 10$ because $(10)(10) = 100$ $\sqrt{169} = 13$ because $(13)(13) = 169$

Topic	Procedure	Examples
Approximating the square root of a number that is not a perfect square, p. 438	1. If a calculator with a square root key is available, enter the number and then press the $\boxed{\sqrt{x}}$ or $\boxed{\sqrt{}}$ key. The approximate value will be displayed. 2. If using a square root table, find the number n then look for the square root of that number. The approximate value will be correct to the nearest thousandth. | Number, n	Square Root of That Number, $\sqrt{n}$
---	---
31	5.568
32	5.657
33	5.745
34	5.831		1. Find on a calculator: (a) $\sqrt{13}$ (b) $\sqrt{182}$ Round to the nearest thousandth. (a) 13 $\boxed{\sqrt{x}}$ 3.60555127 rounds to 3.606 (b) 182 $\boxed{\sqrt{x}}$ 13.49073756 rounds to 13.491 2. Find from a square root table (a) $\sqrt{31}$ (b) $\sqrt{33}$ (c) $\sqrt{34}$ To the nearest thousandth, the approximate values are (a) $\sqrt{31} = 5.568$ (b) $\sqrt{33} = 5.745$ (c) $\sqrt{34} = 5.831$
Finding the hypotenuse of a right triangle when given the length of each leg, p. 441	Hypotenuse $= \sqrt{(\text{leg})^2 + (\text{leg})^2}$ 	Find the hypotenuse of a triangle with sides of 9 m and 12 m, respectively. hypotenuse $= \sqrt{(12)^2 + (9)^2} = \sqrt{144 + 81} = \sqrt{225}$ $= 15$ m
Finding the leg of a right triangle when given the length of the other leg and the hypotenuse, p. 441	Leg $= \sqrt{(\text{hypotenuse})^2 - (\text{leg})^2}$	Find the leg of a right triangle. The hypotenuse is 14 in. and one leg is 12 in. Round your answer to nearest thousandth. Leg $= \sqrt{(14)^2 - (12)^2} = \sqrt{196 - 144} = \sqrt{52}$ Using a calculator or a square root table, leg ≈ 7.211 in.
Solving applied problems involving the Pythagorean theorem, p. 442	1. Read the problem carefully. 2. Draw a sketch. 3. Label the two sides that are given. 4. If the hypotenuse is unknown, use hypotenuse $= \sqrt{(\text{leg})^2 + (\text{leg})^2}$ 5. If one leg is unknown, use leg $= \sqrt{(\text{hypotenuse})^2 - (\text{leg})^2}$	A boat travels 5 mi south and then 3 mi east. How far is it from the starting point? Round your answer to the nearest tenth. Hypotenuse $= \sqrt{(5)^2 + (3)^2} = \sqrt{25 + 9} = \sqrt{34}$ Using a calculator or a square root table, distance is approximately 5.8 mi.
The special 30°–60°–90° right triangle, p. 443	The length of the leg opposite the 30° angle is $\frac{1}{2}$ × the length of the hypotenuse.	Find y. $y = \frac{1}{2}(26) = 13$ m
The special 45°–45°–90° right triangle, p. 443	The sides opposite the 45° angles are equal. The hypotenuse is $\sqrt{2}$ × the length of either leg.	Find z. $z = \sqrt{2}(13) = (1.414)(13) = 18.382$ m

A student found that there were a total of 120 personal computers owned by students in the dormitory. The following circle graph displays the distribution of manufacturers of these computers. Use the graph to answer questions 1–8.

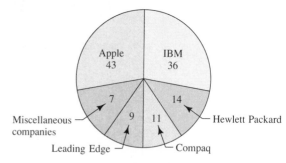

1. How many personal computers were manufactured by IBM?

2. How many personal computers were manufactured by Apple?

3. How many personal computers were manufactured by Leading Edge or Compaq?

4. How many personal computers were manufactured by Hewlett-Packard or miscellaneous companies?

5. What is the *ratio* of the number of computers manufactured by IBM to the number of computers manufactured by Leading Edge?

6. What is the *ratio* of the number of computers manufactured by Hewlett-Packard to the number of computers manufactured by miscellaneous companies?

7. What *percent* of the 120 computers are manufactured by Leading Edge?

8. What *percent* of the computers are manufactured by IBM?

Nancy and Wally Worzowski's family monthly budget of $2400 is displayed in the following circle graph. Use the graph to answer questions 9–16.

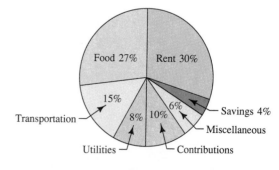

9. What percent of the budget is alloted for transportation?

10. What percent of the budget is alloted for savings?

11. What percent of the budget is used up by the food and rent categories?

12. What percent of the budget is used up by the transportation, utilities, and savings categories?

13. Of the total $2400, how much money per month is budgeted for utilities?

14. Of the total $2400, how much money per month is budgeted for transportation?

15. Of the total $2400, how much money per month is budgeted for the rent and savings categories?

16. Of the total $2400, how much money per month is budgeted for the transportation and food categories?

The following double-bar graph illustrates the number of customers at Reid's Steak House for each quarter for the years 1990 and 1991. Use the graph to answer questions 17–24.

17. How many customers came to the restaurant in the second quarter of 1990?

18. How many customers came to the restaurant in the third quarter of 1991?

19. When did the restaurant have the greatest number of customers?

20. When did the restaurant have the fewest number of customers?

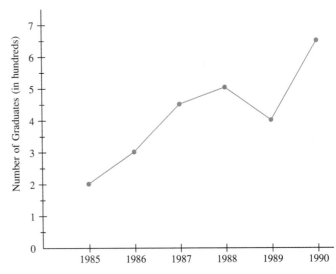

21. By how much did the number of customers increase from the first quarter of 1990 to the first quarter of 1991?

22. By how much did the number of customers increase from the fourth quarter of 1990 to the fourth quarter of 1991?

23. During the third quarter (July, August, September) there was road construction in front of the restaurant in both 1990 and 1991. Does the graph indicate the possibility that this might have caused a drop in the number of customers?

24. In the second quarter (April, May, June), the owner spent less on advertising in 1991 than in 1990. Does the graph suggest that this change might have caused a drop in the number of customers?

The following line graph shows the number of graduates of Williams University during the last 6 years. Use the graph to answer questions 25–32.

25. How many Williams University students graduated in 1989?

26. How many Williams University students graduated in 1988?

27. How many Williams University students graduated in 1990?

28. How many Williams University students graduated in 1987?

29. How many more Williams University students graduated in 1986 than 1985?

30. How many more Williams University students graduated in 1988 than in 1987?

31. Between what two years did the number of graduates decline?

32. Between what two years did the number of graduates increase by the greatest amount?

The following comparison line graph shows the number of ice cream cones purchased at the Junction Ice Cream Stand during a five-month period in 1989 and in 1990. Use this graph to answer questions 33–40.

33. How many ice cream cones were purchased in July 1990?

34. How many ice cream cones were purchased in August 1989?

35. How many more ice cream cones were purchased in May 1989 than in May 1990?

36. How many more ice cream cones were purchased in August 1990 than in August 1989?

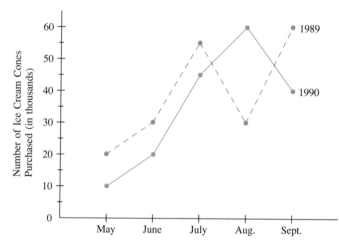

37. How many more ice cream cones were purchased in July 1989 than in June 1989?

38. How many more ice cream cones were purchased in September 1989 than in August 1989?

39. In the location of the ice cream stand, August 1989 was cold and rainy. The months of May, June, July, and September of 1989 were warm and sunny. What trend on the graph do you think is dependent on the weather?

40. July and August of 1990 were warm and very sunny, while May and June of 1990 were cloudy in the location of the ice cream stand. What trend on the graph do you think is dependent on the weather?

The state highway department made a survey of the bridges over all of its county and state highways. The number of bridges of each age interval is displayed in the following histogram. Use the histogram to answer questions 41–46.

41. How many bridges are between 40 and 59 years old?

42. How many bridges are between 60 and 79 years old?

43. The greatest number of bridges in the state are between _____ and _____ years old.

44. The highway Commissioner has ordered an immediate inspection of all bridges older than 79 years old. How many bridges are involved?

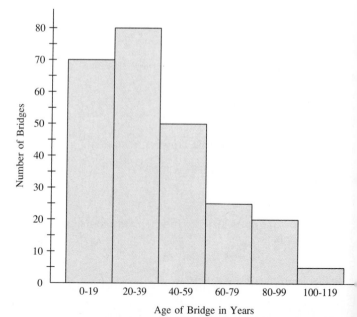

45. All bridges less than 40 years old have had a recent inspection by the highway department. How many bridges were involved?

46. How many more bridges in the state are in the category 60–79 years old than are in the category 80–99 years old?

Complete a table to determine the frequency of each class interval of the following data: Each number tells how many defective televisions were identified daily in the production of 400 new color television sets by a major manufacturer during the last 28 days.

3	5	8	2	0	1	7
13	6	3	4	1	8	12
0	2	16	5	7	13	14
12	10	17	5	4	0	3

Number of Defects

	Class Interval	Tally	Frequency
47.	0–3		
48.	4–7		
49.	8–11		
50.	12–15		
51.	16–19		

52. Construct a histogram using the table prepared in problems 47–51.

53. Based on the data of problems 47–51 how often were between 0 and 7 defective television sets identified in the production?

*Find the **mean value** (average) for each of the following items.*

54. The last seven-day maximum temperature readings in Los Angeles in July were: 86°, 83°, 88°, 95°, 97°, 100°, 81°.

55. The amount of groceries purchased by the Michael Stallard family each week for the last seven weeks was: $87, $105, $89, $120, $139, $160, $98.

56. The number of college textbooks purchased by each man of Jenkins House during his four years of college was: 176, 200, 191, 157, 142, 210, 175, 189.

57. The number of women students enrolled in the school of engineering at Westwood University during each of the last 10 years was: 151, 140, 148, 156, 183, 201, 205, 228, 231, 237.

58. The number of cars parked in the Central City garage for each of the last five days was: 1327, 1561, 1429, 1307, 1481.

59. The number of employees throughout the nation employed by Freedom Rent a Car for the last six years was: 882, 913, 1017, 1592, 1778, 1936.

*Find the **median value** for each of the following.*

60. The score on a recent mathematics exam: 69, 57, 100, 87, 93, 65, 77, 82, 88.

61. The number of students taking Abnormal Psychology for the fall semester for the last nine years at Elmson College: 77, 83, 91, 104, 87, 58, 79, 81, 88.

62. The cost of eight trucks purchased by the Highway Department: $28,500; $29,300; $21,690; $35,000; $37,000; $43,600; $45,300; $38,600.

63. The cost of 10 houses recently purchased at Stillwater: $98,000; $150,000; $120,000; $139,000; $170,000; $156,000; $135,000; $144,000; $154,000; $126,000.

64. The number of cups of coffee consumed by each of the students of the 7:00 A.M. Biology III class during the last semester: 38, 19, 22, 4, 0, 1, 5, 9, 18, 36, 43, 27, 21, 19, 25, 20.

65. The number of deliveries made each day by the Northfield House of Pizza: 21, 16, 0, 3, 19, 24, 13, 18, 9, 31, 36, 25, 28, 14, 15, 26.

66. The eight tests taken by Wong Yin in Calculus last semester were 96, 98, 88, 100, 31, 89, 94, 98. Which is a better measure of his usual score, the *mean* or the *median?* Why?

67. The eight salespersons of People's Dodge sold the following number of cars last month: 13, 16, 8, 4, 5, 19, 15, 18, 39, 12. Which is a better measure of the usual sales of these salespersons, the *mean* or the *median?* Why?

Evaluate exactly.

68. $\sqrt{81}$

69. $\sqrt{64}$

70. $\sqrt{100}$

71. $\sqrt{121}$

72. $\sqrt{144}$

73. $\sqrt{225}$

74. $\sqrt{36} + \sqrt{0}$

75. $\sqrt{25} + \sqrt{1}$

76. $\sqrt{9} + \sqrt{4}$

77. $\sqrt{25} + \sqrt{49}$

Approximate using a square root table or a calculator with a square root key. Round your answer to the nearest thousandth when necessary.

78. $\sqrt{35}$

79. $\sqrt{45}$

80. $\sqrt{88}$

81. $\sqrt{76}$

82. $\sqrt{171}$

83. $\sqrt{180}$

Find the unknown side of the right triangle. If the answer cannot be obtained exactly, use a square root table or a calculator with a square root key. When necessary round your answer to the nearest hundredth.

84.

85.

86.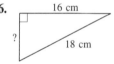

87.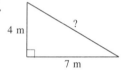

In problems 88–91, round your answer to the nearest tenth.

88. Find the distance between the centers of the holes of a metal plate with the dimensions labeled in this sketch.

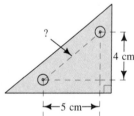

89. A building ramp has the following dimensions. Find the length of the ramp.

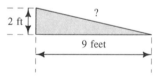

90. A shed is built with the following dimensions. Find the distance from the peak of the roof to the horizontal support brace.

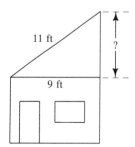

91. Find the width of a door if it is 6 ft tall and the diagonal line measures 7 ft.

The following circle graph displays the highway department budget for the town of Brentwood for the current year. The total budget is $430,000. Use the graph to answer questions 1–6.

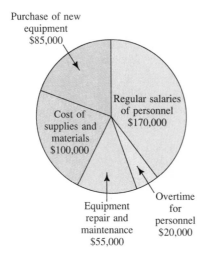

Purchase of new equipment $85,000

Cost of supplies and materials $100,000

Regular salaries of personnel $170,000

Equipment repair and maintenance $55,000

Overtime for personnel $20,000

1. What category takes the least amount of the budget?

2. What category takes the most amount of the budget?

3. How much money is alloted for equipment repair and maintenance?

4. How much money is alloted for the purchase of new equipment?

5. What is the ratio of money alloted for the purchase of new equipment to the money alloted for equipment repair and maintenance?

6. What is the ratio of money alloted for regular salaries of personnel to the total amount of money in the budget?

A local superior court analyzed the results of the most recent 150 criminal cases brought before the court in the following circle graph. Use the graph to answer questions 7–12.

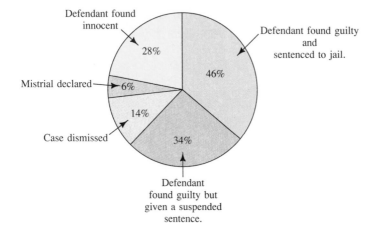

Defendant found innocent 28%

Defendant found guilty and sentenced to jail. 46%

Mistrial declared 6%

Case dismissed 14%

Defendant found guilty but given a suspended sentence. 34%

7. What percent of the cases were dismissed?

8. What percent of the cases had the defendant found guilty but given a suspended sentence?

1. _____

2. _____

3. _____

4. _____

5. _____

6. _____

7. _____

8. _____

9. _____

10. _____

11. _____

12. _____

13. _____

14. _____

15. _____

16. _____

9. In what percent of the cases was the defendant found guilty (consider all possibilities)?

10. In what percent of the cases was there no finding of guilt or innocence (consider all possibilities)?

11. Of the 150 criminal cases, *how many* resulted in a mistrial being declared?

12. Of the 150 criminal cases, *how many* resulted in the defendant being found innocent?

The following bar graph shows the out-of-state student population at Lyman State University for several recent years. Use the graph to answer questions 13–16.

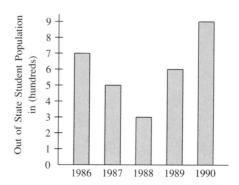

13. What was the smallest population of out-of-state students at the university?

14. How many out-of-state students attended the university in 1987?

15. According to this bar graph, by how much did the out-of-state student population increase between 1988 and 1989?

16. According to this bar graph, by how much did an out-of-state student population decrease between 1987 and 1988?

The following double-bar graph illustrates the number of new houses built in two cities for selected years. Use the graph to answer questions 17–20.

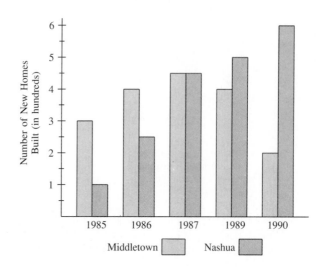

17. How many new houses were built in Nashua in 1986?

18. How many new houses were built in Middletown in 1989?

19. In what year were the *same number* of houses built in Nashua as in Middleton?

20. In what year was the difference between the number of new homes built in Nashua and the number of new homes built in Middleton the greatest?

The following line graph shows the Delta Company's profits during the last 5 years. Use the graph to answer questions 21–24.

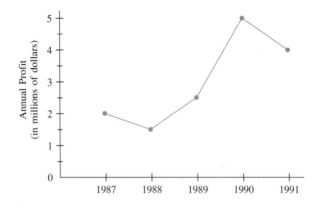

21. What was the profit in 1989?

22. What was the profit in 1988?

23. Between what two years was the most noticeable *decrease* in profit?

24. Between what two years was the most noticeable *increase* in profit?

The wildlife management team at two state wildlife preserves collected data on the population of deer at the two locations. The following comparison line graph summarizes the data. Use the graph to answer questions 25–30.

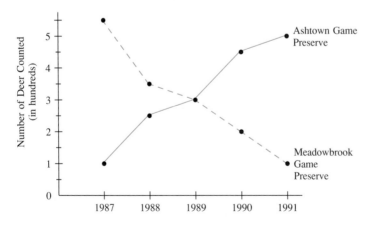

25. How many deer were counted at the Ashtown Game Preserve in 1988?

26. How many deer were counted at the Meadowbrook Game Preserve in 1987?

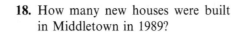

27. _____

28. _____

29. _____

30. _____

31. _____

32. _____

33. _____

34. _____

35. _____

36. _____

27. In what year did the two game preserves have the same count for the number of deer present?

28. In what year was the largest count of deer at the Meadowbrook Game Preserve?

29. The wildlife management teams feel that most of the shift in the population of deer is due to deer moving their natural habitat from the Meadowbrook Game Preserve to the Ashtown Game Preserve. Does this comparison line graph support that opinion?

30. Between what two years did the Meadowbrook Game Preserve experience the greatest *decrease* in the number of deer counted?

During a 1-hour period, a state police officer made the following observations of the number of cars traveling at a certain speed on a state turnpike. The posted speed limit is 65 mi/hr. Use the graph to answer questions 31–36.

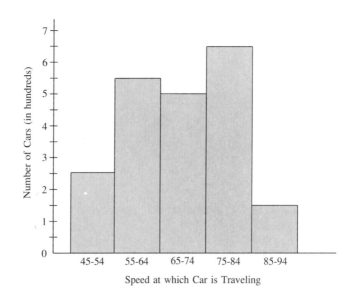

31. How many cars traveled between 55 and 64 mi/hr?

32. How many cars traveled between 65 and 74 mi/hr?

33. How many cars traveled faster than 74 mi/hr?

34. How many cars traveled slower than 65 mi/hr?

35. If the police are able to apprehend and give a speeding ticket to 90% of all drivers exceeding 84 mi/hr, how many tickets will they give out on this day?

36. If the police are able to apprehend and give a speeding ticket to 60% of all drivers exceeding 74 mi/hr, how many tickets will they give out on this day?

Find the **mean value** *(average) for each item.*

1. The last seven-day maximum daily temperature readings in Houston for August: 88°, 91°, 96°, 89°, 94°, 99°, 101°.

2. The last seven-day maximum daily temperature readings for Caribou, Maine for August: 39°, 45°, 47°, 52°, 38°, 60°, 48°.

3. The ages of 10 students in the 2:00 P.M. Basic Mathematics class at the college: 21, 19, 18, 36, 52, 43, 28, 30, 17, 18.

4. The number of audio tapes purchased by 10 students of Smithsville College during the last year: 12, 13, 26, 18, 9, 3, 4, 1, 42, 27.

5. The weekly salary of eight employees of Northshore Grinding Company: $303, $310, $560, $800, $260, $240, $385, $410.

6. The daily number of airplane flights out of Springfield Airport for the last eight days: 201, 203, 216, 245, 212, 230, 277, 260.

Find the **median value** *for each item.*

7. The scores on the last introduction to mathematics test: 88, 43, 60, 99, 24, 73, 86, 91, 90.

8. The scores on the first history test: 99, 57, 82, 77, 39, 69, 85, 84, 93.

9. The cost of each of the last eight new cars sold at Hillside Chevrolet: $15,600; $18,200; $19,365; $20,118; $12,300; $9,345; $7,520; $14,250.

10. The cost of room, board, and tuition for a year at the schools Robert applied to last year: $10,000; $2,945; $8,600; $23,100; $17,200; $13,150; $9980; $3945.

11. One share of stock in the company where Bob works was worth the following amounts during the first business day of each month of last year: $85, $92, $106, $105, $99, $112, $152, $88, $79, $78, $84, $89.

12. The number of hours of television watched by each member of the 12 members of Hale House during the fall semester: 120, 43, 21, 2, 36, 100, 150, 123, 77, 5, 85, 37.

Evaluate exactly.

13. $\sqrt{49}$ 14. $\sqrt{25}$ 15. $\sqrt{36}$ 16. $\sqrt{64}$

1. _____

2. _____

3. _____

4. _____

5. _____

6. _____

7. _____

8. _____

9. _____

10. _____

11. _____

12. _____

13. _____

14. _____

15. _____

16. _____

17.

18.

19.

20.

21.

22.

23.

24.

25.

26.

27.

28.

29.

30.

31.

32.

33.

34.

17. $\sqrt{1}$ **18.** $\sqrt{0}$ **19.** $\sqrt{144}$ **20.** $\sqrt{169}$

21. $\sqrt{9} + \sqrt{25}$ **22.** $\sqrt{81} + \sqrt{4}$ **23.** $\sqrt{100} + \sqrt{64}$ **24.** $\sqrt{121} + \sqrt{16}$

Approximate using the square root table on page 578 or by using a calculator with a square root key. Round your answer to the nearest thousandth when necessary.

25. $\sqrt{18}$ **26.** $\sqrt{23}$ **27.** $\sqrt{157}$ **28.** $\sqrt{161}$

Find the unknown side of the right triangle. If the answer cannot be obtained exactly, use a square root table or a calculator with a square root key. When necessary round to the nearest thousandth.

29.

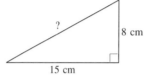

30.

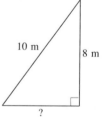

31.

32.

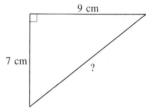

In problems 33 and 34, round your answer to the nearest tenth.

33. An airplane travels 8 mi north and then 5 mi west. How far is the plane from the starting point?

34. A support wire holds a telephone pole at a point 6 m above the ground. The support wire is 9 m long. How far is the bottom of the support wire from the base of the telephone pole?

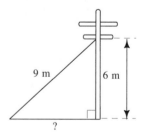

A State Highway Safety Commission recently reported the results of inspecting 200,000 automobiles. The following circle graph depicts the percent of automobiles that passed and the percent that had one or more safety violations. Use this graph to answer questions 1–5.

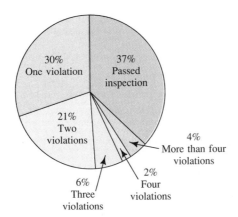

1. What percent of the automobiles passed inspection?

2. What percent of the automobiles had two safety violations?

3. What percent of the automobiles had more than two safety violations?

4. If 200,000 automobiles were inspected, *how many* of them had one safety violation?

5. If 200,000 automobiles were inspected, *how many* of them had either two violations or three violations?

The following double-bar graph indicates the number of cars sold at Boley's Chrysler during each quarter of 1989 and 1990. Use the graph to answer questions 6–11.

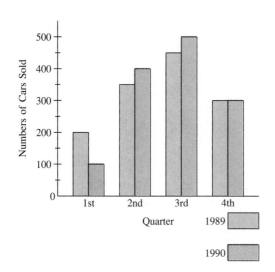

6. How many cars were sold in the second quarter of 1989?

7. How many cars were sold in the third quarter of 1990?

1. _____

2. _____

3. _____

4. _____

5. _____

6. _____

7. _____

8. _____

9. _____

10. _____

11. _____

12. _____

13. _____

14. _____

15. _____

16. _____

8. When were the fewest number of cars sold?

9. During which quarter were more cars sold in 1989 than in 1990?

10. How many more cars were sold in the second quarter of 1990 than in the second quarter of 1989?

11. How many more cars were sold in the third quarter of 1989 than in the fourth quarter of 1989?

A research study by 10 midwestern universities having a medical school produced the following preliminary information on the life expectancy of American men at various ages based on those who smoke and those who do not. The following comparison line graph depicts the approximate results. Use the graph to answer questions 12–16.

12. Approximately how long is a 45-year-old American man expected to live if he smokes?

13. Approximately how long is a 55-year-old American man expected to live if he does not smoke?

14. According to this graph, approximately how many more years is a 25-year-old man expected to live if he does not smoke compared to a 25-year-old man who does smoke?

15. According to this graph, at what age is the difference between life expectancy of a smoker and a non-smoker the greatest?

16. According to this graph, at what age is the difference between life expectancy of a smoker and a non-smoker the smallest?

The following histogram prepared by a consumer research group tells us how many color television sets lasted for a certain period of years. Use the histogram to answer questions 17–20.

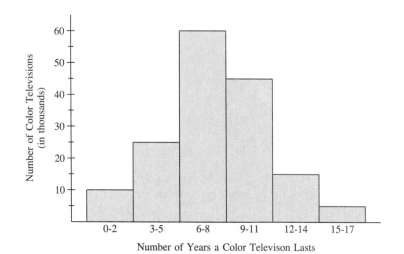

Number of Years a Color Televison Lasts

17. How many color television sets lasted 6–8 years?

18. How many color television sets lasted 3–5 years?

19. How many color television sets lasted more than 11 years?

20. How many color television sets lasted 9–14 years?

A chemistry student had the following scores on eight quizzes in her chemistry class: 20, 17, 16, 13, 19, 18, 15, 12.

21. Find the *mean* quiz score.

22. Find the *median* quiz score.

The monthly salary for seven workers at Richardson's Print Shop is as follows: $374, $450, $600, $720, $410, $930, $800.

23. Find the *median* monthly salary.

24. Find the *mean* monthly salary.

Evaluate exactly.

25. $\sqrt{81}$

26. $\sqrt{64}$

27. $\sqrt{1} + \sqrt{121}$

28. $\sqrt{0} + \sqrt{49}$

17. _____

18. _____

19. _____

20. _____

21. _____

22. _____

23. _____

24. _____

25. _____

26. _____

27. _____

28. _____

29. _____

30. _____

31. _____

32. _____

33. _____

34. _____

35. _____

Approximate using a square root table or a calculator with a square root key. Round your answer to the nearest thousandth when necessary.

29. $\sqrt{54}$

30. $\sqrt{120}$

31. $\sqrt{187}$

Find the unknown side of the right triangle. Use a calculator or a square root table to approximate square roots to the nearest thousandth.

32.
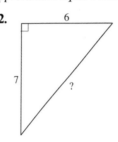

33.

34. Find the distance between the centers of the holes drilled in a rectangular metal plate with the dimensions labeled in this sketch.

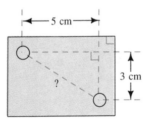

35. A 15-ft-tall ladder is placed so that it reaches 12 ft up on the wall of a house. How far is the base of the ladder from the wall of the house?

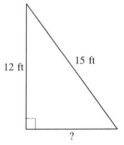

Approximately one-half of this test is based on Chapter 8 material. The remainder is based on material covered in Chapters 1–7.

1. Add: 1376 + 2804 + 9003 + 7642.

2. Multiply: 3004 × 26.

3. Subtract: $7\frac{1}{5} - 3\frac{3}{8}$.

4. Divide: $10\frac{4}{5} \div 3\frac{1}{2}$.

5. Round to the nearest hundredth: 2864.3719.

6. Subtract: $\begin{array}{r} 200.58 \\ -\,127.93 \end{array}$

7. Divide: 52.0056 ÷ 0.72.

8. Find n: $\dfrac{7}{n} = \dfrac{35}{3}$.

9. Of every 2030 cars manufactured, three have major engine defects. If in total 26,390 of these cars were manufactured, approximately how many would have major engine defects?

10. What is 1.3% of 25?

11. 72% of what number is 252?

12. Convert 198 cm to m.

13. Convert 18 yd to ft.

14. Find the area of a circle with radius of 3 in. Round your answer to the nearest tenth.

1. _____

2. _____

3. _____

4. _____

5. _____

6. _____

7. _____

8. _____

9. _____

10. _____

11. _____

12. _____

13. _____

14. _____

15. _____

15. Find the perimeter of a square with a side of 17 in.

The 12,000-member student body of Mason University consists of five groups: freshmen, sophomores, juniors, seniors, and graduate students. The distribution by category is displayed on the following graph.

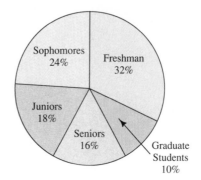

16. _____

16. What _percent_ are either juniors or seniors?

17. How _many_ of the 12,000 students are freshmen?

17. _____

The following double-bar graph indicates the quarterly profits for Dedalon Corporation for 1990 and 1991.

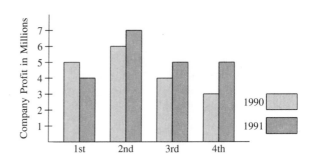

18. What was the quarterly profit for Dedalon Corporation in the fourth quarter of 1990?

18. _____

19. How much greater was the profit of Dedalon Corporation in the second quarter of 1991 than in the second quarter of 1990?

19. _____

The following comparison line graph depicts the annual rainfall in Dixville compared to the annual rainfall in Weston for five specific years.

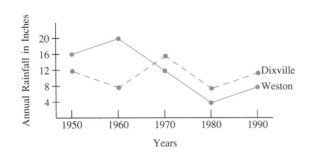

20. How many inches of rain fell in Dixville in 1970?

21. In what years was the annual rainfall in Weston greater than the annual rainfall in Dixville?

The following histogram depicts the number of students in the math course Basic Mathematics who fall in various age groups.

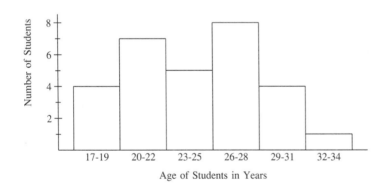

22. How many students are 26–28 years in age?

23. How many students are less than 26 years in age?

24. _____

25. _____

26. _____

27. _____

28. _____

29. _____

30. _____

The following are the hourly salaries of six employees of Delson's Pizza: $5.00, $4.50, $3.95, $4.90, $7.00, $13.65.

24. Find the *mean* hourly salary.

25. Find the *median* hourly salary.

26. Evaluate exactly: $\sqrt{36} + \sqrt{25}$.

27. Approximate to the nearest thousandth using a table or a calculator: $\sqrt{57}$.

Find the unknown side of the right triangle. Use a calculator or square root table if necessary. Round answers to nearest thousandth.

28.

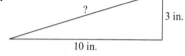

29.

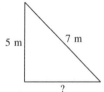

30. A boat travels 12 mi south and then 7 mi east. How far is it from its original starting point? Round your answer to nearest tenth of a mile.

Signed Numbers

With the advent of new technology used in law enforcement, police officers are increasingly met with the need to know more mathematics.

PRETEST CHAPTER 9

If you are familiar with the topics in this chapter, take this test now. Check your answers with those in the back of the book. If an answer was wrong or you couldn't do a problem, study the appropriate section of the chapter.

If you are not familiar with the topics in this chapter, don't take this test now. Instead, study the examples, work the practice problems, and then take the test.

This test will help you identify those concepts that you have mastered and those that need more study.

Section 9.1 Add.

1. $-8 + (-15)$ 2. $-12 + 7$
3. $8.2 + (-6.1)$ 4. $6 + (-7) + 3 + (-4)$
5. $\dfrac{4}{15} + \left(-\dfrac{3}{5}\right)$ 6. $-\dfrac{1}{12} + \left(-\dfrac{1}{2}\right)$
7. $-3.3 + (-6.1)$ 8. $-1.8 + 4.2$

Section 9.2 Subtract.

9. $11 - 16$ 10. $-12 - 19$
11. $\dfrac{5}{19} - \left(-\dfrac{3}{19}\right)$ 12. $-17 - (-3)$
13. $-5.3 - (-7.1)$ 14. $3.9 - 7.7$
15. $14 - (-14)$ 16. $\dfrac{3}{4} - \left(-\dfrac{1}{5}\right)$

Section 9.3 Multiply or divide.

17. $(-2)(-5)$ 18. $-20 \div (-10)$
19. $-36 \div 3$ 20. $(3)(-2)(5)(-1)\left(-\dfrac{1}{2}\right)$
21. $\dfrac{52}{-2}$ 22. $\dfrac{-\dfrac{2}{3}}{-\dfrac{3}{4}}$
23. $(-6)(-3)(-2)$ 24. $100 \div (-4)$

Section 9.4 Perform each operation in the proper order.

25. $27 \div (-3) + 26 \div (-2)$ 26. $12 - 3(4) + (-2) \div (-2)$
27. $5 + (-8) + 2(-3)$ 28. $5(-7) \div (-10)$
29. $7 - (-6) + 10 \div (-2)$ 30. $5(-2) + 3(-3) - (-4)$
31. $\dfrac{7 + 6 - 2}{(-8)(-3) - 2}$ 32. $\dfrac{40 \div (-8) + 1}{(2)(-5) + 2(-1)}$

□ After studying this section, you will be able to:

1 *Add two signed numbers with the same sign*

2 *Add two signed numbers with opposite signs*

3 *Add three or more signed numbers*

9.1 ADDITION OF SIGNED NUMBERS

Think of a line in which each point is paired with a number. This **number line** has a point matched with zero and with each of the whole numbers, fractions, and decimals we have been using thus far in Chapters 1–8. In this chapter we enlarge the set of numbers we work with. We include "new numbers" on the other side of zero on the number line.

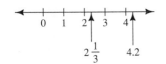 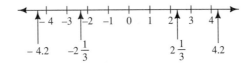

The numbers to the right of zero are positive. The numbers to the left of zero are negative. All these numbers are called **signed numbers**.

Why enlarge our set of numbers? You'll see some reasons in the next two chapters. In this chapter you'll learn how to compute with these "new numbers," and in the following chapter you'll see some of the many practical problems that can now be solved with them in a branch of mathematics called algebra.

1 When we look at this thermometer we see a temperature reading $-20°F$.

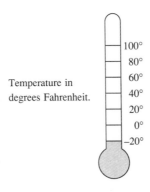

Temperature in degrees Fahrenheit.

In this report of operating profit of four airlines we see examples of both positive and negative numbers. 100, 200, 300 are examples of positive numbers. They represent money gained. -100, -200, -300 are negative numbers. They represent money lost.

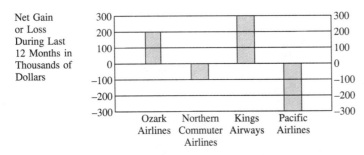

Net Gain or Loss During Last 12 Months in Thousands of Dollars

In the last 12 months a value of $-100,000$ is recorded for Northern Commuter Airlines, which means that it lost \$100,000 during the year. A value of $-300,000$ is recorded for Pacific Airlines. It lost \$300,000 during the year.

In addition to showing positive numbers *up* on a thermometer or profit-and-loss graph, and negative numbers *down*, we can show these positive numbers to the *right* and negative numbers to the *left*. Usually, numbers are displayed on a **number line**. **Positive numbers** are to the right of zero on the number line. **Negative numbers** are to the left of zero on the number line. Zero is considered neither positive or negative.

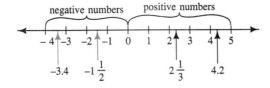

Positive numbers could be written with a plus sign—for example, 2 could be written $+2$—but this is not usually done. Negative numbers, such as -3 or -2, must always have the negative sign so that we know it is a negative number. In the number 2, the *sign* of the number is positive (although it is not written). In the number -2, the *sign* of the number is negative.

> The **absolute value** of a number is the distance between that number and zero on the number line.

This distance is always a positive number, regardless of the direction we travel on the number line. From our definition we see that the absolute value of any number will be a positive number or zero. When we find an absolute value we use the symbol "| |" around the number. The distance from 0 to 5 is 5. Thus $|5| = 5$. This is read "the absolute value of 5 is 5." The distance from 0 to -5 is 5. Thus $|-5| = 5$. This is read "the absolute value of -5 is 5."

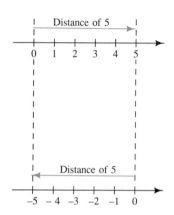

Consider these examples of absolute values:

$$|6| = 6 \qquad |-3| = 3$$

$$|7.2| = 7.2 \qquad \left|-\frac{1}{5}\right| = \frac{1}{5}$$

$$|0| = 0 \qquad |-26| = 26$$

Do you see a pattern? The absolute value of a positive number is always the number itself. The absolute value of zero is, of course, zero. The absolute value of a negative number is that number without the negative sign.

The idea of absolute value will help us work with signed numbers. Let's consider numbers with the same sign. The numbers 7 and 5 have the same sign. They both have a positive sign (not written). The numbers -6 and -2 have the same sign. They both have a negative sign.

Suppose that we earn $52 one day and earn $38 the next day. To learn what our 2-day total is, we add the positive numbers.

We earn

$$(+\$52) + (+\$38) = +\$90$$

The money is coming in, and the plus sign records a gain.

Suppose that we consider money spent as negative dollars. If we spend $52 one day ($-\$52$) and we spend $38 the next day ($-\$38$), we must add two negative numbers. What is our financial position?

$$(-\$52) + (-\$38) = -\$90$$

We have spent $90. The negative sign tells us the direction of the money: out!

Addition Rule for Two Numbers with Same Sign

To add two numbers with the same sign:
1. Add the absolute value of the numbers.
2. Use the common sign in the answer.

Example 1

(a) Add: $7 + 5$. **(b)** Add: $-6 + (-2)$.

(a)
$$\begin{array}{r} 7 \\ +\,5 \\ \hline 12 \end{array}$$

We add the absolute value of the numbers 7 and 5. The positive sign, although not written, is common to both numbers. The answer is a positive 12. (The $+$ sign is not written.)

(b)
$$\begin{array}{r} -6 \\ +\,-2 \\ \hline -8 \end{array}$$

We add the absolute value of the numbers 6 and 2 without regard to sign. Why is the answer -8?

We use a negative sign in our answer because the negative sign is common to both numbers. ■

Practice Problem 1

(a) Add: $9 + 14$.

(b) Add: $-7 + (-3)$. ■

Example 2 Add: $-3.2 + (-5.6)$.

$$
\begin{array}{r}
-3.2 \\
+ \ -5.6 \\
\hline
-8.8
\end{array}
$$ We add the absolute value of the numbers 3.2 and 5.6.

We use a negative sign in our answer because we added two negative numbers. ■

Practice Problem 2 Add: $-4.5 + (-1.9)$. ■

These rules can be applied to fractions as well.

Example 3 Add: $-\dfrac{3}{13} + \left(-\dfrac{5}{13}\right)$.

$$
\begin{array}{r}
-\dfrac{3}{13} \\[2mm]
+ \ -\dfrac{5}{13} \\[1mm]
\hline
-\dfrac{8}{13}
\end{array}
$$ The common sign here is a negative sign. ■

Practice Problem 3 Add: $-\dfrac{2}{15} + \left(-\dfrac{12}{15}\right)$. ■

Example 4

(a) Add: $\dfrac{5}{18} + \dfrac{1}{3}$.

(b) Add: $-\dfrac{1}{7} + \left(-\dfrac{3}{5}\right)$.

(a) The LCD = 18. The first fraction already has the LCD.

$$
\begin{array}{r}
\dfrac{5}{18} = \dfrac{5}{18} \\[2mm]
+ \ \dfrac{1}{3} \cdot \dfrac{6}{6} = + \dfrac{6}{18} \\[1mm]
\hline
\dfrac{11}{18}
\end{array}
$$

We add two positive numbers, so the answer is positive.

(b) The LCD = 35.

$$\dfrac{1}{7} \cdot \dfrac{5}{5} = \dfrac{5}{35}$$

Because $\dfrac{1}{7} = \dfrac{5}{35}$ it follows that $-\dfrac{1}{7} = -\dfrac{5}{35}$.

$$\dfrac{3}{5} \cdot \dfrac{7}{7} = \dfrac{21}{35}$$

Because $\dfrac{3}{5} = \dfrac{21}{35}$ it follows that $-\dfrac{3}{5} = -\dfrac{21}{35}$. Thus

$$
\begin{array}{r}
-\dfrac{1}{7} \\[2mm]
+ \ -\dfrac{3}{5} \\
\hline
\end{array}
$$ is equivalent to $$
\begin{array}{r}
-\dfrac{5}{35} \\[2mm]
+ \ -\dfrac{21}{35} \\[1mm]
\hline
-\dfrac{26}{35}
\end{array}
$$

We add two negative numbers, so the answer is negative. ■

Practice Problem 4

(a) Add: $\dfrac{5}{12} + \dfrac{1}{4}$.

(b) Add: $-\dfrac{1}{6} + \left(-\dfrac{2}{7}\right)$. ■

Example 5 A company lost $50,000 in January and lost $30,000 in February. What is the profit or loss situation for the company after two months?

$$
\begin{array}{rl}
-\$50,000 & \text{January loss}\\
+\ -\$30,000 & \text{February loss}\\
\hline
-\$80,000 & \blacksquare
\end{array}
$$

Practice Problem 5 The company lost $65,000 in March and lost $40,000 in April. What is the profit or loss statement for the company after these two months? ■

2 Sometimes we need to add numbers with different signs, like $6 + (-2)$ or $-12 + 4$.

Addition Rule for Two Numbers with Different Signs

To add two numbers with different signs:
1. Find the difference between the absolute value of the larger number and the absolute value of the smaller number.
2. Place the sign of the number with the larger absolute value in front of the difference.

Example 6

(a) Add: $8 + (-10)$.

(b) Add: $-8 + 10$.

(a)
$$
\begin{array}{r}
8\\
+\ -10\\
\hline
-2
\end{array}
$$

The absolute value of 8 is 8: $|8| = 8$. The absolute value of -10 is 10: $|-10| = 10$. The difference between the two absolute values 10 and 8 is 2. Because the absolute value 10 is larger than the absolute value 8, we place the sign of the number whose absolute value is 10 in front of the difference. In this example, the number whose absolute value is 10 is -10, so we place a negative sign in front of the difference. The answer is negative.

(b)
$$
\begin{array}{r}
-8\\
+\ \ 10\\
\hline
2
\end{array}
$$

In this case, $|-8| = 8$ and $|10| = 10$. The difference between 10 and 8 is 2. Because 10 is larger in absolute value and 10 is positive, the answer is positive. (The $+$ sign is not written.) ■

Practice Problem 6

(a) Add: $7 + (-12)$.

(b) Add: $-7 + 12$. ■

Example 7

(a) Add: $-16.6 + 12.3$.

(b) Add: $16.6 + (-12.3)$.

In both cases we see that 16.6 is larger than 12.3 and that the difference between 16.6 and 12.3 is 4.3.

(a)
$$
\begin{array}{r}
-16.6\\
+\ \ \ 12.3\\
\hline
-4.3
\end{array}
$$
$\uparrow$

Because the sign of the number with the larger absolute value is negative, the answer is negative.

(b)
$$\begin{array}{r} 16.6 \\ +\ -12.3 \\ \hline 4.3 \end{array}$$
↑

Because the sign of the number with the larger absolute value is positive, the answer is positive. ■

Practice Problem 7
(a) Add: $-20.8 + 15.2$ **(b)** Add: $20.8 + (-15.2)$. ■

The commutative property of addition holds for signed numbers also. Thus $a + b = b + a$ when a and b are positive or negative.

Commutative Property of Addition

For any real numbers a, b,
$$a + b = b + a$$

Example 8 Last night the temperature dropped to $-14°F$. From that low, today the temperature rose $33°F$. What was the highest temperature today?

We want to add $-14°F$ and $33°F$. Because addition is commutative, it does not matter if we add $-14 + 33$ or $33 + (-14)$. Do you see why?

$$\begin{array}{r} 33°F \\ +\ -14°F \\ \hline 19°F \end{array}$$
↑ The 33 is larger than 14. The difference between 33 and 14 is 19. The number with the larger absolute value is positive so the answer is positive. ■

Practice Problem 8 Last night the temperature dropped to $-19°F$. From that low, today the temperature rose $28°F$. What was the highest temperature today? ■

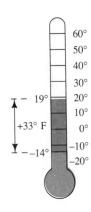

3 We can add three or more numbers at once using these rules.

Example 9 Add: $24 + (-16) + (-10)$.

We go from left to right, adding two numbers at a time.

Step 1:
$$\begin{array}{r} 24 \\ +\ -16 \\ \hline 8 \end{array}$$
↑ First we add $24 + (-16)$. The difference between 24 and 16 is 8. Because the number with the larger absolute value is positive, the answer is positive.

Step 2:
$$\begin{array}{r} 8 \\ +\ -10 \\ \hline -2 \end{array}$$
↑ Next we take the sum, which is 8, and add it to -10. The number with the larger absolute value is negative: $|-10| = 10$. The difference between 10 and 8 is 2. Because 10 is larger than 8, we will have a negative sign in the answer. ■

Practice Problem 9 Add: $36 + (-21) + (-18)$. ■

We may write the numbers to be added in any order. And it does not matter which two numbers are added first. This is called the associative property of signed numbers.

> **Associative Property of Addition**
>
> For any three real numbers a, b, c
> $$(a + b) + c = a + (b + c)$$

If there are many numbers to add, it may be easier to add the positive numbers and the negative numbers separately and then combine the results.

Example 10 Add: $-8 + 3 + 5 + (-7) + (-4) + 6.$

Let us add separately the positive numbers and the negative numbers.

$$
\begin{array}{rr}
-8 & 3 \\
-7 & 5 \\
+\ -4 & +\ 6 \\
\hline
-19 & 14
\end{array}
$$

Now we add together $-19 + 14$, which have opposite signs. The number with the larger absolute value is -19.

$$
\begin{array}{r}
-19 \\
+\quad 14 \\
\hline
-5
\end{array}
$$

> The difference between 19 and 14 is 5.
> 19 is larger than 14, so the answer is negative. ∎

Practice Problem 10 Add: $-5 + 3 + (-2) + 8 + (-7) + 9.$ ∎

Example 11 The results of a five-month operation of a new company is listed in this table. What is the overall profit or loss over the five-month period?

NET OPERATIONS
Profit/Loss Statement in Dollars

Month	Profit	Loss
January	30,000	
February		-50,000
March		-10,000
April	20,000	
May	15,000	

First we will add separately the positive numbers and the negative numbers.

$$
\begin{array}{rr}
30,000 & \\
20,000 & -50,000 \\
+\ 15,000 & +\ -10,000 \\
\hline
65,000 & -60,000
\end{array}
$$

Now we add the positive number $\$65,000$ and the negative number $-\$60,000.$

$$
\begin{array}{r}
\$65,000 \\
+\ -\$60,000 \\
\hline
\$\ 5,000
\end{array}
$$

The company had an overall profit of $5000 for the five-month period. ∎

Practice Problem 11 The results of the next five-month operation of a new company are listed in the table. What is the overall profit or loss over the five-month period?

NET OPERATIONS
Profit/Loss Statement in Dollars

Month	Profit	Loss
June		− 20,000
July	30,000	
August	40,000	
September		− 5,000
October		− 35,000

∎

To Think About

?

Let us verify the commutative law of addition for signed numbers. For any signed numbers a and b,

$$a + b = b + a$$

Is this true when $a = -3.0$ and $b = -5.5$?

$$-3.0 + (-5.5) \overset{?}{=} -5.5 + (-3.0)$$
$$-8.5 = -8.5 \checkmark \quad \text{Yes}$$

Is this true when $a = 6.5$ and $b = -8.0$?

$$6.5 + (-8.0) \overset{?}{=} -8.0 + 6.5$$
$$-1.5 = -1.5 \checkmark \quad \text{Yes}$$

It is a little harder to verify the associative law of addition for signed numbers. The associative law says that for any signed numbers a, b, c,

$$(a + b) + c = a + (b + c)$$

The two numbers inside the parentheses must be added first. Verify that it is true when $a = 6.2$, $b = -3.4$, $c = -7.0$. Notice that we use brackets for the outer parentheses.

$$[6.2 + (-3.4)] + (-7.0) \overset{?}{=} 6.2 + [(-3.4) + (-7.0)]$$
$$2.8 + (-7.0) \overset{?}{=} 6.2 + (-10.4)$$
$$-4.2 = -4.2 \checkmark \quad \text{Yes}$$

Problems 55–58 will give you a chance to verify the laws for additional values. ∎

EXERCISES 9.1

Add each pair of signed numbers, which have the same sign.

1. $8 + 6$

2. $7 + 9$

3. $-2 + (-5)$

4. $-8 + (-3)$

5. $-3.6 + (-2.1)$

6. $-5.2 + (-3.7)$

7. $8.9 + 7.6$

8. $5.8 + 2.7$

9. $\dfrac{1}{5} + \dfrac{2}{7}$

10. $\dfrac{2}{3} + \dfrac{1}{4}$

11. $-\dfrac{1}{18} + \left(-\dfrac{5}{6}\right)$

12. $-\dfrac{1}{16} + \left(-\dfrac{3}{4}\right)$

Add pairs of signed numbers, which have opposite signs.

13. $12 + (-3)$ **14.** $15 + (-6)$ **15.** $-8 + 3$ **16.** $-9 + 6$

17. $-17 + 12$ **18.** $-21 + 15$ **19.** $-36 + 58$ **20.** $-42 + 57$

21. $-9.3 + 6.5$ **22.** $-7.2 + 4.4$ **23.** $\frac{1}{12} + \left(-\frac{3}{4}\right)$ **24.** $\frac{1}{15} + \left(-\frac{3}{5}\right)$

Add.

25. $\frac{7}{9} + \left(-\frac{2}{9}\right)$ **26.** $\frac{11}{13} + \left(-\frac{4}{13}\right)$ **27.** $-15 + (-3)$ **28.** $-18 + (-6)$

29. $1.5 + (-2.2)$ **30.** $3.6 + (-4.1)$ **31.** $\frac{3}{14} + \frac{2}{7}$ **32.** $\frac{7}{15} + \frac{3}{10}$

33. $-15 + (-23)$ **34.** $-18 + (-31)$ **35.** $-5.5 + (-2.1)$ **36.** $-7.7 + (-3.2)$

37. $5 + (-8) + 2$ **38.** $15 + (-7) + 4$ **39.** $-6 + 3 + (-12) + 4$ **40.** $-2 + 15 + (-5) + 4$

41. $-7 + 6 + (-2) + 5 + (-3) + (-5)$ **42.** $-2 + 1 + (-12) + 7 + (-4) + (-1)$

43. $5 + (-8) + 7 + 20 + (-8) + (-6)$ **44.** $2 + (-10) + 6 + 13 + (-10) + (-4)$

Find the profit or loss situation for a company after the following reports.

45. A $36,000 loss in January followed by a $32,000 loss in February

46. A $20,000 loss in September followed by a $46,000 loss in October

47. A $35,000 profit in May followed by a $40,000 loss in June

48. A $60,000 profit in January followed by a $28,000 loss in February

49. A $10,000 loss in April, a $12,000 loss in May, and a $27,000 profit in June

50. A $16,000 loss in December, a $9000 loss in January, and a $20,000 profit in February

51. Last night the temperature was $-8°$F. Today it rose $18°$F. What was the new temperature?

52. Last night the temperature was $-15°$F. Today it rose $23°$F. What was the new temperature?

53. Today the temperature was $-2°$F. Then the temperature dropped $14°$F. What was the new temperature?

54. Today the temperature was $-6°$F. Then the temperature dropped $17°$F. What was the new temperature?

Verify the commutative law for addition of signed numbers:

55. When $a = -8.6$ and $b = 3.5$

56. When $a = -9.2$ and $b = 10.0$

Verify the associative law for addition of signed numbers:

$$a + (b + c) = (a + b) + c$$

57. When $a = 7$, $b = -3$, $c = -5$

58. When $a = 6$, $b = -8$, $c = 1$

Add.

59. $\left(-\dfrac{1}{5}\right) + \left(-\dfrac{2}{3}\right) + \left(\dfrac{4}{25}\right) + \left(-\dfrac{1}{9}\right)$

60. $\left(-\dfrac{1}{7}\right) + \left(-\dfrac{5}{21}\right) + \left(\dfrac{3}{14}\right) + \left(-\dfrac{3}{7}\right)$

Cumulative Review Problems

61. Use $V = \dfrac{4\pi r^3}{3}$ to find the volume of a sphere of radius 6 ft. Use $\pi = 3.14$ and round your answer to the nearest tenth.

62. Use $V = \dfrac{Bh}{3}$ to find the volume of a pyramid whose rectangular base measures 9 m by 7 m and whose height is 10 m.

 Calculator Problems

Add.

63. $122.345 + (-89.23)$

64. $-96.721 + 64.302$

For Extra Practice Examples and Exercises, turn to page 500.

Solutions to Odd-Numbered Practice Problems

1. (a) $\begin{array}{r} 9 \\ +\ 14 \\ \hline 23 \end{array}$ **(b)** $\begin{array}{r} -7 \\ +\ -3 \\ \hline -10 \end{array}$ **3.** $-\dfrac{2}{15} + \left(-\dfrac{12}{15}\right) = -\dfrac{14}{15}$ **5.** $\begin{array}{r} -\$65,000 \\ +\ -\$40,000 \\ \hline -\$105,000 \end{array}$ The company lost a total of \$105,000 in March and April.

7. (a) $\begin{array}{r} -20.8 \\ +\ \ 15.2 \\ \hline -5.6 \end{array}$ **(b)** $\begin{array}{r} 20.8 \\ +\ -15.2 \\ \hline 5.6 \end{array}$ **9.** $\begin{array}{r} 36 \\ +\ -21 \\ \hline 15 \end{array}$ Then we add $\begin{array}{r} 15 \\ +\ -18 \\ \hline -3 \end{array}$

11. $\begin{array}{r} \$30,000 \\ +\ \$40,000 \\ \hline \$70,000 \end{array}$ $\begin{array}{r} -\$20,000 \\ -\$5,000 \\ +\ -\$35,000 \\ \hline -\$60,000 \end{array}$ $\begin{array}{r} \$70,000 \\ +\ -\$60,000 \\ \hline \$10,000 \end{array}$

The company had an overall profit of \$10,000 in the five-month period.

Answers to Even-Numbered Practice Problems

2. -6.4 **4. (a)** $\dfrac{2}{3}$ **(b)** $-\dfrac{19}{42}$ **6. (a)** -5 **(b)** 5 **8.** $9°F$ **10.** 6

1 Subtract one signed number from another

2 Solve problems involving subtraction with more than two signed numbers

3 Solve simple applied problems that involve the subtraction of one signed number from another

9.2 SUBTRACTION OF SIGNED NUMBERS

Remember that with whole numbers we can write

$$8 - 5 = \text{some number}$$

How did we know if we had the "right number" as a result of this subtraction? The "right number" plus 5 adds up to 8. By memorizing facts or by trial and error, we arrive at 3 as the correct answer. With signed numbers we might have any of the following subtractions:

$$8 - 5 = ?$$

$$8 - (-5) = ?$$

$$-8 - (5) = ?$$

$$-8 - (-5) = ?$$

In this section we learn how to do subtraction much more quickly than by memorizing facts or by trial and error. We'll see that *every* time we have a subtraction problem we can replace it with a certain addition problem. And since we know how to do addition of signed numbers from the previous section, we have a known method to follow.

1 We want to examine a way to subtract signed numbers. It is important to be able to find $6 - (-3)$ and $-7 - (-5)$, for example. First we need to be clear about the meaning of the word **opposite**. The opposite of a positive number is a negative number with the same absolute value. For example, the opposite of 7 is -7. The opposite of a negative number is a positive number with the same absolute value. For example, the opposite of -9 is 9. If a number is the opposite of another number, these two numbers are at an equal distance from zero on the number line.

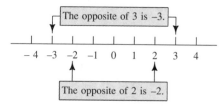

The sum of a number and its opposite is zero.

$$-3 + 3 = 0 \qquad 2 + (-2) = 0$$

Let's think about how a checking account works. Suppose that you deposit $25 and the bank adds a service charge of $5 for a new checkbook. Your account changes

$$\$25 + (-\$5) = \$20$$

Suppose instead that you deposit $25 and the bank adds no charge. The next day, you write a check for $5. The result of these two transitions is

$$\$25 - \$5 = \$20$$

Note that your account has grown by the same amount of money ($20) in both cases. We see that adding a negative amount to a certain number, such as, $25 + (-$5)$ is equivalent to subtracting a positive amount from that number $25 - $5 = 20.

Subtracting is equivalent to adding the opposite.

Subtractions	Adding the Opposite
$25 - 5 = 20$	$25 + (-5) = 20$
$19 - 6 = 13$	$19 + (-6) = 13$
$7 - 3 = 4$	$7 + (-3) = 4$
$15 - 5 = 10$	$15 + (-5) = 10$

If this pattern holds, we could also say

$$-7 - 3 = -10 \qquad \text{because} \qquad -7 + (-3) = -10$$

Let's define a rule that covers the subtraction of signed numbers.

> **Subtraction of Signed Numbers**
>
> To subtract signed numbers, add the opposite of the second number to the first number.

Thus to do a subtraction problem, we first change it to an equivalent addition problem in which the first number does not change but the second number is replaced by its opposite. Then we follow the rules of *addition* for signed numbers.

Example 1 Subtract:

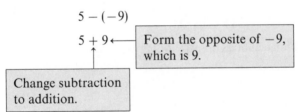

$$-8 - (-2)$$
$$-8 + 2 \quad\longleftarrow\quad \text{Form the opposite of } -2, \text{ which is 2.}$$

Change subtraction to addition.

Using the rules of addition, for two numbers with opposite signs, we add

$$-8 + 2 = -6 \quad\blacksquare$$

Practice Problem 1 Subtract: $-10 - (-5)$. ∎

Example 2 Subtract:

$$5 - (-9)$$
$$5 + 9 \quad\longleftarrow\quad \text{Form the opposite of } -9, \text{ which is 9.}$$

Change subtraction to addition.

Using the rules of addition, for two numbers with the same sign, we have

$$5 + 9 = 14 \quad\blacksquare$$

Practice Problem 2 Subtract: $4 - (-7)$. ∎

Example 3 Subtract:

$$7 - 8$$
$$7 + (-8) \quad\longleftarrow\quad \text{Form the opposite of 8, which is } -8.$$

Change subtraction to addition.

Now we use the rules of addition for two numbers with opposite signs:

$$7 + (-8) = -1 \quad\blacksquare$$

Practice Problem 3 Subtract: $5 - 12$. ∎

Example 4 Subtract:

$$-12 - 16$$

$$-12 + (-16) \longleftarrow \boxed{\text{Form the opposite of 16, which is } -16.}$$

$$\underset{\big|}{\overset{\big|}{}}$$

$\boxed{\text{Change subtraction to addition.}}$

Now we follow the rules of addition of two numbers with the same sign:

$$-12 + (-16) = -28 \quad \blacksquare$$

Practice Problem 4 Subtract: $-11 - 17$.   ■

Sometimes the numbers we subtract are fractions or decimals.

Example 5 Subtract:

$$5.6 - (-8.1)$$

$$5.6 + 8.1 \longleftarrow \boxed{\text{Form the opposite of } -8.1, \text{ which is } 8.1}$$

$$\underset{\big|}{\overset{\big|}{}}$$

$\boxed{\text{Change subtraction to addition.}}$

Now add

$$5.6 + 8.1 = 13.7 \quad \blacksquare$$

Practice Problem 5 Subtract: $3.6 - (-9.5)$.   ■

Example 6 Subtract: $-\dfrac{6}{11} - \left(-\dfrac{1}{22}\right)$.

First we change subtraction to adding the opposite.

$$-\frac{6}{11} + \frac{1}{22}$$

Next we recognize the LCD is 22 and write $\dfrac{6}{11}$ as an equivalent fraction with the LCD in the denominator.

$$\frac{6}{11} \cdot \frac{2}{2} = \frac{12}{22}$$

$\dfrac{6}{11} = \dfrac{12}{22}$; therefore, $-\dfrac{6}{11} = -\dfrac{12}{22}$. Thus

$$-\frac{6}{11} + \frac{1}{22} = -\frac{12}{22} + \frac{1}{22} = -\frac{11}{22} \quad \blacksquare$$

Practice Problem 6 Subtract: $-\dfrac{5}{8} - \left(-\dfrac{5}{24}\right)$. Reduce your answer.   ■

At this point you should be ready to do several subtraction problems quickly. Remember that in performing subtraction of two numbers:

1. The first number does not change.
2. The subtraction sign is changed to addition.
3. We write the opposite of the second number.
4. We find the result of this addition problem.

Remember, when you see $7 - 10$, the sign $-$ means subtraction. Then you should think: "$7 + (-10)$." You should think of replacing the subtraction problem with an addition problem.

If you see $-3 - 19$, think: $-3 + (-19)$.
If you see $6 - (-3)$, think: $6 + (+3)$.

Try to think of each subtraction problem as an equivalent problem of adding the opposite. See if you can follow each of the following in Example 7, then quickly try to do all parts of Practice Problem 7.

Example 7 Subtract.

(a) $7 - 10$ **(b)** $-3 - 19$ **(c)** $6 - (-3)$

(d) $-4 - (-2)$ **(e)** $-\dfrac{1}{2} - \left(-\dfrac{1}{3}\right)$ **(f)** $2.7 - (-5.2)$

(a) $7 - 10 = 7 + (-10) = -3$ **(b)** $-3 - 19 = -3 + (-19) = -22$
(c) $6 - (-3) = 6 + 3 = 9$ **(d)** $-4 - (-2) = -4 + 2 = -2$

(e) $-\dfrac{1}{2} - \left(-\dfrac{1}{3}\right) = -\dfrac{1}{2} + \dfrac{1}{3} = -\dfrac{3}{6} + \dfrac{2}{6} = -\dfrac{1}{6}$

(f) $2.7 - (-5.2) = 2.7 + 5.2 = 7.9$ ■

Practice Problem 7 Subtract.
(a) $8 - 14$ **(b)** $-6 - 13$ **(c)** $20 - (-5)$

(d) $-9 - (-3)$ **(e)** $-\dfrac{1}{5} - \left(-\dfrac{1}{2}\right)$ **(f)** $3.6 - (-5.5)$ ■

2 If a problem involves both addition and subtraction, work from left to right and perform each operation.

Example 8 Perform the following set of operations working from left to right:

$-8 - (-3) + (-5) = -8 + 3 + (-5)$ — First we change subtracting a -3 to adding a 3.

$= -5 + (-5)$ — The sum of $-8 + 3 = -5$ using the rules for adding signed numbers with opposite signs.

$= -10$ — The sum of $-5 + (-5) = -10$ using the rules for adding signed numbers with the same sign. ■

Practice Problem 8 Perform the following set of operations working from left to right:

$$-5 - (-9) + (-14)$$ ■

3 Solving Applied Problems That Involve the Subtraction of One Signed Number from Another

Suppose that we want to find the difference in altitude between two points. We would subtract the lower altitude from the higher altitude in the illustration at right. A location of land below sea level is considered to have a negative altitude reading. The Dead Sea is 1286 ft below sea level as shown in the figure below.

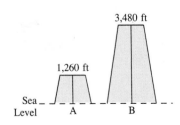

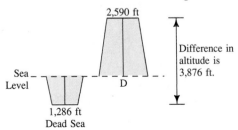

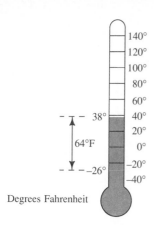

140°
120°
100°
80°
60°
40°
20°
0°
-20°
-40°

38°

64°F

-26°

Degrees Fahrenheit

The difference in altitude between mountain D and the Dead Sea is

$$2590 - (-1286) = 2590 + 1286 = 3876 \text{ feet}$$

Example 9 Find the difference in temperature between 38°F in the day in Anchorage, Alaska, and −26°F at night.

We subtract the lower temperature from the higher temperature:

$$38 - (-26) = 38 + 26 = 64$$

The difference is 64°F. ∎

Practice Problem 9 Find the difference in temperature between 31°F in the day in Fairbanks, Alaska, and −37°F at night. ∎

To Think About

Just why does this "add the opposite of the second number to the first number" rule for subtraction work? Why can we say that subtracting a signed number is the same as adding the opposite of that number? We once learned subtraction as "taking away." Are there any examples in our daily life of how "taking away" is the same as "adding the opposite." Why is

$$-40 - (-100) = -40 + 100?$$

Can these rules work in areas of business and finance? Suppose your checking account is overdrawn by $40. The balance is $−40. Upon investigation at the bank you find that another person's check for $100 has been incorrectly charged to your account. The bank wants to correct that error and "take away" or "subtract" from your account the charge of $100. In other words, the bank wants to perform the operation

$$-40 - (-100)$$

However, the error appears on the bank records and cannot be "erased." The bank can make an equivalent adjustment that has the same effect as "erasing." It adds $100 to your account and calls it a "credit adjustment." So in order to do $-40 - (-100)$ the bank performs $-40 + 100$ and obtains your new balance of $60.00. We see how $-40 - (-100) = -40 + 100 = 60$. Subtracting a negative 100 is equivalent to adding a positive 100. ∎

EXERCISES 9.2

In each case subtract signed numbers by adding the opposite of the second number to the first number.

1. $5 - 8$

2. $7 - 12$

3. $-5 - 9$

4. $-6 - 4$

5. $-19 - (-14)$

6. $-23 - (-17)$

7. $3 - (-21)$

8. $8 - (-17)$

9. $18 - 24$

10. $15 - 20$

11. $-12 - (-15)$

12. $-17 - (-30)$

13. $150 - 210$

14. $330 - 350$

15. $300 - (-256)$

16. $420 - (-300)$

17. $-58 - 32$

18. $-71 - 25$

19. $-45 - (-85)$

20. $-62 - (-90)$

21. $-2.5 - 4.2$

22. $-6.1 - 1.8$

23. $10.6 - 3.5$

24. $9.8 - 2.7$

25. $-10.9 - (-2.3)$

26. $-6.8 - (-2.9)$

27. $5.1 - (-3.6)$

28. $7.7 - (-4.3)$

29. $\dfrac{1}{4} - \left(-\dfrac{3}{4}\right)$

30. $\dfrac{7}{8} - \left(-\dfrac{2}{8}\right)$

31. $-\dfrac{5}{6} - \dfrac{1}{3}$

32. $-\dfrac{2}{8} - \dfrac{1}{4}$

33. $-\dfrac{5}{12} - \left(-\dfrac{1}{4}\right)$

34. $-\dfrac{7}{14} - \left(-\dfrac{1}{7}\right)$

35. $\dfrac{7}{11} - \dfrac{1}{2}$

36. $\dfrac{5}{13} - \dfrac{1}{3}$

Perform each set of operations working from left to right.

37. $2 - (-8) + 5$

38. $7 - (-3) + 9$

39. $-5 - 6 - (-11)$

40. $-3 - 12 - (-5)$

41. $7 - (-2) - (-8)$

42. $6 - (-3) - (-5)$

43. $-10 - (-4) - 3$

44. $-15 - (-7) - 2$

45. $9 - 3 - 2 - 6$

46. $12 - 5 - 4 - 8$

Use your knowledge of signed numbers to answer the following questions.

47. Find the difference in altitude between a mountain 4350 ft high and a desert valley 660 ft below sea level.

48. Find the difference in altitude between a mountain 3480 ft high and a desert valley 890 ft below sea level.

49. Find the difference in temperature in Anchorage, Alaska, between $16°F$ in the day and $-39°F$ at night.

50. Find the difference in temperature in Fairbanks, Alaska, between $27°F$ in the day and $-33°F$ at night.

A company's profit and loss statement in dollars for the last five months is shown below:

Month	Profit	Loss
January	16,500	
February		$-28,200$
March		$-7,500$
April	41,200	
May		$-13,400$

51. What is the difference between the loss in March and the loss in February?

52. What is the difference between the loss in March and the loss in May?

53. What is the difference between the profit in January and the loss in March?

54. What is the difference between the profit in April and the loss in May?

To Think About

55. Give an example of how a bank might want to do the calculation $50 - (-80)$ and would accomplish this by adding $50 + 80$.

56. Give an example of how a bank might want to do the calculation $-100 - 50$ and would actually accomplish this by adding $-100 + (-50)$

Evaluate.

57. $\frac{1}{4} + \left(-\frac{1}{12}\right) - \left(-\frac{2}{3}\right) - \frac{5}{6} - \frac{1}{2}$

58. $-\frac{1}{30} + \left(-\frac{2}{3}\right) - \left(-\frac{5}{6}\right) - \frac{7}{10} + \frac{7}{15}$

Cumulative Review Problems

Perform in order.

59. $20 \times 2 \div 10 + 4 - 3$

60. $5 \times 7 + 6 \times 3 - 8$

Calculator Problems

Subtract.

61. $-111.236 - (-146.78)$

62. $-86.44 - 21.2$

For Extra Practice Examples and Exercises, turn to page 501.

Solutions to Odd-Numbered Practice Problems

1. $-10 - (-5) = -10 + 5 = -5$ **3.** $5 - 12 = 5 + (-12) = -7$ **5.** $3.6 - (-9.5) = 3.6 + 9.5 = 13.1$
7. (a) $8 - 14 = 8 + (-14) = -6$ **(b)** $-6 - 13 = -6 + (-13) = -19$ **(c)** $20 - (-5) = 20 + 5 = 25$

(d) $-9 - (-3) = -9 + 3 = -6$ **(e)** $-\frac{1}{5} - \left(-\frac{1}{2}\right) = -\frac{1}{5} + \frac{1}{2} = -\frac{2}{10} + \frac{5}{10} = \frac{3}{10}$ **(f)** $3.6 - (-5.5) = 3.6 + 5.5 = 9.1$

9. $31 - (-37) = 31 + 37 = 68$ The difference is $68°F$.

Answers to Even-Numbered Practice Problems

2. 11   **4.** -28 **6.** $-\frac{5}{12}$ **8.** -10

☐ After studying this section, you will be able to:

1 *Multiply or divide two signed numbers*

2 *Multiply three or more signed numbers*

9.3 MULTIPLICATION AND DIVISION OF SIGNED NUMBERS

We are familiar with multiplying two positive numbers. For example, $(2)(8) = 16$ gives us no problem. But how do we handle negative numbers in multiplication? What would a negative number *mean* in multiplication? Well, if you work for 2 hours at 8 dollars an hour, $(2)(8) = 16$ tells you what you have gained. But if you pay someone for working 2 hours at 8 dollars an hour, $(2)(-8) = -16$ tells you that you have $16 less than what you started with. In this next section we learn how to multiply and divide signed numbers.

1 In higher-level mathematics, the operation of multiplication is usually indicated by a dot, or by one or two sets of parentheses. Thus to indicate 3×5 we usually write $3 \cdot 5$, $(3)(5)$ or $3(5)$. The number to the left of a parentheses tells you how many times to multiply that number by the quantity inside the parentheses.

Example 1 Evaluate.

(a) $(7)(8)$ **(b)** $3(12)$ **(c)** $2(6)(3)$ **(d)** $(4)(8)(2)$

(a) $(7)(8) = 56$ **(b)** $3(12) = 36$
(c) $2(6)(3) = 12(3) = 36$ **(d)** $(4)(8)(2) = 32(2) = 64$ ∎

Practice Problem 1 Evaluate.
(a) $(6)(9)$ **(b)** $7(12)$ **(c)** $3(5)(8)$ **(d)** $(6)(2)(7)$ ∎

Let us try to think of a simple example in everyday life of why we might multiply signed numbers like

$$(2)(3) = ?$$
$$(-2)(3) = ?$$
$$(2)(-3) = ?$$
$$(-2)(-3) = ?$$

What logical pattern could we find in the answers?

Suppose that water is flowing *into* a tank at the rate of 3 gallons a minute (3). Two minutes from now (2) there will be 6 gallons more in the tank

$$(2)(3) = 6$$

$$\boxed{\text{positive number}} \times \boxed{\text{positive number}} = \boxed{\text{positive number}}$$

Two minutes ago (-2) there were 6 gallons less (-6) in the tank.

$$(-2)(3) = -6$$

$$\boxed{\text{negative number}} \times \boxed{\text{positive number}} = \boxed{\text{negative number}}$$

Now suppose water is flowing *out* of the tank at the rate of 3 gallons per minute, (-3). Two minutes from now (2) there will be 6 gallons less in the tank (-6).

$$(2)(-3) = -6$$

$$\boxed{\text{positive number}} \times \boxed{\text{negative number}} = \boxed{\text{negative number}}$$

Two minutes ago (-2) there were 6 gallons more in the tank $(+6)$.

$$(-2)(-3) = +6$$

$$\boxed{\text{negative number}} \times \boxed{\text{negative number}} = \boxed{\text{positive number}}$$

These patterns we observe allow us to state the following rule:

Multiplication and Division Rule for Two Numbers with the Same Sign

To multiply or divide two numbers with the same sign, multiply or divide the absolute values. The sign of the result is positive.

We now consider the multiplication of two signed numbers.

Example 2 Multiply.

(a) $(-8)(-3)$ **(b)** $-5(-6)$ **(c)** $\left(-\dfrac{1}{2}\right)\left(-\dfrac{3}{5}\right)$ **(d)** $(0.6)(1.3)$

In each case we are multiplying two numbers with the same sign. We will always obtain a positive number.

(a) $(-8)(-3) = 24$

(b) $-5(-6) = 30$

(c) $\left(-\dfrac{1}{2}\right)\left(-\dfrac{3}{5}\right) = \dfrac{3}{10}$

(d) $(0.6)(1.3) = 0.78$ ■

Practice Problem 2 Multiply.

(a) $(-8)(-8)$ **(b)** $-10(-6)$ **(c)** $\left(-\dfrac{1}{3}\right)\left(-\dfrac{2}{7}\right)$ **(d)** $(1.2)(0.4)$ ■

Whenever we divide two numbers with the same sign, the result is a positive number.

Example 3 Divide.

(a) $\dfrac{-20}{-10}$

(b) $(-50) \div (-2)$

(c) $(-9.9) \div (-3.0)$

(d) $\dfrac{\frac{2}{3}}{\frac{1}{4}}$

(a) $\dfrac{-20}{-10} = 2$

(b) $(-50) \div (-2) = 25$

(c) $(-9.9) \div (-3.0) = 3.3$

(d) $\dfrac{2}{3} \div \dfrac{1}{4} = \left(\dfrac{2}{3}\right)\left(\dfrac{4}{1}\right) = \dfrac{8}{3}$ or $2\dfrac{2}{3}$ ■

Practice Problem 3 Divide.

(a) $\dfrac{-16}{-8}$

(b) $-78 \div (-2)$

(c) $(-1.2) \div (-0.5)$

(d) $\dfrac{\frac{3}{5}}{\frac{1}{10}}$ ■

Multiplication Rule for Two Numbers with Opposite Signs

To multiply two numbers with opposite signs, multiply the absolute values. The result is negative.

Example 4 Multiply.

(a) $2(-8)$ **(b)** $(-3)(25)$ **(c)** $\left(-\dfrac{1}{3}\right)\left(\dfrac{2}{5}\right)$ **(d)** $0.6(-2.4)$

In each case we are multiplying two signed numbers with opposite signs. We will always get a negative number for an answer.

(a) $2(-8) = -16$

(b) $(-3)(25) = -75$

(c) $\left(-\dfrac{1}{3}\right)\left(\dfrac{2}{5}\right) = -\dfrac{2}{15}$

(d) $0.6(-2.4) = -1.44$ ■

Practice Problem 4 Multiply.

(a) $(-8)(5)$ **(b)** $3(-60)$ **(c)** $\left(-\dfrac{1}{5}\right)\left(\dfrac{3}{7}\right)$ **(d)** $0.5(-6.7)$ ■

> **Division Rule for Two Numbers with Opposite Signs**
>
> To divide two numbers with opposite signs, divide the absolute values. The result is negative.

Example 5 Divide.

(a) $-20 \div 5$ **(b)** $36 \div (-18)$ **(c)** $\dfrac{-20.8}{2.6}$ **(d)** $\dfrac{\frac{3}{5}}{-\frac{9}{13}}$

(a) $-20 \div 5 = -4$ **(b)** $36 \div (-18) = -2$ **(c)** $\dfrac{-20.8}{2.6} = -8$

(d) $\dfrac{3}{5} \div \left(-\dfrac{9}{13}\right) = \dfrac{\overset{1}{\cancel{3}}}{5}\left(-\dfrac{13}{\underset{3}{\cancel{9}}}\right) = -\dfrac{13}{15}$ ■

Practice Problem 5 Divide.

(a) $-50 \div 25$ **(b)** $49 \div (-7)$ **(c)** $\dfrac{21.12}{-3.2}$ **(d)** $\dfrac{-\frac{2}{7}}{\frac{4}{13}}$ ■

🄴 Multiplying More Than Two Numbers

When multiplying more than two numbers, multiply any two numbers first, then multiply the result by another number. Continue until each factor has been used once.

Example 6 Multiply: $5(-2)(-3)$.

$$5(-2)(-3) = -10(-3) \qquad \text{First multiply } 5(-2) = -10$$
$$= 30 \qquad\qquad\quad \text{Then multiply } -10(-3) = 30 \quad ■$$

Practice Problem 6 Multiply: $(-6)(3)(-4)$. ■

Example 7 Multiply: $6\left(-\dfrac{1}{2}\right)(-4)(-2)$.

First we note that in multiplying the first two numbers that

$$\dfrac{6}{1}\left(-\dfrac{1}{2}\right) = \dfrac{\overset{3}{\cancel{6}}}{1}\left(-\dfrac{1}{\underset{1}{\cancel{2}}}\right) = -\dfrac{3}{1} = -3$$

Then we continue multiplication.

$$(-3)(-4)(-2) = 12(-2) = -24 \quad ■$$

Practice Problem 7 Multiply: $4(-8)\left(-\dfrac{1}{4}\right)(-3)$. ■

If zero is a factor in any multiplication problem, the product is zero.

Example 8 Multiply: $(7)(-9)(8)(-3)(0)(-4)$.

Because zero times any number results in zero, we have

$$(7)(-9)(8)(-3)(0)(-4) = 0 \quad ■$$

Practice Problem 8 Multiply: $(-3)\left(-\dfrac{1}{5}\right)(-2)(0)(4)\left(-\dfrac{2}{7}\right)$. ■

Is there a mathematical pattern to our sign rules? Do the rules seem to follow a logical order? Do you sometimes wonder why a positive number multiplied by a negative number yields a negative number? Do you wonder why a negative number multiplied by a negative number yields a positive number?

Let's look at a list of numbers that are multiplied and see what pattern we observe.

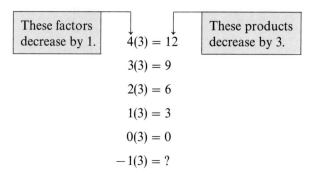

These factors decrease by 1. $4(3) = 12$ These products decrease by 3.

$$3(3) = 9$$
$$2(3) = 6$$
$$1(3) = 3$$
$$0(3) = 0$$
$$-1(3) = ?$$

As the left column of factors decreases by 1 for each row, the right column of products decreases by 3 for each row. What is 3 less than 0? It is -3. We could continue the pattern and see logical support for the idea that a negative number times a positive number yields a negative number.

Now let us carry the process one step further. Look at the following list of numbers:

These factors decrease by 1. $4(-3) = -12$ These products increase by 3.

$$3(-3) = -9$$
$$2(-3) = -6$$
$$1(-3) = -3$$
$$0(-3) = 0$$
$$(-1)(-3) = ?$$

As the left column of factors decreases by 1, the right column of products increases by 3. What number is 3 greater than zero? It is 3. Thus

$$(-1)(-3) = 3$$

We could continue the pattern and see support for the idea that a negative number times a negative number yields a positive number. ∎

EXERCISES 9.3

Multiply.

1. $7(-2)$ **2.** $6(-8)$ **3.** $(-5)(-3)$ **4.** $(-2)(-20)$ **5.** $(-5)(4)$

6. $(-8)(7)$ **7.** $(2)(8)$ **8.** $(5)(9)$ **9.** $(-20)(-3)$ **10.** $(-40)(-6)$

11. $(-20)(8)$ **12.** $(-15)(12)$ **13.** $(2.5)(-0.6)$ **14.** $(7.5)(-0.6)$ **15.** $(-1.5)(-2.5)$

16. $(-4.0)(-3.8)$ **17.** $\left(-\dfrac{2}{5}\right)\left(\dfrac{3}{7}\right)$ **18.** $\left(-\dfrac{4}{9}\right)\left(-\dfrac{1}{3}\right)$ **19.** $\left(-\dfrac{4}{12}\right)\left(-\dfrac{3}{23}\right)$ **20.** $\left(-\dfrac{8}{11}\right)\left(\dfrac{22}{31}\right)$

Divide.

21. $20 \div (-10)$ **22.** $36 \div (-4)$ **23.** $121 \div 11$ **24.** $169 \div 13$ **25.** $-70 \div (-5)$

26. $-60 \div (-3)$ **27.** $-16 \div 8$ **28.** $-28 \div 7$ **29.** $\dfrac{48}{-6}$ **30.** $\dfrac{52}{-13}$

31. $\dfrac{-72}{-8}$ **32.** $\dfrac{-54}{-6}$ **33.** $\dfrac{1}{2} \div \left(-\dfrac{3}{5}\right)$ **34.** $-\dfrac{2}{7} \div \dfrac{3}{4}$ **35.** $\dfrac{-\dfrac{4}{5}}{-\dfrac{7}{10}}$

36. $\dfrac{-\dfrac{26}{15}}{-\dfrac{13}{7}}$ **37.** $49.2 \div (-6)$ **38.** $27.2 \div (-4)$ **39.** $\dfrac{-55.8}{-9}$ **40.** $\dfrac{-45.5}{-7}$

Multiply.

41. $7(-8)(-2)$ **42.** $3(-5)(-9)$ **43.** $(-2)(-4)(-5)$ **44.** $(-7)(-2)(-3)$

45. $2(-8)(3)\left(-\dfrac{1}{3}\right)$ **46.** $7(-2)(-5)\left(\dfrac{1}{7}\right)$ **47.** $(-5)(2)(-1)(-3)$ **48.** $(-8)(-2)(3)(-1)$

49. $8(-3)(-5)(0)(-2)$ **50.** $9(-6)(-4)(-3)(0)$

 To Think About

51. Bob multiplied -12 times a mystery number that was not negative and he did not get a negative answer. What was the mystery number?

52. Fred divided a mystery number that was not negative by -3 and did not get a negative answer. What was the mystery number?

In problems 53 and 54, the letters a, b, c, *and* d *represent numbers that may be either positive or negative. The first three numbers are multiplied to obtain the fourth number. This can be stated by the equation* (a)(b)(c) = d.

53. In the case where a, c, and d are all negative numbers, is the number b positive or negative?

54. In the case where a is positive but b and d are negative, is the number c positive or negative?

Perform each operation. Simplify your answer.

55. $\left(-\dfrac{2}{3}\right)\left(-\dfrac{3}{4}\right)\left(-\dfrac{5}{6}\right)\left(-\dfrac{7}{8}\right)\left(-\dfrac{9}{10}\right)$

56. $(-62.8712) \div (0.0763)$

Cumulative Review Problems

57. Find the area of a parallelogram with height of 6 in. and a base of 15 in.

58. Find the area of a trapezoid with height of 12 m and bases of 18 m and 26 m, respectively.

 Calculator Problems

Multiply.

59. $(2.76)(-3.21)(1.09)$

60. $(-0.024)(0.001)(-0.61)$

Divide.

61. $(-3288) \div (-0.213)$

62. $(0.0712) \div (421)$

For Extra Practice Examples and Exercises, turn to page 501.

Solutions to Odd-Numbered Practice Problems

1. (a) $(6)(9) = 54$ **(b)** $(7)(12) = 84$ **(c)** $3(5)(8) = 15(8) = 120$ **(d)** $(6)(2)(7) = (12)(7) = 84$

3. (a) $\dfrac{-16}{-8} = 2$ **(b)** $-78 \div (-2) = 39$ **(c)** $\dfrac{-1.2}{-0.5} = 2.4$ **(d)** $\dfrac{3}{5} \div \dfrac{1}{10} = \dfrac{3}{5} \cdot \dfrac{\overset{2}{\cancel{10}}}{1} = \dfrac{6}{1} = 6$

5. (a) $-50 \div 25 = -\dfrac{50}{25} = -2$ **(b)** $49 \div (-7) = \dfrac{49}{-7} = -7$ **(c)** $\dfrac{21.12}{-3.2} = -6.6$

$$3.2_{\wedge}\overline{)21.1_{\wedge}2}\ \ \overset{6.6}{}$$
$$\underline{192}$$
$$192$$
$$\underline{192}$$
$$0$$

(d) $-\dfrac{2}{7} \div \dfrac{4}{13} = -\dfrac{\overset{1}{\cancel{2}}}{7} \cdot \dfrac{13}{\cancel{4}_2} = -\dfrac{13}{14}$

7. $4(-8)\left(-\dfrac{1}{4}\right)(-3) = (-32)\left(-\dfrac{1}{4}\right)(-3) = (-\overset{8}{\cancel{32}})\left(-\dfrac{1}{\cancel{4}_1}\right)(-3) = (8)(-3) = -24$

Answers to Even-Numbered Practice Problems

2. (a) 64 **(b)** 60 **(c)** $\dfrac{2}{21}$ **(d)** 0.48   **4. (a)** -40 **(b)** -180 **(c)** $-\dfrac{3}{35}$ **(d)** -3.35 **6.** 72   **8.** 0

☐ After studying this section, you will be able to:

1 *Do combination problems that involve two or more of the following operations: addition, subtraction, multiplication, or division of signed numbers*

9.4 ORDER OF OPERATIONS WITH SIGNED NUMBERS

If we have several computations to do, such as

$$(-48) \div (-8) + 20 \div (-10)$$

there could be confusion. One person, A, could interpret the problem to mean, "go in order, left to right, doing every operation as you come to it." Another person, B, could interpret the problem to mean, "do the operations in any order." Their answers would differ. So, to avoid confusion, there is an agreed-upon order in which the problem should be done. This is called the **order of operations with signed numbers**.

As a matter of fact, *neither* of the methods thought of by A or B is the actual agreed-upon order of operations in use today—as you will see in the section that follows.

1 As we discussed in Section 1.7, it is important to perform operations in the correct order. If there are no grouping symbols or exponents, we need to remember that:

---

1. Multiplication and division are done first, working from left to right.
2. Then addition and subtraction are done from left to right.

Example 1 Perform the indicated operations in the proper order.

$$-6 \div (-2)(5)$$

Multiplication and division are of equal priority. So we work from left to right and first divide

$$-6 \div (-2) = 3$$

Now we have $(3)(5) = 15$. ■

Practice Problem 1 Perform the indicated operations in the proper order.

$$20 \div (-5)(-3) \quad ■$$

Example 2 Perform the indicated operations in the proper order.

$$7 + 6(-2)$$

Multiplication and division must be done first.

$$7 + (-12) \qquad \text{Because } 6(-2) = -12.$$
$$= -5 \qquad \text{Adding } 7 + (-12) = -5. \quad ■$$

Practice Problem 2 Perform the indicated operations in the proper order.

$$9 + 20 \div (-4) \quad ■$$

When subtraction of a number is indicated, we think of it as adding the opposite of that number.

Example 3 Perform the indicated operations in the proper order.

$$9 \div 3 - 16 \div (-2)$$

$$9 \div 3 - 16 \div (-2) = 3 - (-8) \qquad \text{Performing indicated divisions.}$$
$$= 3 + 8 \qquad \text{Changing subtracting a negative 8}$$
$$= 11 \quad ■ \qquad \text{to adding a positive 8.}$$

Practice Problem 3 Perform the indicated operations in the proper order.

$$25 \div (-5) + 16 \div (-8) \quad ■$$

Example 4 Perform the indicated operations in the proper order.

$$4 - 2(7) - 5$$

$$4 - 2(7) - 5 = 4 - 14 - 5 \qquad \text{Performing the multiplication.}$$
$$= 4 + (-14) + (-5) \qquad \text{Transforming subtraction to adding the opposite.}$$
$$= -10 + (-5)$$
$$= -15 \quad ■$$

Practice Problem 4   Perform the indicated operations in the proper order.

$$13 + 7(-6) - 8 \quad ∎$$

If a fraction has operations written in the numerator or in the denominator or both, then these operations must be done first. Then the fraction may be reduced or the division carried out.

Example 5   Perform the indicated operation in the proper order.

$$\frac{5 + 6 - 2}{-8 + 3 - 7}$$

In the numerator we have $5 + 6 - 2 = 11 - 2 = 9$

In the denominator we have $-8 + 3 - 7 = -5 - 7 = -12$

Thus we have

$$\frac{5 + 6 - 2}{-8 + 3 - 7} = \frac{9}{-12} = -\frac{3}{4}$$

Notice that we have reduced $\frac{9}{12}$ to $\frac{3}{4}$, and we have placed a negative sign in front of the $\frac{3}{4}$. Do you see why?   ∎

Practice Problem 5   Perform the indicated operations in the proper order.

$$\frac{12 - 2 - 3}{21 + 3 - 10} \quad ∎$$

Example 6   Perform the indicated operations in the proper order.

$$\frac{7(-2) + 4}{8 \div (-2)(5)}$$

In the numerator we have $7(-2) + 4 = -14 + 4 = -10$

In the denominator we have $8 \div (-2)(5) = (-4)(5) = -20$. Finally, we are able to write

$$\frac{7(-2) + 4}{8 \div (-2)(5)} = \frac{-10}{-20} = \frac{10}{20} = \frac{1}{2}$$

Note that the answer is positive since a negative number divided by a negative number gives a positive result.   ∎

Practice Problem 6   Perform the indicated operations in the proper order.

$$\frac{9(-3) - 5}{2(-4) \div (-2)} \quad ∎$$

Example 7   The Fahrenheit temperature for the last five days of January was $-18°$, $-13°$, $-3°$, $9°$, $-15°$. What was the average temperature?

First we add the temperatures, then divide by 5

$$\frac{-18 + (-13) + (-3) + 9 + (-15)}{5} = \frac{-40}{5} = -8$$

The average temperature was $-8°$F.   ∎

Practice Problem 7   A company had profits/loss reports as follows: January $-\$10,000$, February $+\$30,000$, March $-\$18,000$, April $-\$6,000$. Find the average monthly profit or loss for this time period.   ∎

Perform the operations in the proper order.

1. $7 + (-2) + (-3)$

2. $6 + (-8) + (-5)$

3. $9 + 7(-5)$

4. $-21 + 3(8)$

5. $16 + 32 \div (-4)$

6. $17 - (-20) \div 5$

7. $24 \div (-3) + 16 \div (-4)$

8. $(-48) \div (-8) + 20 \div (-10)$

9. $3(-4) + 5(-2) - (-3)$

10. $6(-3) + 8(-1) - (-2)$

11. $-18 \div 2 + 9$

12. $-20 \div 2 + 10$

13. $5 - 30 \div 3$

14. $7 - 16 \div 4$

15. $48 \div 12(-3)$

16. $30 \div 6(7)$

17. $5(5) - 5$

18. $8(-8) + 8$

19. $3(-4) + 6(-2) - 3$

20. $-7(3) + 5(-4) + 2$

21. $8(-7) - 2(-3)$

22. $10(-5) - 4(12)$

23. $16 - 4(8) + 18 \div (-9)$

24. $20 - 3(-2) + (-20) \div (-5)$

25. $16 \div (-2)(3) + 5 - 6$

26. $-30 \div (-5)(2) - 7 + 6$

In problems 27–36, simplify the numerator and denominator first using the proper order of operations. Then reduce the fraction if possible.

27. $\dfrac{8 + 6 - 12}{3 - 6 + 5}$

28. $\dfrac{20 - 3 - 7}{6 + 5 - 1}$

29. $\dfrac{5(-3) + 1}{1 - 3 - 5}$

30. $\dfrac{7 + 3 - 5}{2(-4) + 3}$

31. $\dfrac{-16 \div (-2)}{3(-4) - 4}$

32. $\dfrac{2(-3) - 5 + 1}{10 \div (-5)}$

33. $\dfrac{4 - (-3) - 5}{20 \div (-10)}$

34. $\dfrac{8 + (-8) - (-3)}{24 \div (-6)}$

35. $\dfrac{7 - (-1)}{9 - 9 \div (-3)}$

36. $\dfrac{4 - (-3) + 1}{2 - 2 \div (-2)}$

37. Find the average Fahrenheit temperature for five days if the temperatures were $-16°$, $-20°$, $-13°$, $5°$, $26°$.

38. Find the average Fahrenheit temperature for five days if the temperatures were $-4°$, $-18°$, $20°$, $5°$, $-31°$.

? To Think About

Perform the operations in the proper order.

39. $\dfrac{1}{2} \div \left(-\dfrac{2}{3}\right)\left(-\dfrac{3}{7}\right) + \left(-\dfrac{5}{14}\right)$

40. $(0.3)(-2.9)(-3.5) + (50.6) \div (-2.0)$

41. A telephone wire that is 3840 m long would be how long measured in kilometers?

42. A container with 36.8 g of protein would be equivalent to how many milligrams of protein?

For Extra Practice Examples and Exercises, turn to page 502.

Solutions to Odd-Numbered Practice Problems

1. $20 \div (-5)(-3) = (-4)(-3) = 12$
Working from left to right, we first evaluate the division. Multiplying the two negative numbers to obtain 12.
3. $25 \div (-5) + 16 \div (-8) = -5 + (-2) = -7$
Working from left to right, we perform each division. Adding the two negative numbers to obtain -7.

5. Evaluating the numerator, we have

$$12 - 2 - 3 = 10 - 3 = 7$$

Next we evaluate the denominator.

$$21 + 3 - 10 = 24 - 10 = 14$$

Thus we finally have

$$\frac{12 - 2 - 3}{21 + 3 - 10} = \frac{7}{14} = \frac{1}{2}$$

7. $\dfrac{-10,000 + 30,000 - 18,000 - 6,000}{4} = -\dfrac{4000}{4} = -\$1000$

Average loss is $1000 per month.

Answers to Even-Numbered Practice Problems

2. 4   **4.** -37 **6.** -8

EXTRA PRACTICE: EXAMPLES AND EXERCISES

Section 9.1

Add the following pairs of signed numbers that have the same sign.

Example $-8.7 + (-2.6)$

$$\begin{array}{r} -8.7 \\ + \ -2.6 \\ \hline -11.3 \end{array}$$

1. $4 + 5$

2. $-3.6 + (-6.2)$

3. $-\dfrac{1}{3} + \left(-\dfrac{3}{5}\right)$

4. $\dfrac{3}{18} + \dfrac{1}{6}$

5. $-7.7 + (-3.9)$

Example $-63 + 41$

$$\begin{array}{r} -63 \\ + \ \ 41 \\ \hline -22 \end{array}$$

6. $-15 + 7$

7. $-7 + 21$

8. $54 + -(18)$

9. $5.9 + (-6.3)$

10. $-\dfrac{1}{12} + \dfrac{5}{6}$

Example $-2 + 3 + (-7) + (-8) + 4$

$$\begin{array}{r} -2 \\ -7 \\ + \ -8 \\ \hline -17 \end{array} \qquad \begin{array}{r} 3 \\ +4 \\ \hline 7 \end{array} \qquad \begin{array}{r} -17 \\ + \ \ \ 7 \\ \hline -10 \end{array}$$

11. $-9 + 4 + (-5) + (-7) + 3$

12. $6.9 + (-7.1) + (-2.1) + 8 + (-1)$

13. $8 + (-6) + 9 + (-2) + (-3)$

14. $-7 + (-5) + 4 + (-1) + 10$

15. $9.1 + (-6.4) + 5.8 + (-2.6) + (-3.5)$

Find the profit or loss situation for a company after the following reports.

Example   A $28,000 profit in February followed by a $31,000 loss in March

Profit $28,000   loss $-$$31,000

$$\begin{array}{r} \$28,000 \\ + \ -\$31,000 \\ \hline -\$3,000 \end{array}$$

The company lost $3000.

16. A $22,000 loss in June followed by a $12,000 loss in July

17. A $14,000 profit in August followed by a $32,000 loss in September

18. A $15,000 loss in September, a $11,000 loss in October, and a $40,000 profit in November

19. Last night the temperature was $-2°$F. Today it rose $16°$F. What was the new temperature?

20. Last night the temperature was $-12°$F. Today it rose $31°$F. What was the new temperature?

Example $-3.8 - (-2.6)$

$$-3.8 + 2.6 \qquad \begin{array}{r} -3.8 \\ + \;\; 2.6 \\ \hline -1.2 \end{array}$$

1. $3 - 5$

2. $-6 - (-9)$

3. $-9.9 - (-8.3)$

4. $78 - 92$

5. $-25 - 31$

Example $-\dfrac{7}{8} - \dfrac{1}{4}$

$$\text{LCM} = 8 \qquad \frac{1\,(2)}{4\,(2)} = \frac{2}{8}$$

$$-\frac{7}{8} - \frac{1}{4} = -\frac{7}{8} - \frac{2}{8} = -\frac{7}{8} + \left(-\frac{2}{8}\right) = -\frac{9}{8} \quad \text{or} \quad -1\frac{1}{8}$$

6. $-\dfrac{2}{3} - \dfrac{5}{6}$

7. $\dfrac{3}{4} - \dfrac{9}{10}$

8. $\dfrac{3}{14} - \left(-\dfrac{1}{7}\right)$

9. $\dfrac{1}{2} - \dfrac{1}{4}$

10. $-\dfrac{3}{8} - \left(-\dfrac{1}{4}\right)$

Perform the following set of operations working from left to right.

Example $-6 - 8 - (-12)$

$$-6 + (-8) - (-12)$$
$$-14 - (-12)$$
$$-14 + 12 = -2$$

11. $-4 - 7 - (-13)$

12. $7 - (-9) - (-3)$

13. $7 - 5 - 3 - 8$

14. $12 - 7 - (-5)$

15. $9 - 3 - 8 - (-1)$

A company's profit and loss statement for the last six months is shown below.

Month	Profit	Loss
June		-$21,300
July	$12,000	
August	32,000	
September		-16,500
October	8,500	
November		-13,700

Example

1. What is the difference between the loss in June and the loss in September?

2. What is the difference between the profit in October and the loss in November?

1. Loss in June, $-\$21,000$; loss in September, $-\$16,500$

$$\begin{array}{lll} \textit{Difference} & \text{loss June} - \text{loss September} \\ & (-21,000) - \;\;(-16,500) & = (-\$4500) \end{array}$$

2. Profit in October, $\$8500$; loss in November, $(-\$13,700)$

$$\text{Profit October} - \text{loss November}$$
$$8500 \quad - \quad (-13,700) \quad = 8500 + (13,700)$$
$$= \$22,200$$

16. What is the difference between the loss in September and the loss in November?

17. What is the difference between the profit in August and the loss in September?

18. Find the difference in altitude between a mountain 5270 ft high and a desert valley 760 ft below sea level.

19. Find the difference in altitude between a mountain 4700 ft high and a desert valley 380 ft below sea level.

20. Find the difference in temperature in Anchorage, Alaska, between $12°F$ in the day and $-31°F$ at night.

Multiply the signed numbers.

Example $\left(\dfrac{3}{8}\right)\left(-\dfrac{2}{7}\right)$

$$\frac{(3)}{(8)} \frac{(-2)}{(7)} = \frac{(3)(\overset{-1}{-2})}{(8)(7)} = -\frac{3}{28}$$

1. $\left(-\dfrac{2}{3}\right)\left(\dfrac{4}{7}\right)$

2. $\left(\dfrac{5}{12}\right)\left(-\dfrac{2}{5}\right)$

3. $\left(-\dfrac{1}{2}\right)\left(-\dfrac{3}{5}\right)$

4. $\left(-\dfrac{6}{11}\right)\left(-\dfrac{1}{3}\right)$

5. $\left(\dfrac{2}{5}\right)\left(-\dfrac{5}{9}\right)$

Example $-50 \div 5$

$$-50 \div 5 = -10$$

6. $56 \div (-7)$

7. $-26 \div 13$

8. $-45 \div -9$

9. $52 \div 12$

10. $-32 \div -4$

Example $\dfrac{-\dfrac{35}{21}}{-\dfrac{7}{42}}$

$$\frac{-35}{21} \div \frac{-7}{42} = \left(\frac{-\overset{5}{35}}{21}\right)\left(\frac{\overset{2}{42}}{-7}\right) = 10$$

11. $\dfrac{-\dfrac{8}{13}}{\dfrac{16}{26}}$

12. $\dfrac{-\dfrac{4}{5}}{-\dfrac{10}{12}}$

13. $\dfrac{3}{7} \div -\dfrac{4}{5}$

14. $\dfrac{4}{17} \div \dfrac{34}{19}$

15. $\dfrac{-\dfrac{35}{12}}{\dfrac{2}{3}}$

Example $\quad (3)(9)(-7)(2)\left(-\dfrac{1}{4}\right)$

$(3)(9) = 27 \qquad (27)(-7)(2)\left(-\dfrac{1}{4}\right)$

$(27)(-7) = -189 \qquad (-189)(2)\left(-\dfrac{1}{4}\right)$

$(-189)(2) = -378 \qquad (-378)\left(-\dfrac{1}{4}\right)$

$$\dfrac{378}{4} = \dfrac{189}{2} \text{ or } 94\dfrac{1}{2}$$

16. $(8)(-5)(4)$

17. $(-10)(-2)(4)(-1)$

18. $(2)(-18)\left(-\dfrac{1}{2}\right)(12)$

19. $(-8)\left(\dfrac{1}{4}\right)(-6)(0)$

20. $\left(-\dfrac{1}{3}\right)(21)(-5)(-2)$

Section 9.4

Perform each operation in the proper order.

Example $\quad (6)(-10) - (-3)$

$$(6)(-10) - (-3)$$
$$-60 - (-3)$$
$$-60 + 3 = -57$$

1. $3 + (-7) + (-2)$

2. $(2)(-6) - 10$

3. $3 + (-2)(5)$

4. $-18 + (3)(-6)$

5. $(5)(-7) - (-2)$

Example $\quad 16 - 35 \div -7$

$$16 - 35 \div -7 = 16 + (-35) \div -7$$
$$16 + 5$$
$$21$$

6. $-30 \div 2 + (-6)$

7. $25 - (-5) \div -15$

8. $-18 \div 3 - 6$

9. $-20 \div 5 - 2$

10. $32 \div (-8) - 6$

Example $\quad (2)(-3) + (5)(-2) \div 5$

$$(2)(-3) + (5)(-2) \div 5$$
$$-6 + (-10) \div 5$$
$$-6 + -2 = -8$$

11. $-30 \div (2)(-6)$

12. $5 - 1 \div 4$

13. $(6)(-3) + (-5)(2) \div 4$

14. $35 - (4)(-3) + 3 \div (-5)$

15. $-22 \div (11)(2) + (-5) - 6$

Example $\quad \dfrac{3 + 9 - 4}{6 - 8 + 3}$

$$\dfrac{3 + 9 - 4}{6 - 8 + 3} = \dfrac{12 - 4}{-2 + 3} = \dfrac{8}{1} = 8$$

16. $\dfrac{6(-4) + 2}{2 - 4 - 3}$

17. $\dfrac{5 + 6 - 8}{9 - 2 + 6}$

18. $\dfrac{-18 \div (-9)}{8(-2) - 4}$

19. $\dfrac{4(-3) - 8 + 2}{36 \div (-4)}$

20. $6 + (-6) - (-2)$

CHAPTER ORGANIZER

Topic	Procedure	Examples
Absolute value, p. 476	The absolute value of a number is the distance between the number on the number line and zero.	$\lvert -6 \rvert = 6 \qquad \lvert 3 \rvert = 3 \qquad \lvert 0 \rvert = 0$
Adding signed numbers with the same sign, p. 476	To add two numbers with the same sign: 1. Add the absolute value of the numbers. 2. Use the common sign in the answer.	$12 + 5 = 17$ $-6 + (-8) = -14$ $-5.2 + (-3.5) = -8.7$ $-\dfrac{1}{7} + \left(-\dfrac{3}{7}\right) = -\dfrac{4}{7}$
Adding signed numbers with opposite signs, p. 478	To add two numbers with different signs: 1. Find the difference between the absolute value of the larger number and the absolute value of the smaller number. 2. Place the sign of the number with the larger absolute value in front of the difference.	$14 + (-8) = 6$ $-14 + 8 = -6$ $-3.2 + 7.1 = 3.9$ $\dfrac{5}{13} + \left(-\dfrac{8}{13}\right) = -\dfrac{3}{13}$

Topic	Procedure	Examples
Subtracting signed numbers, p. 485	To subtract signed numbers, add the opposite of the second number to the first number.	$-9 - (-3) = -9 + 3 = -6$ $5 - (-7) = 5 + 7 = 12$ $8 - 12 = 8 + (-12) = -4$ $-4 - 13 = -4 + (-13) = -17$ $-\dfrac{1}{12} - \left(-\dfrac{5}{12}\right) = -\dfrac{1}{12} + \dfrac{5}{12} = \dfrac{4}{12} = \dfrac{1}{3}$
Multiplying or dividing signed numbers with the same sign, p. 491	To multiply or divide two numbers with the same sign, multiply or divide the absolute values. The sign of the result is positive.	$(0.5)(0.3) = 0.15$ $(-6)(-2) = 12$ $\dfrac{-20}{-2} = 10$ $\dfrac{-\dfrac{1}{3}}{-\dfrac{1}{7}} = \left(-\dfrac{1}{3}\right)\left(-\dfrac{7}{1}\right) = \dfrac{7}{3}$
Multiplying or dividing signed numbers with opposite signs, p. 492	To multiply or divide two numbers with opposite signs, multiply or divide the absolute values. The result is negative.	$7(-3) = -21$ $(-6)(4) = -24$ $(-36) \div (2) = -18$ $\dfrac{41.6}{-8} = -5.2$
Combined operations, with addition, subtraction, multiplication, and division, p. 497	1. First multiplication and division are done from left to right. 2. Then addition and subtraction are done from left to right.	Perform the operations in the proper order. **(a)** $-12 + 30 \div (-5) = -12 + (-6) = -18$ **(b)** $-21 \div 3 - 26 \div (-2) = -7 - (-13)$ $\qquad\qquad = -7 + 13 = 6$
Simplifying fractions with combined operations in numerator and denominator, p. 498	1. Perform the operations in the numerator. 2. Perform the operations in the denominator. 3. Reduce the fraction.	Perform the operations in the proper order. $\dfrac{7(-4) - (-2)}{8 - (-5)}$ The numerator is $7(-4) - (-2) = -28 - (-2) = -26$ The denominator is $8 - (-5) = 8 + 5 = 13$ Thus the fraction becomes $-\dfrac{26}{13} = -2$

REVIEW PROBLEMS CHAPTER 9

9.1 *Add.*

1. $17 + (-6)$ **2.** $-12 + (-3)$ **3.** $-15 + (-7)$ **4.** $16 + (-3)$

5. $-20 + 5$ **6.** $-18 + 4$ **7.** $-3.6 + (-5.2)$ **8.** $-8.6 + (-1.1)$

9. $-\dfrac{1}{5} + \left(-\dfrac{1}{3}\right)$ **10.** $-\dfrac{2}{7} + \dfrac{5}{14}$ **11.** $20 + (-14)$ **12.** $-80 + 60$

13. $7 + (-2) + 9 + (-3)$ **14.** $5 + (-4) + 6 + (-10)$

15. $8 + (-7) + (-6) + 3 + (-2) + 8$ **16.** $4 + (-10) + 6 + (-3) + (-8) + 7$

9.2 *Subtract.*

17. $12 - 16$ **18.** $18 - 36$ **19.** $-2 - 7$ **20.** $-5 - 6$ **21.** $-36 - (-21)$ **22.** $-21 - (-28)$

23. $12 - (-7)$ **24.** $14 - (-3)$ **25.** $1.6 - 3.2$ **26.** $-5.2 - 7.1$ **27.** $-\dfrac{2}{5} - \left(-\dfrac{1}{3}\right)$ **28.** $\dfrac{1}{4} - \left(-\dfrac{3}{8}\right)$

Perform the operations from left to right.

29. $5 - (-2) - (-6)$ **30.** $-15 - (-3) + 9$ **31.** $9 - 8 - 6 - 4$ **32.** $-7 - 8 - (-3)$

9.3 *Multiply or divide.*

33. $7(-2)$ **34.** $6(-9)$ **35.** $(-10)(-5)$ **36.** $(-16)(-3)$ **37.** $\left(-\dfrac{2}{7}\right)\left(-\dfrac{1}{5}\right)$

38. $\left(-\dfrac{9}{17}\right)\left(\dfrac{2}{3}\right)$ **39.** $(5.2)(-1.5)$ **40.** $(-3.6)(-1.2)$ **41.** $-60 \div (-20)$ **42.** $-18 \div (-3)$

43. $\dfrac{-36}{4}$ **44.** $\dfrac{-20}{10}$ **45.** $\dfrac{-13.2}{-2.2}$ **46.** $\dfrac{-17.4}{5.8}$ **47.** $\dfrac{-\frac{2}{5}}{\frac{4}{7}}$

48. $\dfrac{-\frac{1}{3}}{-\frac{7}{9}}$ **49.** $3(-5)(-2)$ **50.** $6(-8)(-1)$ **51.** $4(-7)(-8)\left(-\dfrac{1}{2}\right)$ **52.** $3(-6)(-5)\left(-\dfrac{1}{5}\right)$

9.4 *Perform the operations in the proper order.*

53. $7(-8) - (-2)$ **54.** $12 \div (-3) + 5$ **55.** $7 - 7(-1)$ **56.** $6 - 5(-2)(-1)$ **57.** $8 - (-30) \div 6$

58. $26 + (-28) \div 4$ **59.** $2(-6) + 3(-4) - (-13)$ **60.** $-49 \div (-7) + 3(-2)$ **61.** $36 \div (-12) + 50 \div (-25)$

62. $15 - (-30) \div 15$ **63.** $50 \div 25(-4)$ **64.** $-70 \div 35(-3)$ **65.** $9(-9) + 9$

66. $4 - 3(-4)$ **67.** $8 \div (-4)(2) + 3(-7) - 5$ **68.** $(-2)(-6) + 3(-2) + 10 \div (-5)$

In problems 69–76, simplify the numerator and denominator first using the proper order of operations. Then reduce the fraction if possible.

69. $\dfrac{5 - 9 + 2}{3 - 5}$ **70.** $\dfrac{5(-2) + 3 + 2}{17 - 6 - 1}$ **71.** $\dfrac{20 \div (-5) - (-6)}{(2)(-2)(-5)}$ **72.** $\dfrac{6 - (-3) - 2}{5 - 2 \div (-1)}$

73. $\dfrac{7 + (-7) - (-4)}{(-3)(-4) + (-2)}$ **74.** $\dfrac{-18 \div (-9)}{(3)(-4) - 6(-3)}$ **75.** $\dfrac{9 - (-3) + 4 + 30 \div (-15)}{2 - (-6) + 4(-3) - 3}$ **76.** $\dfrac{(2)(4)(-3) + 5(-4)}{2(-4) + 18 \div (-6)}$

In problems 1–11, add the signed numbers.

1. $-13 + 6$ **2.** $-9 + (-2)$ **3.** $3.7 + (-5.6)$ **4.** $\dfrac{1}{4} + \left(-\dfrac{2}{3}\right)$

5. $-100 + (-30)$ **6.** $-85 + 65$

7. $\dfrac{1}{5} + \left(-\dfrac{3}{25}\right)$ **8.** $-3.4 + (-2.2)$

9. $5 + 9 + (-4)$ **10.** $8 + (-2) + (-7) + 11$

11. $-12 + (-3) + 4 + (-7)$ **12.** Last night the temperature was $-13°F$. Today it rose $16°F$. What is the new temperature?

Find the profit or loss situation during the two-month period specified for a company having the following financial reports.

13. A $30,000 loss in October followed by a $40,000 loss in November

14. A $50,000 loss in January followed by a $25,000 profit in February

Subtract.

15. $17 - 20$ **16.** $-5 - (-3)$ **17.** $\dfrac{2}{5} - \left(-\dfrac{1}{3}\right)$

1. _____

2. _____

3. _____

4. _____

5. _____

6. _____

7. _____

8. _____

9. _____

10. _____

11. _____

12. _____

13. _____

14. _____

15. _____

16. _____

17. _____

18. _____

19. _____

20. _____

21. _____

22. _____

23. _____

24. _____

25. _____

26. _____

27. _____

28. _____

29. _____

30. _____

31. _____

32. _____

33. _____

34. _____

18. $-4.2 - 6.7$ **19.** $-35 - (-27)$ **20.** $40 - (-72)$

21. $9.9 - 12.5$ **22.** $\dfrac{1}{4} - \left(-\dfrac{5}{6}\right)$ **23.** $150 - (-150)$

24. $-200 - 316$ **25.** Find the difference in temperature in Copper Center, Alaska, between $-13°F$ in the day and $-40°F$ at night.

26. Find the difference in temperature in Anchorage, Alaska, between $18°F$ in the day and $-26°F$ at night.   **27.** Find the difference in altitude between a mountain 3965 ft high and a desert valley 365 ft below sea level.

Perform each set of operations working from left to right.

28. $7 + (-6) + (-2)$ **29.** $-50 + (-30) + 10$

30. $7 - (-3) + 10$ **31.** $-16 - (-12) - 3$

32. $-2.6 + (-3.5) + 20.2$ **33.** $5.3 + 7.2 - 15.8$

34. $16 - (-8) + 5 - (-4)$

Multiply or divide.

1. $(-7)(-6)$

2. $4(-9)$

3. $\dfrac{-50}{-25}$

4. $\dfrac{-39}{3}$

5. $48 \div (-16)$

6. $-28 \div (-7)$

7. $\dfrac{1}{3}\left(-\dfrac{1}{5}\right)$

8. $\left(-\dfrac{2}{7}\right)\left(-\dfrac{3}{4}\right)$

9. $(-1.2)(-3)$

10. $(-5.6)(2)$

11. $(5)(-3)(-2)$

12. $(-7)(-6)(-2)$

13. $\dfrac{-\frac{2}{5}}{4}$

14. $\dfrac{-\frac{3}{5}}{7}$

15. $\dfrac{-\frac{4}{7}}{-\frac{1}{3}}$

16. $\dfrac{\frac{8}{9}}{-\frac{4}{5}}$

1. _____

2. _____

3. _____

4. _____

5. _____

6. _____

7. _____

8. _____

9. _____

10. _____

11. _____

12. _____

13. _____

14. _____

15. _____

16. _____

17. _____

18. _____

19. _____

20. _____

21. _____

22. _____

23. _____

24. _____

25. _____

26. _____

27. _____

28. _____

29. _____

30. _____

31. _____

32. _____

33. _____

34. _____

35. _____

36. _____

17. $\dfrac{150.5}{-5}$

18. $\dfrac{-52.5}{3}$

19. $(-2.5)(-6.5)$

20. $(-3.8)(-1.5)$

21. $(-2)(4)(-3)$

22. $(-6)(-1)(-5)$

23. $(7)(-6)(3)(0)(-9)$

24. $(3)(-5)(-1)(-2)$

25. $3(-4)\left(-\dfrac{1}{2}\right)(5)$

26. $\left(\dfrac{2}{7}\right)(-14)(-1)(3)$

Perform the operations in the proper order.

27. $6 + 7(-3)$

28. $-5 + 2(-4)$

29. $5(-2) + (-3)(-4)$

30. $-25 \div (-5) + 10$

31. $-49 \div 7(3) - 4$

32. $5 - (-30) \div 6$

33. $\dfrac{8 + 2 - 6}{3(-2) + 2}$

34. $\dfrac{3(-4) + 5}{-10 \div (-5) + 12}$

35. $\dfrac{6 - (-1) + 1}{4 - 8 \div (-2)}$

36. $\dfrac{2(-3) - 3(2)}{(-5)(-3) - 3}$

Add.

1. $-15 + 8$

2. $-36 + (-10)$

3. $7.5 + (-3.8)$

4. $-2 + (-5) + 10 + (-6)$

5. $6 + (-8) + 1 + (-4)$

6. $-\dfrac{1}{3} + \left(-\dfrac{5}{6} \right)$

7. $-\dfrac{7}{18} + \left(-\dfrac{1}{9} \right)$

8. $5.2 + (-4.7)$

Subtract.

9. $-23 - 4$

10. $21 - 16$

11. $\dfrac{3}{5} - \left(-\dfrac{2}{3} \right)$

12. $-20 - (-9)$

13. $-1.5 - (-6.5)$

14. $5.9 - 2.6$

15. $\dfrac{1}{14} - \left(-\dfrac{3}{7} \right)$

16. $35 - (-35)$

1. _____

2. _____

3. _____

4. _____

5. _____

6. _____

7. _____

8. _____

9. _____

10. _____

11. _____

12. _____

13. _____

14. _____

15. _____

16. _____

17. _____

18. _____

19. _____

20. _____

21. _____

22. _____

23. _____

24. _____

25. _____

26. _____

27. _____

28. _____

29. _____

30. _____

31. _____

32. _____

Multiply or divide.

17. $(-10)(-3)$ **18.** $49 \div (-7)$ **19.** $-30 \div (-6)$

20. $(6)(-1)(-2)(-3)\left(\dfrac{1}{3}\right)$ **21.** $\dfrac{-26}{-13}$ **22.** $\dfrac{-\dfrac{2}{7}}{\dfrac{4}{5}}$

23. $(-9)(-2)(-3)$ **24.** $84 \div (-4)$

Perform the operations in the proper order.

25. $9 - 3(-7)$ **26.** $(-36) \div (-9) + 3$

27. $15 \div (-5) + 38 \div (-19)$ **28.** $9(-2) + 5(-6)$

29. $5 - 8 - (-3) + 5(-7)$ **30.** $-16 \div (-8) - 3(-2)$

31. $\dfrac{6 + 1 - 5}{(-6)(4) + (-5)(-4)}$ **32.** $\dfrac{8 + 30 \div (-6)}{5 - (-4)}$

Approximately one-half of this test is based on Chapter 9 material. The remainder is based on material covered in Chapters 1–8.

Do each problem. Simplify your answers.

1. Subtract: 28,981
            − 16,598

2. Divide: $36 \overline{)4572}$

3. Add: $3\frac{1}{4} + 8\frac{2}{3}$.

4. Multiply: $1\frac{5}{6} \times 2\frac{1}{2}$.

5. Round to the nearest thousandth: 9.812456.

6. Add: $5.82 + 38.964 + 0.571 + 9.305 + 8.8$.

7. Multiply: 12.89×5.12.

8. Find n: $\dfrac{n}{8} = \dfrac{56}{7}$.

9. For every 156 parts manufactured, there are seven defects. If 2808 parts are manufactured, how many defects would you expect?

10. What is 0.8% of 38?

11. 12% of what number is 480?

12. Convert 94 km to m.

13. Convert 180 in to yd.

14. Find the area of a circle with radius 5 m. Round your answer to the nearest tenth.

1. _____

2. _____

3. _____

4. _____

5. _____

6. _____

7. _____

8. _____

9. _____

10. _____

11. _____

12. _____

13. _____

14. _____

15. _____

16. _____

17. _____

18. _____

19. _____

20. _____

21. _____

22. _____

23. _____

24. _____

25. _____

26. _____

27. _____

28. _____

29. _____

30. _____

15. The following histogram depicts the ages of students at Wolfville College.
(a) How many students are between ages 23 and 25?
(b) How many students are older than 19 years?

16. Evaluate: $\sqrt{36} + \sqrt{49}$.

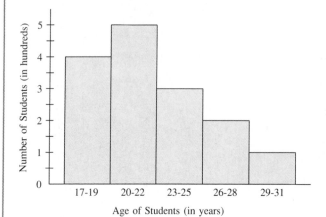

Add.

17. $-1.2 + (-3.5)$

18. $-\dfrac{1}{4} + \dfrac{2}{3}$

Subtract.

19. $7 - 18$

20. $-8 - (-3)$

Multiply or divide.

21. $(5)(-3)(-1)(-2)(2)$

22. $\dfrac{-\dfrac{3}{7}}{-\dfrac{5}{14}}$

Perform the operations in the proper order.

23. $6 - 3(-4)$

24. $(-20) \div (-2) + (-6)$

25. $5 - (-7) + 36 \div (-18)$

26. $-8 - (3)(-7) + 5$

Simplify the numerator and denominator in each problem. Then reduce the fraction if possible. Perform all operations in the proper order.

27. $\dfrac{6 - 5 + (-8)}{(-5)(-3) + (-1)}$

28. $\dfrac{22 \div (-11)(-8)}{1 - 7(-2)}$

29. $\dfrac{(-2)(-1) + (-4)(-3)}{1 + (-4)(2)}$

30. $\dfrac{7 - 8 + 3 - (-10)}{-3(8)}$

Introduction to Algebra

Computer programmers use algebra to write computer programs of all types.

PRETEST CHAPTER 10

If you are familiar with the topics in this chapter, take this test now. Check your answers with those in the back of the book. If an answer was wrong or you couldn't do a problem, study the appropriate section of the chapter.

If you are not familiar with the topics in this chapter, don't take this test now. Instead, study the examples, work the practice problems, and then take the test.

This test will help you identify those concepts that you have mastered and those that need more study.

Section 10.1 Collect like terms.

1. $5x - 8x$

2. $-12y - 16y + 3y$

3. $a + 2b - 3a - 6b$

4. $5x + y - 2 - 8x + 3y - 5$

5. $3a + 6 - 5b - 8 - 7b - 9a$

6. $7x - 4w + 3z - 8w$

Section 10.2 Simplify.

7. $(5)(6x - 7y)$

8. $(-2)(x + 3y - 5)$

9. $(-3)(2.5x - 1.6y + 2z - 4)$

10. $(3)(2a - b) - (6)(3a + b)$

Section 10.3 Solve for the variable.

11. $7x - 8 = 20$

12. $15 - 2x = 6x - 1$

13. $(3)(x - 1) = 9 - (2)(x - 4)$

14. $2x - 3 = (3)(4 - x)$

15. $7x + 6 - 5x = -12 + 3x - 4$

16. $8x - 9 = (-2)(x + 3 - 4x)$

Section 10.4 Translate the English sentence into an equation using the variables indicated for each problem.

17. The computer weighs 9 kg more than the printer. Use c to represent the weight of the computer and p to represent the weight of the printer.

18. The length of the rectangle is 11 m longer than double the width. Use l to represent the length and w the width of the rectangle.

Write algebraic expressions for each of the specified quantities using the same variable.

19. Mount Hood is 1758 m shorter than Mount Ararat in Turkey. Use the variable a to describe the height of each mountain in meters.

20. Angle A is double angle B. Angle C is 35° smaller than angle B. Use the variable b to describe the number of degrees in each of the three angles.

Section 10.5 Solve each problem by using an equation. Show what variable or expression you use to represent each quantity.

21. Hector drove 898 mi in three days. He drove 56 more miles on Tuesday than Monday. He drove 37 fewer miles on Wednesday than on Monday. How many miles did he drive on each of the three days?

22. A rectangular field has a perimeter of 92 mi. The length is 5 m less than double the width. Find the dimensions of the field.

23. An 18-ft board is cut into two pieces. One piece is 2.5 ft longer than the other. Find the length of each piece.

24. The three employees of Mountainland Surveying Company have a combined income of $82,000 per year. The surveyor earns triple the salary of the company secretary. The surveyor's assistant earns $12,000 more than the company secretary. What is the annual salary of each of the three people?

10.1 VARIABLES AND LIKE TERMS

After studying this section, you will be able to:

1 *Understand what a variable is*

2 *Collect like terms containing a variable*

In algebra we reason and solve problems by means of symbols. A *variable* is a symbol, usually a letter of the alphabet, that stands for a number. We can *use* it even though we may not know exactly the value of that number. However, by working with the variable in a logical manner, we can often figure out what that value is.

The "logical manner" of working with symbols can be described as making up the "rules of algebra." In this section, we'll learn about *like terms*—exactly when and how you can combine them and when terms cannot be combined further.

As we progress through this chapter we'll learn more logical rules, until, by the end of the chapter, we'll be able to solve some challenging applied problems.

1 The Concept of a Variable

When we do not know the value of a number, we use a letter to represent that number. A letter that represents a number is called a **variable**. In the formula for the area of a circle the equation $A = \pi r^2$ contains the variable r. The r represents the value of the radius. The letter A is a variable. The A represents the value of the area. π is not a variable. Its value is known. It is a number that is represented by a decimal that goes on forever without a noticeable pattern.

When we want to solve a proportion such as

$$\frac{3}{n} = \frac{27}{16}$$

we represent the unknown value by n. The n is the variable in the proportion equation.

Example 1 Name the variables in the following equations:

(a) $A = lw$
(b) $p = 2w + 2l$
(c) $V = \dfrac{4\pi r^3}{3}$

(a) $A = lw$ The variables are A, l, and w.
(b) $p = 2w + 2l$ The variables are p, w, and l.
(c) $V = \dfrac{4\pi r^3}{3}$ The variables are V and r.   ■

Practice Problem 1 Name the variables.

(a) $A = \dfrac{bh}{2}$
(b) $C = \pi d$
(c) $V = lwh$ ■

The product of a number times a variable can be written with a multiplication sign. To multiply three times n we can write $3 \times n$. However, it is usually written as $3n$. Whenever a number is written just to the left of a letter, with no $+$ or $-$ sign between the number and the letter, it indicates that the number is multiplied by the variable. Thus $4ab$ means 4 times a times b. A number just to the left of a parentheses also means multiplication. Thus $3(n + 8)$ means $3 \times (n + 8)$. So a formula can be written with a multiplication sign, $\times$, or without: $A = \dfrac{b \times h}{2}$ can be written without a multiplication sign as $A = \dfrac{bh}{2}$.

Example 2 Write the formula without a multiplication sign.

(a) $V = \dfrac{B \times h}{3}$
(b) $A = \dfrac{h \times (B + b)}{2}$

(a) $V = \dfrac{Bh}{3}$
(b) $A = \dfrac{h(B + b)}{2}$ ■

Practice Problem 2 Write the formula without a multiplication sign.
(a) $p = 2 \times w + 2 \times l$
(b) $A = \pi \times r^2$ ■

❷ Like Terms

We have seen many examples of adding and subtracting like quantities. This is called **combining** like quantities. A carpenter might perform the following:

$$20 \text{ m} - 3 \text{ m} = 17 \text{ m}$$

$$7 \text{ in} + 9 \text{ in} = 16 \text{ in}$$

We cannot combine quantities that are not the same. We cannot add 5 yd + 7 gal. We cannot subtract 12 lbs − 3 in.

Similarly, when using variables, we can add or subtract only when the same variable is used. For example, we can add $4a + 5a = 9a$, but we cannot add $4a + 5b$.

A **term** is a number, a variable, or a product of a number and one or more variables separated in an expression by a + sign or a − sign. In the expression $2x + 4y + (-1)$ there are three terms, $2x$, $4y$, and -1. **Like terms** have identical variables and identical exponents. So in the expression $3x + 4y + (-2x)$, the two terms $3x$ and $(-2x)$ are called like terms. The terms $2x^2y$ and $5xy$ are not like terms since the exponents for the variable x are not the same. To combine like terms you combine the numbers, called the numerical coefficients, which are directly in front of the terms by using the rules for adding signed numbers.

Example 3   Add: $5x + 7x$.

We add $5 + 7 = 12$. Thus

$$5x + 7x = 12x$$

The reason we can do this is by a property of signed numbers called the distributive property. Using this property we can write

$$5x + 7x = (5 + 7)x = 12x$$

We discuss this property in detail in Section 10.2.   ■

Practice Problem 3   Add: $9x + 2x$.   ■

When we combined signed numbers as coefficients we try to add them mentally without rewriting the problem from subtraction to addition. Thus to combine $7x - 9x$ we think $\boxed{7x + (-9x)}$ and we write the answer, $-2x$. However, you may find this extra step to be really necessary to be sure that you understand what you are doing. Your instructor may ask you to write out this step as part of your work. In the following example we show this extra step inside a $\boxed{}$ box. You should determine from your instructor if he or she feels that this step is necessary.

Example 4   Combine.

(a) $14x - 20x$ **(b)** $3x + 7x - 15x$ **(c)** $9x - 12x - 11x$

(a) $14x - 20x = \boxed{14x + (-20x)} = -6x$

(b) $3x + 7x - 15x = \boxed{3x + 7x + (-15x)} = 10x + (-15x) = -5x$

(c) $9x - 12x - 11x = \boxed{9x + (-12x) + (-11x)} = -3x + (-11x) = -14x$ ■

Practice Problem 4   Combine.
(a) $26x - 18x$ **(b)** $8x - 22x + 5x$ **(c)** $19x - 7x - 12x$ ■

A variable without a numerical coefficient is understood to have a coefficient of 1. $7x + y$ means $7x + 1y$. $5a - b$ means $5a - 1b$.

Example 5 Combine.

(a) $3x - 8x + x$
(b) $12x - x - 20.5x$

(a) $3x - 8x + x = 3x - 8x + 1x = -5x + 1x = -4x$
(b) $12x - x - 20.5x = 12x - 1x - 20.5x = 11.0x - 20.5x = -9.5x$ ∎

Practice Problem 5 Combine.
(a) $9x - 12x + x$
(b) $5.6x - 8x + 10x$ ∎

> Numbers cannot be combined with variable terms.

Example 6 Combine.

(a) $3 - 10x + 5 - 18x$
(b) $7.8 - 2.3x + 9.6x - 10.8$

In each case we combine the numbers separately and the variable terms separately. It may help to use the commutative and associative properties first.

(a) $3 - 10x + 5 - 18x$
$= 3 + 5 - 10x - 18x$
$= 8 - 28x$

(b) $7.8 - 2.3x + 9.6x - 10.8$
$= 7.8 - 10.8 - 2.3x + 9.6x$
$= -3 + 7.3x$ ∎

Practice Problem 6 Combine.
(a) $-8 + 2x - 9 - 15x$
(b) $17.5 - 6.3x - 8.2x + 10.5$ ∎

There may be more than one letter in a problem. However, only like terms may be combined. Fortunately, there are mathematical properties stating that we can rearrange the terms in the problem as we want and combine the like terms in any order.

To Think About

?

Why can we rearrange the order of the terms? Why can we add the terms in any fashion? The commutative property of addition says that $4 + 5 = 5 + 4$. For any two real numbers we can write the commutative property of addition as $a + b = b + a$.

The associative property of addition says that when we add $3 + 5 + 7$ we can add $3 + 5$ first or we can add $5 + 7$ first. The answer will be the same. We write this 15 as follows:

$$(3 + 5) + 7 = 3 + (5 + 7)$$

We can see this is true since

$$8 + 7 = 3 + 12$$

The two numbers inside the parentheses are combined first, but the answer is the same. We can write the associative property of addition as $(a + b) + c = a + (b + c)$.

Suppose that we want to add $5x + 7x + 8x$. We could add $5x + 7x$ first or $7x + 8x$ first. That is, $(5x + 7x) + 8x = 5x + (7x + 8x)$. This is true by the associative property of addition. We could rearrange the order of the terms so that

$$(5x + 7x) + 8x = (7x + 5x) + 8x$$

This is true by the commutative property of addition. Some further work on this appears in problems 45 and 46. ∎

Example 7 Combine.

(a) $5x + 2y + 8 - 6x + 3y - 4$
(b) $2n + p + q - 8n + 5p - 6q$

For convenience we will rearrange the problem to place like terms next to each other. This is an optional step; you do not need to do this.
(a) $5x - 6x + 2y + 3y + 8 - 4 = -1x + 5y + 4$ or $-x + 5y + 4$
(b) $2n - 8n + 1p + 5p + 1q - 6q = -6n + 6p - 5q$ ∎

Practice Problem 7 Combine.

(a) $2w + 3z - 12 - 5w - z - 16$ (b) $-3a + 2b + c - 7a - 5b - 2c$ ■

The order of the terms in an answer is not important in this type of problem. The answer to Example 7(a) could have been $5y + 4 - x$ or $4 + 5y - x$. Often we give the answer with the letters in alphabetical order.

EXERCISES 10.1

Name the variables in each equation.

1. $N = 2bh$

2. $N = 3bh$

3. $S = 2\pi r^2$

4. $S = 3\pi r^3$

5. $p = \dfrac{4ab}{3}$

6. $p = \dfrac{7ab}{4}$

Write each equation without a multiplication sign.

7. $p = 4 \times s^2$

8. $p = 2 \times w + 2 \times l$

9. $A = \dfrac{b \times h}{2}$

10. $A = b \times h$

11. $V = \dfrac{4 \times \pi \times r^3}{3}$

12. $V = l \times w \times h$

13. $H = 2 \times a - 3 \times b$

14. $F = 4 \times p - 2 \times q$

Combine like terms.

15. $9x + 8x$

16. $2x + 12x$

17. $-15x + 20x$

18. $-18x + 30x$

19. $2x - 8x + 5x$

20. $9x + 3x - 7x$

21. $6x - 12x - 15x$

22. $4x - 3x - 21x$

23. $x + 3x + 8 - 7$

24. $5x - x + 9 + 16$

25. $1.3x + 10 - 2.4x - 3.6$

26. $5.2x + 12.3 - 7.4x - 16.8$

27. $-6.5x + 1.2x + 3 - 8.6$

28. $9x - 14.6x - 3.2 + 1.8$

29. $-13x + 7 - 19x - 10$

30. $5 - 30x - 7 - 15x$

31. $4x + 3y - 8 - 2x - 6y + 4$

32. $9x + 3y - 5 + 6x - 9y - 2$

33. $5a - 3b - c + 8a - 2b - 6c$

34. $10a + 5b - 7c - a - 8b - 3c$

35. $-7s - 2r + 3 - s - r - 8$

36. $-8s - 9r + 3 + 7s - 8r - 10$

37. $8n - 12p + q + 3q - 8p - 9p$

38. $15n - p + 3q - 14n - 7p - 10q$

39. $7x - 9.5x + 3.6 - 12x - 14x - 8$

40. $2x - 6.5x + 5.3 - 10x - 22x - 11$

41. $7x - 8y + 3 + 8y - 3 - 7x$

42. $14x - 3y + 12 - 14x - 12 + 3y$

43. $5.2x - 3.4y + 6.8 - 7.2x - 9.5y$

44. $3x - 8.6y + 5.5 - 9.6x + 3.2y$

In problems 45 and 46, give a reason for each step.

45. $(9x + 3) + 5x$
  $= 9x + (3 + 5x)$ Step 1:
  $= 9x + (5x + 3)$ Step 2:
  $= (9x + 5x) + 3$ Step 3:
  $= (9 + 5)x + 3$ Step 4:
  $= 14x + 3$ Step 5:

46. $(-2p + 7) + 8p$
  $= -2p + (7 + 8p)$ Step 1:
  $= -2p + (8p + 7)$ Step 2:
  $= (-2p + 8p) + 7$ Step 3:
  $= (-2 + 8)p + 7$ Step 4:
  $= 6p + 7$ Step 5:

Combine like terms.

47. $-17.8x + 2.3y + 5.8w - 12.2 - 9.6x + 13y - 9.2w - 15.3$

48. $28.6x - 5.2y - 9.8w + 27.6 + 3.4x + 19y - 12.6w - 24.9$

Cumulative Review Problems

Solve for n. Round to the nearest hundredth if necessary.

49. $\dfrac{n}{6} = \dfrac{12}{15}$

50. $\dfrac{n}{9} = \dfrac{36}{40}$

51. $6n = 18$

52. $5.5n = 46.75$

 Calculator Problems

Combine like terms.

53. $3.25x + 2.206y - 12.66x + 6.22y$

54. $8.123y + 0.0211 - 2.66y - 1.011$

For Extra Practice Examples and Exercises, turn to page 545.

Solutions to Odd-Numbered Practice Problems

1. (a) The variables are A, b, and h. **(b)** The variables are C and d. **(c)** The variables are V, l, w, and h.
3. $9x + 2x = 11x$ **5. (a)** $9x - 12x + x = -3x + 1x = -2x$ **(b)** $5.6x - 8x + 10x = -2.4x + 10x = 7.6x$
7. (a) $2w - 5w + 3z - 1z - 12 - 16 = -3w + 2z - 28$ **(b)** $-3a - 7a + 2b - 5b + 1c - 2c = -10a - 3b - c$

Answers to Even-Numbered Practice Problems

2. (a) $p = 2w + 2l$ **(b)** $A = \pi r^2$ **4. (a)** $8x$ **(b)** $-9x$ **(c)** 0 **6. (a)** $-17 - 13x$ **(b)** $28 - 14.5x$

10.2 THE DISTRIBUTIVE PROPERTIES

☐ After studying this section, you will be able to:

1 Remove parentheses using the distributive properties

2 Simplify expressions by removing parentheses and collecting like terms

What do we mean by the word "property" in mathematics (for example, the commutative *property* of addition, and the associative *property* of addition)? A property is an essential characteristic. A *property* of addition is an *essential characteristic* of addition. In this section we learn about the distributive property (in two forms), a property with far-reaching implications in algebra. In this section we use this property to simplify expressions. The shortcut direction "remove parentheses" means "use the distributive property." We learn how to do that here.

1 Sometimes we encounter expressions like $4(x + 3)$. Here we have two operations: multiplication of the 4 times the quantity $(x + 3)$, and addition of $x + 3$.

We'd like to be able to simplify the expression so that no parentheses appear. How can we do this? We cannot use the commutative property of addition or the associative property of addition: these deal only with addition, and here we have both multiplication and addition.

There is another property that we can use, which we have briefly mentioned before, called the **distributive property**. It says that

$$4(x + 3) \qquad \text{is equal to} \qquad 4(x) + 4(3)$$

That is, the multiplication of the numerical coefficient 4 can be "distributed over" the expression $x + 3$ by multiplying the 4 by each of the terms in the parentheses. In the expression that results, no parentheses appear. In this process we could say that "multiplication is distributive over addition." In this way we have a new expression, $4x + 12$, equivalent to the one we started with but with no parentheses.

Using variables, we can write the distributive property two ways.

Distributive Properties of Multiplication over Addition

If a, b, and c are signed numbers, then

$$a(b + c) = ab + ac \qquad \text{and} \qquad (b + c)a = ba + ca$$

A numerical example shows that the distributive property works.

$$(7)(4 + 6) = (7)(4) + (7)(6)$$
$$(7)(10) = 28 + 42$$
$$70 = 70$$

Example 1 Simplify.

(a) $(4)(x + 3)$ **(b)** $(-3)(x + 3y)$

We will use $a(b + c) = ab + ac$

(a) $(4)(x + 3) = 4x + (4)(3)$
$\qquad\qquad\qquad = 4x + 12$

(b) $(-3)(x + 3y) = -3x + (-3)(3)(y) = -3x + (-9y)$
$\qquad\qquad\qquad\qquad\qquad\qquad = -3x - 9y$ ∎

Practice Problem 1 Simplify.
(a) $(7)(x + 5)$ **(b)** $(-4)(x + 2y)$ ∎

Sometimes the distributive property is used with three terms within the parentheses. If a parentheses is used inside a parentheses, we will often change the outside ( ) to a bracket [ ] notation.

Example 2 Simplify.

(a) $(7)(2x - y + 7)$ **(b)** $(-5)(x + 2y - 8)$ **(c)** $(2)(1.5x + 3.6y + 7)$

(a) $(7)(2x - y + 7) = (7)[2x + (-1y) + 7] = 7(2x) + 7(-1y) + 7(7)$
$\qquad\qquad\qquad\qquad\qquad = 14x + (-7y) + 49 = 14x - 7y + 49$

(b) $(-5)(x + 2y - 8) = (-5)[x + 2y + (-8)] = -5(x) + (-5)(2y) + (-5)(-8)$
$\qquad\qquad\qquad\qquad\qquad = -5x + (-10y) + 40 = -5x - 10y + 40$

(c) $(2)(1.5x + 3.6y + 7) = 2(1.5x) + 2(3.6y) + 2(7) = 3x + 7.2y + 14$ ∎

In this example every step is shown in detail. You may find that you do not need to write so many steps.

Practice Problem 2 Simplify.

(a) $(8)(3x - y - 6)$ **(b)** $(-5)(x + 4y + 5)$ **(c)** $(3)(2.2x + 5.5y + 6)$ ∎

Sometimes the number is to the right of the parentheses, so we use

$$(b + c)a = ba + ca$$

Example 3 Simplify.

(a) $(5x + y)(2)$ **(b)** $(3x - y - 4)(3)$

(a) $(5x + y)(2) = (5x)2 + (y)(2)$
$$= 10x + 2y$$
(b) $(3x - y - 4)(3) = [3x + (-y) + (-4)](3)$
$$= (3x)(3) + (-y)(3) + (-4)(3)$$
$$= 9x + (-3y) + (-12)$$
$$= 9x - 3y - 12 \quad ∎$$

Notice in Example 3 that we write our final answer with the numerical coefficient to the *left* of the variable. We would not leave $(y)(2)$ as an answer but would write $2y$.

Practice Problem 3 Simplify.

(a) $(x + 3y)(8)$ **(b)** $(2x + 3y - 7)(2)$ ∎

The parentheses may contain four terms.

Example 4 Simplify.

(a) $(3)(4.5x + 3.2y - z - 2.5)$ **(b)** $(-7)(x - 5y + 10z - 10)$

(c) $\frac{2}{3}\left(x + \frac{1}{2}y - \frac{1}{4}z + \frac{1}{5}\right)$

(a) $(3)[4.5x + 3.2y + (-1z) + (-2.5)]$
$$= 13.5x + 9.6y + (-3z) + (-7.5) = 13.5x + 9.6y - 3z - 7.5$$
(b) $(-7)[1x + (-5y) + 10z + (-10)]$
$$= -7x + 35y + (-70z) + 70 = -7x + 35y - 70z + 70$$

(c) $\frac{2}{3}\left(x + \frac{1}{2}y - \frac{1}{4}z + \frac{1}{5}\right) = \left(\frac{2}{3}\right)(x) + \left(\frac{2}{3}\right)\left(\frac{1}{2}y\right) + \left(\frac{2}{3}\right)\left(-\frac{1}{4}z\right) + \left(\frac{2}{3}\right)\left(\frac{1}{5}\right)$

$$= \frac{2}{3}x + \frac{1}{3}y - \frac{1}{6}z + \frac{2}{15} \quad ∎$$

Practice Problem 4 Simplify.

(a) $(8)(1.2x + 2.4y - 3z - 1.5)$ **(b)** $(-9)(x + 3y - 4z + 5)$

(c) $\frac{3}{2}\left(\frac{1}{2}x - \frac{1}{3}y + 4z - \frac{1}{2}\right)$ ∎

2 After removing parentheses we may have a chance to collect like terms. The direction "simplify" means remove parentheses, collect like terms, and leave the answer in as simple, and correct, a form as possible.

Example 5 Simplify $(2)(x + 3y) + (3)(4x + 2y)$

$(2)(x + 3y) + (3)(4x + 2y) = 2x + 6y + 12x + 6y$ Using the distributive property.
$$= 14x + 12y \quad\quad \text{Collecting like terms.} \quad ∎$$

Practice Problem 5 Simplify $(3)(2x + 4y) + (2)(5x + y)$ ∎

Example 6 Simplify $(-3)(2x + 4) + (2)(-3 + 5x)$

$(-3)(2x + 4) + (2)(-3 + 5x) = -6x - 12 - 6 + 10x$ Using the distributive property.

$$= 4x - 18 \quad\quad\quad\quad \text{Collecting like terms.} \quad ∎$$

Practice Problem 6 Simplify $(-4)(x - 5) + (3)(-1 + 2x)$ ∎

Example 7 Simplify $\quad (2)(x - 3y) + (5)(2x + 6)$

$(2)(x - 3y) + (5)(2x + 6) = 2x - 6y + 10x + 30 \qquad$ Using the distributive property.

$= 12x - 6y + 30 \qquad$ Collecting like terms. ∎

Notice that in the final step of Example 7 only the x terms could be combined. There are no other like terms.

Practice Problem 7 Simplify $\quad (-3)(x + 2) + (4)(2x - 3y)$ ∎

To Think About

Is there any relationship between geometry and the distributive property? Can we show a simple realistic example of how the distributive property would work with geometric shapes? Yes. We can use the area of rectangles to illustrate the distributive property. Suppose that we take two rectangles of width 4 in × 2 in and 4 in × 5 in, respectively. If we combine the area of the two rectangles we have $4 \times 2 + 4 \times 5 = 8 + 20 = 28$ in². This is equivalent to finding the area of one rectangle formed by joining these two smaller rectangles together.
In this case we have shown that

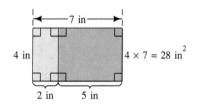

(a)   (b)

$$4 \times 2 + 4 \times 5 = 4 \times (2 + 5)$$

which is a specific numerical example of

$$ab + ac = a(b + c)$$

In general, for any rectangles with sides a, b, and c we can show that

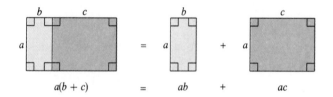

The area of the large $=$ the sum of the areas of the two smaller
rectangle $a(b + c)$ rectangles that make up the large rectangle.

Some similar questions are left for your study in problems 39–42. ∎

EXERCISES 10.2   PAUL

Simplify.

1. $(5)(x + 7)$ **2.** $(8)(2x + 1)$ **3.** $(9)(3x + 2)$ **4.** $(12)(x + 3)$

5. $(-2)(x + y)$ **6.** $(-3)(x + y)$ **7.** $(-7)(1.5x - 3y)$ **8.** $(-6)(2.5x - y)$

9. $(-10)(-3x + 7y)$ **10.** $(-12)(-2x + 8y)$ **11.** $(2)(x + 3y - 5)$ **12.** $(3)(2x + y - 6)$

13. $(-4)(a + b + 2c)$

14. $(-3)(3a - b + c)$

15. $(-8)(2a - 5b + c)$

16. $(-7)(-3a - b + 2c)$

17. $(15)(-12a + 2.2b + 6.7)$

18. $(14)(-10a + 3.2b + 4.5)$

19. $(4)(a - 3b + 5c + 7)$

20. $(5)(2a + b - 4c + 8)$

21. $(-2)(1.3x - 8.5y - 5z + 12)$

22. $(-3)(1.4x - 7.6y - 9z - 4)$

23. $\dfrac{1}{2}\left(2x - 3y + 4z - \dfrac{1}{2}\right)$

24. $\dfrac{1}{3}\left(-3x + \dfrac{1}{2}y + 2z - 3\right)$

Simplify. Be sure to collect like terms.

25. $(3)(2x + y) + (2)(x - y)$

26. $(4)(x + 2y) + (3)(x - y)$

27. $(5)(3x - 1) + (6)(x - 8)$

28. $(7)(2x - 1) + (2)(x - 3)$

29. $(-2)(a - 3b) + (3)(-a + 4b)$

30. $(-2)(3a - b) + (4)(-a + 3b)$

31. $(6)(3x + 2y) - (4)(x + 7)$

32. $(7)(2x + 3y) - (5)(x + 6)$

33. $(1.5)(x + 2.2y) + (3)(2.2x + 1.6y)$

34. $(2.4)(x + 3.5y) + (2)(1.4x + 1.9y)$

35. $(3)(a + b + 2c) - (4)(3a - b + 2c)$

36. $(2)(2a + b + c) - (5)(2a - b + c)$

37. The area of a trapezoid is $A = \dfrac{h(B + b)}{2}$. Write this formula without parentheses and without a multiplication sign.

38. The surface area of a cylinder is $S = 2\pi r(h + r)$. Write this formula without parentheses and without a multiplication sign.

? To Think About

Multiplication is distributive over subtraction. That is, for all signed numbers a, b, c, $a(b - c) = ab - ac$.

39. Illustrate this property by using the area of two rectangles.

40. Illustrate this property by using the area of two parallelograms.

41. Verify that this property is true for $a = 3$, $b = 2$, $c = -6$.

42. Verify that this property is true for $a = 4$, $b = -3$, $c = 8$.

Simplify.

43. $\left(\dfrac{1}{3}\right)\left(2x + \dfrac{1}{4}y + 6\right) - \left(\dfrac{1}{4}\right)\left(3x - \dfrac{2}{3}y + 2\right)$

44. $(1.6)(2.5x - 3.8y + 5.2) - (3.8)(4.6x - 7.2y - 9.8)$

Cumulative Review Problems

Round all answers to the nearest tenth. Use $\pi = 3.14$ when necessary.

45. Find the circumference of a circle with diameter of 7 in.

46. Find the circumference of a circle with a radius of 6 in.

47. Find the area of the shaded figure.

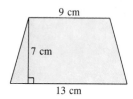

48. Find the volume of a cylinder of radius 4 in and height 6 in.

 Calculator Problems

Simplify.

49. $(3.22)(x - 2.06)$

50. $(-0.0122)(x + 0.316)$

For Extra Practice Examples and Exercises, turn to page 546.

Solutions to Odd-Numbered Practice Problems

1. (a) $(7)(x + 5) = 7x + 35$ **(b)** $(-4)(x + 2y) = -4x + (-8y) = -4x - 8y$
3. (a) $(x + 3y)(8) = 8x + 24y$ **(b)** $(2x + 3y - 7)(2) = [2x + 3y + (-7)](2)$
$$= 4x + 6y + (-14) = 4x + 6y - 14$$
5. $(3)(2x + 4y) + (2)(5x + y) = 6x + 12y + 10x + 2y = 16x + 14y$
7. $(-3)(x + 2) + (4)(2x - 3y) = -3x - 6 + 8x - 12y = 5x - 12y - 6$

Answers to Even-Numbered Practice Problems

2. (a) $24x - 8y - 48$ **(b)** $-5x - 20y - 25$ **(c)** $6.6x + 16.5y + 18$
4. (a) $9.6x + 19.2y - 24z - 12$ **(b)** $-9x - 27y + 36z - 45$ **(c)** $\frac{3}{4}x - \frac{1}{2}y + 6z - \frac{3}{4}$ **6.** $2x + 17$

After studying this section, you will be able to:
1 *Solve equations without parentheses*
2 *Solve equations with parentheses*

10.3 SOLVING EQUATIONS

One of the most important skills in algebra is that of "solving an equation." There are many kinds of equations and many techniques for solving them. But all rest on the same idea. Starting from an equation with a variable whose value is unknown, we transform the equation into a simpler, equivalent equation by choosing a logical step. By looking at this second equation we may be able to see the solution. However, we often have to continue by again choosing a logical step until the solution is finally apparent: the variable equals a certain number.

In this section we'll learn some of the "logical steps" to choose in solving an equation successfully.

1 Solve Equations without Parentheses

Earlier we solved equations like

$$3n = 75$$

by dividing each side of the equation by 3.

$$\frac{3n}{3} = \frac{75}{3}$$

$$n = 25$$

This is one of two important procedures used to solve equations.

Division Property of Equations

You may divide each side of an equation by the same number to obtain an equivalent equation.

In this chapter we use the new notation and write expressions like $3 \times n$ as $3n$.

Example 1   Solve for n.

(a) $6n = 72$ **(b)** $12n = 30$

(a) $6n = 72$ The variable n is multiplied by 6.

$\dfrac{6n}{6} = \dfrac{72}{6}$ Dividing each side by 6.

$n = 12$ $(72 \div 6 = 12)$

(b) $12n = 30$ The variable n is multiplied by 12.

$\dfrac{12n}{12} = \dfrac{30}{12}$ Divide each side by 12.

$n = 2.5$ $(30 \div 12 = 2.5)$ ■

Practice Problem 1   Solve for n.
(a) $8n = 104$ **(b)** $5n = 36$ ■

Sometimes the coefficient of the variable is a negative number. In solving problems of this type we therefore need to divide each side of the equation by that negative number.

Example 2   Solve for the variable.

(a) $-3n = 51$ **(b)** $-11x = -55$

(a) $-3n = 51$ The coefficient of n is -3.

$\dfrac{-3n}{-3} = \dfrac{51}{-3}$ Dividing each side of the equation by -3.

$n = -17$ Watch your signs! $[51 \div (-3) = -17]$

(b) $-11x = -55$ The coefficient of x is -11.

$\dfrac{-11x}{-11} = \dfrac{-55}{-11}$ Dividing each side of the equation by -11.

$n = 5$ Watch your signs! $[-55 \div (-11) = 5]$ ■

Practice Problem 2   Solve for the variable.
(a) $-7n = 35$ **(b)** $-9n = -108$ ■

In solving an equation such as $3x + 18 = 27$, we want to have all the variable terms on one side and all the numbers on the other side. To obtain this format we may need to add a number to each side of the equation. This is the second of two key procedures to solve equations.

Addition Property of Equations

You may add the same number to each side of an equation to obtain an equivalent equation.

In solving an equation we want to replace a given equation by another, equivalent one, that is somehow simpler. We keep doing this until finally the variable is alone on one side of the equation.

The exact steps for changing the way the equation looks (but not changing its meaning) will be different for each equation. We know we *may* add a number to both sides of the equation. But what number *should* we choose to add? We always add the opposite of the number we want to "move to the other side." To solve $7x + (-2) = 12$ we would add $+2$ to each side. To solve $3x + 4 = 12$ we would add -4 to each side.

Example 3 Solve for x. $\quad 3x + 18 = 27$

We want only x terms on the left and only numbers on the right.

$$3x + 18 + (-18) = 27 + (-18)$$

Adding the opposite of 18, which is negative 18, to each side.

$$3x = 9$$

Simplifying.

$$\frac{3x}{3} = \frac{9}{3}$$

Dividing each side by 3.

$$x = 3 \quad \blacksquare$$

Practice Problem 3 Solve for x. $\quad 5x + 13 = 33 \quad \blacksquare$

Example 4 Solve for x. $\quad 9x - 5 = -41$

$$9x + (-5) = -41$$

Subtracting 5 is equivalent to adding -5.

$$9x + (-5) + 5 = -41 + 5$$

Adding the opposite of -5, which is 5, to each side.

$$9x = -36$$

Simplifying.

$$\frac{9x}{9} = \frac{-36}{9}$$

Dividing each side by 9.

$$x = -4 \quad \blacksquare$$

Practice Problem 4 Solve for x. $\quad 7x - 8 = -50 \quad \blacksquare$

To get all the x terms on one side of the equation, it is necessary in some problems to add an x term to each side of the equation.

Example 5 Solve for x.

(a) $8x = 5x - 21$

(b) $2x = -6x + 64$

(a) $\qquad 8x = 5x - 21$

We want to "move $5x$ to the other side." The opposite of $5x$ is $-5x$.

$$8x + (-5x) = 5x + (-5x) - 21$$

Adding $-5x$ to each side.

$$3x = -21$$

Simplifying.

$$\frac{3x}{3} = \frac{-21}{3}$$

Dividing each side by 3.

$$x = -7$$

(b) $\qquad 2x = -6x + 64$

We want to move $-6x$ to the other side. The opposite of $-6x$ is $6x$.

$$2x + 6x = -6x + 6x + 64$$

Adding $6x$ to each side.

$$8x = 64$$

Simplifying.

$$\frac{8x}{8} = \frac{64}{8}$$

Dividing each side by 8.

$$x = 8 \quad \blacksquare$$

Practice Problem 5 Solve for x.
(a) $7x = 4x - 33$

(b) $4x = -8x + 42 \quad \blacksquare$

In our goal to obtain only x terms on one side and only numbers on the other side, it may be necessary to first add a numerical value to each side of the equation. Then we may also need to add an x term to each side of the equation.

Example 6 Solve for x. $2x + 9 = 5x - 3$

$2x + 9 = 5x + (-3)$ Writing subtraction as adding.
We want to get all numbers on one side.
The opposite of 9 is -9.

$2x + 9 + (-9) = 5x + (-3) + (-9)$ Adding -9 to each side.

$2x = 5x + (-12)$ We want to get all x terms on the left.
The opposite of $5x$ is $-5x$.

$2x + (-5x) = 5x + (-5x) + -12$ Adding $-5x$ to each side.

$-3x = -12$ Simplifying.

$\dfrac{-3x}{-3} = \dfrac{-12}{-3}$ Dividing each side by -3.

$x = 4$ Simplifying. ■

Practice Problem 6 Solve for x. $4x - 7 = 9x + 13$ ■

Checking Your Solution

To check your answer, go back to the original equation. Replace the variable by the value you obtained. Do all the calculations. You will obtain a true statement of equality (such as $5 = 5$) if your solution is correct.

Example 7 Check to determine if $x = 4$ is the correct solution to

$$2x + 9 = 5x - 3$$

$$(2)(4) + 9 \stackrel{?}{=} (5)(4) - 3$$

$$8 + 9 \stackrel{?}{=} 20 - 3$$

$$17 = 17 \checkmark \quad \text{It checks.}$$

Thus $x = 4$ is the correct solution. ■

Practice Problem 7 Check to determine if $x = -4$ is the correct solution to

$$4x - 7 = 9x + 13$$ ■

If there are like terms on one side of the equation, these should be combined first. *Then* proceed with the step of adding a value to each side of the equation.

Example 8 *Solve* for x and *check* your solution.

$$-5 + 2x + 8 = 7x + 23$$

$2x + 3 = 7x + 23$ Collecting like terms on one side
of the equation.

$2x + 3 + (-3) = 7x + 23 + (-3)$ Adding -3 to each side.

$2x = 7x + 20$ We want to get all x terms on the left.

$2x + (-7x) = 7x + (-7x) + 20$ Adding $-7x$ to each side.

$-5x = 20$ The coefficient of x is -5.

$\dfrac{-5x}{-5} = \dfrac{20}{-5}$ Dividing each side by -5.

$x = -4$

Check: $-5 + (2)(-4) + 8 \stackrel{?}{=} (7)(-4) + 23$

$$-5 + (-8) + 8 \stackrel{?}{=} -28 + 23$$

$$-5 = -5 \checkmark \quad ■$$

Practice Problem 8 *Solve* for x and *check* your solution.

$$4x - 23 = 3x + 7 - 2x \quad \blacksquare$$

The solution to a problem may be a fraction.

Example 9 Solve for x.

$$9x + 3 - 2x = 18 + 4x - 4$$

$7x + 3 = 14 + 4x$	Collecting like terms on each side of the equation.
$7x + 3 + (-3) = 14 + (-3) + 4x$	Adding -3 to each side.
$7x = 11 + 4x$	The opposite of $4x$ is $-4x$.
$7x + (-4x) = 11 + 4x + (-4x)$	Adding $-4x$ to each side.
$3x = 11$	Simplifying.
$\dfrac{3x}{3} = \dfrac{11}{3}$	Dividing each side by 3.
$x = \dfrac{11}{3}$	Leave the answer in fractional form. $\blacksquare$

Practice Problem 9 Solve for x.

$$7x - 5 + 2x = 2 + 3x + 7 \quad \blacksquare$$

2 If a problem contains one or more parentheses, remove them using the distributive property. Then collect like terms on each side of the equation. Then solve.

Example 10 Solve for x.

$$(2)(x + 3) = 5x - 8 + 10x$$

$2x + 6 = 5x - 8 + 10x$	Removing parentheses by using the distributive property.
$2x + 6 = 15x - 8$	Adding like terms on the right side of the equation.
$2x + 6 + (-6) = 15x + (-8) + (-6)$	Adding -6 to each side.
$2x = 15x + (-14)$	The opposite of $15x$ is $-15x$.
$2x + (-15x) = 15x + (-15x) + (-14)$	Adding $-15x$ to each side.
$-13x = -14$	Simplifying.
$\dfrac{-13x}{-13} = \dfrac{-14}{-13}$	Dividing each side by -13.
$x = \dfrac{14}{13}$	Leave answer in fractional form. (Note that the answer is positive!) $\blacksquare$

Practice Problem 10 Solve for x.

$$(3)(x - 2) + 5x = 7x + 10 \quad \blacksquare$$

We now list a procedure that may be used to help you remember all the steps we are using to solve equations. See if you can use this procedure to solve Example 11.

You have probably noticed that steps 3 and 4 are interchangeable. You can do step 3 and then step 4 or step 4 before step 3.

Example 11   Solve for y. $8y - 3(2 - 3y) = (6)(3 - 4y) + 17$

$8y - 6 + 9y = 18 - 24y + 17$	Removing parentheses.
$17y - 6 = 35 - 24y$	Collecting like terms.
$17y + (-6) = 35 + (-24y)$	Writing as addition.
$17y + (-6) + 6 = 35 + 6 + (-24y)$	Adding 6 to each side.
$17y = 41 + (-24y)$	Simplifying.
$17y + 24y = 41 + (-24y) + 24y$	Adding 24y to each side.
$41y = 41$	Simplifying.
$\dfrac{41y}{41} = \dfrac{41}{41}$	Dividing each side by 41.
$y = 1$	Simplifying.

Check: Can you verify this answer?   ■

Practice Problem 11   Solve for y. $(5)(4 + y) = (3)(3y - 1) - 9$ ■

To Think About   ?

We have always placed the variables on the left and the numbers on the right in solving our equations. What would happen if we put the variables on the right and the numbers on the left? We would obtain the same answer. Let us examine Example 6 again.

$$2x + 9 = 5x + (-3)$$

Suppose that we put the numbers on the left

$2x + 9 = 5x + (-3)$	
$2x + 9 + 3 = 5x + (-3) + 3$	Adding 3 to each side.
$2x + 12 = 5x$	Simplifying.
$2x + (-2x) + 12 = 5x + (-2x)$	Adding $-2x$ to each side.
$12 = 3x$	Simplifying.
$\dfrac{12}{3} = \dfrac{3x}{3}$	Dividing each side by 3.
$4 = x$	Simplifying.

We obtain exactly the same result as before. The correct value is $x = 4$.   ■

Solve for the variable.

1. $7x = -28$

2. $5x = -45$

3. $-6x = -102$

4. $-4x = -68$

5. $5x + 8 = 3$

6. $3x + 10 = 1$

7. $12x - 30 = 6$

8. $15x - 10 = 35$

9. $9x - 3 = -7$

10. $6x - 9 = -12$

11. $2x = 7x + 25$

12. $4x = 6x + 14$

13. $-9x = 3x - 10$

14. $-7x = 2x + 11$

15. $16 - 3x = 2x + 1$

16. $5x + 8 = 8x - 7$

17. $8 + x = 3x - 6$

18. $9 - 8x = 3 - 2x$

19. $7 + 3x = 6x - 8$

20. $2x - 7 = 3x + 9$

21. $5 + 2y = 7 + 5y$

22. $12 + 5y = 9 - 3y$

23. $4y + 5 = 2y - 9$

24. $4y + 7 = 6y - 7$

25. $(5)(x + 4) = 4x + 15$

26. $(4)(x + 2) = 8x + 12$

Check to see if the given answer is a solution to the equation.

27. Is $x = -1$ a solution to $-4 - 8x = -2 - 6x$?

28. Is $x = -5$ a solution to $-2x + 7 = -4x - 3$?

29. Is $x = 2$ a solution to $3 - 4x = 5 - 3x$?

30. Is $x = 5$ a solution to $3x + 2 = -2x - 23$?

Solve for the variable. Then check your solution.

31. $(3)(x + 4) + 2x = 7$

32. $2x + (3)(x - 4) = 3$

33. $9 + (6)(x - 3) = 3x$

34. $8 + (2)(x + 9) = 4x$

35. $(2)(y + 3)$
$= (4)(y + 5) - 7$

36. $13 + (7)(2y - 1)$
$= (5)(y + 6)$

37. $(2)(x - 7)$
$= (4)(x + 1) - 26$

38. $2x - (4)(x + 1)$
$= 3x + 1$

? **To Think About**

39. (a) Solve $7 + 3x = 6x - 8$ by collecting x terms on the left.
(b) Solve by collecting x terms on the right.
(c) Which method is easier? Why?

40. (a) Solve $8 + x = 2x - 6 + x$ by collecting x terms on the left.
(b) Solve by collecting x terms on the right.
(c) Which method is easier? Why?

Solve for x.

41. $(3)(x + 0.2) - (2)(x + 0.25) = (2)(x + 0.3) - 0.5$

42. $(0.2)(x + 3) - (2)(0.5x - 0.75) = (0.3)(x + 2) - 2.9$

Cumulative Review Problems

Use π = 3.14. Round your answer to the nearest tenth.

43. Find the volume of a sphere with radius of 12 cm.

44. Find the area of the shaded region.

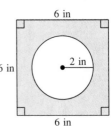

Calculator Problems

Solve for the variable. Round your answer to the nearest ten thousandth.

45. $8.01x - 4.223 = 0.0217x - 7.7$

46. $4.3y - 5.7201 = -3.2y + 6.782$

For Extra Practice Examples and Exercises, turn to page 546.

Solutions to Odd-Numbered Practice Problems

1. (a) $8n = 104$
$$\frac{8n}{8} = \frac{104}{8}$$
$$n = 13$$

(b) $5n = 36$
$$\frac{5n}{5} = \frac{36}{5}$$
$$n = 7.2$$

3.
$$5x + 13 = 33$$
$$5x + 13 + (-13) = 33 + (-13)$$
$$5x = 20$$
$$\frac{5x}{5} = \frac{20}{5}$$
$$x = 4$$

5. (a)
$$7x = 4x - 33$$
$$7x + (-4x) = 4x + (-4x) - 33$$
$$3x = -33$$
$$\frac{3x}{3} = \frac{-33}{3}$$
$$x = -11$$

(b)
$$4x = -8x + 42$$
$$4x + 8x = -8x + 8x + 42$$
$$12x = 42$$
$$\frac{12x}{12} = \frac{42}{12}$$
$$x = 3.5$$

7. Check $x = -4$ as a solution to
$$4x - 7 = 9x + 13$$
$$(4)(-4) - 7 \overset{?}{=} 9(-4) + 13$$
$$-16 - 7 \overset{?}{=} -36 + 13$$
$$-23 = -23 \checkmark \quad \text{It checks.}$$
$$x = -4 \text{ is the solution.}$$

9.
$$7x - 5 + 2x = 2 + 3x + 7$$
$$9x + (-5) = 9 + 3x$$
$$9x + (-5) + 5 = 9 + 3x + 5$$
$$9x = 14 + 3x$$
$$9x + (-3x) = 14 + 3x + (-3x)$$
$$6x = 14$$
$$\frac{6x}{6} = \frac{14}{6}$$
$$x = \frac{7}{3} \quad \text{or} \quad 2\frac{1}{3}$$

11.
$$(5)(4 + y) = (3)(3y - 1) - 9$$
$$20 + 5y = 9y - 3 - 9$$
$$20 + 5y = 9y - 12$$
$$20 + (-20) + 5y = 9y + (-12) + (-20)$$
$$5y = 9y + (-32)$$
$$5y + (-9y) = 9y + (-9y) + (-32)$$
$$-4y = -32$$
$$\frac{-4y}{-4} = \frac{-32}{-4}$$
$$y = 8$$

Answers to Even-Numbered Practice Problems

2. (a) $n = -5$ **(b)** $n = 12$ **4.** $x = -6$ **6.** $x = -4$ **8.** $x = 10$

Check: $(4)(10) - 23 \overset{?}{=} (3)(10) + 7 - (2)(10)$
$$17 = 17 \checkmark$$

10. $x = 16$

10.4 TRANSLATING ENGLISH TO ALGEBRA

In the preceding section you learned to solve an equation that had been given to you. But to solve actual applied problems, you usually need to know not only how to solve an equation but also how to *write one* from a verbal description of the problem.

To help you write your own mathematical equations, in this section we practice *translating*. We translate English expressions into algebraic expressions. In the next section we'll apply this translation skill to equations and to a variety of problems.

☐ After studying this section, you will be able to:

1 Translate simple comparisons in English into mathematical equations using two given variables

2 Write algebraic expressions for several quantities using a given variable

1 The relationship between two or more objects described in an English sentence can often be conveniently expressed as a short equation using variables. For example, if we say in English "Bob's salary is $1000 greater than Fred's salary," we can express the mathematical relationship by the equation

$$b = 1000 + f$$

where b represents Bob's salary and f represents Fred's salary.

The following chart presents the mathematical symbols generally used in translating English phrases into equations.

The English Phrase:	Is Usually Represented by the Symbol:
greater than increased by more than added to taller than total of sum of	$+$
less than decreased by smaller than fewer than shorter than difference of	$-$
double	$2\times$
product of times	$\times$
triple	$3\times$
ratio of quotient of	$\div$
is was has has the value of costs weighs equals represents amounts to	$=$

Example 1 Translate the English sentence into an equation using variables. Use r to represent Roberto's weight and j to represent Juan's weight.

Roberto's weight is 42 lb more than Juan's weight.

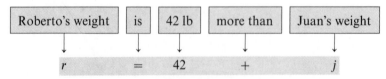

Roberto's weight	is	42 lb	more than	Juan's weight
r	$=$	42	$+$	j

The equation $r = 42 + j$ could also be written as

$$r = j + 42$$

Both are correct translations because addition is commutative. ■

Practice Problem 1 Translate the English sentence into an equation using variables. Use t to represent Tom's height and a to represent Abdul's height.

Tom's height is 7 in. more than Abdul's height. ■

When translating the phrases "less than" or "fewer than" it is important to watch out that the number is subtracted *from* the variable. Words like "costs" "weighs," or "has the value of" act like an equality sign (=).

Example 2 Translate the English sentence into an equation using variables. Use *c* to represent the cost of the chair in dollars and *s* to represent the cost of the sofa in dollars.

The chair costs $200 less than the sofa

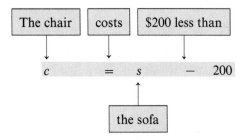

Note the order of the equation. We subtract 200 from *s*, so we have $s - 200$. It would be incorrect to write $200 - s$. Do you see why? ■

Practice Problem 2 Translate the English sentence into an equation using variables. Use *n* to represent the number of students in the noon class and *m* to represent the number of students in the morning class.

The noon class has 24 students less than the morning class. ■

Words such as "drove" or "carried" sometimes are translated with an equality symbol.

Example 3 Translate the English sentence into an equation using variables. Use *t* to represent the number of miles driven on Tuesday and *m* to represent the number of miles driven on Monday.

On Tuesday Yvonne drove 80 mi more than she did on Monday

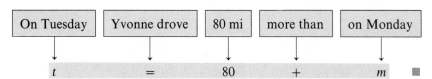

Practice Problem 3 Translate the English sentence into an equation with variables. Use *t* to represent the number of boxes carried on Thursday and *f* to represent the number of boxes carried on Friday.

On Thursday, Adrianne carried five more boxes into the dorm than she did on Friday. ■

Example 4 Translate the English sentence into an equation with variables. Use *l* to represent the length and *w* to represent the width.

The length of the rectangle is 3 inches more than double the width.

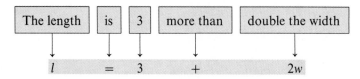

Note here that if the width is *w*, then to double the width is multiplying the width by 2. We may write this as $2 \times w$ or as $2(w)$ or more simply as $2w$. ■

Practice Problem 4 Translate the English sentence into an equation with variables. Use l to represent the length and w to represent the width.

The length of the rectangle is 7 m more than triple the width. ■

2 In each of the examples so far we have used two *different variables*. Now we'll learn how to write algebraic expressions for several quantities using the *same variable*. In the next section we'll use this skill to write and solve equations.

A mathematical expression that contains a variable is often called an **algebraic expression**.

Example 5 Write algebraic expressions for Bob's salary and Fred's salary. Fred's salary is $150 more than Bob's salary. Use the letter b.

Let b = Bob's salary.

Let $b + 150$ = Fred's salary.

Notice that Fred's salary is compared to Bob's salary. Thus it is logical to let Bob's salary be b and then express Fred's salary as $150 more than Bob's. ■

Practice Problem 5 Write algebraic expressions for Sally's trip and Melinda's trip. Melinda's trip is 380 mi longer than Sally's trip. Use the letter s. ■

Example 6 Write algebraic expressions for the size of each of two angles of a triangle. Angle B of the triangle is 34° less than angle A. Use the letter A.

Let A = the number of degrees in angle A.

Let $A - 34$ = the number of degrees in angle B. ■

Practice Problem 6 Write algebraic expressions for the height of each of two buildings. Larson Center is 126 ft shorter than McCormick Hall. Use the letter m. ■

Often in algebra when we write expressions for one or two unknown quantities we will use the letter x.

Example 7 Write algebraic expressions for the number of students in each of two classes. The Monday class has 17 more students than the Tuesday class. Use the letter x.

Let x = the number of students in the Tuesday class.

Let $x + 17$ = the number of students in the Monday class. ■

Practice Problem 7 Write an algebraic expression for the number of dollars it costs to make each of two purchases. A sofa costs $215 more than a chair. Use the letter x. ■

Example 8 Write algebraic expressions for the length of each of three sides of a triangle. The second side is 4 in longer than the first. The third side is 7 in shorter than triple the length of the first side. Use the letter x.

Since the other two sides are compared to the first side, we start with writing an expression for the first side.

Let x = the length of the first side.

Let $x + 4$ = the length of the second side.

Let $3x - 7$ = the length of the third side. ■

Practice Problem 8 Write an algebraic expression for the length of each of three sides of a triangle. The second side is double the length of the first side. The third side is 6 in. longer than the first side. Use the letter x. ■

Translate the English sentence into an equation using the variables indicated for each problem.

1. Ricardo's height is 6 inches more than Charlie's height. Use r for Ricardo's height and c for Charlie's height.

2. Marcia's car is 4 feet longer than Alice's car. Use m for Marcia's car and a for Alice's car.

3. The box with the computer weighs 29 lb more than the box for the printer. Use c for the weight of the box with the computer and p for the weight of the box with the printer.

4. The large cereal box contains 7 oz more than the small box of cereal. Use l for the number of ounces in the large cereal box and s for the number of ounces in the small cereal box.

5. The chair cost $95 dollars less than the table. Use c for the cost of the chair and t for the cost of the table.

6. The noon class has 17 students less than the morning class. Use n to represent the number of students in the noon class and m to represent the number of students in the morning class.

7. On Wednesday we drove 136 mi less than the number of miles we drove on Tuesday. Use w for the number of miles we drove on Wednesday and t for the number of miles we drove on Tuesday.

8. During the spring semester the dorm had 17 fewer students than it did in the fall semester. Use s for the number of students in the spring semester in the dorm and f for the number of students in the fall semester in the dorm.

9. The temperature on Monday was 19 degrees cooler than the temperature on Tuesday. Let m be the number of degrees in the temperature on Monday and t be the number of degrees in the temperature on Tuesday.

10. The sailboat anchor weighs 5 lb less than the motorboat anchor. Use s for the number of pounds the sailboat anchor weighs and m for the number of pounds the motorboat anchor weighs.

11. The length of the rectangle is 7 m longer than double the width. Use l for the length and w for the width of the rectangle.

12. The length of the rectangle is 5 m longer than triple the width. Use l for the length and w for the width of the rectangle.

13. The length of the second side of a triangle is 2 inches shorter than triple the length of the first side of the triangle. Use s to represent the dimension of the second side and f to represent the dimension of the first side.

14. The length of the second side of a triangle is 4 inches shorter than double the length of the first side of the triangle. Use s to represent the dimension of the second side and f to represent the dimension of the first side.

15. The cost of the average sedan in the 1980s is $1000 more than triple the cost of the average sedan in the 1950s. Use f to represent the cost of the average sedan in the 1950s and e to represent the cost of the average sedan in the 1980s.

16. The cost of the average station wagon in the 1980s is $2500 more than double the cost of the average station wagon in the 1960s. Use s to represent the cost of the average station wagon in the 1960s and e to represent the cost of the average station wagon in the 1980s.

17. The total of Wally's salary and Tom's salary is $800 per week. Use w for Wally's salary and t for Tom's salary.

18. The West campus has more students than the East campus. The difference in enrollment between the two campus locations is 2500 students. Use w for number of students at West campus and e for number of students at East campus.

19. The product of hourly wages by the time worked results in $500. Let h = number of dollars per hour paid for hourly wages and t = number of hours worked on time records.

20. The ratio of men to women at Central College is 5 to 3. Let m = the number of men and w = the number of women.

Write algebraic expressions for each quantity using the given variable.

21. Jim's salary is $600 more than Charlie's salary. Use the letter c.

22. The truck's weight is 1500 lb more than the car's weight. Use the letter c.

23. Barbara's car trip was 386 mi longer than Julie's car trip. Use the letter j.

24. The cost of the television set was $212 more than the cost of the compact disc player. Use the letter p.

25. Angle A of the triangle is 46° less than angle B. Use the letter b.

26. The top of the box is 38 cm shorter than the side of the box. Use the letter s.

27. Mount Everest is 4430 m taller than Mount Whitney. Use the letter w.

28. Mount McKinley is 1802 m taller than Mount Ranier. Use the letter r.

29. Wally has taken twice as many biology classes as Ernesto. Use the letter e.

30. Juanita has worked for the company three times as long as Charo. Use the letter c.

31. The length of the rectangle is 12 m longer than triple the width. Use the letter x.

32. The length of the rectangle is 7 m longer than double the width. Use the letter x.

33. The second angle of a triangle is double the first. The third angle of the triangle is 14° smaller than the first. Use the letter x.

34. The second angle of a triangle is triple the first. The third angle of the triangle is 36° larger than the first. Use the letter x.

?

To Think About

35. Bob's salary is $30 more than Fred's salary. Mike's salary is $90 more than Bob's salary. Joe's salary is $10 more than Mike's salary. Let x represent Joe's salary. How would you represent the salary of each of the other three men?

36. Walden's father owns a motorcycle, car, boat, and a truck. The car weighs 500 lb more than the motorcycle. The boat weighs 600 lb more than the car. The truck weighs 1200 lb more than the boat. Let x represent the weight of the truck. How would you represent the weight of the other three items?

Cumulative Review Problems

Perform the operations in the proper order.

37. $-6 - (-7)(2)$

38. $24 \div (-6) + 5$

39. $5 - 5 + 8 - (-4) + 2 - 15$

40. $(2)(-3)(-1)(3)(-1)$

For Extra Practice Examples and Exercises, turn to page 547.

Solutions to Odd-Numbered Practice Problems

1.

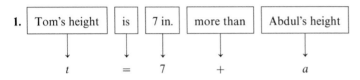

Tom's height	is	7 in.	more than	Abdul's height
t	$=$	7	$+$	a

3.

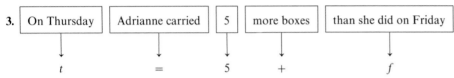

On Thursday	Adrianne carried	5	more boxes	than she did on Friday
t	$=$	5	$+$	f

5. Use the letter s. Melinda's trip is 380 mi longer than Sally's trip.
Melinda's trip is compared to Sally's trip so we start with Sally's trip.

Let $s = $ the number of miles in Sally's trip.
Let $s + 380 = $ the number of miles in Melinda's trip.

7. Use the letter x. A sofa costs $215 more than a chair.

Let $x = $ the cost of the chair.
Let $x + 215 = $ the cost of the sofa.

Answers to Even-Numbered Practice Problems

2. $n = m - 24$ **4.** $l = 3w + 7$ **6.** Let $m = $ the height of McCormick hall in feet.
Let $m - 126 = $ the height of Larson Center in feet.
8. Let $x = $ the length of the first side of the triangle in inches.
Let $2x = $ the length of the second side of the triangle in inches.
Let $x + 6 = $ the length of the third side of the triangle in inches.

10.5 PROBLEM SOLVING

The cost of three cars totals $36,300. The station wagon costs $5000 less than double the cost of the sedan. The sports car costs $4000 less than triple the cost of the station wagon. How much does each vehicle cost?

To solve such a complex problem of comparisons, we use each of the skills we've learned in this chapter.

▣ Applied Problems Involving Comparisons

As we saw in the preceding section, when two values are being compared it is helpful to let a variable represent the quantity to which things are being compared. For example, if a blue box weighs 3 lb more than a red box, the blue box is *compared to* the red box. We let $x = $ the weight of the red box and then we can describe the weight of the blue box as $x + 3$.

After studying this section, you will be able to:

1 Solve problems involving comparisons

2 Solve problems involving geometric formulas

3 Solve problems involving rates and percent

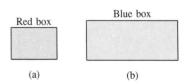

Red box

Blue box

(a)   (b)

This general approach is useful to solve applied problems that compare two items.

Example 1   A 12-ft board is cut into two pieces. The longer piece is 3.5 ft longer than the shorter piece. What is the length of each piece?

Since the longer piece is *compared to* the shorter piece, we let the variable represent the shorter piece. Let x = the length of the shorter piece. The longer piece is 3.5 ft longer than the shorter piece. Let $x + 3.5$ = the length of the longer piece. Then we write an equation. The sum of the two pieces is 12 ft.

$$x + x + 3.5 = 12$$

$$2x + 3.5 = 12 \qquad \text{Collecting like terms.}$$

$$2x + 3.5 + (-3.5) = 12 + (-3.5) \qquad \text{Adding } -3.5 \text{ to each side.}$$

$$2x = 8.5 \qquad \text{Collecting like terms.}$$

$$\frac{2x}{2} = \frac{8.5}{2} \qquad \text{Dividing each side by 2.}$$

$$x = 4.25 \qquad \text{Performing the division by 2.}$$

The shorter piece is 4.25 ft long.

$$x + 3.5 = \text{the longer piece}$$

$$4.25 + 3.5 = 7.75$$

The longer piece is 7.75 ft long.

Check: We verify solutions to word problems by making sure that all the calculated values satisfy the original conditions. Do the two pieces add up to 12 ft?

$$4.25 + 7.75 \overset{?}{=} 12$$

$$12 = 12 \checkmark \quad \text{Yes}$$

Is one piece 3.5 ft longer than the other?

$$7.75 \overset{?}{=} 3.5 + 4.25$$

$$7.75 = 7.75 \checkmark \quad \text{Yes} \qquad \blacksquare$$

Practice Problem 1   An 18-ft board is cut into two pieces. The longer piece is 4.5 ft longer than the shorter piece. What is the length of each piece?   ■

If one item is less in quantity than another, we use subtraction. We usually let the variable represent the larger quantity.

Example 2   José is a store manager. The assistant manager earns $7600 less annually than does José. The sum of José's annual salary and the assistant manager's annual salary is $42,000. How much does each earn?

Let x = José's annual salary.

Let $x - 7600$ = the assistant manager's annual salary.

José's salary x	$+$	assistant manager's salary $x - 7600$	$=$	$42,000 total annual salary of the two people.

Now we write an equation.

$$x + x - 7600 = 42,000$$

$2x - 7600 = 42,000$	Collecting like terms.
$2x + (-7600) + 7600 = 42,000 + 7600$	Adding 7600 to each side.
$2x = 49,600$	Collecting like terms.
$\dfrac{2x}{2} = \dfrac{49,600}{2}$	Dividing each side by 2.
$x = 24,800$	Performing the division.

José earns $24,800 annually. The assistant manager earns $7600 less.

$$x - 7600 = 24,800 - 7600 = 17,200$$

The assistant manager earns $17,200 annually.

Check: Does the assistant manager earn $7600 less than José?

$$24,800 - 7600 \stackrel{?}{=} 17,200$$

$$17,200 = 17,200 \checkmark \quad \text{Yes}$$

Is the sum of the two salaries 42,000?

$$24,800 + 17,200 \stackrel{?}{=} 42,000$$

$$42,000 = 42,000 \checkmark \quad \text{Yes} \quad \blacksquare$$

Practice Problem 2 Mike and Linda bought a sofa. They also bought an easy chair that cost $186 less. The total cost of the two items was $649. How much did each item cost? ∎

Sometimes three items are compared. Let a variable represent the quantity to which things are compared. Then write an expression for the other two quantities.

Example 3 Professor Jones is teaching 332 students in three sections of general psychology this semester. His noon class has 23 students more than his 8:00 A.M. class. His 2:00 P.M. class has 36 students less than his 8:00 A.M. class. How many students are in each class?

All class measurements are compared to the enrollment in his 8:00 A.M. class.
Let x = the number of students in the 8:00 A.M. class.
The noon class has 23 more students than the 8:00 A.M. class.
Let $x + 23$ = the number of students in the noon class.
The 2:00 P.M. class has 36 fewer students than the 8:00 A.M. class.
Let $x - 36$ = the number of students in the 2:00 P.M. class.

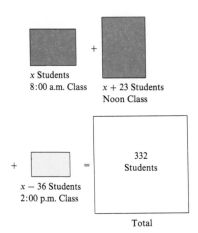

x Students
8:00 a.m. Class

x + 23 Students
Noon Class

x − 36 Students
2:00 p.m. Class

332 Students

Total

$$x + x + 23 + x - 36 = 332$$

$3x - 13 = 332$	Collecting like terms.
$3x + (-13) + 13 = 332 + 13$	Adding 13 to each side.
$\dfrac{3x}{3} = \dfrac{345}{3}$	Dividing each side by 3.
$x = 115$	8:00 A.M. class
$x + 23 = 115 + 23 = 138$	noon class
$x - 36 = 115 - 36 = 79$	2:00 P.M. class

Thus there are 115 students in the 8:00 A.M. class, 138 students in the noon class, and 79 students in the 2:00 P.M. class.

Check: Does the number of students in each class total 332?

$$115 + 138 + 79 \stackrel{?}{=} 332$$

$$332 = 332 \checkmark \quad \text{Yes}$$

Does his noon class have 23 students more than his 8:00 A.M. class?

$$138 \stackrel{?}{=} 23 + 115$$

$$138 = 138 \checkmark \quad \text{Yes}$$

Does his 2:00 P.M. class have 36 students less than his 8:00 A.M. class?

$$79 \stackrel{?}{=} 115 - 36$$

$$79 = 79 \checkmark \quad \text{Yes} \quad \blacksquare$$

Practice Problem 3 The city airport had 349 departures on Monday, Tuesday, and Wednesday. There were 29 more departures on Tuesday than on Monday. There were 16 fewer departures on Wednesday than on Monday. How many departures occurred on each day? ∎

2 Some applied problems have to do with perimeters and other geometric properties of two-dimensional figures. When something is doubled, we multiply by 2. If something is x units, then double that value is $2x$. Triple that value is $3x$. Four times that value is $4x$. Make sure you see the difference between multiplying a number by x and adding a number to x.

Example 4 A farmer wishes to fence in a rectangular field with 804 ft of fence. The length is to be 3 ft longer than *double the width*. How long and how wide is the field?

The perimeter of a rectangle is given by $P = 2w + 2l$.

Let w = the width

The length is 3 ft *longer than double the width.*

$$\text{Length} = 3 + 2w$$

Thus $2w + 3$ = the length.

Now we substitute into the perimeter equation the given facts and expressions.

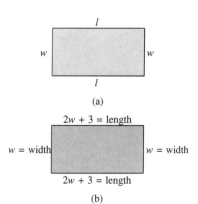

$$2w + 2l = P$$

$2w + (2)(2w + 3) = 804$	Substituting.
$2w + 4w + 6 = 804$	Using the distributive property.
$6w + 6 = 804$	Collecting like terms.
$6w + 6 + (-6) = 804 + (-6)$	Adding -6 to each side.
$6w = 798$	Simplifying.
$\dfrac{6w}{6} = \dfrac{798}{6}$	Dividing each side by 6.
$w = 133$	Performing the division.

The width is 133 ft.

The length $= 2w + 3$. When $w = 133$ we have

$$(2)(133) + 3 = 266 + 3 = 269$$

Thus the length is 269 ft.

Check: Is the length 3 ft longer than double the width?

$$269 \stackrel{?}{=} 3 + (2)(133)$$

$$269 \stackrel{?}{=} 3 + 266$$

$$269 = 269 \checkmark \quad \text{Yes}$$

Is the perimeter 804 ft?

$$(2)(133) + (2)(269) \stackrel{?}{=} 804$$

$$266 + 538 \stackrel{?}{=} 804$$

$$804 = 804 \checkmark \quad \text{Yes} \quad \blacksquare$$

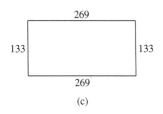

(c)

Practice Problem 4 What is the length and width of a rectangular field that has a perimeter of 772 ft and a length that is 8 ft longer than double the width? ∎

Example 5 The perimeter for a triangular rug section is 21 ft. The second side is double the length of the first side. The third side is 3 ft longer than the first side. Find the length of the three sides of the rug.

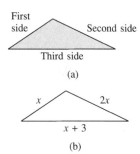

Let x = the length of the first side.

Let $2x$ = the length of the second side.

Let $x + 3$ = the length of the third side.

The distance around the three sides totals 21 ft.

Thus

$x + 2x + x + 3 = 21$	The perimeter equation.
$4x + 3 = 21$	Collecting like terms.
$4x + 3 + (-3) = 21 + (-3)$	Adding -3 to each side.
$4x = 18$	Simplifying.
$\dfrac{4x}{4} = \dfrac{18}{4}$	Dividing each side by 4.
$x = 4.5$	Performing the division.

The first side is 4.5 ft long.

$$2x = (2)(4.5) = 9 \text{ ft}$$

The second side is 9 ft long.

$$x + 3 = 4.5 + 3 = 7.5 \text{ ft}$$

The third side is 7.5 ft long.

Check: Do the three sides add up to a perimeter of 21 ft?

$$4.5 + 9 + 7.5 \stackrel{?}{=} 21$$

$$21 = 21 \checkmark \quad \text{Yes}$$

Is the second side double the length of the first side?

$$9 \stackrel{?}{=} (2)(4.5)$$

$$9 = 9 \checkmark \quad \text{Yes}$$

Is the third side 3 ft longer than the first side?

$$7.5 \stackrel{?}{=} 3 + 4.5$$

$$7.5 = 7.5 \checkmark \quad \text{Yes} \quad \blacksquare$$

Practice Problem 5   The perimeter of a triangle is 36 m. The second side is double the first side. The third side is 10 m longer than the first side. Find the length of each side. Check your solutions.   ■

Example 6   A triangle has three angles, A, B, and C. Angle C is triple angle B. Angle A is 105° larger than angle B. Find the measure of each angle. Check your answer.

Let $\quad$ $x =$ the number of degrees in angle B.

Let $\quad$ $3x =$ the number of degrees in angle C.

Let $x + 105 =$ the number of degrees in angle A.

The sum of the interior angles of a triangle is 180°. Thus we can write

$$x + 3x + x + 105 = 180$$

$$5x + 105 = 180$$

$$5x + 105 + (-105) = 180 + (-105)$$

$$5x = 75$$

$$\frac{5x}{5} = \frac{75}{5}$$

$$x = 15$$

Angle B measures 15°.

$$3x = (3)(15) = 45°$$

Angle C measures 45°.

$$x + 105 = 15 + 105 = 120°$$

Angle A measures 120°.

$\quad$ *Check:* Do the angles total 180°?

$$15 + 45 + 120 \overset{?}{=} 180°$$

$$180 = 180 \checkmark \quad \text{Yes}$$

Is angle C triple angle B? $\qquad$ $45 \overset{?}{=} (3)(15)$

$$45 = 45 \checkmark \quad \text{Yes}$$

Is angle A 105° larger than angle B? $\qquad$ $120 \overset{?}{=} 105 + 15$

$$120 = 120 \checkmark \quad \text{Yes} \quad ■$$

Practice Problem 6   Angle C of a triangle is triple angle A. Angle B is 30° less than angle A. Find the measure of each angle.   ■

❸ Percent and Rate Problems

Example 7   This month's salary for a encyclopedia saleswoman was $3000. This includes her base monthly salary of $1800 plus a 5% commission on total sales. Find the total sales for the month.

Let $\quad$ $s =$ the amount of total sales.

Then $0.05s =$ the amount of commission earned from the sales.

Total salary $3000	=	base salary of $1800	+	5% commission on total sales

$$3000 = 1800 + 0.05s$$

$$1200 = 0.05s$$

$$\frac{1200}{0.05} = \frac{0.05s}{0.05}$$

She sold $24,000 worth of encyclopedias.

$$24,000 = s$$

Check: Does 5% of $24,000 added to $1800 yield a salary of $3000?

$$(0.05)(24,000) + 1800 \overset{?}{=} 3000$$

$$1200 + 1800 \overset{?}{=} 3000$$

$$3000 = 3000 \checkmark \quad \text{Yes} \quad \blacksquare$$

Practice Problem 7 A car salesman earns $1000 a month plus a 3% commission on the total sales price of the cars he sells. Last month he earned $3250. What was the total sales price of the cars he sold? ■

EXERCISES 10.5

Solve each problem by using an equation. Show what you set the variable equal to.

1. A 16-ft board is cut into two pieces. The longer piece is 5.5 ft longer than the shorter piece. What is the length of each piece?

2. A 20-ft board is cut into two pieces. The longer piece is 4.5 ft longer than the shorter piece. What is the length of each piece?

3. In two days Ramone Sanchez drove 560 mi to a vacation campground. He drove 97 more miles the first day than the second day. How far did he drive each day?

4. In two days Felicia drove 615 mi to visit an aunt. She drove 88 more miles the first day than the second day. How far did she drive each day?

5. Mike's car weighs 1420 lb less than his truck. The two vehicles together weigh 5340 lb. How much does each weigh separately?

6. Marlene earns $191 less per month than Barbara. Together they earn $2520 per month. How much does each girl earn?

7. Last year 426 students took general psychology. A total of 110 more students took it in the spring than in the fall. A total of 86 fewer students took it in the summer than in the fall. How many students took it during each semester?

8. The freshman class has 1202 students. A total of 128 more students live on campus than live in nearby off-campus housing. A total of 69 fewer students live at home and commute than live in nearby off-campus housing. How many students are there in each category?

Solve each problem by using an equation. Show what you set the variable equal to. Check your answers.

9. A rectangular field is 7 yd longer than double the width. The perimeter is 1094 yd. What are the dimensions of the field?

10. A rectangular field is 6 yd longer than double the width. The perimeter is 972 yd. What are the dimensions of the field?

11. The perimeter of a rectangle is 52 cm. The length is 2 cm less than triple the width. What are the dimensions of the rectangle?

12. The perimeter of a rectangle is 64 cm. The length is 4 cm less than triple the width. What are the dimensions of the rectangle?

13. The college has a triangular piece of land with a perimeter of 99 m. The length of the second side is double the first side. The length of the third side is 27 m longer than the first side. Find the length of each side.

14. Next to City Hall is a triangular piece of land with a perimeter of 177 m. The length of the second side is double the first side. The length of the third side is 17 m longer than the first side. Find the length of each side.

15. A triangular machine part has a perimeter of 144 mm. The second side is 10 mm longer than the first side. The third side is 7 mm shorter than the first side. Find the length of each side.

16. A triangular machine part has a perimeter of 173 mm. The second side is 6 mm longer than the first side. The third side is 7 mm shorter than the first side. Find the length of each side.

17. A triangle has three angles, *A*, *B*, and *C*. Angle *B* is triple angle *A*. Angle *C* is 40° larger than angle *A*. Find the measure of each angle.

18. A triangle has three angles, *A*, *B*, and *C*. Angle *B* is triple angle *A*. Angle *C* is 35° smaller than angle *A*. Find the measure of each angle.

19. A saleswoman has a base monthly salary of $2000 plus a 2% commission on total sales. Last month her total salary was $2800. What was the amount of her total sales?

20. A salesman has a base monthly salary of $1500 plus a 3% commission on total sales. Last month his total salary was $2400. What was the amount of his total sales?

21. A realtor charges $100 to place a rental listing plus 12% of the yearly rent. An apartment in Central City was rented by the realtor for one year. She charged the landowner $820. How much does the apartment cost to rent for one year?

22. A realtor charges $50 to place a rental listing plus 9% of the yearly rent. An apartment in the town of West Longmeadow was rented by the realtor for one year. He charged the landowner $482. How much does the apartment cost to rent for one year?

23. The population of Bennington this year is 24,720. This was a 3% increase over the population last year. What was the population last year?

24. The population of Riverhead this year is 21,840. This was a 4% increase over the population last year. What was the population last year?

 To Think About

25. The cost of three machine parts is $72.83. The cost of the second part is $3 less than double the cost of the first part. The cost of the third part is $5 more than triple the cost of the second part. How much does each part cost?

26. The cost of three vehicles is $36,300. The cost of a station wagon is $5000 less than double the cost of a *sedan*. The cost of a sports car is $4000 less than triple the cost of the *station wagon*. How much does each vehicle cost?

Cumulative Review Problems

27. What is 16.5% of 350?

28. What percent of 20 is 12?

29. 38% of what number is 190?

30. What is 0.52% of 4000?

31. A rectangular field is 8.2066 yd longer than double the width. The perimeter is 2876.39 yd. What are the dimensions of the field.

32. Joan earns $17.65 less per week than Heather. Together they earn $572.69. How much does each girl earn?

For Extra Practice Examples and Exercises, turn to page 548.

Solutions to Odd-Numbered Practice Problems

1. Let x = length of shorter piece
$x + 4.5$ = length of longer piece
$$x + x + 4.5 = 18$$
$$2x + 4.5 = 18$$
$$3x + 4.5 + (-4.5) = 18 + (-4.5)$$
$$2x = 13.5$$
$$x = 6.75$$
$x + 4.5 = 6.75 + 4.5 = 11.25$
Shorter piece = 6.75 ft.
Longer piece = 11.25 ft.

3. Let x = number of departures on Monday
$x + 29$ = number of departures on Tuesday
$x - 16$ = number of departures on Wednesday
$$x + x + 29 + x - 16 = 349$$
$$3x + 13 = 349$$
$$3x + 13 + (-13) = 349 + (-13)$$
$$3x = 336$$
$$x = 112$$
$x + 29 = 141$ $x - 16 = 96$
There were 112 departures on Monday, 141 on Tuesday, and 96 on Wednesday.

5. Let x = length of the first side
$2x$ = length of the second side
$x + 10$ = length of the third side
$$x + 2x + x + 10 = 36$$
$$4x + 10 = 36$$
$$4x + 10 + (-10) = 36 + (-10)$$
$$4x = 26$$
$$\frac{4x}{4} = \frac{26}{4}$$
$$x = 6.5$$
$$2x = (2)(6.5) = 13$$
$$x + 10 = 16.5$$
The length of the sides is: first side 6.5 m, second side 13 m, and third side 16.5 m.
   Check: Does $6.5 + 13 + 16.5 = 36$?   Yes
   Does $13 = (2)(6.5)$?   Yes
   Does $16.5 = 10 + 6.5$?   Yes

7. Let c = total sales price of the cars he sells in one month
   Then $0.03c$ = the amount of commission earned from the sales

total salary	=	base salary of $1000 a month	+	3% commission on total car sales

$$3250 = 1000 + 0.03c$$
$$2250 = 0.03c$$
$$\frac{2250}{0.03} = \frac{0.03c}{0.03}$$
$$75,000 = c$$
The saleman sold $75,000 worth of cars in one month.
   Check: Does 3% of $75,000 added to 1000 yield a monthly salary of $3250?
$$(0.03)(75,000) + 1000 \overset{?}{=} 3250$$
$$2250 + 1000 \overset{?}{=} 3250$$

Answers to Even-Numbered Practice Problems

2. The sofa costs $417.50.
The easy chair costs $231.50.

4. The width is 126 ft.
The length is 260 ft.

6. Angle A is 42°.
Angle B is 12°.
Angle C is 126°.

EXTRA PRACTICE: EXAMPLES AND EXERCISES

Section 10.1

Name the variables in each equation.

Example $s = \dfrac{3n}{2 + a}$

   The variables are s, n, and a.

1. $L = r^2 + 3k$

2. $w = 2hL$

3. $t = 3r + s$

4. $c = 2st$

5. $s = s + b + 2c$

Write each equation without a multiplication sign.

Example $I = P \times R \times T$

$$I = PRT$$

6. $A = \dfrac{h \times (b + B)}{2}$

7. $E = M \times C^2$

8. $d = r \times t$

9. $s = a \times b \times c$

10. $x = 2 \times y \times z$

Combine like terms.

Example $-3x + 6x - 2x$

$$\underbrace{-3x + 6x}_{\uparrow} - 2x =$$
$$3x \qquad - 2x = 1x = x$$

11. $6x + 2x$ **12.** $-12x + 30x$

13. $5x - 8x + 7x$ **14.** $-2x + 8x - 6x$

15. $-4x + 7x - 2x$

Example $-2x + 5y + 9 + 3y - 6 + 9x - z$
$$-2x + 9x + 5y + 3y + 9 - 6 - z$$
$$7x + 8y + 3 - z$$

16. $x + 7 + 6x - 3$

17. $-4.5a + 1.2b - 6 + 3.2a - 1.1b$

18. $8 - 20x - 5 - 12x$

19. $-4x + 7y - 2 + 8x - 6y + z + 5$

20. $-6s - 9r + 2 + 4s - 2r$

Section 10.2

Simplify.

Example $(5)(6 + 2x)$
$$(5)(6 + 2x) = 30 + 10x$$

1. $(4)(2 + 3x)$ **2.** $(7)(3x - 4)$

3. $(11)(x + 1)$ **4.** $(3)(5 - 9x)$

5. $(5)(1.3x + 2.5)$

Example $(-8)(2.3x - 3.1y)$
$$(-8)(2.3x - 3.1y) = 18.4x + 24.8y$$

6. $(-6)(1.7x - 3.4y)$

7. $(9)(3x + 2y)$

8. $(-5)(4a - 3c)$

9. $(-1)(24t + 1)$

10. $(-10)(9.4r - 3.7t)$

Example $(-12)(2a + b - c - 3d)$
$$(-12)(2a + b - c - 3d) = -24a - 12b + 12c + 36d$$

11. $(-7)(6a - 2b + c - 5d)$

12. $(-4)(2.5x - 3.1y + 1.1)$

13. $(9)(4.1s - 2.2t - 5)$

14. $(-1)(5x - 2b - c - d)$

15. $(-3)(-6a - 2b - 4c + 9d)$

Example $(-3)(x - 2y) + (2)(-x - 4y)$
$$(-3)(x - 2y) + (2)(-x - 4y) = -3x + 6y - 2x - 8y = -5x - 2y$$

16. $(-2)(x + 5y) - (4)(-2x - 3y)$

17. $(9)(3x - 8y) + (6)(-x - y)$

18. $(-6)(5a - 7b) - (5)(-a + b)$

19. $(4.1)(s + t) - (3.8)(2s - 7t)$

20. $(3)(-a + b - 6c) - (9)(4a - 2b + c)$

Section 10.3

Solve for the variable.

Example $8x + 3 = 16$
$$8x + 3 = 16$$
$$8x + 3 + (-3) = 16 + (-3)$$
$$8x = 13$$
$$\frac{8x}{8} = \frac{13}{8} \qquad x = \frac{13}{8} \text{ or } 1\frac{5}{8}$$

1. $3x = -21$ **2.** $-5x = 35$

3. $9x - 10 = 80$ **4.** $-2x + 6 = 16$

5. $6x - 1 = -9$

Example $3x = 9x + 20$
$$3x + (-9x) = 9x + (-9x) + 20$$
$$-6x = 20$$
$$\frac{-6x}{-6} = \frac{20}{-6} \qquad x = -\frac{10}{3} \text{ or } -3\frac{1}{3}$$

6. $5x = 2x + 10$

7. $-8y = 7y - 4$

8. $6y - 5 = 3y + 9$

9. $3x + 2 = 5x - 2$

10. $8 - 3x = 5 - 4x$

Example $(3)(y + 1) = 6y - 2$
$$(3)(y + 1) = 6y - 2$$
$$3y + 3 = 6y - 2$$
$$3y + 3 + (-3) = 6y + (-2) + (-3)$$
$$3y = 6y + (-5)$$
$$3y + (-6y) = 6y + (-6y) + (-5)$$
$$-3y = -5 \qquad y = \frac{-5}{-3} = \frac{5}{3}$$

11. $(5)(x + 2) = 4x - 9$ **12.** $(11)(s - 6) = 6s + 2$

13. $(-4)(y + 1) = 3y - 2$ **14.** $(8)(x - 1) = -3x + 7$

15. $(2)(y - 7) = 4y + 2$

Solve for the variable. Then check your solution in each case.

Example $(3)(x + 2) = (2)(x - 5) - 6$
$$(3)(x + 2) = (2)(x - 5) - 6$$
$$3x + 6 = 2x - 10 - 6$$
$$3x + 6 = 2x + (-10) + (-6)$$
$$3x + 6 + (-6) = 2x + (-16) + (-6)$$
$$3x = 2x + (-22)$$
$$3x + (-2x) = 2x + (-2x) + (-22)$$
$$1x = -22 \qquad x = -22$$

Check:

$$(3)(x + 2) = (2)(x - 5) - 6$$
$$(3)(-22 + 2) \overset{?}{=} (2)(-22 - 5) - 6$$
$$(3)(-20) \overset{?}{=} (2)(-27) - 6$$
$$-60 \overset{?}{=} -54 - 6$$
$$-60 = -60 \checkmark$$

16. $(2)(x + 1) + 6 = (3)(x - 4) + 2$
17. $(4)(x + 1) = (3)(x - 4) + 1$
18. $(3)(x - 5) + 2 = (2)(x + 1)$
19. $(6)(x + 3) - 1 = (2)(x - 1) + 3$
20. $(3)(2x + 1) = (2)(x + 5)$

Section 10.4

Translate the English sentence into an equation using variables. Use the letters indicated for each problem.

Example The temperature Sunday was 10 degrees cooler than the temperature Monday. Use *s* to equal the temperature on Sunday and *m* to equal the temperature on Monday.

Temperature Sunday	was	temperature Monday	ten degrees less than
s	$=$	m	-10

$$s = m - 10$$

1. Leslie's height is 5 in. more than Angela's height. Use *l* for Leslie's height and *a* for Angela's height.

2. The box with the television set weighs 15 lb more than the box with the VCR. Use *t* for the weight of the box with the television set and *v* for the weight of the box with the VCR.

3. The desk cost $65 dollars less than the chair. Use *d* for the cost of the desk and *c* for the cost of the chair.

4. The temperature on Tuesday was 13 degrees cooler than the temperature on Wednesday. Let *t* be the number of degrees in temperature on Tuesday and *w* be the number of degrees in temperature on Wednesday.

5. The lawn mower weighs 3 lb less than the edger. Use *l* for the number of pounds the lawn mower weighs and *e* for the number of pounds the edger weighs.

Example The length of the second side of a triangle is 4 inches shorter than double the length of the first side of the triangle. Use *s* to represent the dimensions of the second side and *f* to represent the dimensions of the first side.

Length of second side	is	4 inches shorter	than double the first side
$\downarrow$	$\downarrow$		
s	$=$	$2f$	-4

$$s = 2f - 4$$

6. The length of the rectangle is 3 m longer than triple the width. Use *l* for the length and *w* for the width of the rectangle.

7. The length of the rectangle is 8 m longer than double the width. Use *l* for length and *w* for the width of the rectangle.

8. The length of the second side of a triangle is 3 in. shorter than triple the length of the first side of the triangle. Use *s* to represent the dimensions of the second side and *f* to represent the dimension of the first side.

9. The length of the second side of a triangle is 5 in. shorter than double the length of the first side of the triangle. Use *s* to represent the dimensions of the second side and *f* to represent the dimension of the first side.

10. John is 2 in. more than double the height of Dave. Use *J* to represent the height of John and *D* to represent the height of Dave.

Describe each quantity specified using the same variable.

Example Julie's salary is $300 more than Charlene's salary. Use *c* to describe each salary.

$$c = \text{Charlene's salary}$$
$$c + 300 = \text{Julie's salary}$$

11. Sara's car trip was 260 mi longer than Janet's car trip. Use the letter *j* to describe the length of each trip.

12. The cost of the washer was $50 more than the cost of the dryer. Use the letter *D* to describe the cost of each item.

13. Angle *A* of the triangle is 23° less than angle *B*. Use the letter *b* to describe the number of degrees in each of the two angles.

14. The top of a desk is 21 cm shorter than the side of the desk. Use the letter *s* to describe each of these two dimensions.

15. Juanita weighs 16 lb less than Rosa. Use the letter *r* to describe the number of pounds each of the girls weigh.

Example Juanita has worked for the company three times as long as Wayne. Use the letter *w* to describe the length of time each person has worked for the company.

$$w = \text{the length of time Wayne worked for the company}$$
$$3w = \text{the length of time Juanita worked for the company}$$

16. Alice has taken twice as many English classes as Justin. Use the letter *j* to describe the number of English courses taken by each person.

17. Les has worked for the company twice as long as Bob. Use the letter *b* to describe the length of time each person has worked for the company.

18. The length of the rectangle is 15 m longer than triple the width. Use the letter *x* to describe each dimension of the rectangle.

19. The length of the rectangle is 8 m longer than double the width. Use the letter *x* to describe each dimension of the rectangle.

20. The second angle of a triangle is double the first. The third angle of the triangle is 21° smaller than the first. Use the letter *x* to describe each angle of the triangle.

Example In two days Jerome drove 465 mi. He drove 57 more miles the first day than the second day. How far did he drive each day?

Let x = the number of miles Jerome drove
the *second* day

Let $x + 57$ = the number miles Jerome drove the first day

miles first day + miles second day = total miles

$$x + 57 \quad + \quad x \quad = \quad 465$$

$$2x + 57 = 465$$

$$2x + 57 + (-57) = 465 + (-57)$$

$$2x = 408$$

$$\frac{2x}{2} = \frac{408}{2} = 204$$

Jerome drove 204 mi the second day and $204 + 57 = 261$ mi the first day.

1. In two days José drove 385 mi to a friend's house. He drove 86 more miles the first day than the second day. How far did he drive each day?

2. In two days Dawn drove 462 mi to visit her parents. She drove 96 more miles the second day than the first day. How far did she drive each day?

3. In two days Nina drove 595 mi to Chicago. She drove 72 more miles the first day than the second day. How far did she drive each day?

4. Jean's annual salary is $1250 more than Daniel's annual salary. Together they earn $42,500 annually. What is the salary of each person?

5. George and Dale work for the same company. Dale is a new employee and earns $82 less per month than George. Together they earn $1850 per month. What is the monthly salary of each person?

Example A restaurant had twice as many customers in December as in November. They had 2000 more customers in January than in November. Over the three months 22,800 customers came to the restaurant. How many came each month?

Let x = the number of customers in November.
Let $2x$ = the number of customers in December.
Let $x + 2000$ = the number of customers in January.

Number of customers in:

$$\text{Nov.} + \text{Dec.} + \text{Jan.} = 22,800$$

$$x + 2x + x + 2000 = 22,800$$

$$4x + 2000 = 22,800$$

$$4x + 2000 + (-2000) = 22,800 + (-2000)$$

$$4x = 20,800$$

$$\frac{4x}{4} = \frac{20,800}{4}$$

$$x = 5200$$

Number of customers in: Nov. = $x = 5200$
Dec. = $2x = (2)(5200) = 10,400$
Jan. = $x + 2000 = 5200 + 2000$
$= 7200$

Check:
Number of customers in: Nov. + Dec. + Jan. $= 22,800$
$5200 + 10,400 + 7200 = 22,800$

6. Last year 395 students took biology. A total of 95 more students took it in the spring than in the fall. A total of 75 fewer students took it in the summer than in the fall. How many students took it during each semester?

7. The community college has 1704 students. A total of 115 more students live on campus than live in nearby off-campus housing. A total of 55 fewer students live at home and commute than live in nearby off-campus housing. How many students are there in each category?

8. On a three-day trip Anna drove 856 mi. On Tuesday she traveled 23 mi less than she did on Monday. On Wednesday she traveled 53 mi more than she did on Monday. How far did she drive each day? Round your answer to nearest tenth.

9. A total fast-food restaurant had twice as many customers in June as in May. They had 2500 more customers in July than in May. Over the three months 35,700 customers came to the restaurant. How many came each month?

10. Alex drove 963 mi during three days of travel. He drove 90 more miles on Friday than on Thursday. He drove 27 fewer miles on Saturday than on Thursday. How many miles did he drive each day?

Example A triangular machine part has a perimeter of 122 mm. The second side is 4 mm longer than the first side. The third side is 8 mm shorter than the first side. Find the length of each side.

$$\text{Perimeter} = \text{side } 1 + \text{side } 2 + \text{side } 3 = 122 \text{ mm.}$$

Let x = the length of the first side.
Let $x + 4$ = the length of the second side.
Let $x - 8$ = the length of the third side.

Perimeter:

$$x + x + 4 + x - 8 = 122$$

$$3x + 4 - 8 = 122$$

$$3x - 4 = 122$$

$$3x + (-4) + 4 = 122 + 4$$

$$3x = 126$$

$$\frac{3x}{3} = \frac{126}{3} = 42 = x \qquad \text{length of first side}$$

Side 1 = 42 mm

Side 2 = $x + 4 = 42 + 4 = 46$ mm

Side 3 = $x - 8 = 42 - 8 = 34$ mm

11. The perimeter of a rectangle is 48 cm. The length is 4 cm less than triple the width. What are the dimensions of the rectangle?

12. The perimeter of a rectangle is 68 m. The length is 2 m less than triple the width. What are the dimensions of the rectangle?

13. A triangular piece of land has a perimeter of 102 m. The length of the second side is double the first side. The length of the third side is 25 m longer than the first side. Find the length of each side.

14. The perimeter of a triangle is 120 m. The length of the second side is double the first side. The length of the third side is 12 m longer than the first side. Find the length of each side.

15. A triangular machine has a perimeter of 176 mm. The second side is 25 mm longer than the first side. The third side is 5 mm shorter than the first side. Find the length of each side.

Example A triangle has three angles, A, B, and C. Angle B is triple angle A. Angle C is $15°$ smaller than angle A. Find the measure of each angle.

$$\text{Angle } A + \text{angle } B + \text{angle } C = 180°.$$

Let $x = $ the number of degrees in angle A.
Let $3x = $ the number of degrees in angle B.
Let $x - 15 = $ the number of degrees in angle C.

$$x + 3x + x - 15 = 180$$
$$5x - 15 = 180$$
$$5x + (-15) + 15 = 180 + 15$$
$$5x = 195$$
$$\frac{5x}{5} = \frac{195}{5} = 39° = \text{angle } A$$

$$\text{Angle } B = 3x = (3)(39) = 117°$$

$$\text{Angle } C = x - 15 = 39 - 15 = 24°$$

Check: Angle A + angle B + angle C = 180
$$39 + 117 + 24 = 180$$

16. A triangle has three angles, A, B, and C. Angle B is triple angle C. Angle A is $40°$ larger than angle C. Find the measure of each angle.

17. A triangle has three angles, E, F, and G. Angle E is double angle F. Angle G is $20°$ larger than angle F. Find the measure of each angle.

18. A triangle has three angles, A, B, and C. Angle B is triple angle A. Angle C is $20°$ smaller than angle A. Find the measure of each angle.

19. A triangle has three angles, A, B, and C. Angle C is triple the measure of angle B. Angle A is $65°$ larger than angle B. Find the measure of each angle.

20. A triangle has three angles, A, B, and C. Angle B is double the measure of angle A. Angle C is $30°$ smaller than angle A. Find the measure of each angle.

CHAPTER ORGANIZER

Topic	Procedure	Examples
Combining like terms, p. 516	If the terms are like terms, you combine the numerical coefficients directly in front of the variables.	Combine like terms. (a) $7x - 8x + 2x = -1x + 2x = x$ (b) $3a - 2b - 6a - 5b = -3a - 7b$ (c) $a - 2b + 3 - 5a = -4a - 2b + 3$
The distributive properties, p. 520	$a(b + c) = ab + ac$ and $(b + c)a = ba + ca$	$(5)(x - 4y) = 5x - 20y$ $(3)(a + 2b - 6) = 3a + 6b - 18$ $(-2x + y)(7) = -14x + 7y$
Problems involving parentheses and like terms, p. 521	1. Remove the parentheses using the distributive property. 2. Collect like terms.	Simplify: $(2)(4x - y) - (3)(-2x + y) = 8x - 2y + 6x - 3y = 14x - 5y$
Solving equations, p. 524	1. Remove any parentheses. 2. Collect like terms on each side of the equation. 3. Add the appropriate value to both sides of the equation to obtain all numbers on one side. 4. Add the appropriate value to both sides of the equation to obtain all variable terms on the other side. 5. Divide both sides of the equation by the numerical coefficient of the variable. 6. Check by substituting your answer back into the original equation.	Solve for x: $5x - (2)(6x - 1) = (3)(1 + 2x) + 12$ $5x - 12x + 2 = 3 + 6x + 12$ $-7x + 2 = 15 + 6x$ $-7x + 2 + (-2) = 15 + (-2) + 6x$ $-7x = 13 + 6x$ $-7x + (-6x) = 13 + 6x + (-6x)$ $-13x = 13$ $\dfrac{-13x}{-13} = \dfrac{13}{-13}$ $x = -1$ *Check:* For $x = -1$ $5x - (2)(6x - 1) = (3)(1 + 2x) + 12$ $(5)(-1) - (2)[6(-1) - 1] \overset{?}{=} (3)[1 + (2)(-1)] + 12$ $-5 - (2)[-6 - 1] \overset{?}{=} (3)[1 - 2] + 12$ $-5 - (2)[-7] \overset{?}{=} (3)[-1] + 12$ $-5 + 14 \overset{?}{=} -3 + 12$ $9 = 9 \checkmark$ It checks.

Topic	Procedure	Examples
Translating an English sentence to an equation, p. 532	When translating English to an equation: | The English Phrase:	Is Usually Represented by the Symbol:
---	---
greater than increased by more than added to taller than total of sum of	+
less than smaller than fewer than shorter than difference of	−
product of times	×
double	$2\times$
triple	$3\times$
ratio of quotient of	÷
is costs has the value of weighs has was equals represents amounts to	=		Translate a comparison in English into an equation using two given variables. Use t to represent Thursday's temperature and w to represent Wednesday's temperature. The temperature Thursday was 12 degrees higher than the temperature on Wednesday. Temperature Thursday was 12° higher than Temperature Wednesday t = 12 + w
Writing algebraic expressions for several quantities, p. 534	1. Use a variable to describe the quantity that other quantities are compared to. 2. Write an expression in terms of that variable for each of the other quantitites.	Write algebraic expressions for the size of each angle of a triangle. The second angle of a triangle is 7° less than the first angle. The third angle of a triangle is double the first angle. Use the letter x. Since other angles are compared to the first angle, we let the variable x represent that angle. Let x = the number of degrees in the first angle. Let $x - 7$ = the number of degrees in the second angle. Let $2x$ = the number of degrees in the third angle.

Solving applied problems using equations, p. 537	1. Read over the problem carefully. Find out what is asked for. Draw a picture.
	2. Write down the numbers and formulas that are to be used in solving the problem.
	3. Express each quantity in terms of a variable.
	4. Write an equation that uses the given values. Solve the equation.
	5. Check your answer. Does it satisfy the original conditions?
		The perimeter of a field is 62 yd. The field is rectangular in shape. The length of the field is 4 yd more than double the width. Find the dimensions of the field.
		1. We need to find the dimensions of the rectangular field: length and width.
		
		2. The perimeter is 62 yd.
		   $P = 2(\text{width}) + 2(\text{length})$
		4. $P = 2(\text{width}) + 2(\text{length})$
		   $62 = 2w + 2(2w + 4)$
		   $62 = 2w + 4w + 8$
		   $62 = 6w + 8$
		   $54 = 6w$
		    $9 = w$
		
		3. Let w = the width.
		   Let $2w + 4$ = the length.
		5. $62 = (2)(9) + 2(22)$ ✓
		   $22 = (2)(9) + 4$ ✓
		
		The width = 9 yd
		$2w + 4 = (2)(9) + 4 = 22$ yd = length

10.1 *Combine like terms.*

1. $x + 2y - 6x$

2. $5x + 6y - 7x$

3. $a + 8 - 2a + 4$

4. $7 + 2a - 6a - 5$

5. $5x + 2y - 7x - 9y$

6. $3x - 7y + 8x + 2y$

7. $5x - 9y - 12 - 6x$ $- 3y + 18$

8. $7x - 2y - 20 - 5x$ $- 8y + 13$

9. $1.2a + 5.6b - 3 - 4a - 2.2b + 1$

10. $-1.5a + 3.4b - 7 - 6a + 5.6b + 3$

10.2 *Simplify.*

11. $(-3)(5x + y)$

12. $(-4)(2x + 3y)$

13. $(2)(x - 3y + 4)$

14. $(3)(2x - 6y - 1)$

15. $(-8)(3a - 5b - c)$

16. $(-9)(2a - 8b - c)$

17. $(5)(1.2x + 3y - 5.5)$

18. $(6)(1.4x - 2y + 3.4)$

Simplify.

19. $(2)(x + 3y) - (4)(x - 2y)$

20. $(2)(5x - y) - (3)(x + 2y)$

21. $(-2)(a + b) - (3)(2a + 8)$

22. $(-4)(a - 2b) + (3)(5 - a)$

10.3 *Solve for the variable.*

23. $7x + 8 = 43$

24. $9x + 5 = 50$

25. $3 - 2x = 9 - 8x$

26. $8 - 6x = -7 - 3x$

27. $10 + x = 3x - 6$

28. $8x - 7 = 5x + 8$

29. $9x - 3x + 18 = 36$

30. $4 + 3x - 8 = 12 + 5x + 4$

31. $5x - 2 = 27$

32. $(2)(x + 3) = -3x + 10$

33. $2x - 3 - 5x = 13 + (2)(2x - 1)$

34. $(2)(3x - 4) = 7 - 2x + 5x$

35. $5 + 2y + (5)(y - 3) = (6)(y + 1)$

36. $3 + (5)(y + 4) = (4)(y - 2) + 3$

37. The weight of the truck is 3000 lb more than the weight of the car. Use *w* for the weight of the truck and *c* for the weight of the car.

38. The afternoon class had 18 fewer students than the morning class. Use *a* for the number of students in the afternoon class and *m* for the number of students in the morning class.

39. The number of degrees in angle *A* is triple the number of degrees in angle *B*. Use *A* for the number of degrees in angle *A* and *B* for the number of degrees in angle *B*.

40. The length of a rectangle is 3 in shorter than double the width of the rectangle. Use *w* for the width of the rectangle in inches and *l* for the length of the rectangle in inches.

Write algebraic expressions for each of the specified quantities using the same variable.

41. Michael's salary is $2050 more than Roberto's salary. Use the letter *r*.

42. The length of the second side of a triangle is double the length of the first side of the triangle. Use the letter *x*.

43. Nancy has six fewer graduate courses completed than Connie does. Use the letter *c*.

44. The number of books in the new library is 450 more books than double the number of books in the old library. Use the letter *b*.

45. A 60-ft length of pipe is divided into two pieces. One piece is 6.5 ft longer than the other. Find the length of each piece.

46. Two clerks work in a store. The new employee earns $28 less per week than a person hired six months ago. Together they earn $412 per week. What is the weekly salary of each person?

47. A local fast-food restaurant had twice as many customers in March as in February. They had 3000 more customers in April than in February. Over the three months 45,200 customers came to the restaurant. How many came each month?

48. Alfredo drove 856 mi during three days of travel. He drove 106 more miles on Friday than on Thursday. He drove 39 fewer miles on Saturday than on Thursday. How many miles did he drive each day?

49. A rectangle has a perimeter of 72 in. The length is 3 in less than double the width. Find the dimensions of the rectangle.

50. A rectangle has a perimeter of 180 m. The length is 2 m more than triple the width. Find the dimensions of the rectangle.

51. A triangle has a perimeter of 99 m. The second side is 7 m longer than the first side. The third side is 4 m shorter than the first side. How long is each side?

52. A triangle has three angles labeled *A*, *B*, and *C*. Angle *C* is triple the measure of angle *B*. Angle *A* is 74 degrees larger than angle *B*. Find the measure of each angle.

Collect like terms.

1. $5x + 2x - 12x$

2. $7y - 19y + 4$

3. $5a - 9b - 3a - 12b$

4. $a - 3b + 6 - 5b - 2a$

5. $3x - 12x - 16y + 5y - 9y$

6. $x + 2y - 8 - 5x - 12y - 16$

7. $3w + 2x - 4y - 8x + w$

8. $-15w - 13y + w + 2y - 8y$

Simplify.

9. $(3)(x - 6y)$

10. $(2)(2x + 5y)$

11. $(-4)(a + 2b - 3c)$

12. $(-3)(2a + b - 5c)$

13. $(6)(-5x - 6y + 4)$

14. $(8)(7x - 8y - 7)$

15. $(-2)(1.5x - 3.6y + 5)$

16. $(-4)(2.6x + 3.5y - 7)$

1. _____

2. _____

3. _____

4. _____

5. _____

6. _____

7. _____

8. _____

9. _____

10. _____

11. _____

12. _____

13. _____

14. _____

15. _____

16. _____

17. _____

18. _____

19. _____

20. _____

21. _____

22. _____

23. _____

24. _____

25. _____

26. _____

27. _____

28. _____

29. _____

30. _____

31. _____

32. _____

Simplify. Be sure to collect like terms.

17. $(5)(x - 2y) - (3)(2y - x)$

18. $(6)(2x + y) - (7)(-4x + y)$

19. $(5)(6 + 3y) - (2)(x + 5y)$

20. $(7)(4 - 3y) - (5)(2x + y)$

Solve for the variable.

21. $9x + 3 = 30$

22. $5x + 4 = 39$

23. $6x - 4 = -2x - 8$

24. $9x + 3 = -5x - 4$

25. $4x - 2 = 7x - 2$

26. $8x - 5 = 3x - 5$

27. $8 - (2)(3x - 4) = 2x$

28. $20 - (2)(5x + 5) = -10$

29. $(5)(4x + 3) = (6)(3x + 2) - 1$

30. $(5)(2x - 1) - 7 = (4)(2x + 1)$

31. Is $x = 0$ a solution to $(5)(3x - 4) + 5 = -15$?

32. Is $x = 3$ a solution to $10x + 6 - 3 = 12x - 18 + 5$?

Translate the English sentence into an equation using the variables indicated for each problem.

1. Hector's height is 3 in. more than Won Lin's height. Use h for Hector's height and w for Won Lin's height.

2. The chair costs $145 more than the table. Use c for the cost of the chair and t for the cost of the table.

3. Polly traveled 367 fewer miles than Barbara. Use p for the number of miles Polly traveled and b for the number of miles Barbara traveled.

4. During the spring the dorm had 23 fewer students than in the fall. Use s for the number of students in the spring and f for the number of students in the fall.

5. The length of the rectangle is 4 m longer than double the width. Use l for the length and w for the width of the rectangle.

6. The second side of a triangle is 8 m longer than triple the first side of a triangle. Use s to represent the length of the second side of the triangle and f to represent the length of the first side of the triangle.

Write algebraic expressions for each quantity using the given variable.

7. Susan's salary is $1800 more than Jeffrey's salary. Use the letter j.

8. The sofa weighs 230 lb more than the chair. Use the letter c.

9. The top of the box is 19 cm shorter than the side of the box. Use the letter s.

10. Rita has taken twice as many accounting courses as Carlos. Use the letter c.

11. William has owned three times as many cars as Frederick. Use the letter f.

12. Angle A is 13° larger than triple angle B. Use the letter B.

13. Angle B is 14° smaller than double angle C. Use the letter C.

14. The length of the rectangle is 7 m less than double the width. Use the letter x.

1. _____

2. _____

3. _____

4. _____

5. _____

6. _____

7. _____

8. _____

9. _____

10. _____

11. _____

12. _____

13. _____

14. _____

15. _____

16. _____

17. _____

18. _____

19. _____

20. _____

21. _____

22. _____

23. _____

24. _____

25. _____

26. _____

15. The first side of a triangle is 5 ft shorter than triple the length of the second side. Use the letter x.

16. Dr. Kaiser has driven to California twice as often as Dr. Schoenert. Dr. Slater has driven to California one more time than Dr. Schoenert. Use the letter x.

17. The second angle of a triangle is 15° more than the first. The third angle of a triangle is 29° less than the first. Use the letter x.

18. The Williamson family went to Disneyworld. Their vacation cost $560 more than the one taken by the Frydrych family. The Van Dyke family vacation cost $112 less than the one taken by the Frydrych family. Use the letter x.

Solve each applied problem by using an equation. Show what you set the variable equal to.

19. A 20-ft board is cut into two pieces. One piece is 3.5 ft longer than the other. What is the length of each piece?

20. Bob's annual salary is $2640 more than Samuel's annual salary. Together they earn $36,400 annually. What is the salary of each one?

21. On a three-day trip Valerie drove 945 mi. On Tuesday she traveled 18 mi less than she did on Monday. On Wednesday she traveled 69 mi more than she did on Monday. How far did she drive each day?

22. The length of a rectangle is one meter less than double the width. The perimeter of the rectangle is 100 m. Find the length and the width.

23. The perimeter of a triangle is 45 cm. The second side is 3 cm longer than the first side. The third side is 6 m shorter than double the first side. Find the length of each side.

24. A triangle has three angles, labeled A, B, and C. Angle B is triple the measure of angle A. Angle C is 55° larger than angle A. Find the measure of each angle.

25. A saleswoman has a base salary of $2200 per month plus a 2% commission on total sales. Last month her total salary was $3400. What was the amount of her total sales?

26. The population of Chathan this year is 15,450. This was a 3% increase over the population last year. What was the population last year?

Collect like terms.

1. $3a - 8a$

2. $2a - 15a - 6a$

3. $-5x + 2y - 8x - 4y$

4. $x + 2y - 3 - 6x - 7y - 8$

5. $5x - 8y + z + 3z - 6y$

6. $a + b - 3 - 5a + 6b - 12$

Simplify.

7. $(3)(12x - 5y)$

8. $(-4)(x - 2y + 3)$

9. $(-2.5)(x + 2y - 3z - 5)$

10. $(2)(-3a + 2b) - (5)(a - 2b)$

Solve for the variable.

11. $2 - 6x = 20$

12. $(4)(3 - x) = (-5)(2 + 3x)$

1. _____

2. _____

3. _____

4. _____

5. _____

6. _____

7. _____

8. _____

9. _____

10. _____

11. _____

12. _____

13. _____

14. _____

15. _____

16. _____

17. _____

18. _____

19. _____

20. _____

21. _____

22. _____

23. _____

24. _____

13. $-3x - 7 = 2x + 8$

14. $-5x + 4 + 2x = x - 8 + 2x$

15. $5x + 4 - 2x = 3x - 1 + x$

16. $3 - (x + 2) = 5 + (3)(x + 2)$

Translate the English sentence into an equation using the variables indicated for each problem.

17. The second floor of Trabor Laboratory has 18 more classrooms than the first floor. Use s to represent the number of classrooms on the second floor and f to represent the number of classrooms on the first floor.

18. The north field yields 20,000 fewer bushels of wheat than the south field. Use n to represent the number of bushels of wheat in the north field and s to represent the number of bushels of wheat in the south field.

Write algebraic expressions for each quantity using the given variable.

19. The first angle of a triangle is double the second angle. The third angle of a triangle is triple the second angle. Use the variable s.

20. The length of a rectangle is 5 in shorter than double the width. Use the letter w.

Solve each problem by using an equation. Show what variable or expression you use to represent each quantity.

21. The number of acres of land in the old Smithfield farm is five times the number of acres of land in the Prentice farm. Together the two farms have 282 acres. How many acres of land is there on each farm?

22. Wanda earns $900 less per year than Marcia does. The combined income of the two people is $29,500 per year. How much does each person earn?

23. Gina drove 965 mi in three days. She drove 49 more miles on Tuesday than on Monday. She drove 35 fewer miles on Wednesday than on Monday. How many miles did she drive each day?

24. A rectangular field has a perimeter of 106 ft. The length is 5 ft longer than double the width. Find the dimensions of the rectangle.

Approximately one-half of this test is based on Chapter 10 material. The remainder is based on material covered in Chapters 1–9.

Do each problem. Simplify your answers.

1. Add: $456 + 89 + 123 + 79$.

2. Multiply: $\begin{array}{r} 309 \\ \times\ \ 35 \\ \hline \end{array}$

3. Round to the nearest hundred: 45,678,934.

4. Divide: $\dfrac{1}{2} \div \dfrac{1}{4}$.

5. Multiply: $3\dfrac{1}{4} \times 2\dfrac{1}{2}$.

6. Multiply: 9.3×0.0078.

7. Subtract: $34,007.090 - 3456.789$.

8. Find n: $\dfrac{9}{n} = \dfrac{40.5}{72}$.

9. What is 28.5% of $5600?

10. 34% of what number is 1870?

11. Convert 345 mm to m.

12. Convert 10 ft to in.

13. Find the circumference of a circle with diameter of 12 yd. Round your answer to the nearest tenth.

14. Find the area of a triangle that has a base of 13 m and a height of 22 m.

1. _____

2. _____

3. _____

4. _____

5. _____

6. _____

7. _____

8. _____

9. _____

10. _____

11. _____

12. _____

13. _____

14. _____

15. _____

16. _____

17. _____

18. _____

19. _____

20. _____

21. _____

22. _____

23. _____

24. _____

25. _____

26. _____

27. _____

28. _____

Perform the following operations.

15. $4 - 8 + 12 - 32 - 7$

16. $(5)(-2)(3)(-1)$

Collect like terms.

17. $3a - 5b - 12a - 6b$

18. $-4x + 5y - 9 - 2x - 3y + 12$

Simplify.

19. $(-7)(-3x + y - 8)$

20. $(2)(3x - 4y) - (8)(x + 2y)$

Solve for the variable.

21. $5x - 5 = 7x - 13$

22. $7 - 9y - 12 = 3y + 5 - 8y$

23. $x - 2 + 5x + 3 = 183 - x$

24. $9(2x + 8) = 20 - (x + 5)$

Write algebraic expressions for each of the specified quantities, using the given variable.

25. The weight of the computer was 322 lb more than the weight of the printer. Use the letter p.

26. The summer enrollment in algebra was 87 students less than the Fall enrollment. Use the letter f.

Solve each word problem by using an equation. Show what you set the variable equal to.

27. Barbara drove 1081 mi in three days. She drove 48 more miles on Friday than on Thursday. She drove 95 fewer miles on Saturday than on Thursday. How many miles did she drive each day?

28. A rectangle has a perimeter of 98 ft. The length is 8 ft longer than double the width. Find each dimension.

This examination is based on Chapters 1–10 of the book. There are 10 problems covering the context of each chapter.

CHAPTER 1

1. Write in words 82,367

2. Add.   13,428
          + 16,905

3. Add.   19
               23
               16
               45
             + 70

4. Subtract.   89,071
                   − 54,968

Multiply the following.

5.   78
       × 54

6.   2035
       × 107

Divide the following. (Be sure to indicate the remainder if one exists.)

7. $7\overline{)1106}$

8. $26\overline{)15756}$

9. Evaluate. Perform operations in their proper order.
$3^4 + 20 \div 4 \times 2 + 5^2$

10. Melinda traveled 512 miles in her car. The car used 16 gallons of gas on the entire trip. How many miles per gallon did the car achieve?

CHAPTER 2

11. Reduce the fraction $\dfrac{14}{30}$

12. Change to an improper fraction $3\dfrac{9}{11}$

Add the following fractions.

13. $\dfrac{1}{10} + \dfrac{3}{4} + \dfrac{4}{5}$

14. $2\dfrac{1}{3} + 3\dfrac{3}{5}$

15. Subtract $4\dfrac{5}{7} - 2\dfrac{1}{2}$

16. Multiply $1\dfrac{1}{4} \times 3\dfrac{1}{5}$

Divide the following.

17. $\dfrac{7}{9} \div \dfrac{5}{18}$

18. $\dfrac{5\frac{1}{2}}{3\frac{1}{4}}$

1. _____

2. _____

3. _____

4. _____

5. _____

6. _____

7. _____

8. _____

9. _____

10. _____

11. _____

12. _____

13. _____

14. _____

15. _____

16. _____

17. _____

18. _____

19. _____

20. _____

21. _____

22. _____

23. _____

24. _____

25. _____

26. _____

27. _____

28. _____

29. _____

30. _____

31. _____

32. _____

33. _____

34. _____

35. _____

36. _____

37. _____

38. _____

19. Lucinda jogged $1\frac{1}{2}$ miles on Monday, $3\frac{1}{4}$ miles on Tuesday, and $2\frac{1}{10}$ miles on Wednesday. How many miles in all did she jog over the three-day period?

20. A butcher has $11\frac{2}{3}$ pounds of steak. She wishes to place them in several equally-sized packages of steak. Each package will hold $2\frac{1}{3}$ pounds of steak. How many packages can be made?

CHAPTER 3

21. Express as a decimal: $\frac{719}{1000}$

22. Write in reduced fractional notation: 0.86

23. Fill in the blank with $<$, $=$, or $>$
0.315 _____ 0.309

24. Round to the nearest hundredth.
506.3782

25. Add   9.6
        3.82
        1.05
      + 7.3

26. Subtract   3.61
            − 2.853

27. Multiply   1.23
          × 0.4

28. Divide $0.24\,\overline{)\,0.8856}$

29. Write as a decimal $\frac{13}{16}$

30. Evaluate, by doing operations in the proper order,
$0.7 + (0.2)^3 - 0.08\,(0.03)$.

CHAPTER 4

31. Write a rate in simplest form to compare 7000 students to 215 faculty.

32. Is this proportion true or false?
$\frac{12}{15} = \frac{17}{21}$

Solve the following proportions. Round to the nearest tenth if necessary.

33. $\frac{5}{9} = \frac{n}{17}$

34. $\frac{3}{n} = \frac{7}{18}$

35. $\frac{n}{12} = \frac{5}{4}$

36. $\frac{n}{7} = \frac{36}{28}$

Solve each of the following problems by using a proportion. Round your answers to the nearest hundredth if necessary.

37. Bob earned $2000 for painting 3 houses. How much would he earn for painting 5 houses?

38. Two cities that are actually 200 miles apart appear 6 inches apart on the map. Two other cities are 325 miles apart. How far apart will they appear on the same map?

39. Roberta earned $68 last week on her part-time job. She had $5 withheld for federal income tax. Last year she earned $4,000 on her part-time job. Assuming the same rate, how much was withheld for federal income tax last year?

40. Malaga's recipe feeds 18 people and calls for 1.2 pounds of butter. If she wants to feed 24 people, how many pounds of butter does she need?

CHAPTER 5

41. Write as a percent 0.0063.

42. Change $\dfrac{17}{80}$ to a percent.

Round all answers to the nearest tenth if necessary.

43. Write as a decimal 164%

44. What percent of 300 is 52?

45. Find 6.3% of 4800.

46. 145 is 58% of what number?

47. 126% of 3400 is what number?

48. Pauline bought a new car. She got an 8% discount. The car listed for $11,800. How much did she pay for the car?

49. A total of 1260 freshmen were admitted to Central College. This is 28% percent of the student body. How big is the student body?

50. There are 11.28 centimeters of water in the rain gauge this week. Last week the rain guage held 8.40 centimeters of water. What is the percent of increase from last week to this week?

CHAPTER 6

Convert the following. When necessary, express your answers as a decimal rounded to the nearest hundredth.

51. 17 qt = _____ gal

52. 3.25 tons = _____ lb

53. 16 ft = _____ in

54. 5.6 km = _____ m

55. 69.8 g = _____ kg

56. 2.48 ml = _____ L

Round to the nearest hundredth.

57. 12 mi = _____ km

39.	
40.	
41.	
42.	
43.	
44.	
45.	
46.	
47.	
48.	
49.	
50.	
51.	
52.	
53.	
54.	
55.	
56.	
57.	

58. _____

59. _____

60. _____

61. _____

62. _____

63. _____

64. _____

65. _____

66. _____

67. _____

68. _____

69. _____

70. _____

Write in scientific notation.

58. 0.00063182

59. 126,400,000,000

60. Two metal sheets are 0.623 cm and 0.74 cm thick, respectively. An insulating foil is 0.0428 mm thick. When all three layers are placed tightly together, what is the total thickness?

CHAPTER 7

Round answers to the nearest hundredth when necessary. Use $\pi \doteq 3.14$ when necessary.

61. Find the perimeter of a rectangle that is 6 meters long and 1.2 meters wide.

62. Find the perimeter of a trapezoid with sides of 82 cm, 13 cm, 98 cm, and 13 cm.

63. Find the area of a triangle with base 6 feet and height 1.8 feet.

64. Find the area of a trapezoid with bases of 12 meters, and 8 meters, and a height of 7.5 meters.

65. Find the area of a circle with radius 6 meters.

66. Find the circumference of a circle with diameter 18 meters.

67. Find the volume of a right-circular cone with a radius of 4 centimeters and a height of 10 centimeters.

68. Find the volume of a rectangular pyramid with a base of 12 feet by 19 feet and a height of 2.7 feet.

69. Find the area of this object, consisting of a square and a triangle.

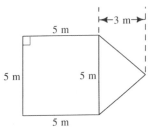

70. In the following pair of similar triangles find n.

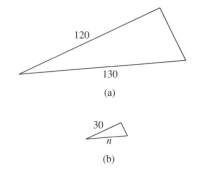

CHAPTER 8

The following double bar graph indicated the quarterly profits for Westar Corporation.

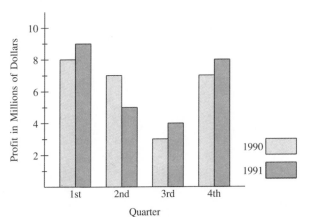

71. What were the profits in the fourth quarter of 1991?

72. How much greater were the profits in the first quarter of 1991 than the first quarter of 1990?

The following line graph depicts the average annual temperature at West Valley for the years 1950, 1960, 1970, 1980, 1990.

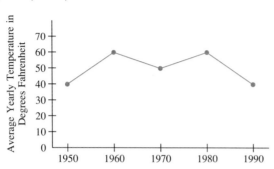

73. What was the average temperature in 1970?

74. In what 10-year period did the average temperature show the greatest decline?

The following histogram shows the number of students in each age category at Center City College.

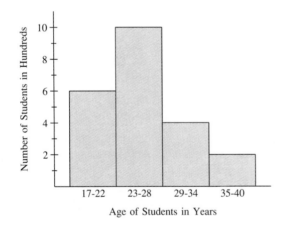

75. How many students are between 17–22 years old?

76. How many students are between 23–34 years old?

77. Find the *mean* and the *median* of the following. 8, 12, 16, 17, 20, 22

78. Evaluate exactly. $\sqrt{49} + \sqrt{81}$

79. Approximate to the nearest thousandth with a calculator or the Square Root table. $\sqrt{123}$

80. Find the unknown side of the right triangle.

? 12 feet

9 feet

71. _____

72. _____

73. _____

74. _____

75. _____

76. _____

77. _____

78. _____

79. _____

80. _____

81. _____
82. _____
83. _____
84. _____
85. _____
86. _____
87. _____
88. _____
89. _____
90. _____
91. _____
92. _____
93. _____
94. _____
95. _____
96. _____
97. _____
98. _____
99. _____
100. _____

CHAPTER 9

Add the following.

81. $-8 + (-2) + (-3)$ **82.** $-\dfrac{1}{4} + \dfrac{3}{8}$

Subtract the following.

83. $9 - 12$ **84.** $-20 - (-3)$

85. Multiply $(2)(-3)(4)(-1)$ **86.** Divide $-\dfrac{2}{3} \div \dfrac{1}{4}$

Perform the following operations in the proper order:

87. $(-16) \div (-2) + (-4)$ **88.** $12 - 3(-5)$

89. $7 - (-3) + 12 \div (-6)$ **90.** $\dfrac{(-3)(-1) + (-4)(2)}{(0)(6) + (-5)(2)}$

CHAPTER 10

Collect like terms.

91. $5x - 3y - 8x - 4y$ **92.** $5 + 2a - 8b - 12 - 6a - 9b$

Simplify the following.

93. $-2(x - 3y - 5)$ **94.** $-2(4x + 2) - 3(x + 3y)$

Solve for the variable.

95. $5 - 4x = -3$ **96.** $5 - 2(x - 3) = 15$

97. $7 - 2x = 10 + 4x$ **98.** $-3(x + 4) = 2(x - 5)$

Solve the following word problems by using an equation.

99. There are 12 more students taking History than Math. There are twice as many students taking Psychology as there are students taking Math. There are 452 students taking these 3 subjects. How many are taking History? How many are taking Math?

100. A rectangle has a perimeter of 106 feet. The length is 5 meters longer than double the width. Find the length and width of the rectangle.

The Use of a Calculator

In this day and age, calculators are readily available at almost any food store, drug store, convenience store, or department store. Many people use the convenience of a calculator to balance their checkbooks or determine their income taxes.

In this section, we will deal with the simple inexpensive calculators and how to use them. (There are more powerful calculators called scientific calculators that will do far more complicated problems with more efficiency. We will not discuss their use in this book.)

Simple calculators have many individual differences. However, they have in common a basic keyboard and several keys. They all have a display that shows the result of a calculation.

The most important features are shown in the following diagram:
The keys $\boxed{0}\boxed{1}\boxed{2}\boxed{3}\boxed{4}\boxed{5}\boxed{6}\boxed{7}\boxed{8}\boxed{9}$ are used to enter numbers. The $\boxed{\cdot}$ key is used to enter a decimal point. The keys $\boxed{+}\boxed{-}\boxed{\times}\boxed{\div}$ are used to enter the operations of addition, subtraction, multiplication, and division. The $\boxed{=}$ key is used to have the calculator display the answer to a calculation or to perform all operations entered thus far.

In addition, your calculator may have many other keys that we have not mentioned or shown in the diagram. For example, if your calculator has a Memory feature, you will see such keys as $\boxed{M+}\boxed{M}$ or $\boxed{MR}$. We will discuss this later in this section.

Display →

Simple Calculations

Let us begin with some calculations. Suppose we wish to find $156 + 298$.

Example 1   Add $156 + 298$.

We first key in the number 156, then press the $\boxed{+}$ key, then enter the number 298, and finally press the $\boxed{=}$ key.

The answer is 454. The calculator will display the answer like this.   ∎

Practice Problem 1   Add $3792 + 5896$.   ∎

Example 2   Subtract $1508 - 963$.

We first enter the number 1508, then press the Subtraction Key $\boxed{-}$, then enter the number 963 and finally press the $\boxed{=}$ key.

The answer is 545. The calculator will display the answer like this.   ∎

Practice Problem 2   Subtract 7930 − 5096.   ■

Example 3   Multiply 196 × 358.

$$196 \boxed{\times} 358 \boxed{=} \quad \boxed{70168}$$

The answer is 70,168.   ■

Practice Problem 3   Multiply 896 × 273.   ■

Example 4   Divide 2054 ÷ 13.

$$2054 \boxed{\div} 13 \boxed{=} \quad \boxed{158.}$$

The answer is 158.   ■

Practice Problem 4   Divide 2352 ÷ 16.   ■

Decimal Problems

Problems involving decimals can be readily done on the calculator. Entering numbers with a decimal point is done by pressing the $\boxed{\cdot}$ key at the appropriate time.

Example 5   Calculate 4.56 × 283.

To enter 4.56 we press the $\boxed{4}$ key, the Decimal Point key, then the $\boxed{5}$ key, and finally the $\boxed{6}$ key.

$$4.56 \boxed{\times} 283 \boxed{=} \quad \boxed{1290.48}$$

The answer is 1290.48. Observe how your calculator displays the decimal point.

Practice Problem 5   Calculate 72.8 × 197.   ■

Example 6   Add 128.6 + 343.7 + 103.4 + 207.5.

$$128.6 \boxed{+} 343.7 \boxed{+} 103.4 \boxed{+} 207.5 \boxed{=} \quad \boxed{783.2}$$

The answer is 783.2. Observe how your calculator displays the answer.   ■

Practice Problem 6   Add 52.98 + 31.74 + 40.37 + 99.82   ■

Example 7   The college purchased 38 filing cabinets that cost $129.49 each. How much did the college pay for them?

We will need to multiply 38 times the cost of each cabinet, which is $129.49.

$$38 \boxed{\times} 129.49 \boxed{=} \quad \boxed{4920.62}$$

The cost is $4920.62. Observe how your calculator displays the answer.   ■

Practice Problem 7   Templeton Dormitory had 126 new study desks purchased for a cost of $107.38 each. What was the total purchase cost?   ■

Combined Operations

You must use extra caution concerning the order of mathematical operations when you are doing a problem on the calculator when two or more different operations are involved.

Example 8   Perform the calculation

$$(3.86)(5.92) + 7.98.$$

We know that the product (3.86)(5.92) must be done first because multiplication is a higher priority than is addition. When the problem is done from left to right the calculator will perform the multiplication first.

First Method

$$3.86 \boxed{\times} 5.92 \boxed{+} 7.98 \boxed{=} \quad \boxed{30.8312}$$

The answer is 30.8312. Did you get the correct answer? Some calculators require an extra step in order to do these operations. If your calculator did not display the correct answer, then try the sequence of steps listed in the second method below.

Second Method

This method is needed for those calculators requiring the entry of the $\boxed{=}$ key after every two numbers with their associated operation is entered.

$$3.86 \boxed{\times} 5.92 \boxed{=} \boxed{+} 7.98 \boxed{=} \quad \boxed{30.8312}$$

Again, the answer is 30.8312. ■

Practice Problem 8   Perform the calculation (7.03)(4.98) + 6.54. ■

Example 9   Perform the calculation 1298.7 − (4.13)(2.99).

Because of the order of operations, the first calculation here that must be done is to find the product (4.13)(2.99). The resulting value will then be subtracted from 1298.7. It is important to remember that multiplication and division problems must always be done before addition and subtraction problems. We will show two methods of solution: One if your calculator has memory, and one if it does not.

First Method

If your calculator does not have any memory that is usable by you (which you can tell if it does not have any $\boxed{M+}$ $\boxed{M}$ or $\boxed{MR}$ keys), we use the following approach: Calculate the product (4.13)(2.99) and write down the result.

$$4.13 \boxed{\times} 2.99 \boxed{=} \quad \boxed{12.3487}$$

We are recording this value on paper so we can use it in the next step. Now we want to subtract 1298.7 − 12.3487

$$1298.7 \boxed{-} 12.3487 \boxed{=} \quad \boxed{1286.3513}$$

The final answer is 1286.3513.

Second Method

If your calculator has a memory that you can use you will be able to store a value for future use. Look to see if you have a $\boxed{M}$ $\boxed{M+}$ or a $\boxed{M \text{ in}}$ key. If so, you will find this memory feature most helpful.
   First enter the $\boxed{AC}$ to clear all your storage registers. Then enter the following:

$$4.13 \boxed{\times} 2.99 \boxed{=} \boxed{M+}$$

   This will add the result of multiplying 4.13 by 2.99 into the memory. Since you have pushed the All Clear key $\boxed{AC}$ the memory did not contain anything else. Thus the product (which is 12.3487) is now in the memory. Now we subtract 12.3487 from the first number 1298.7. We will do this by utilizing the Memory Recall key $\boxed{MR}$

$$1298.7 \boxed{-} \boxed{MR} \boxed{=} \quad \boxed{1286.3513}$$

The final answer is 1286.3513 which agrees with our result using the first method.

■

Practice Problem 9   Calculate 658.3 − (12.8)(3.46). ■

Example 10 Calculate $\dfrac{(9.83)(4.06)}{12.22}$. Round your answer to the nearest hundredth.

Since the fraction bar acts like a grouping symbol we proceed to do the multiplication first and then divide the result by 12.22. So that the method will be usable by all calculators we will enter the $\boxed{=}$ key after the first multiplication.

$$9.83 \boxed{\times} 4.06 \boxed{=} \boxed{\div} 12.22 \boxed{=} \quad \boxed{3.26594108}$$

Your calculator may display fewer digits in the display. We now round the value to the nearest hundredth to obtain 3.27. ∎

Practice Problem 10 Calculate $\dfrac{(36.2)(58.9)}{19.3}$. Round your answer to the nearest tenth. ∎

Applied Problems

Example 11 Find the area of a rectangle with a width of 16.92 centimeters and a length of 47.32 centimeters. Round your answer to the nearest hundredth.

We will use the Area Formula

$$A = lw$$

$$A = (47.32)(16.92)$$

On the calculator we have

$$47.32 \boxed{\times} 16.92 \boxed{=} \quad \boxed{800.6544}$$

Rounding to the nearest hundredth we find the area is 800.65 square centimeters. ∎

Practice Problem 11 Find the area of a rectangle with a width of 36.93 meters and a length of 85.21 meters. Round your answer to the nearest hundredth. ∎

length = 47.32 cm

width = 16.92 cm

Operations With Fractions on a Calculator

A simple calculator is not so readily used for fractions. Many fractions are cumbersome to use in decimal form. It is important to remember that almost all calculator calculations involving fractions will result in a decimal answer that is an *approximation*.

Example 12 Find an approximate value for $\dfrac{3}{7} + \dfrac{5}{48}$. Round your answer to five decimal places.

We will need to find a decimal approximation for $\dfrac{5}{48}$ and write it down or else use the memory feature if your calculator has one.

First Method (No Memory)

$$5 \boxed{\div} 48 \boxed{=} \quad \boxed{0.104166}$$

Write down the number 0.104166. We will now divide the two numbers in the first fraction and add that result to the number we have written down.

$$3 \boxed{\div} 7 \boxed{=} \boxed{+} 0.104166 \boxed{=} \quad \boxed{0.532738}$$

At this stage your calculator may display more or fewer digits than this. Now we round our answer to five decimal places to obtain an *approximate* answer of 0.53274.

Second Method (Using Memory)

We will first divide $5 \div 48$ and store that result in memory. First we press the All Clear key $\boxed{AC}$

$$\boxed{AC} \, 5 \boxed{\div} 48 \boxed{=} \boxed{M+}$$

Now we calculate $3 \div 7$ and add the result to the result stored in memory.

$$3 \;\boxed{\div}\; 7 \;\boxed{=}\; \boxed{+} \;\boxed{MR}\; \boxed{=}\; \boxed{\mathtt{0.532738}}$$

Rounding to five decimal places we have 0.53274 ■

Practice Problem 12   Find an approximate value for

$$\frac{8}{15} + \frac{7}{17}$$

Round your answer to five decimal places.   ■

Solving Percent Problems with a Calculator Using a Percent Key.

Find the amount when the base and the rate are given.

Example 13   What is 65% of 800?

The percent key $\boxed{\%}$ must be used last on most calculators, so we will enter it as $800 \times 65\%$. Use the keystrokes

$$800 \;\boxed{\times}\; 65 \;\boxed{\%}\; \boxed{\mathtt{520.}}$$

The answer is 520. Note that we did not use the equal sign key.   ■

Practice Problem 13   What is 75% of 500?   ■

Find the base when the amount and rate are given.

Example 14   198 is 66% of what number?

We want to solve to find

$$198 = 66\% \times n$$

This means

$$\frac{198}{66\%} = n$$

On a calculator we use the keystrokes

$$198 \;\boxed{\div}\; 66 \;\boxed{\%}\; \boxed{\mathtt{300.}}$$

The answer is 300. Note again we did not use the equal sign key.   ■

Practice Problem 14   57.6 is 24% of what number?   ■

Find the rate when the amount and base are given.

Example 15   What percent is 128 of 250?

We want to solve the problem

$$\frac{128}{250} = n\%$$

We enter on the calculator the following keystrokes

$$128 \;\boxed{\div}\; 250 \;\boxed{\%}\; \boxed{\mathtt{51.2}}$$

The answer is 51.2. Note that we did not use the equal sign key.   ■

Practice Problem 15   What percent is 153 of 170?   ■

Use your calculator to complete each of the following. Your answers may vary slightly due to the characteristics of individual calculators.

Complete the following table.

To Do This Operation	Use These Keystrokes	Record Your Answer Here
1. 8963 + 2784	8963 $\boxed{+}$ 2784 $\boxed{=}$	
2. 15,308 − 7980	15308 $\boxed{-}$ 7980 $\boxed{=}$	
3. 2631 × 134	2631 $\boxed{\times}$ 134 $\boxed{=}$	
4. 70,221 ÷ 89	70221 $\boxed{\div}$ 89 $\boxed{=}$	
5. 5.325 − 4.031	5.325 $\boxed{-}$ 4.031 $\boxed{=}$	
6. 184.68 + 73.98	184.68 $\boxed{+}$ 73.98 $\boxed{=}$	
7. 2004.06 ÷ 7.89	2004.06 $\boxed{\div}$ 7.89 $\boxed{=}$	
8. 1.34 × 0.763	1.34 $\boxed{\times}$ 0.763 $\boxed{=}$	

Write down the answer for the following and then show what problem you have solved

9. 123.45 $\boxed{+}$ 45.9876 $\boxed{+}$ 8765.3 $\boxed{=}$ **10.** 0.0897 $\boxed{\times}$ 234.56 $\boxed{\times}$ 2.5428 $\boxed{=}$

11. 34 $\boxed{\div}$ 8 $\boxed{+}$ 12.56 $\boxed{=}$ **12.** 458 $\boxed{\div}$ 4 $\boxed{-}$ 16.897 $\boxed{=}$

Perform the following calculations using your calculator:

13. 9.467 + 0.563   **14.** 0.347 + 23.457   **15.** 34.89 + 39.6 + 214.897   **16.** 12.567 + 48.31 + 189.38

17. 412,899 − 34,675   **18.** 87,456 − 2,876   **19.** 3,567,089 − 2,876,805   **20.** 8,345,802 − 4,985,004

21. 234 × 4.567   **22.** 1.9876 × 347   **23.** 0.456 × 3.48   **24.** 67,876 × 0.0946

25. 3458 ÷ 2.5   **26.** 9764 ÷ 8   **27.** 12.107524 ÷ 15.86   **28.** 16.06513 ÷ 17.98

Perform the following calculations using your calculator:

29.	**30.**	**31.**	**32.**
1.98	8.92	$103.91	$3986.21
6.34	9.31	$2653.82	$4502.89
+ 7.71	+ 7.79	+ $9804.61	+ $989.30

33.	**34.**	**35.**	**36.**
368,781.5	571,809.6	$1,393,271.86	$8,571,300.76
− 283,617.8	− 539,376.8	− $1,289,663.21	− $4,098,789.39

37. 345.34
   ×   45.7

38. 8954.34
   ×   425.4

39. 0.6314
   ×   3.96

40. 0.0789
   × 12.38

41. $40.36 \overline{) 36202.92}$

42. $52.98 \overline{) 172608.84}$

43. $0.7613 \overline{) 17.12925}$

44. $0.9854 \overline{) 3.59671}$

Perform the following operations in the proper order using your calculator:

45. $4.567 + 87.89 - 2.45 \times 3.3$

46. $4.891 + 234.5 - 0.98 \times 23.4$

47. $7 \div 8 + 3.56$

48. $9 \div 4.5 + 0.6754$

49. $(9.34)(0.345) + 98.345$

50. $(0.628)(398) + 34.4581$

51. $\dfrac{(95.34)(0.9874)}{381.36}$

52. $\dfrac{(0.8759)(45.87)}{183.48}$

Find an approximate value for each of the following. Round your answer to five decimal places.

53. $\dfrac{7}{18} + \dfrac{9}{13}$

54. $\dfrac{5}{22} + \dfrac{1}{31}$

55. $\dfrac{1}{4} + \dfrac{7}{8} + \dfrac{3}{11}$

56. $\dfrac{1}{3} + \dfrac{9}{14} + \dfrac{5}{19}$

57. $\dfrac{3 + 9.2 + 6.41}{7 + 8 - 2.3}$

58. $\dfrac{9.6 + 4.3 + 5}{18 + 9 + 3.2}$

Find an answer to the following applied situations:

59. Find the area of a rectangular lot which is 345.5 feet wide and 568.2 feet long.

60. Find the area of a rectangular state park which is 12.45 miles wide and 18.96 miles long.

61. Alberta traveled 846.8 miles and used 31.5 gallons of gas. How many miles did her car travel on a gallon of gas? Round your answer to the nearest tenth.

62. The Smithville foundary made an extra profit this year of $112,456.08, and the company president decided to divide it equally among the 126 employees. How much money did each employee get?

63. The Hendersons purchased a new car for $14,788.31. They will have to pay a state sales tax of $4\frac{1}{2}\%$. Multiply 0.045 times the cost of the car to find the amount of tax they will have to pay. Round your answer to the nearest cent.

64. The state police purchased 139 new police cruisers at a cost of $17,937.82 each. Find the amount it cost to purchase the new cruisers.

65. In the past year the following items were purchased for Old North Dormitory: Two televisions at $346.89, three sofas at $893.99, and four chairs at $136.89. What was the total bill for these purchases?

66. Bonnie Blair won a World Speed Skating record in 1988 in Calgary Canada by skating 500 meters in 39.10 seconds. What was her speed in meters per second? Round your answer to three decimal places.

67. Florence Griffith-Joyner ran 200 meters in 21.56 seconds in Seoul, Korea in October 1988, winning a world record. What was her speed in meters per second? Round your answer to three decimal places.

68. Bob used about 536 gallons of gas last year. In his area the average price of gasoline is $1.249 per gallon for regular unleaded and $1.439 per gallon for premium unleaded. If he uses premium unleaded gas all year how much more will it cost than if he uses regular unleaded? (Assume that he will use 536 gallons either way.)

69. Wally has a Diesel Rabbit. He used 720 gallons of diesel fuel last year. Joan has a gasoline Rabbit. She used 893 gallons of gas last year. In their area of the country regular unleaded gas costs $1.239 per gallon, and Diesel fuel costs $1.199 per gallon. How much did each person spend on fuel last year? Who spent more? By how much?

Do the following problems on a calculator using the percent key. Where necessary round to the nearest hundredth.

70. What is 19% of 3400?

71. What is 28% of 2300?

72. 30% what number is 180?

73. 40% of what number is 336?

74. What percent of 2000 is 1650?

75. What percent of 7000 is 2660?

76. 5.8% of 1260 is what?

77. 9.2% of 3470 is what?

78. 2400 is 16% of what number?

79. 462 is 14% of what number?

80. 21,600 is what percent of 27,000?

81. 20,800 is what percent of 32,500?

82. What is 148.6% of 12,320?

83. What is 202.9% of 34,160?

84. 0.56% of what number is 32.984?

85. 0.71% of what number is 45.582?

Solutions for Odd-Numbered Practice Problems

1. Add 3792 + 5896
   3792 $+$ 5896 $=$ | 9688|
   The solution is 9688.

3. Multiply 896 × 273
   896 $\times$ 273 $=$ | 244608|
   The solution is 244,608.

5. Multiply 72.8 × 197
   72.8 $\times$ 197 $=$ | 143416|
   The solution is 14,341.6.

7. We multiply 126 times the cost of each one which is $107.38
   126 $\times$ 107.38 $=$ | 1352988|
   The total cost is $13,529.88.

9. Calculate 658.3 − (12.8)(3.46).
   Using Memory you calculate (12.8)(3.46) first and store it in memory.
   AC 12.8 $\times$ 3.46 $=$ M+
   Then you subtract
   658.3 $-$ MR $=$ | 614012|
   The solution is 614.012. Without memory you calculate first (12.8)(3.46) = 44.288 and then subtract that value from 658.3 to obtain the same answer.

11. The area is found by multiplying length times width. Here we multiply 85.21 times 36.93 on the calculator.
    85.21 $\times$ 36.93 $=$ | 3146.8053|
    If we round to the nearest hundredth, we have 3146.81 square meters.

13. What is 75% of 500? The percent key must be used last on most calculators.
    500 $\times$ 75 % | 375|
    The answer is 375. Note no equal key was used.

15. What percent is 153 of 170? We want to solve the problem.
    153 ÷ 170 = n%
    We enter on the calculator the following keystrokes:
    153 $\div$ 170 % | 90|
    The answer is 90%. Note we did not use the equal sign key.

Answers to Even-Numbered Practice Problems

2. 2834   4. 147   6. 224.91   8. 41.5494   10. Rounded to the nearest tenth, 110.5
12. Rounded to five decimal places, 0.94510   14. 240

Tables

TABLE OF BASIC ADDITION FACTS

+	0	1	2	3	4	5	6	7	8	9
0	0	1	2	3	4	5	6	7	8	9
1	1	2	3	4	5	6	7	8	9	10
2	2	3	4	5	6	7	8	9	10	11
3	3	4	5	6	7	8	9	10	11	12
4	4	5	6	7	8	9	10	11	12	13
5	5	6	7	8	9	10	11	12	13	14
6	6	7	8	9	10	11	12	13	14	15
7	7	8	9	10	11	12	13	14	15	16
8	8	9	10	11	12	13	14	15	16	17
9	9	10	11	12	13	14	15	16	17	18

TABLE OF BASIC MULTIPLICATION FACTS

×	0	1	2	3	4	5	6	7	8	9
0	0	0	0	0	0	0	0	0	0	0
1	0	1	2	3	4	5	6	7	8	9
2	0	2	4	6	8	10	12	14	16	18
3	0	3	6	9	12	15	18	21	24	27
4	0	4	8	12	16	20	24	28	32	36
5	0	5	10	15	20	25	30	35	40	45
6	0	6	12	18	24	30	36	42	48	54
7	0	7	14	21	28	35	42	49	56	63
8	0	8	16	24	32	40	48	56	64	72
9	0	9	18	27	36	45	54	63	72	81

TABLE OF PRIME FACTORS

Number	Prime Factors	Number	Prime Factors	Number	Prime Factors	Number	Prime Factors
		51	3×17	101	prime	151	prime
2	prime*	52	$2^2 \times 13$	102	$2 \times 3 \times 17$	152	$2^3 \times 19$
3	prime	53	prime	103	prime	153	$3^2 \times 17$
4	2^2	54	2×3^3	104	$2^3 \times 13$	154	$2 \times 7 \times 11$
5	prime	55	5×11	105	$3 \times 5 \times 7$	155	5×31
6	2×3	56	$2^3 \times 7$	106	2×53	156	$2^2 \times 3 \times 13$
7	prime	57	3×19	107	prime	157	prime
8	2^3	58	2×29	108	$2^2 \times 3^3$	158	2×79
9	3^2	59	prime	109	prime	159	3×53
10	2×5	60	$2^2 \times 3 \times 5$	110	$2 \times 5 \times 11$	160	$2^5 \times 5$
11	prime	61	prime	111	3×37	161	7×23
12	$2^2 \times 3$	62	2×31	112	$2^4 \times 7$	162	2×3^4
13	prime	63	$3^2 \times 7$	113	prime	163	prime
14	2×7	64	2^6	114	$2 \times 3 \times 19$	164	$2^2 \times 41$
15	3×5	65	5×13	115	5×23	165	$3 \times 5 \times 11$
16	2^4	66	$2 \times 3 \times 11$	116	$2^2 \times 29$	166	2×83
17	prime	67	prime	117	$3^2 \times 13$	167	prime
18	2×3^2	68	$2^2 \times 17$	118	2×59	168	$2^3 \times \cdot 3 \times 7$
19	prime	69	3×23	119	7×17	169	13^2
20	$2^2 \times 5$	70	$2 \times 5 \times 7$	120	$2^3 \times 3 \times 5$	170	$2 \times 5 \times 17$
21	3×7	71	prime	121	11^2	171	$3^2 \times 9$
22	2×11	72	$2^3 \times 3^2$	122	2×61	172	$2^2 \times 43$
23	prime	73	prime	123	3×41	173	prime
24	$2^3 \times 3$	74	2×37	124	$2^2 \times 31$	174	$2 \times 3 \times 29$
25	5^2	75	3×5^2	125	5^3	175	$5^2 \times 7$
26	2×13	76	$2^2 \times 19$	126	$2 \times 3^2 \times 7$	176	$2^4 \times 11$
27	3^3	77	7×11	127	prime	177	3×59
28	$2^2 \times 7$	78	$2 \times 3 \times 13$	128	2^7	178	2×89
29	prime	79	prime	129	3×43	179	prime
30	$2 \times 3 \times 5$	80	$2^4 \times 5$	130	$2 \times 5 \times 13$	180	$2^2 \times 3^2 \times 5$
31	prime	81	3^4	131	prime	181	prime
32	2^5	82	2×41	132	$2^2 \times 3 \times 11$	182	$2 \times 7 \times 13$
33	3×11	83	prime	133	7×19	183	3×61
34	2×17	84	$2^2 \times 3 \times 7$	134	2×67	184	$2^3 \times 23$
35	5×7	85	5×17	135	$3^3 \times 5$	185	5×37
36	$2^2 \times 3^2$	86	2×43	136	$2^3 \times 17$	186	$2 \times 3 \times 31$
37	prime	87	3×29	137	prime	187	11×17
38	2×19	88	$2^3 \times 11$	138	$2 \times 3 \times 23$	188	$2^2 \times 47$
39	3×13	89	prime	139	prime	189	$3^3 \times 7$
40	$2^3 \times 5$	90	$2 \times 3^2 \times 5$	140	$2^2 \times 5 \times 7$	190	$2 \times 5 \times 19$
41	prime	91	7×13	141	3×47	191	prime
42	$2 \times 3 \times 7$	92	$2^2 \times 23$	142	2×71	192	$2^6 \times 3$
43	prime	93	3×31	143	11×13	193	prime
44	$2^2 \times 11$	94	2×47	144	$2^4 \times 3^2$	194	2×97
45	$3^2 \times 5$	95	5×19	145	5×29	195	$3 \times 5 \times 13$
46	2×23	96	$2^5 \times 3$	146	2×73	196	$2^2 \times 7^2$
47	prime	97	prime	147	3×7^2	197	prime
48	$2^4 \times 3$	98	2×7^2	148	$2^2 \times 37$	198	$2 \times 3^2 \times 11$
49	7^2	99	$3^2 \times 11$	149	prime	199	prime
50	2×5^2	100	$2^2 \times 5^2$	150	$2 \times 3 \times 5^2$	200	$2^3 \times 5^2$

TABLE OF SQUARE ROOTS

Square Root Values are Rounded to the Nearest Thousandth Unless the Answer Ends in .000

n	$\sqrt{n}$	n	$\sqrt{n}$	n	$\sqrt{n}$	n	$\sqrt{n}$	n	$\sqrt{n}$
1	1.000	41	6.403	81	9.000	121	11.000	161	12.689
2	1.414	42	6.481	82	9.055	122	11.045	162	12.728
3	1.732	43	6.557	83	9.110	123	11.091	163	12.767
4	2.000	44	6.633	84	9.165	124	11.136	164	12.806
5	2.236	45	6.708	85	9.220	125	11.180	165	12.845
6	2.449	46	6.782	86	9.274	126	11.225	166	12.884
7	2.646	47	6.856	87	9.327	127	11.269	167	12.923
8	2.828	48	6.928	88	9.381	128	11.314	168	12.961
9	3.000	49	7.000	89	9.434	129	11.358	169	13.000
10	3.162	50	7.071	90	9.487	130	11.402	170	13.038
11	3.317	51	7.141	91	9.539	131	11.446	171	13.077
12	3.464	52	7.211	92	9.592	132	11.489	172	13.115
13	3.606	53	7.280	93	9.644	133	11.533	173	13.153
14	3.742	54	7.348	94	9.695	134	11.576	174	13.191
15	3.873	55	7.416	95	9.747	135	11.619	175	13.229
16	4.000	56	7.483	96	9.798	136	11.662	176	13.266
17	4.123	57	7.550	97	9.849	137	11.705	177	13.304
18	4.243	58	7.616	98	9.899	138	11.747	178	13.342
19	4.359	59	7.681	99	9.950	139	11.790	179	13.379
20	4.472	60	7.746	100	10.000	140	11.832	180	13.416
21	4.583	61	7.810	101	10.050	141	11.874	181	13.454
22	4.690	62	7.874	102	10.100	142	11.916	182	13.491
23	4.796	63	7.937	103	10.149	143	11.958	183	13.528
24	4.899	64	8.000	104	10.198	144	12.000	184	13.565
25	5.000	65	8.062	105	10.247	145	12.042	185	13.601
26	5.099	66	8.124	106	10.296	146	12.083	186	13.638
27	5.196	67	8.185	107	10.344	147	12.124	187	13.675
28	5.292	68	8.246	108	10.392	148	12.166	188	13.711
29	5.385	69	8.307	109	10.440	149	12.207	189	13.748
30	5.477	70	8.367	110	10.488	150	12.247	190	13.784
31	5.568	71	8.426	111	10.536	151	12.288	191	13.820
32	5.657	72	8.485	112	10.583	152	12.329	192	13.856
33	5.745	73	8.544	113	10.630	153	12.369	193	13.892
34	5.831	74	8.602	114	10.677	154	12.410	194	13.928
35	5.916	75	8.660	115	10.724	155	12.450	195	13.964
36	6.000	76	8.718	116	10.770	156	12.490	196	14.000
37	6.083	77	8.775	117	10.817	157	12.530	197	14.036
38	6.164	78	8.832	118	10.863	158	12.570	198	14.071
39	6.245	79	8.888	119	10.909	159	12.610	199	14.107
40	6.325	80	8.944	120	10.954	160	12.649	200	14.142

Glossary

Absolute value of a number (9.1) The absolute value of a number is the distance between that number and zero on the number line. When we find the absolute value of a number, we use the $|\ \ |$ notation. To illustrate, $|-4| = 4$, $|6| = 6$, $|-20 - 3| = |-23| = 23$, $|0| = 0$.

Addends (1.2) When two or more numbers are added, the numbers being added are called addends. In the problem $3 + 4 = 7$, the numbers 3 and 4 are both addends.

Altitude of a triangle (7.3) The height of a triangle.

Amount of a percent equation (5.3A) The product we obtain when we multiply a percent times a number. In the equation $75 = 50\% \times 150$, the amount is 75.

Angle (7.3) An angle is formed whenever two line segments have a common point of origin.

Area (7.1) The measure of the surface inside a geometric figure. Area is measured in square units, such as square feet.

Associative property of addition (1.2) The property of addition that tells that if three numbers are to be added, it does not matter which two numbers are added first. An example of the associative property is $5 + (1 + 2) = (5 + 1) + 2$. If we add $1 + 2$ first and then add 5 to that, or if we add $5 + 1$ first and then add that result to 2, we will obtain the same result.

Associative property of multiplication (1.4) The property that tells us that if we have three numbers to multiply, it does not matter which two numbers we group together first to multiply, the result will be the same. An example of the associative property of multiplication is $2 \times (5 \times 3) = (2 \times 5) \times 3$.

Base (1.6) The number that is to be repeatedly multiplied in exponent notation. When we write $16 = 2^4$, the number 2 is the base.

Base of a percent equation (5.3A) The quantity we take a percent of. In the equation $80 = 20\% \times 400$, the base is 400.

Billion (1.1) The number 1,000,000,000.

Borrowing (1.3) The renaming of a number in order to facilitate subtraction. When we subtract $42 - 28$ we rename 42 as 3 12. This represents 3 tens and 12 ones. This renaming is called borrowing.

Box (7.5) A three dimensional object whose every side is a rectangle. Another name for a box is a rectangular solid.

Building up a fraction (2.6) To make one fraction into an equivalent fraction by making the denominator and numerator larger numbers. For example, the fraction $\dfrac{3}{4}$ can be built up to the fraction $\dfrac{30}{40}$.

Building fraction property (2.6) For whole numbers a, b, c where neither b or c can be zero

$$\frac{a}{b} = \frac{a}{b} \times 1 = \frac{a}{b} \times \overbrace{\frac{c}{c}} = \frac{a \times c}{b \times c}$$

Cancellation of common factors (2.2) A procedure where a fraction is reduced by dividing the numerator and denominator by a common factor. If we reduce the fraction $\dfrac{35}{42}$ by dividing both numerator and denominator by a common factor of 7 we obtain the fraction $\dfrac{5}{6}$.

Caret (3.5) A symbol $\wedge$ used to indicate the new location of a decimal point when performing division of decimal fractions.

Celsius temperature (6.4) A temperature scale in which water boils at 100 degrees Celsius and freezes as 0 degrees Celsius. To convert Celsius temperature to Fahrenheit we use the formula $F = 1.8 \times C + 32$.

Center of a circle (7.4) The point in the middle of a circle from which all points of the circle are an equal distance.

Centimeter (6.2) A unit of length commonly used in the metric system to measure small distances. 1 centimeter = 0.01 meter.

Circle (7.4) A figure for which all points are at an equal distance from a given point.

Circumference of a circle (7.4) The distance around the rim of a circle.

Commission (5.4) The amount of money a salesperson gets that is a percentage of the value of the sales of that salesperson is called a commission. The commission is obtained by multiplying the commission rate times the value of the sales. If a salesman sold $120,000 of insurance and his commission rate was 0.5% then his commission is $0.5\% \times 120,000 = \$600.00$.

Common denominator (2.7)   Two fractions have a common denominator if the same number appears in the denominator of each fraction. $\frac{3}{7}$ and $\frac{1}{7}$ have a common denominator of 7.

Commutative property of addition (1.2)   The property of addition that tells us that the order in which two numbers are added does not change the sum. An example of the commutative property of addition is $3 + 6 = 6 + 3$.

Commutative property of multiplication (1.4)   The property that tells us that the order in which two numbers are multiplied does not change the value of the answer. An example of the commutative property of multiplication would be $7 \times 3 = 3 \times 7$.

Composite number (2.2)   A composite number is a whole number greater than 1 that can be divided by whole numbers other than itself. The number 6 is a composite number since it can be divided exactly by 2 and 3 (as well as 1 and 6).

Cone (7.5)   A three–dimensional object shaped like an ice cream cone or the sharpened end of a pencil.

Cross-multiplying (4.3)   If you have a proportion such as $\frac{n}{3} = \frac{12}{15}$, then to cross multiply you form products to have $n \times 15 = 3 \times 12$.

Cubic centimeter (6.3)   A metric measurement of volume equal to one milliliter.

Cup (6.1)   Smallest unit of volume in the American system. 2 cups = 1 pint.

Cylinder (7.5)   A three dimensional object shaped like a tin can.

Debit (1.2)   A debit in banking is the removing of money from an account. If you had a savings account and took $300 out of it on Wednesday, we would say that you had a debit of $300 from your account. Often a bank will charge a service charge to your account and use the word debit to mean that they have removed money from your account to cover the charge.

Decimal fraction (3.1)   A fraction whose denominator is a power of ten.

Decimal places (3.4)   The number of digits to the right of the decimal point in a decimal fraction. The number 1.234 has three decimal places while the number 0.129845 has six decimal places. A whole number such as 42 is considered to have zero decimal places.

Decimal point (3.1)   The period that is used when writing a decimal fraction. In the number 5.346, the period between the 5 and the 3 is the decimal point. It separates the whole number from the fractional part that is less than one.

Decimal system (1.1)   Our number system is called the decimal system or base ten system because the value of numbers written in our system is based on tens and ones.

Decimeter (6.2)   A unit of length not commonly used in the metric system. 1 decimeter = 0.1 meter.

Degree (7.3)   A unit used to measure an angle. A degree is $\frac{1}{360}$ of a complete revolution. An angle of 32 degrees is written as 32°.

Dekameter (6.2)   A unit of length not commonly used in the metric system. 1 dekameter = 10 meters.

Denominator (2.1)   The number on the bottom of a fraction. In the fraction $\frac{2}{9}$ the denominator is 9.

Deposit (1.2)   A deposit in banking is the placing of money in an account. If you had a checking account and on Tuesday you placed $124 into that account we say that you made a deposit of $124.

Diameter of a circle (7.4)   A line segment across the circle that passes through the center of the circle. The diameter of a circle is equal to twice the radius of a circle.

Difference (1.3)   The result of performing a subtraction is called the difference. In the problem $9 - 2 = 7$ the number 7 is the difference.

Digits (1.1)   The symbols 0, 1, 2, 3, 4, 5, 6, 7, 8, and 9 are called digits.

Discount (5.4)   The amount of reduction in a price. The discount is a product of the discount rate times the list price. If the list price of a television is $430.00 and it has a discount rate of 35%, then the amount of discount is $35\% \times \$430.00 = \$150.50$. The price would be reduced by $150.50.

Distributive property of multiplication over addition (1.4)   The property illustrated by $5 \times (4 + 3) = (5 \times 4) + (5 \times 3)$. In general, for any numbers a, b, c it is true that $a(b + c) = ab + ac$.

Dividend (1.5)   The number that is being divided by another is called the dividend. If we consider the problem $14 \div 7 = 2$, the number 14 is the dividend.

Divisor (1.5)   The divisor is the number that you divide into another number. In the problem $30 \div 5 = 6$, the number 5 is called the divisor.

Earned run average (4.4)   A ratio formed by finding the number of runs a pitcher would give up in a 9 inning game. If a pitcher had an earned run average of 2, it would mean that on the average he would give up two runs for every nine innings he pitched.

Ellipsoid (7.5)   A three dimensional object with a shape like a cold capsule, a football, or an egg.

Equal fractions (2.2)   Fractions that represents the same number. The fractions $\frac{3}{4}$ and $\frac{6}{8}$ are equal fractions.

Equality test of fractions (2.2)   Two fractions $\frac{a}{b}$ and $\frac{c}{d}$ are equal if the product $a \times d = b \times c$.

Equations (10.3)   Numerical sentences with variables such as $x + 3 = -8$ and $2s + 5s = 34 - 4s$ are called equations.

Equilateral triangle (7.3)   A triangle with three equal sides.

Equivalent equations (10.3)   Equations that have the same solution.

Equivalent fractions (2.2)   Two fractions that are equal.

Expanded notation for a number (1.1)   A number is written in expanded notation if it is written as a sum of hundreds, tens, ones, etc. The expanded notation for 763 is $700 + 60 + 3$.

Exponent (1.6)   The number that indicates the number of times a factor occurs. When we write $8 = 2^3$ the number 3 is the exponent.

Factors (1.4)   Each of the numbers which are multiplied are called factors. In the problem $8 \times 9 = 72$, the numbers 8 and 9 are called factors.

Fahrenheit temperature (6.4)   A temperature scale at which water boils at 212 degrees Fahrenheit and freezes at 32 degrees Fahrenheit. To convert Fahrenheit temperature to Celsius we use the formula $C = \dfrac{5 \times F - 160}{9}$.

Foot (6.1)   American system unit of length. 3 feet = 1 yard.

Fundamental theorem of arithmetic (2.2)   Every composite number has a unique product of prime numbers.

Gallon (6.1)   Largest unit of volume in the American System. 4 quarts = 1 gallon.

Gigameter (6.2)   A metric unit of length equal to 1,000,000,000 meters.

Gram (6.3)   The basic unit of weight in the metric system. A gram is defined as the weight of the water in a box that is 1 centimeter on each side. 1 gram = 1000 milligrams. 1 gram = 0.001 kilograms.

Hectometer (6.2)   A unit of length not commonly used in the metric system. 1 hectometer = 100 meters.

Height (7.2)   The distance between two parallel sides in a four–sided figure such as a parallelogram or a trapezoid.

Height of a cone (7.5)   The distance from the vertex of a cone to the base of a cone.

Height of a pyramid (7.5)   The distance from the point of a pyramid to the base of the pyramid.

Height of a triangle (7.3)   The distance of a line drawn from a vertex perpendicular to the other side or, an extension of the other side. This is sometimes called the altitude of a triangle.

Hemisphere (7.5)   A three–dimensional object created by taking exactly one half of a sphere.

Hexagon (7.2)   A six-sided figure.

Hypotenuse (8.6)   The side opposite the right angle in a right triangle is called the hypotenuse. It is always the longest side of a right triangle.

Improper fraction (2.3)   A fraction in which the numerator is greater than, or equal to, the denominator. The fractions $\dfrac{34}{29}, \dfrac{8}{7}$, and $\dfrac{6}{6}$ are all improper fractions.

Inch (6.1)   The smallest unit of length in the American System. 12 inches = 1 foot.

Inequality symbol (3.2)   The symbol that is used to indicate if a number is greater than another number or less than another number. Since 5 is greater than 3, we would write this with a "greater than" symbol as follows: $5 > 3$. The statement 7 is less than 12 would be written as follows: $7 < 12$.

Interest (5.5)   The money that is paid for the use of money. If you deposit money in a bank, the bank uses that money and pays you interest. If you borrow money, you pay the bank interest for the use of that money. Simple interest is determined by the formula $I = P \times R \times T$. Compound interest is usually determined by a table, a calculator or by a computer.

Invert a fraction (2.5)   To invert a fraction is to interchange the numerator and the denominator. If we invert $\dfrac{5}{9}$ we obtain the fraction $\dfrac{9}{5}$. To invert a fraction is sometimes referred to as to take the reciprocal of a fraction.

Irreducible (2.2)   A fraction that cannot be reduced is called irreducible.

Isosceles triangle (7.3)   A triangle with two sides equal.

Kilogram (6.3)   The most commonly used metric unit of weight. 1 kilogram = 1000 grams.

Kiloliter (6.3)   The metric unit of volume normally used to measure large volumes. 1 kiloliter = 1000 liters.

Kilometer (6.2)   The unit of length commonly used in the metric system to measure large distances. 1 kilometer = 1000 meters.

Least common denominator (LCD) (2.6)   The least common denominator (LCD) of two or more fractions is the smallest number that can be divided without remainder by each of the original denominators. The LCD of $\dfrac{1}{3}$ and $\dfrac{1}{4}$ is 12. The LCD of $\dfrac{5}{6}$ and $\dfrac{4}{15}$ is 30.

Legs of a right triangle (8.6)   The two shortest sides of a right triangle.

Length of a rectangle (7.1)   Each of the longer sides of a rectangle.

Like terms (10.1)   Like terms have identical variables. $-5x$ and $3x$ are like terms. $-7xyz$ and $-12xyz$ are like terms.

Line segment (7.3)   A portion of a straight line that has a beginning and an end.

Liter (6.3)   The standard metric measurement of volume. 1 liter = 1000 milliliters. 1 liter = 0.001 kiloliter.

Mach number (4.1)   A Mach number is the ratio of the velocity of an object to the velocity of sound. If the Mach number of a jet is 2, this would tell us that the jet travels at twice the speed of sound.

Mean (8.4)   The mean of a set of values is the sum of the values divided by the number of values. The mean of the numbers 10, 11, 14, 15 is 12.5. In everyday use when people use the word "average" they usually are referring to the mean.

Median (8.4)   If a set of numbers is arranged in order from smallest to largest, the median is that value that has the same number of values above it as below it. The median of the numbers 3, 7, 8 is 7. If the list contains an even number of items we obtain the median by finding the mean of the two middle numbers. The median of the numbers 5, 6, 10, 11 is 8.

Megameter (6.2)   A metric unit of length equal to 1,000,000 meters.

Meter (6.2)   The basic unit of length in the metric system. 1 meter = 1000 millimeters. 1 meter = 0.001 kilometer.

Metric ton (6.3)   A metric unit of measurement for very heavy weights. 1 metric ton = 1,000,000 grams.

Microgram (6.3)   A unit of weight equal to 0.000001 gram.

Micrometer (6.2)   A metric unit of length equal to 0.000001 meter.

Mile (6.1)   Largest unit of length in the American system. 5280 feet = 1 mile, 1760 yards = 1 mile.

Milligram (6.3)   A metric unit of weight used for very, very small objects. 1 milligram = 0.001 gram.

Milliliter (6.3)   The metric unit of volume normally used to measure small volumes. 1 milliliter = 0.001 liter.

Millimeter (6.2)   A unit of length commonly used in the metric system to measure very small distances. 1 millimeter = 0.001 meter.

Million (1.1)   The number 1,000,000.

Minuend (1.3)   The number being subtracted from in a subtraction problem. In the problem $8 - 5 = 3$ the number 8 is the minuend.

Mixed number (2.3)   A number created by the sum of a whole number greater than 1 and a proper fraction. The numbers $4\frac{5}{6}$ and $1\frac{1}{8}$ are both mixed numbers. Mixed numbers are sometimes referred to as mixed fractions.

Multiplicand (1.4)   The first factor in a multiplication problem. In the problem $7 \times 2 = 14$ the number 7 is the multiplicand.

Multiplier (1.4)   The second factor in a multiplication problem. In the problem $6 \times 3 = 18$ the number 3 is the multiplier.

Nanogram (6.3)   A unit of weight equal to 0.000000001 gram.

Nanometer (6.2)   A metric unit of length equal to 0.000000001 meter.

Negative numbers (9.1)   All of the numbers to the left of zero on the number line. The numbers -1.5, -16, -200.5, -4500 are all negative numbers. All negative numbers are written with a negative sign in front of the digits.

Numerator (2.1)   The number on the top of a fraction. In the fraction $\frac{3}{7}$ the numerator is 3.

Numerical coefficients (10.1)   Numerical coefficients are the numbers in front of the variables in one or more terms. If we look at $-3xy + 12w$, we find that the numerical coefficient of the xy term is -3 while the numerical coefficient of the w term is 12.

Number line (1.6)   A line on which numbers are placed in order from smallest to largest.

Octagon (7.2)   An eight-sided figure.

Odometer (1.8)   A device on an automobile that displays how many miles the car has been driven since it was first put into operation.

Opposite of a signed number (9.2)   The opposite of a signed number is a number that has the same absolute value. The opposite of -5 is 5. The opposite of 7 is -7.

Order of operations (1.7)   An agreed upon prodecure to do a problem with several arithmetic operations in the proper order.

Ounce (6.1)   Smallest unit of weight in the American System. 16 ounces = 1 pound.

Overtime (2.8)   The pay earned by a person if he works more than a certain number of hours per week. In most jobs that pay by the hour, a person will earn $1\frac{1}{2}$ times as much per hour for every hour beyond 40 hours worked in one work week. For example, Carlos earns $6.00 per hour for the first 40 hours in a week and overtime for each additional hour. He would earn $9.00 per hour for all hours in that week beyond 40 hours.

Parallel lines (7.2)   Two straight lines that are always the same distance apart.

Parallelogram (7.2)   A four-sided figure with both pairs of opposite sides parallel.

Parentheses (7.2)   One of several symbols used in mathematics to indicate multiplication. For example, (3)(5) means 3 multiplied by 5.

Percent (5.1)   The word percent means per one hundred. For example, 14 percent means $\frac{14}{100}$.

Percent of decrease (5.4)   The percent that something decreases is determined by dividing the amount of decrease by the original amount. If a tape deck sold for $300 and its price was decreased by $60, the percent of decrease is $\frac{60}{300} = 0.20 = 20\%$.

Percent of increase (5.4)   The percent that something increases is determined by dividing the amount of increase by the original amount. If the population of a town was 5,000 people and the population increased by 500 people, the percent of increase is $\frac{500}{5000} = 0.10 = 10\%$.

Percent proportion (5.3B) The percent proportion is the equation $\frac{a}{b} = \frac{p}{100}$ where a is the amount, b is the base, and p is the percent number.

Percent symbol (5.1) A symbol that is used to indicate percent. To indicate 23 percent we write 23%.

Perfect square (8.5) When a whole number is multiplied by itself the number that is obtained is a perfect square. The numbers 1, 4, 9, 16, 25, 36, 49, 64, 81, and 100 are all perfect squares.

Perimeter (7.1) The distance around a figure.

Perpendicular lines (7.3) Lines that meet at an angle of 90 degrees.

Pi (7.4) Pi is an irrational number that we obtain if we divide the circumference of a circle by the diameter of a circle. It is represented by the symbol π. Accurate to eleven decimal places, the value of Pi is given by 3.14159265359. For most work in mathematics the value of 3.14 is used to approximate the value of Pi.

Picogram (6.3) A unit of weight equal to 0.000000000001 gram.

Pint (6.1) Unit of volume in the American system. 2 pints = 1 quart.

Place holder (1.1) The use of a digit to indicate a place. Zero is a place holder in our number system. It holds a position and shows that there is no other digit in that place.

Place value system (1.1) Our number system is called a place value system because the placement of the digits tells the value of the number. If we use the digits 5 and 4 to write the number 54, the result is different than if we placed them in opposite order and wrote 45.

Positive numbers (9.1) All of the numbers to the right of zero on the number line. The numbers 5, 6.2, 124.186, 5000 are all positive numbers. A positive number such as +5 is usually written without the positive sign.

Pound (6.1) Basic unit of weight in the American System. 2000 pounds = 1 ton. 16 ounces = 1 pound.

Power of ten (1.4) Whole numbers that begin with 1 and end in one or more zeros are called powers of ten. The numbers 10, 100, 1000, etc. are all powers of ten.

Prime factors (2.2) Factors that are prime numbers. If we write 15 as a product of prime factors we would have $15 = 5 \times 3$.

Prime number (2.2) A prime number is a whole number greater than 1 that can only be divided by 1 and itself. The first fifteen prime numbers are 2, 3, 5, 7, 11, 13, 17, 19, 23, 29, 31, 37, 41, 43, 47. The list of prime numbers goes on indefinitely.

Principal (5.5) The amount of money deposited or borrowed on which interest is computed. In the simple interested formula $I = P \times R \times T$ the P stands for the Principal. (The other letters are: I = Interest, R = Interest Rate, and T = the amount of time.)

Product (1.4) The answer in a multiplication problem. In the problem $3 \times 4 = 12$ the number 12 is the product.

Proper fraction (2.3) A fraction in which the numerator is less than the denominator. The fractions $\frac{3}{4}$ and $\frac{15}{16}$ are proper fractions.

Proportion (4.2) A proportion is a true or false statement that two ratios or two rates are equal. The proportion $\frac{3}{4} = \frac{15}{20}$ is true. The proportion $\frac{5}{7} = \frac{7}{9}$ is false.

Pyramid (7.5) A three–dimensional object made up of a geometric figure for a base and triangular sides that meet at a point. Some pyramids are shaped like the great pyramids of Egypt.

Pythagorean theorem (8.6) A statement that for any right triangle the square of the hypotenuse equals the sum of the squares of the two legs of the triangle.

Quadrilateral (7.2) A four-sided geometric figure.

Quadrillion (1.1) The number 1,000,000,000,000,000.

Quart (6.1) Unit of volume in the American System. 4 quarts = 1 gallon.

Quotient (1.5) The answer after performing a division problem. In the problem $60 \div 6 = 10$, the number 10 is the quotient.

Radius of a circle (7.4) A line segment from the center of a circle to any point on the circle. The radius of a circle is equal to one-half of the diameter of a circle.

Rate (4.1) A rate compares two quantities that have different units. Examples of rates are $5.00 an hour or 13 pounds for every 2 inches. In fraction form, these two rates would be written as: $\frac{\$5.00}{1 \text{ hour}}$ and $\frac{13 \text{ pounds}}{2 \text{ inches}}$.

Ratio (4.1) A ratio is a comparison of two quantities that have the same units. To compare 2 to 3 we can express the ratio in three ways: "The ratio of 2 to 3"; We can write 2:3; or we can write the fraction $\frac{2}{3}$.

Ratio in simplest form (4.1) A ratio is in simplest form when the two numbers do not have a common factor.

Rectangle (7.1) A four–sided figure that has four right angles.

Reduced fraction (2.2) A fraction for which the numerator and denominator have no common factor other than 1. The fraction $\frac{5}{7}$ is a reduced fraction. The fraction $\frac{15}{21}$ is not a reduced fraction because both numerator and denominator have a common factor of 3.

Regular octagon (7.2) An eight-sided figure with all sides equal.

Repeating decimals (3.6) Decimals that have a digit or a group of digits that repeat. The decimals 0.33333333333 ... and 1.234234234234 ... are repeating decimals. The pattern of repeating continues indefinitely. Repeating decimals can be written in a form with a bar denoted over the repeating digit. Thus, the decimals above could be written as $0.\overline{3}$ and $1.\overline{234}$.

Remainder (1.5)   When two numbers do not divide exactly, a part is left over. This part is called the remainder. For example, if we say $13 \div 2 = 6$ with a 1 left over, the 1 is called the remainder.

Right angle (7.1) and (7.3)   An angle that measures 90 degrees.

Right triangle (7.3)   A triangle with one 90 degree angle.

Regular hexagon (7.2)   A six-sided figure with all sides equal.

Rounding (1.6)   Rounding is the process of writing a number in an approximate form for convenience. The number 9756 rounded to the nearest hundred is 9800.

Sales tax (5.4)   The sales tax is the amount of tax on a purchase. The sales tax for any item is a product of the sales tax rate times the purchase price. If an item is purchased for $12.00 and the sales tax rate is 5%, the sales tax is $5\% \times 12.00 = \$0.60$.

Scientific notation (6.5)   A positive number is written in scientific notation if it is in the form $a \times 10^n$ where a is a number greater than or equal to 1, but less than 10, and n is an integer. If we write 5678 in scientific notation, we have 5.678×10^3. If we write 0.00825 in scientific notation, we have 8.25×10^{-3}.

Semicircle (7.4)   One-half of a circle. The semicircle usually includes a diameter of a circle connected to one-half of the circumference of a circle.

Sides of an angle (7.3)   The two line segments that meet to form an angle.

Signed numbers (9.1)   All of the numbers on a number line. Numbers like $-33, 2, 5, -4.2, 18.678, -8.432$ are all signed numbers. A negative number always has a negative sign in front of the digits. A positive number such as $+3$ is usually written without the positive sign in front of it.

Simple interest (5.5)   The interest determined by the formula $I = P \times R \times T$ where I = the interest obtained, P = the principal or the amount borrowed or invested, R = the interest rate, usually on an annual rate, and T = the number of time periods, usually years.

Similar triangles (7.6)   Two triangles that have the same shape but not necessarily the same size. The corresponding angles of similar triangles are equal. The corresponding sides of similar triangles have the same ratio.

Solution of an equation (10.3)   A number is a solution of an equation if replacing the variable by the number makes the number sentence always true. The solution of $x - 5 = -20$ is the number -15.

Sphere (7.5)   A three–dimensional object shaped like a perfectly round ball.

Square (7.1)   A rectangle with all four sides equal.

Square root (8.5)   The square root of a number is one of two identical factors of that number. The square root of 9 is 3. The square root of 121 is 11.

Square root sign (8.5)   The symbol $\sqrt{\ }$ is called the square root sign. When we want to find the square root of 25 we write $\sqrt{25}$. The answer is 5.

Standard notation for a number (1.1)   When a number is written in the normal fashion. For example, we would write $70 + 2$ in standard notation as 72.

Subtrahend (1.3)   The number being subtracted is called the subtrahend. In the problem $7 - 1 = 6$, the number 1 is the subtrahend.

Sum (1.2)   The result of an addition of two or more numbers. In the problem $7 + 3 + 5 = 15$, the number 15 is called the sum.

Trapezoid (7.2)   A four-sided figure with at least two parallel sides.

Term (10.1)   A term is a number, a variable, or a product of a number and one or more variables. $5x, 2ab, -43cdef$ are three examples of terms.

Terminating decimals (3.6)   Every fraction can be written as a decimal. If the division process of dividing denominator into numerator ends with a remainder of zero, the decimal is a terminating decimal. Decimals such as 1.28, 0.007856, and 5.123 are called terminating decimals.

Triangle (7.3)   A three-sided figure.

Trillion (1.1)   The number 1,000,000,000,000.

Unit fraction (6.1)   A fraction used to convert one unit to another. For example, to change 180 inches to feet we multiply by the unit fraction $\dfrac{1 \text{ foot}}{12 \text{ inches}}$. Thus we have:

$$180 \text{ inches} \times \boxed{\dfrac{1 \text{ foot}}{12 \text{ inches}}} = 15 \text{ feet}$$

Variable (10.1)   A letter that is used to represent a number is called a variable.

Vertex of an angle (7.3)   The point at which two line segments meet to form an angle.

Vertex of a cone (7.5)   The sharp point of a cone.

Volume (7.5)   The measure of the space inside a three-dimensional object. Volume is measured in cubic units such as cubic feet.

Width of a rectangle (7.1)   Each of the shorter sides of a rectangle.

Whole numbers (1.1)   The whole numbers are the set of numbers 0, 1, 2, 3, 4, 5, 6, 7, 8, 9, 10, 11, 12, ... The set goes on forever. There is no largest whole number.

Word names for whole numbers (1.1)   The notation for a number where each digit is expressed by a word. To write 389 with a word name we would write three hundred eighty-nine.

Zero (1.1)   The smallest whole number. It is normally written 0.

Selected Answers

CHAPTER 1 Whole Numbers

Pretest Chapter 1
1. Sixty-five million, two hundred eighty-seven thousand, twenty-three   **2.** $50,000 + 9,000 + 300 + 60 + 1$ **3.** 1,058,498
4. 244   **5.** 91,926   **6.** 19,435   **7.** 7131   **8.** 63,541   **9.** 1,944,753   **10.** 180   **11.** 1392   **12.** 340,500
13. 50,632   **14.** 20,301   **15.** 8,253 R 2   **16.** 127   **17.** 7^5 **18.** 64   **19.** 157,000   **20.** 48,300   **21.** 50,000,000
22. 80   **23.** 43   **24.** 15   **25.** 1166 miles   **26.** $22.00   **27.** $299

Exercises 1.1
1. $6000 + 700 + 30 + 1$ **3.** $100,000 + 8,000 + 200 + 70 + 6$
5. $20,000,000 + 3,000,000 + 700,000 + 60,000 + 1,000 + 300 + 40 + 5$
7. $100,000,000 + 3,000,000 + 200,000 + 60,000 + 700 + 60 + 8$ **9.** 671   **11.** 9863   **13.** 54,027   **15.** 706,200
17. Fifty-three   **19.** Four hundred sixty-five   **21.** Eight thousand, nine hundred thirty-six
23. One hundred five thousand, two hundred sixty-one
25. Twenty-three million, five hundred sixty-one thousand, two hundred forty-eight
27. Four billion, three hundred two million, one hundred fifty-six thousand, two hundred   **29.** 375   **31.** 56,281
33. 100,079,826   **35.** 7,164,742   **37.** 2,913,000   **39. (a)** 5 **(b)** 2   **41. (a)** 9 **(b)** 8   **43.** 596,000,734,129,029
45. One thousand, nine hundred sixty-five

Exercises 1.2
1.

+	3	5	4	8	0	6	7	2	9	1
2	5	7	6	10	2	8	9	4	11	3
7	10	12	11	15	7	13	14	9	16	8
5	8	10	9	13	5	11	12	7	14	6
3	6	8	7	11	3	9	10	5	12	4
0	3	5	4	8	0	6	7	2	9	1
4	7	9	8	12	4	10	11	6	13	5
1	4	6	5	9	1	7	8	3	10	2
8	11	13	12	16	0	14	15	10	17	9
6	9	11	10	14	6	12	13	8	15	7
9	12	14	13	17	9	15	16	11	18	10

3. 23   **5.** 26   **7.** 87   **9.** 99   **11.** 6579   **13.** 13,336   **15.** 13,951   **17.** 42,739

19. 126   **21.** 1132   **23.** 17,909   **25.** 11,579,426   **27.** 1,135,280,240   **29.** 2,303,820   **31.** 248
33. 20,909   **35.** 818 miles   **37.** $8775   **39.** 855 feet   **41.** 4,753,420 people voted for Mondale.
43. $12,854,835,000   **45. (a)** 1937 good TV's **(b)** 1982 televisions
47. The answer to an addition problem would not be unique.   **49. (a)** $9553 **(b)** $7319   **51.** 12,117,289,695
53. Seventy-six million, two hundred eight thousand, nine hundred forty-one   **55.** 8,724,396   **57.** 243,722,443
59. $202,429

Exercises 1.3
1. 6   **3.** 5   **5.** 1   **7.** 9   **9.** 16   **11.** 9   **13.** 7   **15.** 6   **17.** 9   **19.** 9   **21.** 58   **23.** 12   **25.** 31
27. 444   **29.** 1341   **31.** 11,191   **33.** 553,101   **35.** 108   **37.** 1113   **39.** 1010   **41.** 56,232   **43.** 46   **45.** 37
47. 75   **49.** 372   **51.** 3,296   **53.** 34,092   **55.** 2,314   **57.** 223,116   **59.** Latasha got 182 more votes.
61. The Horizon Enterprise bid was $403,445 lower   **63.** She received $244.   **65.** 662,106 people   **67.** 3,686,434 people
69. $x = 6$ **71.** If a and b represent the same number. For example if $a = 10$ and $b = 10$.   **73.** 21,222,415,864
75. 8,466,084   **77.** 168   **79.** 2894   **81.** $12,039,724

Exercises 1.4

1.

×	6	2	3	8	0	5	7	9	1	4
5	30	10	15	40	0	25	35	45	5	20
7	42	14	21	56	0	35	49	63	7	28
1	6	2	3	8	0	5	7	9	1	4
0	0	0	0	0	0	0	0	0	0	0
6	36	12	18	48	0	30	42	54	6	24
2	12	4	6	16	0	10	14	18	2	8
3	18	6	9	24	0	15	21	27	3	12
8	48	16	24	64	0	40	56	72	8	32
4	24	8	12	32	0	20	28	36	4	16
9	54	18	27	72	0	45	63	81	9	36

3. 84   **5.** 208   **7.** 18,306   **9.** 303,612   **11.** 60   **13.** 522   **15.** 1630   **17.** 7609
19. 189,560   **21.** 762,324   **23.** 1560   **25.** 8,936,100   **27.** 482,000
29. 56,231,400,000   **31.** 8460   **33.** 63,600   **35.** 186,000,000   **37.** 31,682   **39.** 3906
41. 5696   **43.** 26,880   **45.** 41,537   **47.** 69,312   **49.** 148,567   **51.** 490,374
53. 41,830   **55.** 89,496   **57.** 217,980   **59.** 1,427,961   **61.** 4,097,115   **63.** 3900
65. 9210   **67.** 70   **69.** 308   **71.** 500   **73.** 360   **75.** 0   **77.** $168   **79.** $978
81. $2826   **83.** 612 miles   **85.** 1428 VCRs   **87.** $12,461,040,000
89. No, it would not always be true. In our number system $62 = 60 + 2$. But in roman numerals, IV $\neq$ I + V. The digit system in roman numerals involves subtraction. Thus (XII) × (IV) $\neq$ (XII × I) + (XII × V).   **91.** 1,279,922,641,023   **93.** 6,756
95. $139   **97.** 87,897,646   **99.** 2,552,199

Exercises 1.5

1. 7   **3.** 8   **5.** 3   **7.** 8   **9.** 3   **11.** 9   **13.** 9   **15.** 7   **17.** 5   **19.** 8   **21.** 8   **23.** 0   **25.** 8   **27.** 6
29. 1   **31.** 9 R 1   **33.** 9 R 4   **35.** 25 R 3   **37.** 23 R 4   **39.** 14   **41.** 37   **43.** 322 R 1   **45.** 127 R 1
47. 362   **49.** 1357 R 4   **51.** 1757 R 5   **53.** 2478 R 3   **55.** 5 R 5   **57.** 5 R 7   **59.** 7   **61.** 160 R 10
63. 57 R 16   **65.** 615 R 11   **67.** 210 R 8   **69.** 202 R 7   **71.** 5   **73.** 5 R 32   **75.** 27
77. Each employee receives $1586.   **79.** 64 miles per hour   **81.** $17,652 per car   **83.** $171 for each monthly payment
85. a and b must represent the same number. For example if $a = 12$ and $b = 12$.   **87.** 9875   **89.** 5504   **91.** 406,195
93. 8807

Exercises 1.6

1. 6^4 **3.** 2^5 **5.** 12^3 **7.** 1^7 **9.** 35^1 **11.** 27   **13.** 25   **15.** 1000   **17.** 1   **19.** 32   **21.** 64   **23.** 169
25. 512   **27.** 625   **29.** 1   **31.** 128   **33.** 8000   **35.** 8   **37.** 100,000   **39.** 196   **41.** 80   **43.** 70   **45.** 100
47. 1300   **49.** 1600   **51.** 4000   **53.** 14,000   **55.** 95,000   **57.** 790,000,000   **59.** 90,000
61. (a) 268,000   **(b)** 268,400   **63. (a)** 17,200,000   **(b)** 17,240,000   **65.** 531,441   **67.** $70,000 + 6000 + 300 + 20 + 5$
69. 9688   **71.** 46,656

Exercises 1.7

1. 37   **3.** 38   **5.** 7   **7.** 59   **9.** 420   **11.** 19   **13.** 114   **15.** 60   **17.** 144   **19.** 96   **21.** 7   **23.** 51
25. 16   **27.** 2   **29.** 18   **31.** Yes, because multiplication is commutative.   **33.** 2769
35. $100,000 + 50,000 + 6000 + 300 + 10 + 2$ **37.** Two hundred sixty-one million, seven hundred sixty-three thousand, two
39. 1156   **41.** 695   **43.** 90,000   **45.** 578,652

Exercises 1.8

1. 14,295 people   **3.** 572 miles   **5.** 6000 sheets   **7.** 6 cents per ounce   **9.** 1519 miles   **11.** 160 minutes
13. 3,932,160 vibrations   **15.** 17,602 kilowatt-hours   **17.** $317   **19.** Her balance will be $185.   **21.** $14,040
23. 25 miles per gallon   **25.** $8 for each student contribution   **27.** 343

Extra Practice: Examples and Exercises

Section 1.1   **1.** 37,043   **3.** 27,940   **5.** 830,426   **7.** Eight thousand, nine hundred twenty-one
9. Thirty-two million, five hundred eight thousand, two hundred fifty-four   **11.** 251   **13.** 36,320   **15.** 500,023,651
17. 0   **19.** 8   *Section 1.2*   **1.** 15   **3.** 46   **5.** 56   **7.** 258   **9.** 3567   **11.** 5901   **13.** 43,315   **15.** 4,926,638
17. $7664   **19 (a)** $4,218,558   **(b)** $6,520,243   *Section 1.3*   **1.** 6342   **3.** 3130   **5.** 2442   **7.** 3128   **9.** 5059
11. 7278   **13.** 41,681   **15.** 4229   **17.** 2222   **19.** 69,114   **21.** 106   **23.** 206 miles   **25.** $879,621
Section 1.4   **1.** 68   **3.** 9396   **5.** 1,293,369   **7.** 3115   **9.** 974,616   **11.** 9262   **13.** 297,738   **15.** 8,181,312
Section 1.5   **1.** 5 R 1   **3.** 6 R 3   **5.** 8 R 4   **7.** 392 R 3   **9.** 265 R 5   **11.** 251 R 11   **13.** 67 R 2   **15.** 605 R 8
17. 22 R 5   **19.** 43   *Section 1.6*   **1.** 5^4 **3.** 13^6 **5.** 52^1 **7.** 144   **9.** 1   **11.** 346,000   **13.** 450,000
15. 3,920,000   **17.** 948,700,000   **19.** 156,800,000   *Section 1.7*   **1.** 26   **3.** 22   **5.** 21   **7.** 32   **9.** 8   **11.** 28
13. 1   **15.** 23   *Section 1.8*   **1.** $761   **3.** $982   **5.** $11   **7.** 7 cents per ounce   **9.** 9 cents per ounce
11. $12 per share   **13.** 140 minutes   **15.** 5,754,240 times   **17.** $21,420   **19.** $1680

Review Problems Chapter 1

1. Three hundred seventy-six   **3.** One hundred nine thousand, two hundred seventy-six   **5.** $4000 + 300 + 60 + 4$
7. $1,000,000 + 300,000 + 5,000 + 100 + 20 + 8$ **9.** 924   **11.** 1,328,828   **13.** 130   **15.** 690   **17.** 400   **19.** 1598
21. 14,703   **23.** 17   **25.** 27   **27.** 159   **29.** 3026   **31.** 224,757   **33.** 36   **35.** 0   **37.** 18   **39.** 105   **41.** 144
43. 240   **45.** 2,612,100   **47.** 83,200,000   **49.** 864   **51.** 4050   **53.** 25,524   **55.** 87,822   **57.** 543,510
59. 150,000   **61.** 7,200,000   **63.** 2,000,000,000   **65.** 2   **67.** 14   **69.** 0   **71.** 7   **73.** 7   **75.** Undefined   **77.** 7
79. 8   **81.** 125   **83.** 258   **85.** 25,874   **87.** 36,958   **89.** 15,986 R 2   **91.** 7 R 21   **93.** 31 R 15
95. 38 R 30   **97.** 258   **99.** 54   **101.** 13^2 **103.** 8^5 **105.** 64   **107.** 125   **109.** 49   **111.** 216   **113.** 5670
115. 15,310   **117.** 12,000   **119.** 676,000   **121.** 5,700,000   **123.** 11   **125.** 16   **127.** 107   **129.** 17   **131.** 44
133. 175 words   **135.** $59,470   **137.** $7028   **139.** $64 per share   **141.** $334   **143.** 25 miles per gallon
145. $5041

Quiz Sections 1.1–1.4

1. Ten thousand, thirty-six   **2.** Forty-two million, three hundred ten thousand, fifty   **3.** $9000 + 300 + 60 + 7$
4. $100,000 + 4000 + 700 + 60$ **5.** 64,986   **6.** 20,152,804   **7.** 189   **8.** 900   **9.** 139,294   **10.** 4,105,504   **11.** 652
12. 3319   **13.** 23,192   **14.** 48,090,583   **15.** 867   **16.** 0   **17.** 640   **18.** 953,778   **19.** 8016   **20.** 238,952
21. 378   **22.** 58,400

Quiz Sections 1.5–1.8

1. 10,410   **2.** 2481 R 1   **3.** 820 R 1   **4.** 336   **5.** 8^4 **6.** 125   **7.** 870   **8.** 58,600   **9.** 670,000
10. 20,000,000   **11.** 6   **12.** 69   **13.** 33   **14.** 14   **15.** $542   **16.** $3687   **17.** 364 miles   **18.** 251 miles
19. $212   **20.** $3916

Test Chapter 1

1. Twenty-six million, five thousand, nine hundred eighty-six   **2.** $30,000 + 8000 + 500 + 90 + 8$ **3.** 2,361,054
4. 586   **5.** 834   **6.** 323,238   **7.** 3419   **8.** 48,441   **9.** 5,136,646   **10.** 240   **11.** 2870   **12.** 72,240
13. 210,815   **14.** 2052 R 3   **15.** 3652   **16.** 1582   **17.** 12^3 **18.** 243   **19.** 26,500   **20.** 1,670,000
21. 9,800,000   **22.** 35   **23.** 288   **24.** 59   **25.** $2189   **26.** 419 feet   **27.** $135   **28.** $152

CHAPTER 2   Fractions

Pretest Chapter 2

1. $\frac{7}{12}$ **2.** Answers will vary.   **3.** $\frac{19}{136}$ **4.** $\frac{2}{3}$ **5.** $\frac{4}{5}$ **6.** $\frac{1}{6}$ **7.** $\frac{5}{8}$ **8.** $\frac{3}{14}$ **9.** $\frac{11}{8}$ **10.** $\frac{37}{7}$ **11.** $33\frac{1}{3}$

12. $6\frac{3}{4}$ **13.** $2\frac{2}{15}$ **14.** $\frac{10}{21}$ **15.** $\frac{2}{3}$ **16.** $\frac{299}{3}$ or $99\frac{2}{3}$ **17.** $\frac{33}{35}$ **18.** $\frac{1}{2}$ **19.** $\frac{24}{5}$ or $4\frac{4}{5}$ **20.** $\frac{15}{4}$ or $3\frac{3}{4}$ **21.** 6

22. 35   **23.** 66   **24.** 72   **25.** $\frac{67}{70}$ **26.** $\frac{93}{20}$ or $4\frac{13}{20}$ **27.** $\frac{4}{3}$ or $1\frac{1}{3}$ **28.** $\frac{81}{28}$ or $2\frac{25}{28}$ **29.** $\frac{49}{72}$ **30.** $9\frac{1}{8}$ miles

31. $20\frac{7}{8}$ tons   **32.** 16 students

Exercises 2.1

1. N: 5, D: 7   **3.** N: 2, D: 3   **5.** N: 1, D: 12   **7.** $\frac{1}{2}$ **9.** $\frac{5}{6}$ **11.** $\frac{2}{3}$ **13.** $\frac{5}{6}$ **15.** $\frac{1}{4}$ **17.** $\frac{3}{10}$ **19.** $\frac{5}{8}$ **21.** $\frac{4}{7}$

23. $\frac{7}{8}$ **25.** $\frac{1}{5}$ **27.** ▢▢▢▢   **29.** ▢▢▢▢▢▢▢▢▢▢

31. ▢▢▢▢▢▢▢▢▢▢▢▢▢▢▢▢▢▢▢▢   **33.** $\frac{2}{9}$ **35.** $\frac{101}{3651}$ **37.** $\frac{31}{63}$ **39.** $\frac{21}{40}$

41. $\frac{12}{23}$ **43.** The amount of money each of 6 business owners gets if the business has a profit of $0   **45. (a)** $\frac{90}{195}$ **(b)** $\frac{22}{195}$

47. 241   **49.** 119,944

Exercises 2.2

1. 5, 11, 19, 41   **3.** 3×5 **5.** 2×3 **7.** 7^2 **9.** 2^6 **11.** $2^2 \times 5$ **13.** $3^2 \times 5$ **15.** 3×5^2 **17.** 2×3^3
19. $2^2 \times 3 \times 7$ **21.** 2×7^2 **23.** Prime   **25.** 3×19 **27.** Prime   **29.** 7×11 **31.** Prime   **33.** Prime
35. $2^4 \times 7$ **37.** 7×23 **39.** The product is $2^4 \times 19$. Yes, they are the same by the fundamental theorem of Arithmetic.

41. $\frac{2}{3}$ **43.** $\frac{3}{5}$ **45.** $\frac{2}{3}$ **47.** $\frac{7}{8}$ **49.** $\frac{3}{5}$ **51.** $\frac{3}{4}$ **53.** $\frac{2}{7}$ **55.** $\frac{3}{5}$ **57.** $\frac{11}{12}$ **59.** $\frac{5}{13}$ **61.** $\frac{5}{8}$ **63.** $\frac{4}{5}$ **65.** $\frac{17}{30}$

67. Yes   **69.** Yes   **71.** No   **73.** No   **75.** Yes   **77.** $\frac{223}{234}$ **79.** 164,050

Exercises 2.3

1. $\frac{7}{2}$ **3.** $\frac{14}{3}$ **5.** $\frac{17}{7}$ **7.** $\frac{53}{10}$ **9.** $\frac{55}{8}$ **11.** $\frac{65}{3}$ **13.** $\frac{95}{2}$ **15.** $\frac{169}{6}$ **17.** $\frac{131}{12}$ **19.** $\frac{79}{10}$ **21.** $\frac{201}{25}$ **23.** $\frac{407}{2}$

25. $\frac{494}{3}$ **27.** $\frac{119}{15}$ **29.** $\frac{138}{25}$ **31.** $1\frac{2}{5}$ **33.** $2\frac{1}{4}$ **35.** $2\frac{1}{2}$ **37.** $3\frac{3}{8}$ **39.** 5   **41.** $9\frac{5}{9}$ **43.** $2\frac{2}{13}$ **45.** $3\frac{3}{16}$

47. $9\frac{1}{3}$ **49.** $17\frac{1}{2}$ **51.** 13   **53.** $33\frac{8}{9}$ **55.** 6   **57.** $2\frac{2}{3}$ **59.** $3\frac{1}{6}$ **61.** $12\frac{1}{5}$ **63.** $\frac{16}{5}$ **65.** $\frac{16}{9}$ **67.** $\frac{65}{11}$

69. $1\frac{80}{113}$ **71.** $2\frac{27}{82}$ **73.** $1\frac{2}{11}$ **75.** $287\frac{16}{31}$ **77.** No. 157 is prime and is not a factor of 9810.   **79.** $\frac{547}{4}$ **81.** 260,247

Exercises 2.4

1. $\frac{3}{35}$ **3.** $\frac{15}{52}$ **5.** 1 **7.** $\frac{24}{35}$ **9.** $\frac{5}{12}$ **11.** $\frac{21}{8}$ or $2\frac{5}{8}$ **13.** $\frac{21}{5}$ or $4\frac{1}{5}$ **15.** $\frac{5}{2}$ or $2\frac{1}{2}$ **17.** $\frac{4}{15}$ **19.** $\frac{1}{2}$

21. $\frac{14}{5}$ or $2\frac{4}{5}$ **23.** $\frac{55}{12}$ or $4\frac{7}{12}$ **25.** 15 **27.** $\frac{161}{50}$ or $3\frac{11}{50}$ **29.** 0 **31.** $\frac{154}{3}$ or $51\frac{1}{3}$ **33.** $\frac{53}{15}$ or $3\frac{8}{15}$ **35.** $12\frac{5}{13}$

37. $14\frac{1}{4}$ square miles **39.** 305 miles **41.** $47\frac{1}{4}$ yards **43.** $11\frac{13}{16}$ ounces **45.** \$3483 **47.** $8\frac{17}{24}$

49. (a) $\frac{5}{8}$ of the voters (b) 7500 people voted. **51.** $\frac{9}{595}$ **53.** 529 cars **55.** 1752 lines

Exercises 2.5

1. $\frac{4}{35}$ **3.** $\frac{6}{5}$ or $1\frac{1}{5}$ **5.** $\frac{9}{10}$ **7.** $\frac{5}{4}$ or $1\frac{1}{4}$ **9.** $\frac{4}{5}$ **11.** $\frac{16}{7}$ or $2\frac{2}{7}$ **13.** $\frac{4}{5}$ **15.** 1 **17.** $\frac{9}{7}$ or $1\frac{2}{7}$ **19.** $\frac{36}{7}$ or $5\frac{1}{7}$

21. 0 **23.** Cannot be done **25.** $\frac{3}{4}$ **27.** $\frac{4}{5}$ **29.** $\frac{36}{175}$ **31.** 16 **33.** $\frac{7}{32}$ **35.** 4500 **37.** $\frac{2}{15}$ **39.** $\frac{7}{40}$

41. $a = 2, b = 3, c = 4, d = 6; \frac{2}{3} \div \frac{4}{6} = \frac{4}{6} \div \frac{2}{3}$. In general if $\frac{a}{b} = \frac{c}{d}$, then it is true. **43.** $\frac{13}{2}$ or $6\frac{1}{2}$ **45.** $\frac{7}{12}$ **47.** $\frac{2}{3}$ **49.** 4

51. $\frac{15}{7}$ or $2\frac{1}{7}$ **53.** 3 **55.** $\frac{63}{32}$ or $1\frac{31}{32}$ **57.** $\frac{23}{50}$ **59.** 12 **61.** $\frac{2}{3}$ **63.** $\frac{5}{21}$ **65.** $\frac{5}{16}$ mile

67. 60 miles per hour **69.** 23 shirts **71.** $\frac{1014}{217}$ or $4\frac{146}{217}$

73. Thirty-nine million, five hundred seventy-six thousand, three hundred four **75.** 1099

Exercises 2.6

1. 10 **3.** 28 **5.** 40 **7.** 18 **9.** 60 **11.** 16 **13.** 90 **15.** 60 **17.** 105 **19.** 40 **21.** 12 **23.** 180

25. 144 **27.** 84 **29.** 120 **31.** 5 **33.** 45 **35.** 20 **37.** 14 **39.** 96 **41.** 39 **43.** $\frac{21}{36}$ and $\frac{20}{36}$

45. $\frac{24}{200}$ and $\frac{35}{200}$ **47.** $\frac{36}{432}$ and $\frac{57}{432}$ **49.** (a) 42 (b) $\frac{30}{42}$ and $\frac{7}{42}$ **51.** (a) 48 (b) $\frac{20}{48}$ and $\frac{3}{48}$

53. (a) 90 (b) $\frac{81}{90}$ and $\frac{65}{90}$ **55.** (a) 60 (b) $\frac{35}{60}$ and $\frac{46}{60}$ **57.** (a) 12 (b) $\frac{10}{12}, \frac{11}{12}, \frac{9}{12}$ **59.** (a) 56 (b) $\frac{3}{56}, \frac{49}{56}, \frac{40}{56}$

61. (a) 63 (b) $\frac{5}{63}, \frac{12}{63}, \frac{56}{63}$ **63.** (a) 16 (b) $\frac{3}{16}, \frac{12}{16}, \frac{6}{16}$ **65.** 168 **67.** 357 **69.** 178 R 3

Exercises 2.7

1. $\frac{5}{8}$ **3.** $\frac{11}{15}$ **5.** $\frac{18}{23}$ **7.** $\frac{17}{44}$ **9.** $\frac{11}{10}$ or $1\frac{1}{10}$ **11.** $\frac{9}{20}$ **13.** $\frac{37}{100}$ **15.** $\frac{26}{175}$ **17.** $\frac{31}{24}$ or $1\frac{7}{24}$ **19.** $\frac{7}{20}$ **21.** $\frac{1}{4}$

23. $\frac{8}{35}$ **25.** $\frac{9}{16}$ **27.** $\frac{11}{60}$ **29.** 0 **31.** $\frac{8}{5}$ or $1\frac{3}{5}$ **33.** $\frac{11}{30}$ **35.** $41\frac{4}{5}$ **37.** $6\frac{7}{12}$ **39.** $16\frac{2}{3}$ **41.** $73\frac{37}{40}$ **43.** $5\frac{1}{2}$

45. $1\frac{7}{15}$ **47.** $4\frac{41}{60}$ **49.** $8\frac{8}{15}$ **51.** $102\frac{5}{8}$ **53.** $8\frac{1}{12}$ hours **55.** $\$13\frac{3}{4}$ per share

57. (a) $30\frac{5}{8}$ kilometers (b) $15\frac{1}{2}$ kilometers **59.** $\frac{2607}{40}$ or $65\frac{7}{40}$ **61.** $11\frac{44}{45}$ **63.** 111,303 **65.** 512,012

Exercises 2.8

1. $16\frac{1}{4}$ tons **3.** $6\frac{5}{8}$ miles **5.** $\$15\frac{3}{4}$ **7.** 36 pieces **9.** $1\frac{9}{16}$ inches **11.** \$312 **13.** $134\frac{3}{8}$ nautical miles

15. $275\frac{5}{8}$ gallons **17.** (a) 41 ties (b) \$1 (c) \$287 **19.** (a) $13\frac{3}{8}$ ounce (b) $\frac{7}{8}$ ounce less than the expected amount

21. (a) 5485 bushels (b) $11,998\frac{7}{16}$ cubic feet (c) $9598\frac{3}{4}$ bushels **23.** 44,245 **25.** 45,441

Extra Practice: Examples and Exercises

Section 2.1 **1.** $\frac{3}{5}$ **3.** $\frac{6}{8}$ **5.** $\frac{3}{5}$ **7.**

9. ▯▯▯▯▯▯▯▯▯▯▯▯▯▯▯▯▯▯   **11.** $\dfrac{5}{13}$ **13.** $\dfrac{3}{5}$ **15.** $\dfrac{6}{11}$ **17.** $\dfrac{19}{63}$ **19.** $\dfrac{27}{57}$

Section 2.2   **1.** $2^4 \times 3$ **3.** $2^2 \times 5 \times 7$ **5.** $3 \times 5 \times 7$ **7.** $\dfrac{3}{7}$ **9.** $\dfrac{8}{9}$ **11.** $\dfrac{4}{7}$ **13.** $\dfrac{3}{7}$ **15.** $\dfrac{4}{13}$ **17.** Yes   **19.** No

Section 2.3   **1.** $\dfrac{31}{7}$ **3.** $\dfrac{377}{17}$ **5.** $\dfrac{95}{3}$ **7.** 5   **9.** $11\dfrac{7}{9}$ **11.** $5\dfrac{1}{3}$ **13.** $2\dfrac{1}{2}$ **15.** $9\dfrac{4}{7}$ **17.** 3   **19.** 7

Section 2.4   **1.** $\dfrac{2}{45}$ **3.** $\dfrac{5}{7}$ **5.** $\dfrac{9}{5}$ *or* $1\dfrac{4}{5}$ **7.** $\dfrac{1}{3}$ **9.** 11   **11.** 6   **13.** 6   **15.** $\dfrac{27}{2}$ *or* $13\dfrac{1}{2}$ **17.** 161 square miles

19. $206\dfrac{2}{3}$ miles   ***Section 2.5***   **1.** $\dfrac{10}{21}$ **3.** Undefined   **5.** $\dfrac{2}{5}$ **7.** $\dfrac{27}{5}$ **9.** $\dfrac{19}{2}$ *or* $9\dfrac{1}{2}$ **11.** $\dfrac{35}{44}$ **13.** $2\dfrac{11}{27}$ **15.** $1\dfrac{1}{8}$

17. $\dfrac{5}{24}$ **19.** $23\dfrac{1}{3}$ ***Section 2.6***   **1.** 4   **3.** 21   **5.** 28   **7.** 150   **9.** 50   **11.** 30   **13.** 180   **15.** 180

17. LCD = 21; $\dfrac{9}{21}, \dfrac{5}{21}$ **19.** LCD = 80; $\dfrac{45}{80}, \dfrac{12}{80}$ ***Section 2.7***   **1.** $\dfrac{5}{7}$ **3.** $\dfrac{14}{29}$ **5.** $\dfrac{8}{41}$ **7.** $\dfrac{5}{7}$ **9.** $\dfrac{5}{24}$ **11.** $\dfrac{18}{35}$

13. $46\dfrac{7}{12}$ **15.** $56\dfrac{11}{20}$ **17.** $5\dfrac{11}{12}$ **19.** $4\dfrac{2}{15}$ **21.** $11\dfrac{4}{11}$ ***Section 2.8***   **1.** 19 tons   **3.** $4\dfrac{7}{12}$ miles

5. (a) $15\dfrac{1}{12}$ ounces   (b) $\dfrac{5}{12}$ ounce   **7.** $\$38\dfrac{1}{8}$ **9.** $222\dfrac{1}{2}$ gallons   **11.** 22   **13.** 12 gallons   **15.** 35 bags

17. $\$354\dfrac{1}{2}$ **19.** $\$74\dfrac{1}{4}$

Review Problems Chapter 2

1. $\dfrac{5}{12}$ **3.** Sketches will vary.   **5.** $\dfrac{6}{31}$ **7.** $2 \times 3 \times 7$ **9.** $2^3 \times 3 \times 7$ **11.** $2 \times 3 \times 13$ **13.** $\dfrac{1}{4}$ **15.** $\dfrac{7}{12}$ **17.** $\dfrac{7}{8}$

19. $\dfrac{35}{8}$ **21.** $2\dfrac{5}{7}$ **23.** $3\dfrac{3}{11}$ **25.** $1\dfrac{29}{48}$ **27.** $\dfrac{7}{15}$ **29.** $\dfrac{4}{63}$ **31.** $\dfrac{817}{40}$ *or* $20\dfrac{17}{40}$ **33.** $486\dfrac{17}{20}$ square inches

35. $\dfrac{15}{14}$ *or* $1\dfrac{1}{14}$ **37.** 1920   **39.** $\dfrac{25}{6}$ *or* $4\dfrac{1}{6}$ **41.** 0   **43.** $186\dfrac{2}{3}$ calories   **45.** 98   **47.** 90   **49.** $\dfrac{33}{72}$

51. $\dfrac{187}{198}$ **53.** $\dfrac{13}{12}$ *or* $1\dfrac{1}{12}$ **55.** $\dfrac{11}{40}$ **57.** $\dfrac{209}{48}$ *or* $4\dfrac{17}{48}$ **59.** $\dfrac{857}{12}$ *or* $71\dfrac{5}{12}$ **61.** $\$8\dfrac{3}{4}$ **63.** 15 pieces

65. 30 words per minute   **67.** $1\dfrac{1}{16}$ inches   **69.** $\$242$

Quiz Sections 2.1–2.5

1. $\dfrac{3}{5}$ **2.** $\dfrac{7}{8}$ **3.** Sketches will vary.   **4.** Sketches will vary.   **5.** $\dfrac{6}{17}$ **6.** $2^3 \times 5$ **7.** $3^2 \times 5^2$ **8.** $\dfrac{2}{3}$ **9.** $\dfrac{3}{4}$

10. $\dfrac{3}{8}$ **11.** $11\dfrac{3}{4}$ **12.** $5\dfrac{3}{7}$ **13.** $\dfrac{22}{5}$ **14.** $\dfrac{95}{4}$ **15.** $\dfrac{15}{77}$ **16.** $\dfrac{5}{12}$ **17.** $\dfrac{35}{3}$ *or* $11\dfrac{2}{3}$ **18.** $\dfrac{17}{4}$ *or* $4\dfrac{1}{4}$ **19.** $\dfrac{21}{16}$ *or* $1\dfrac{5}{16}$

20. $\dfrac{4}{3}$ *or* $1\dfrac{1}{3}$ **21.** $\dfrac{21}{2}$ *or* $10\dfrac{1}{2}$ **22.** $\dfrac{63}{40}$ *or* $1\dfrac{23}{40}$ **23.** $\$44$ **24.** 12 dresses

Quiz Sections 2.6–2.8

1. 30   **2.** 20   **3.** 36   **4.** 60   **5.** 160   **6.** 84   **7.** 48   **8.** 54   **9.** $\dfrac{23}{40}$ **10.** $\dfrac{5}{4}$ *or* $1\dfrac{1}{4}$ **11.** $\dfrac{23}{35}$

12. $\dfrac{59}{57}$ *or* $1\dfrac{2}{57}$ **13.** $\dfrac{161}{20}$ *or* $8\dfrac{1}{20}$ **14.** $14\dfrac{2}{3}$ **15.** $\dfrac{19}{60}$ **16.** $\dfrac{103}{200}$ **17.** $\dfrac{2}{3}$ **18.** $\dfrac{283}{48}$ *or* $5\dfrac{43}{48}$ **19.** $10\dfrac{1}{12}$ inches

20. $3\dfrac{7}{8}$ teaspoons   **21.** $1\dfrac{5}{12}$ hours   **22.** $6\dfrac{7}{20}$ miles   **23.** $1\dfrac{1}{4}$ feet   **24.** The stock went down $\$4\dfrac{7}{8}$.   **25.** $64\dfrac{4}{5}$ miles

26. $11\dfrac{1}{4}$ feet

Test Chapter 2

1. $\dfrac{5}{7}$ **2.** $\dfrac{307}{364}$ **3.** $\dfrac{3}{16}$ **4.** $\dfrac{7}{13}$ **5.** $\dfrac{11}{2}$ *or* $5\dfrac{1}{2}$ **6.** $\dfrac{39}{5}$ **7.** $8\dfrac{9}{13}$ **8.** 15   **9.** $\dfrac{16}{27}$ **10.** 20   **11.** $\dfrac{40}{27}$ *or* $1\dfrac{13}{27}$

12. $\dfrac{25}{18}$ *or* $1\dfrac{7}{18}$ **13.** $\dfrac{18}{5}$ *or* $3\dfrac{3}{5}$ **14.** $\dfrac{26}{33}$ **15.** 36   **16.** 42   **17.** 72   **18.** $\dfrac{56}{72}$ **19.** $\dfrac{5}{36}$ **20.** $\dfrac{13}{20}$ **21.** $\dfrac{29}{20}$ *or* $1\dfrac{9}{20}$

22. $\dfrac{211}{12}$ or $17\dfrac{7}{12}$ **23.** $7\dfrac{23}{24}$ **24.** 10 square yards   **25.** 7 packages   **26.** $\dfrac{3}{10}$ mile   **27.** $14\dfrac{3}{8}$ miles

28. (a) 38 oranges   **(b)** 912 cents or $9.12

Cumulative Test Chapters 1–2

1. Eighty-four million, three hundred sixty-one thousand, two hundred eight   **2.** 869   **3.** 719,220   **4.** 2075
5. 17,216   **6.** 4788   **7.** 202,896   **8.** 4307 R 1   **9.** 369   **10.** 49   **11.** 6,037,000   **12.** 38   **13.** $174   **14.** $306

15. $\dfrac{55}{84}$ **16.** $\dfrac{7}{13}$ **17.** $\dfrac{75}{4}$ **18.** $14\dfrac{2}{7}$ **19.** $\dfrac{527}{48}$ or $10\dfrac{47}{48}$ **20.** $\dfrac{12}{35}$ **21.** 39   **22.** $\dfrac{31}{54}$ **23.** $\dfrac{71}{8}$ or $8\dfrac{7}{8}$

24. $\dfrac{113}{15}$ or $7\dfrac{8}{15}$ **25.** $\dfrac{13}{28}$ **26.** $23\dfrac{1}{8}$ tons   **27.** $24\dfrac{3}{5}$ miles per gallon   **28.** $8\dfrac{1}{8}$ cups of sugar, $5\dfrac{5}{6}$ cups of flour

---

CHAPTER 3   Decimals

Pretest Chapter 3

1. Thirty six and five hundred twenty-four thousandths   **2.** 0.1234   **3.** $1\dfrac{39}{100}$ **4.** $\dfrac{93}{200}$ **5.** 2.69, 2.7, 2.701, 2.71

6. 158.3   **7.** 0.381   **8.** 13.43   **9.** 27.436   **10.** 33.37   **11.** 8.186   **12.** 0.7404   **13.** 2864.3   **14.** 7.918
15. 0.129   **16.** 12.8   **17.** 0.3125   **18.** $0.1\overline{36}$ or $0.1363636\ldots$ **19.** 0.39   **20.** 22.9 miles per gallon   **21.** $464.94
22. $4.80

Exercises 3.1

1. Fifty-seven hundredths   **3.** One and six tenths   **5.** One hundred twenty-four thousandths

7. Twenty eight and seven ten thousandths   **9.** Eighty-seven and $\dfrac{36}{100}$ dollars

11. One thousand two hundred thirty-six and $\dfrac{8}{100}$ dollars   **13.** Ten thousand and $\dfrac{76}{100}$ dollars   **15.** 0.5   **17.** 0.76

19. 0.771   **21.** 0.0009   **23.** 8.7   **25.** 84.13   **27.** 1.019   **29.** 126.0571   **31.** $\dfrac{9}{50}$ **33.** $3\dfrac{3}{5}$ **35.** $\dfrac{121}{1000}$ **37.** $12\dfrac{5}{8}$

39. $7\dfrac{3}{2000}$ **41.** $307\dfrac{603}{5000}$ **43.** $\dfrac{187}{10,000}$ **45.** $8\dfrac{27}{2500}$ **47.** $289\dfrac{47}{125}$ **49.** $\dfrac{3}{20,000}$ **51.** $\dfrac{3}{20,000,000}$ **53.** 818

55. 56,800

Exercises 3.2

1. >   **3.** =   **5.** <   **7.** >   **9.** <   **11.** >   **13.** <   **15.** >   **17.** 12.6, 12.65, 12.8   **19.** 0.003, 0.005, 0.0053
21. 1.1, 1.79, 1.8, 1.81   **23.** 26.003, 26.033, 26.034, 26.04   **25.** 18.006, 18.060, 18.065, 18.066, 18.606   **27.** 5.7   **29.** 29.5
31. 197.1   **33.** 2176.8   **35.** 26.03   **37.** 5.77   **39.** 156.12   **41.** 2786.72   **43.** 1.061   **45.** 0.0913   **47.** 5.00761
49. 0.007537   **51.** 129   **53.** $7813   **55.** $10,098   **57.** $56.98   **59.** $5783.72
61. You should only consider one digit to the right of the decimal place that you wish to round to. 86.23498 is closer to 86.23

than to 86.24.   **63.** 0.0059, 0.006, 0.0519, $\dfrac{6}{100}$, 0.0601, 0.0612, 0.062, $\dfrac{6}{10}$, 0.61   **65.** $12\dfrac{1}{8}$ **67.** 692 miles

Exercises 3.3

1. 72.8   **3.** 1215.55   **5.** 5.531   **7.** 136.844   **9.** 323.9   **11.** 23.00   **13.** 36.7287   **15.** 67.42   **17.** 1112.21
19. 19.86 M   **21.** 47.6 gallons   **23.** $12.31   **25.** $427.12   **27.** 3.9   **29.** 27.17   **31.** 64.18
33. 108.118   **35.** 0.02151   **37.** 4.6465   **39.** 6.737   **41.** 1189.07   **43.** $8306.00   **45.** 2.156 centimeters short
47. 856.2 miles   **49.** $37.14   **51.** 1.46 cm   **53.** $601,409.07 revenue shortage   **55.** 5.0799   **57.** 13,197.6825
59. 20.288   **61.** 276.7268   **63.** 95.87454

Exercises 3.4

1. 0.12   **3.** 0.06   **5.** 0.00196   **7.** 0.00711   **9.** 0.0003   **11.** 0.7416   **13.** 2555.52   **15.** 0.013244   **17.** 1524.8839
19. 4911.3   **21.** $234.00   **23.** $288.00   **25.** 87.74 square inches   **27.** $651.60   **29.** $0.77   **31.** 514.8 miles
33. 28.6   **35.** 23.6   **37.** 128,650   **39.** 12,798.6   **41.** 28,056,020   **43.** 7,634,900   **45.** 671.8   **47.** 71.63
49. $1964   **51.** 2980 meters   **53.** $36,405,000.00
55. Rule: Move the decimal point one place to the left for every zero.   **(a)** 0.013684   **(b)** 0.02587193   **(c)** 0.000000061834
57. $62,279.00   **59.** 98   **61.** 125 R 4   **63.** 430.03378   **65.** 2,384,979.84

Exercises 3.5

1. 2.18   **3.** 0.0369   **5.** 17.62   **7.** 0.0565   **9.** 0.0029   **11.** 12.2   **13.** 64.3   **15.** 8.01   **17.** 21   **19.** 340
21. 5.3   **23.** 2.3   **25.** 33.8   **27.** 30.96   **29.** 11.53   **31.** 24.92   **33.** 29.770   **35.** 12.246   **37.** 50   **39.** 13
41. $9.71 per month   **43.** 17.5 miles per gallon   **45.** $0.75 per pound   **47.** 25 payments

49. The error was to pack the box two pens short.   **51.** 74   **53.** 0.017712   **55.** $1\frac{31}{40}$ **57.** $\frac{91}{12}$ or $7\frac{7}{12}$

59. 280.38925   **61.** 75,184

Exercises 3.6

1. 0.25   **3.** 0.625   **5.** 0.4375   **7.** 0.35   **9.** 0.62   **11.** 1.75   **13.** 2.125   **15.** 1.025   **17.** 0.6   **19.** 0.5
21. $0.7\overline{2}$ **23.** $0.\overline{72}$ **25.** $0.9\overline{3}$ **27.** $0.\overline{15}$ **29.** $1.13\overline{8}$ **31.** $2.2\overline{7}$ **33.** 0.571   **35.** 0.952   **37.** 0.146   **39.** 1.424
41. 0.160   **43.** 0.944   **45.** 1.214   **47.** 0.474   **49.** 6.01   **51.** 255.72   **53.** 376   **55.** 156.0664   **57.** 21.414
59. (a) 0.16   (b) $0.0149\overline{49}$ (c) The repeating patterns line up differently.   **61.** Yes. Because $2.\overline{0} - 1.\overline{9} = 0$ **63.** 20.836
65. 312 square feet   **67.** $377 was deposited.   **69.** 0.586930

Exercises 3.7

1. $409.69   **3.** 2724.8 miles   **5.** 9751.691 square meters   **7.** 23 packages   **9.** $7.74 left
11. 20.2 miles per gallon   **13.** 22 days   **15.** $731.85 total bill   **17.** $743.60   **19.** $426.55 balance   **21.** $583.88

23. $\frac{79}{70}$ or $1\frac{9}{70}$ **25.** $\frac{5}{9}$ **27.** $14,697,388.00

Extra Practice: Examples and Exercises

Section 3.1   **1.** Ninety-six hundredths   **3.** Nineteen and eight hundred seven thousandths

5. Twenty-nine and three ten-thousandths   **7.** 0.04   **9.** 98.17   **11.** $\frac{13}{20}$ **13.** $\frac{39}{250}$ **15.** $5\frac{17}{10,000}$ *Section 3.2*   **1.** >

3. <   **5.** >   **7.** 3.98   **9.** 4703.28   **11.** 2.071   **13.** 15.00841   **15.** 133   **17.** $8110   **19.** $30,249.51
Section 3.3   **1.** 16.167   **3.** 17.12   **5.** 103.977   **7.** $203.22   **9.** 1.8   **11.** 13.591   **13.** 8.3335   **15.** 22.84201
17. 44.015779   *Section 3.4*   **1.** 70.5486   **3.** 37.746   **5.** 2649.036   **7.** 3998.12   **9.** 13,498.12   **11.** 1009.123
13. 56.8   **15.** 54,933,001,202   **17.** 7.84266758   **19.** 0.042768   *Section 3.5*   **1.** 14.26   **3.** 19.85   **5.** 0.6588
7. 59.58   **9.** 33.597   **11.** 80   **13.** 400   **15.** 1235   **17.** 74   **19.** 21   *Section 3.6*   **1.** 0.75   **3.** 0.583   **5.** 0.25
7. 0.254   **9.** 0.525   **11.** $4.6\overline{1}$ **13.** $12.1\overline{6}$ **15.** $2.1\overline{6}$ **17.** 37.65   **19.** 198.4744   *Section 3.7*   **1.** $19.84 left
3. 441.4 miles   **5.** $471.98   **7.** $657.05   **9.** $932.25

Review Problems Chapter 3

1. Thirteen and six hundred seventy-two thousandths   **3.** 0.7   **5.** 1.523   **7.** $\frac{17}{100}$ **9.** $26\frac{22}{25}$ **11.** >

13. 0.901, 0.918, 0.98, 0.981   **15.** 0.6   **17.** 1.100   **19.** 6.639   **21.** (a) 8.5 gallons   (b) 9 gallons   **23.** 90.739
25. 0.0228   **27.** 0.000364   **29.** 86.4   **31.** 78.6   **33.** 0.613   **35.** $0.90   **37.** 0.00258   **39.** 232.9   **41.** 0.059
43. $0.2\overline{7}$ **45.** $1.8\overline{3}$ **47.** 0.786   **49.** 3.6   **51.** 1.152   **53.** $9.13   **55.** $368.08   **57.** $2170.30

Quiz Sections 3.1–3.4

1. Sixty-two hundredths   **2.** One hundred thirty-six and four hundred thirteen thousandths   **3.** 0.0037   **4.** 12.049

5. 1.2   **6.** 18.62   **7.** 7.070   **8.** 28.0764   **9.** 0.06   **10.** $\frac{591}{1000}$ **11.** 71.084   **12.** 103.9   **13.** 11.2361   **14.** 0.41

15. 1976.531   **16.** 1823.6   **17.** 20,075   **18.** 0.544   **19.** 10.73   **20.** 73.8556   **21.** 0.7, 0.711, 0.714, 0.74, 0.741
22. 1.0063, 1.063, 1.163, 1.613, 1.629, 1.63

Quiz Sections 3.5–3.7

1. 12.562   **2.** 3.654   **3.** 3.6   **4.** 30.7   **5.** 36.8   **6.** 0.026   **7.** 0.045   **8.** 0.46875   **9.** 2.875   **10.** 1.714
11. 0.765   **12.** $0.2\overline{7}$ **13.** $0.59\overline{0}$ **14.** 1.6224 square centimeters   **15.** $227.55   **16.** 24.4 miles per gallon
17. $130.05 more   **18.** $0.01 more per pound

Test Chapter 3

1. Two hundred sixty-three thousandths   **2.** 0.7899   **3.** $5\frac{31}{50}$ **4.** $\frac{77}{200}$ **5.** 1.09, 1.9 1.903, 1.91   **6.** 69.98

7. 0.0792   **8.** 37.65   **9.** 44.29   **10.** 0.0989   **11.** 23.282   **12.** 0.3168   **13.** 583.6   **14.** $1.1\overline{3}$ **15.** 0.15625
16. 42.645   **17.** $23.80   **18.** 24.7 miles per gallon   **19.** 2.27 centimeters

Cumulative Test Chapters 1–3

1. Thirty-eight million, fifty-six thousand, nine hundred fifty-four   **2.** 479,587   **3.** 54,480   **4.** 39,463   **5.** 258   **6.** 16

7. $\frac{3}{8}$ **8.** $7\frac{1}{2}$ **9.** $\frac{9}{35}$ **10.** $\frac{7}{6}$ or $1\frac{1}{6}$ **11.** 16   **12.** $\frac{33}{10}$ or $3\frac{3}{10}$ **13.** 0.571   **14.** 2.01, 2.1, 2.11, 2.12, 20.1

15. 26.080   **16.** 19.54   **17.** 8.639   **18.** 1.136   **19.** 36,512.3   **20.** 1.058   **21.** 0.8125   **22.** 13.597

23. 456 miles   **24.** $195.57

CHAPTER 4 Ratio and Proportion

Pretest Chapter 4

1. $\frac{15}{17}$ **2.** $\frac{7}{33}$ **3.** $\frac{34}{7}$ **4.** $\frac{31}{34}$ **5. (a)** $\frac{13}{42}$ **(b)** $\frac{3}{35}$ **6.** $\frac{\$13}{4 \text{ cabinets}}$ **7.** $\frac{29 \text{ gallons}}{36 \text{ square feet}}$

8. $\frac{31.5 \text{ miles}}{1 \text{ hour}}$ or 31.5 miles/hour   **9.** $\frac{\$39}{1}$ radio or $39/radio   **10.** $\frac{23}{50} = \frac{69}{150}$ **11.** $\frac{54}{66} = \frac{27}{33}$ **12.** True   **13.** False

14. $n = 16$ **15.** $n = 15$ **16.** $n = 5$ **17.** $n = 5$ **18.** $n = 480$ **19.** 6.75 cups   **20.** 220.5 miles   **21.** 165 miles

22. 63 defective bulbs

Exercises 4.1

1. $\frac{2}{3}$ **3.** $\frac{7}{6}$ **5.** $\frac{5}{11}$ **7.** $\frac{2}{3}$ **9.** $\frac{15}{16}$ **11.** $\frac{2}{3}$ **13.** $\frac{8}{5}$ **15.** $\frac{2}{3}$ **17.** $\frac{3}{10}$ **19.** $\frac{10}{17}$ **21.** $\frac{43}{60}$ **23.** $\frac{9}{1}$ **25.** $\frac{13}{19}$

27. $\frac{165}{285} = \frac{11}{19}$ **29.** $\frac{35}{165} = \frac{7}{33}$ **31.** $\frac{205}{1225} = \frac{41}{245}$ **33.** $\frac{450}{205} = \frac{90}{41}$ **35.** $\frac{72,000}{128,000} = \frac{9}{16}$ **37.** $\frac{\$40}{3 \text{ chairs}}$ **39.** $\frac{41 \text{ pounds}}{9 \text{ people}}$

41. $\frac{62 \text{ gallons}}{125 \text{ sq ft}}$ **43.** $\frac{410 \text{ revolutions}}{1 \text{ mile}}$ or 410 rev/mile   **45.** $\frac{9 \text{ miles}}{4 \text{ hours}}$ **47.** $13/hr   **49.** 16 mi/gallon   **51.** 245 gal/hour

53. 230 words/page   **55.** 517 km/hour   **57.** $6/share   **59.** $4/radio

61. (a) $0.11 and $0.09/ounce   **(b)** 2 cents per ounce or $0.02 per ounce   **63.** Increased by Mach 0.2   **65.** $10.50/sq yd

67. $2\frac{5}{8}$ **69.** $\frac{5}{46}$ **71.** $24.53

Exercises 4.2

1. $\frac{48}{32} = \frac{3}{2}$ **3.** $\frac{9}{26} = \frac{18}{52}$ **5.** $\frac{20}{36} = \frac{5}{9}$ **7.** $\frac{27}{15} = \frac{9}{5}$ **9.** $\frac{44}{60} = \frac{22}{30}$ **11.** $\frac{45}{135} = \frac{9}{27}$ **13.** $\frac{5.5}{10} = \frac{11}{20}$ **15.** True

17. True   **19.** False   **21.** True   **23.** False   **25.** True   **27.** True   **29.** False   **31.** False   **33.** True

35. True   **37.** True   **39.** False   **41.** Yes   **43.** No

45. (a) True **(b)** True **(c)** For most students it is faster to multiply than to reduce fractions   **47.** True

49. 23.1405   **51.** 402.408   **53.** True

Exercises 4.3

1. 8   **3.** 42   **5.** 16   **7.** 13   **9.** 26   **11.** 5.6   **13.** 5   **15.** 7   **17.** 5.5   **19.** 20   **21.** 16   **23.** 7.5   **25.** 32

27. 24   **29.** 7   **31.** 66   **33.** 8.8   **35.** 20   **37.** 25   **39.** 18   **41.** 14   **43.** 132   **45.** 50   **47.** $n \approx 2.9$

49. 9.4   **51.** 18   **53.** 2.8 grams   **55.** $121.60   **57.** 8.64 acres   **59.** $3\frac{5}{8}$ **61.** $10\frac{8}{9}$ **63.** $n \approx 17.54$ **65.** 76

67. $n \approx 24.4$

Exercises 4.4

1. 105 defective ones   **3.** 309 faculty members   **5.** 10.5 quarts   **7.** 616 francs   **9.** $n \approx 14.9$ miles per hour

11. 171.6 feet   **13.** $n \approx 103$ minutes   **15.** $n \approx 75$ miles   **17.** 2 ohms   **19.** 6.25 cups or $6\frac{1}{4}$ cups   **21.** 55 games

23. $n \approx 1.35$ earned runs   **25.** No. Should be $\frac{49}{60} = \frac{n}{88}$ **27.** No. Should be $\frac{n}{88} = \frac{49}{60}$ **29.** $10,367.50   **31.** $79,900

33. 56,200   **35.** 56.1   **37.** 0.0762   **39.** 3410 people will respond.

Extra Practice: Examples and Exercises

Section 4.1 **1.** $\frac{2}{5}$ **3.** $\frac{3}{7}$ **5.** $\frac{7}{6}$ **7.** $\frac{3}{4}$ **9.** $\frac{5}{8}$ **11.** $\frac{12 \text{ dollars}}{1 \text{ piece}}$ **13.** $\frac{1027 \text{ revolutions}}{4 \text{ miles}}$ **15.** $\frac{33 \text{ gallons}}{70 \text{ square feet}}$

17. $\frac{17.8 \text{ miles}}{\text{gallon}}$ **19.** $\frac{262.9 \text{ kilometers}}{\text{hour}}$ *Section 4.2* **1.** $\frac{3}{21} = \frac{1}{7}$ **3.** $\frac{35}{95} = \frac{7}{19}$ **5.** $\frac{6.5}{10} = \frac{13}{20}$ **7.** False   **9.** True

11. False   **13.** True   **15.** False   *Section 4.3*   **1.** 8   **3.** 7   **5.** 5   **7.** 4.2   **9.** 10.9   **11.** 2   **13.** 13.2
15. 5.83   **17.** 4 liters   **19.** 22 tons   *Section 4.4*   **1.** 390 faculty   **3.** 1960 students   **5.** 192.5 pounds

7. 45 feet   **9.** 126 minutes   **11.** 128 miles   **13.** $10\frac{1}{2}$ cups   **15.** 48 games will need to be won.

Review Problems Chapter 4
1. $\frac{11}{5}$ **3.** $\frac{4}{5}$ **5.** $\frac{25}{62}$ **7.** $\frac{52}{147}$ **9.** $\frac{2}{5}$ **11.** $\frac{3}{4}$ **13.** $\frac{7}{43}$ **15.** $\frac{5\text{ gallons}}{9\text{ people}}$ **17.** $\frac{47\text{ vibrations}}{4\text{ seconds}}$ **19.** $\frac{\$17.00}{\text{share}}$
21. $\frac{\$13.50}{\text{square yard}}$ **23.** $\frac{12}{48}=\frac{7}{28}$ **25.** $\frac{7.5}{45}=\frac{22.5}{135}$ **27.** $\frac{136}{17}=\frac{408}{51}$ **29.** False   **31.** True   **33.** False   **35.** True
37. True   **39.** 23   **41.** 31   **43.** 33   **45.** 7   **47.** $n \approx 3.9$ **49.** 3   **51.** 87   **53.** 16.8   **55.** 12 gallons
57. $n \approx 12.6$ horsepower   **59.** 15 gallons   **61.** 9 nurses   **63.** 2016 French francs   **65.** 600 miles
67. 67,200 feet   **69.** 120 feet

Quiz Sections 4.1–4.2
1. $\frac{3}{4}$ **2.** $\frac{1}{9}$ **3.** $\frac{6}{7}$ **4.** $\frac{42}{23}$ **5.** $\frac{4}{3}$ **6.** $\frac{6}{37}$ **7.** $\frac{3}{37}$ **8.** $\frac{\$67}{3\text{ chairs}}$ **9.** $\frac{772\text{ students}}{37\text{ faculty}}$ **10.** $\frac{69\text{ kilometers}}{2\text{ liters}}$
11. $\frac{3\text{ pounds}}{2\text{ people}}$ **12.** $\frac{315\text{ revolutions}}{2\text{ miles}}$ **13.** $\frac{178\text{ words}}{\text{page}}$ **14.** 19.5 miles per gallon   **15.** $25.92 per share   **16.** $\frac{11}{7}=\frac{55}{35}$
17. $\frac{27.5}{33}=\frac{10}{12}$ **18.** $\frac{34}{50}=\frac{51}{45}$ **19.** $\frac{238}{357}=\frac{38}{57}$ **20.** False   **21.** True   **22.** True   **23.** False   **24.** True

Quiz Sections 4.3–4.4
1. 8   **2.** 8.5   **3.** 3   **4.** 48.6   **5.** 162   **6.** 90   **7.** 45   **8.** 400   **9.** $52.50   **10.** 2964 miles   **11.** 17 gallons
12. 63 servings   **13.** 80 shares   **14.** 21 cars   **15.** 969 students   **16.** 1325 miles

Test Chapter 4
1. $\frac{4}{17}$ **2.** $\frac{12}{35}$ **3.** $\frac{101\text{ miles}}{3\text{ gallons}}$ **4.** $\frac{480\text{ square feet}}{7\text{ pounds}}$ **5.** $\frac{3.4\text{ tons}}{\text{day}}$ **6.** $\frac{\$5.28}{\text{hour}}$ **7.** $\frac{288.89\text{ feet}}{\text{pole}}$ **8.** $\frac{\$85.13}{\text{share}}$
9. $\frac{19}{31}=\frac{57}{93}$ **10.** $\frac{12}{17}=\frac{18}{25.5}$ **11.** $\frac{420\text{ miles}}{18\text{ gallons}}=\frac{350\text{ miles}}{15\text{ gallons}}$ **12.** $\frac{3\text{ tablespoons}}{16\text{ people}}=\frac{9\text{ tablespoons}}{48\text{ people}}$ **13.** True   **14.** True
15. True   **16.** True   **17.** 6   **18.** 16   **19.** 8   **20.** 33.6   **21.** 36 women   **22.** 16.8 grams   **23.** 60 inches of snow
24. $1.30   **25.** 9 eggs   **26.** 61.90 pounds   **27.** 196 miles   **28.** $350   **29.** 6 quarts   **30.** 706.67 kilometers

Cumulative Test Chapters 1–4
1. Twenty six million, five hundred ninety-seven thousand, eighty nine   **2.** 411   **3.** 13,936   **4.** 68   **5.** $\frac{43}{40}$ or $1\frac{3}{40}$
6. $\frac{27}{35}$ **7.** $\frac{117}{8}$ or $14\frac{5}{8}$ **8.** $\frac{17}{6}$ or $2\frac{5}{6}$ **9.** 163.58   **10.** 8.2584   **11.** 0.179586   **12.** 14   **13.** $\frac{3}{1}$ **14.** $\frac{\$0.03}{1\text{ banana}}$
15. $\frac{4\text{ yen}}{1\text{ peso}}$ **16.** True   **17.** True   **18.** 3   **19.** 2   **20.** 128   **21.** 9   **22.** $n \approx 3.4$ **23.** 8.33 inches   **24.** $750.00
25. 5 pounds

CHAPTER 5   Percent

Pretest Chapter 5
1. 17%   **2.** 34%   **3.** 49.6%   **4.** 58.3%   **5.** 125%   **6.** 783%   **7.** 0.7%   **8.** 0.5%   **9.** 16%   **10.** 24%
11. 12.3%   **12.** 17.6%   **13.** $5\frac{1}{4}\%$ **14.** $2\frac{1}{8}\%$ **15.** 80%   **16.** 7.5%   **17.** 105%   **18.** 118.75%   **19.** 42.86%
20. 85.71%   **21.** 91.3%   **22.** 73.68%   **23.** 320%   **24.** 225%   **25.** 0.67%   **26.** 0.75%   **27.** $\frac{13}{50}$ **28.** $\frac{17}{50}$
29. $\frac{51}{100}$ **30.** $\frac{43}{100}$ **31.** $\frac{5}{4}$ **32.** $\frac{7}{4}$ **33.** $\frac{23}{300}$ **34.** $\frac{4}{75}$ **35.** $\frac{9}{16}$ **36.** $\frac{39}{80}$ **37.** 36.8   **38.** 44.72
39. 94.44%   **40.** 69.23%   **41.** 2000   **42.** 1350   **43.** 76.47%   **44.** $3600   **45.** $12.60   **46.** 20% decrease

Exercises 5.1

1. 45% **3.** 7% **5.** 80% **7.** 11% **9.** 78% **11.** 4% **13.** 371% **15.** 5.3% **17.** 0.6% **19.** 0.51
21. 0.07 **23.** 0.2 **25.** 0.436 **27.** 0.003 **29.** 0.0072 **31.** 1.82 **33.** 4.55 **35.** 74% **37.** 30%
39. 8% **41.** 56.3% **43.** 0.2% **45.** 0.57% **47.** 135% **49.** 272% **51.** 27%

53. $36\% = 36$ percent $= 36$ "per one hundred" $= 36 \times \dfrac{1}{100} = \dfrac{36}{100} = 0.36$. The rule is using the fact that 36% means 36 per one

hundred. **55. (a)** 555.62 **(b)** $\dfrac{55562}{10^2}$ **(c)** $\dfrac{27781}{50}$ **57.** $\dfrac{14}{25}$ **59.** 0.6875

Exercises 5.2

1. $\dfrac{11}{25}$ **3.** $\dfrac{7}{100}$ **5.** $\dfrac{21}{100}$ **7.** $\dfrac{11}{20}$ **9.** $\dfrac{3}{4}$ **11.** $\dfrac{1}{5}$ **13.** $\dfrac{29}{200}$ **15.** $\dfrac{22}{125}$ **17.** $\dfrac{81}{125}$ **19.** $\dfrac{57}{80}$ **21.** $1\dfrac{19}{25}$ **23.** $3\dfrac{2}{5}$

25. $\dfrac{33}{400}$ **27.** $\dfrac{13}{600}$ **29.** $\dfrac{99}{800}$ **31.** $\dfrac{117}{200}$ **33.** 75% **35.** 33.33% **37.** 31.25% **39.** 28% **41.** 27.5%

43. 58.33% **45.** 360% **47.** 283.33% **49.** 412.5% **51.** 42.86% **53.** 93.75% **55.** 38% **57.** $83\dfrac{1}{3}\%$

59. $68\dfrac{3}{4}\%$ **61.** $\dfrac{5}{12}, 0.4167, 41.67\%$ **63.** $\dfrac{3}{50}, 0.06, 6\%$ **65.** $\dfrac{2}{5}, 0.4, 40\%$ **67.** $\dfrac{69}{200}, 0.345, 34.5\%$ **69.** $\dfrac{3}{200}, 0.015, 1.5\%$

71. $\dfrac{5}{9}, 0.5556, 55.56\%$ **73.** $\dfrac{1}{32}, 0.03125, 3\dfrac{1}{8}\%$ **75.** 15.375% **77.** $\dfrac{463}{1600}$ **79.** 5.625 **81.** 88

Exercises 5.3A

1. $n = 38\% \times 500$ **3.** $75\% \times n = 9$ **5.** $17 = n \times 85$ **7.** $n = 128\% \times 4000$ **9.** $n \times 400 = 15$ **11.** $156 = 130\% \times n$
13. 150 **15.** 912 **17.** 1200 **19.** 1300 **21.** 60% **23.** 28% **25.** 50.4 **27.** 68 **29.** 12% **31.** 3.28
33. 64% **35.** 445 **37.** 55% **39.** 39.6 **41.** 1.44 **43.** 471.96 **45.** 57.6 **47.** 2.448 **49.** 2834

Exercises 5.3B

1. $p = 75, b = 660, a = 495$ **3.** $p = 42, b = 400, a = n$ **5.** $p = 49, b = n, a = 2450$ **7.** $p = p, b = 50, a = 30$
9. $p = p, b = 25, a = 10$ **11.** $p = 160, b = n, a = 400$ **13.** 91 **15.** 300 **17.** 16.4% **19.** 3.64 **21.** 25%
23. 180 **25.** 43% **27.** 47.5 **29.** 2.8 **31.** 153.615 **33.** 118.8 **35.** 15.82% **37.** 10.91

Exercises 5.4

1. 80% **3.** 22% **5.** $7.56 **7.** $4.40 **9.** $4.70 **11.** 6000 parts **13.** 289 people **15.** $42.40 **17.** 25%
19. $60,000 **21.** 810 workers **23.** $35.00 **25.** 15% **27.** 40% **29. (a)** $7.70 **(b)** $117.70
31. (a) $655.20 **(b)** $15,724.80 **33. (a)** $148.50 **(b)** $346.50 **35.** $21,000 **37.** 3% **39. (a)** $200 **(b)** $2700
41. (a) $253 **(b)** $2553 **(c)** $63.25 **43.** $28,709.19 **45.** 1,698,000 **47.** 1.63 **49.** 0.0556 **51.** $977.13
53. $7021.88

Extra Practice: Examples and Exercises

Section 5.1 **1.** 0.7% **3.** 9% **5.** 355% **7.** 87% **9.** 2% **11.** 0.0589 **13.** 0.28 **15.** 3.55 **17.** 0.56%

19. 125% *Section 5.2* **1.** $\dfrac{143}{400}$ **3.** $\dfrac{13}{25}$ **5.** $1\dfrac{13}{20}$ or $\dfrac{33}{20}$ **7.** $\dfrac{3}{40}$ **9.** $\dfrac{9}{400}$ **11.** 75% **13.** 56.25% **15.** 285.71%

17. 66.67% **19.** 72.73% *Section 5.3* **1.** 39.47% **3.** 63.43% **5.** 36.89% **7.** 23.87 **9.** 136.5 **11.** 194.12
13. 468.75 **15.** 12,000 **17.** 19.5 **19.** 25% *Section 5.4* **1.** 440 people **3.** $570.00 **5.** 12.5% **7.** $21,000
9. 2.5% **11. (a)** $568.50 **(b)** $18,381.50 **13. (a)** $173.75 **(b)** $521.25 **15. (a)** $645 **(b)** $1505
17. (a) $78 **(b)** $1278 **19. (a)** $1260 **(b)** $4760 **(c)** $210

Review Problems Chapter 5

1. 87% **3.** 27.6% **5.** 7.13% **7.** 252% **9.** 103.6% **11.** 0.6% **13.** 0.29% **15.** 72% **17.** 19.5%

19. 0.24% **21.** $4\dfrac{1}{12}\%$ **23.** 317% **25.** 64% **27.** 90% **29.** 45.45% **31.** 225% **33.** 442.86% **35.** 190%

37. 0.38% **39.** 0.002 **41.** 0.219 **43.** 1.66 **45.** 0.32125 **47.** $\dfrac{41}{50}$ **49.** $\dfrac{37}{20}$ **51.** $\dfrac{41}{250}$ **53.** $\dfrac{5}{16}$ **55.** $\dfrac{1}{2000}$

57. $\dfrac{3}{5}, 0.6, 60\%$ **59.** $\dfrac{3}{8}, 0.375, 37.5\%$ **61.** $\dfrac{2}{250}, 0.008, 0.8\%$ **63.** 288 **65.** 15 **67.** 40% **69.** 57.5 **71.** 160

73. 20% **75.** 13 **77.** 464.29% **79.** 51 students **81.** 4.17% **83.** $11,200 **85.** 15% **87.** $1200
89. 4.09% **91. (a)** $319 **(b)** $1276

Quiz Sections 5.1–5.2

1. 46% **2.** 93% **3.** 2.3% **4.** 5.7% **5.** 482% **6.** 503% **7.** 0.2% **8.** 0.9% **9.** 17% **10.** 29%

11. 28.6%   **12.** 16.4%   **13.** $7\frac{3}{8}$%   **14.** $4\frac{3}{4}$%   **15.** 85%   **16.** 50%   **17.** 180%   **18.** 106.25%   **19.** 71.43%

20. 57.14%   **21.** 32.14%   **22.** 39.53%   **23.** 775%   **24.** 960%   **25.** 0.63%   **26.** 0.6%   **27.** $\frac{21}{25}$ **28.** $\frac{7}{20}$

29. $\frac{77}{100}$ **30.** $\frac{51}{100}$ **31.** $\frac{33}{20}$ **32.** $\frac{21}{10}$ **33.** $\frac{33}{800}$ **34.** $\frac{13}{400}$ **35.** $\frac{13}{60}$ **36.** $\frac{13}{75}$

Quiz Sections 5.3–5.4
1. 325   **2.** 147   **3.** 88.89%   **4.** 58.33%   **5.** 1300   **6.** 278.57   **7.** 90   **8.** 150   **9.** 2.8%   **10.** 2.75%
11. $12.25   **12.** $23,520   **13.** 21%   **14. (a)** $119.76   **(b)** $379.24   **15. (a)** $840   **(b)** $105   **16.** $22,000

Test Chapter 5
1. 56%   **2.** 3%   **3.** 0.8%   **4.** 12.7%   **5.** 135%   **6.** 83%   **7.** 5.6%   **8.** $2\frac{1}{6}$%   **9.** 42.5%   **10.** 20%

11. 120%   **12.** 225%   **13.** 18.52%   **14.** 213.6%   **15.** $\frac{33}{20}$ **16.** $\frac{37}{400}$ **17.** 31.73   **18.** 160   **19.** 71.43%

20. 100   **21.** 5000   **22.** 52%   **23.** 592.2   **24.** 1.78%   **25.** $4101   **26. (a)** $94.16   **(b)** $333.84   **27.** 90.43%
28. 9.43% decrease   **29.** 13,000 registered voters   **30. (a)** $140   **(b)** $560

Cumulative Test Chapters 1–5
1. 2241   **2.** 8444   **3.** 5292   **4.** 89   **5.** $\frac{67}{12}$ or $5\frac{7}{12}$ **6.** $\frac{1}{12}$ **7.** $\frac{35}{12}$ or $2\frac{11}{12}$ **8.** $\frac{5}{21}$ **9.** 5731.7   **10.** 34.118

11. 1.686   **12.** 0.368   **13.** $\frac{3 \text{ pounds}}{5 \text{ square feet}}$ **14.** True   **15.** $n = 24$ **16.** 673 faculty   **17.** 2.3%   **18.** 46.8%

19. 198%   **20.** 3.75%   **21.** 2.43   **22.** 0.0675   **23.** 17.76%   **24.** 114.58   **25.** 300   **26.** 718.2   **27.** $8370
28. 3200 students   **29.** 11.31% increase   **30.** $352

CHAPTER 6   Measurement

Pretest Chapter 6
1. 204   **2.** 56   **3.** 3520   **4.** 6400   **5.** 1320   **6.** 24   **7.** 5320   **8.** 4680   **9.** 98.6   **10.** 0.027   **11.** 529.6
12. 0.123   **13.** 2376 m   **14.** 94.262 m   **15.** 3820   **16.** 3.162   **17.** 0.0563   **18.** 4800   **19.** 0.568   **20.** 8900
21. 4.73   **22.** 1.28   **23.** 59.52   **24.** 1826.78   **25.** 39.69   **26.** 103.4   **27.** 4.86×10^5 **28.** 2.0×10^{-5}
29. 0.00593   **30.** 260,000,000   **31.** 55.8 feet   **32. (a)** 95°F   **(b)** No.   **33.** 12.2 miles farther   **34.** 22.5 gallons/hr
35. 7.2×10^{15} atoms   **36.** $45

Exercises 6.1
1. 12   **3.** 1760   **5.** 2000   **7.** 4   **9.** 2   **11.** 60   **13.** 4   **15.** 7   **17.** 2   **19.** 12,320   **21.** 108   **23.** 123
25. 6.25   **27.** 11   **29.** 26,000   **31.** 36   **33.** 28   **35.** 9   **37.** 62   **39.** 64   **41.** 35   **43.** 11   **45.** 264
47. 4200   **49.** 64,800   **51.** $135   **53.** $2.88   **55.** 1152 land miles   **57.** 12,038,400,000 inches   **59.** 39   **61.** 6.0
63. 41550.72   **65.** 1049.92   **67.** 2520

Exercises 6.2
1. False   **3.** True   **5.** True   **7.** False   **9.** True   **11.** 30   **13.** 270   **15.** 1.98   **17.** 0.497   **19.** 5236
21. 370   **23.** 4200   **25.** 0.328   **27.** 0.482   **29.** 200,000   **31.** 0.078   **33.** 538,600   **35.** 9.6, 0.096
37. 3.582, 0.003582   **39.** 0.0032, 0.0000032   **41.** b **43.** c **45.** 3626 m   **47.** 463 cm   **49.** 406.71 m
51. 1.8068 cm or 18.068 mm   **53.** 3,816,000,000,000 meters   **55.** $341.01
57. Thirty eight million, five hundred sixteen thousand, two hundred forty-three
59. Two thousand, one hundred seventy-three ten thousandths   **61.** 61.2%   **63.** 1263.3592 km

Exercises 6.3
1. True   **3.** False   **5.** True   **7.** False   **9.** True   **11.** 64,000   **13.** 5300   **15.** 0.0189   **17.** 0.752
19. 2,430,000   **21.** 82   **23.** 0.005261   **25.** 74,000   **27.** 0.162   **29.** 0.027   **31.** 6.328   **33.** 2920   **35.** 17,000
37. 0.00032   **39.** 0.007896   **41.** 5,900,000   **43.** 0.007, 0.000007   **45.** 0.128, 0.000128   **47.** 0.522, 0.000522
49. 6.822, 0.006822   **51.** b **53.** a **55.** 41.582 L   **57.** 9.803 t   **59.** 260,160 mg   **61.** 0.005632
63. (a) 714.29 mg   **(b)** 0.26 kg   **65.** 20%   **67.** $321.30   **69.** 13.187 L

Exercises 6.4
1. 2.14 m   **3.** 22.86 cm   **5.** 15.26 yd   **7.** 28.89 yd   **9.** 8.06 m   **11.** 21.94 m   **13.** 132.02 km   **15.** 82 ft
17. 6.90 in.   **19.** 218 yd   **21.** 26.53 L   **23.** 1061.2 L   **25.** 21.76 L   **27.** 5.02 gal   **29.** 4.77 qt

31. 14.53 kg **33.** 198.45 g **35.** 39.6 lb **37.** 4.45 oz **39.** 5.45 feet **41.** 502.92 cm **43.** $\dfrac{31 \text{ mi}}{\text{hr}}$ **45.** $\dfrac{96.6 \text{ km}}{\text{hr}}$
47. 0.51 in. **49.** 104°F **51.** 185°F **53.** 53.6°F **55.** 55°C **57.** 75.56°C **59.** 30°C **61.** 180.6448 sq cm
63. $\dfrac{32.7273 \text{ mi}}{\text{hr}}$ **65.** 47 **67.** 91 **69.** 508.55 gal **71.** 307.31 kg

Exercises 6.5
1. 2.6×10^1 **3.** 1.37×10^2 **5.** 7.163×10^3 **7.** 1.2×10^2 **9.** 5.0×10^2 **11.** 2.63×10^4 **13.** 1.99×10^5
15. 1.71×10^6 **17.** 1.2×10^7 **19.** 6.7×10^{-1} **21.** 3.98×10^{-1} **23.** 2.79×10^{-3} **25.** 4.0×10^{-1}
27. 1.5×10^{-3} **29.** 1.6×10^{-5} **31.** 5.31×10^{-6} **33.** 7.0×10^{-4} **35.** 16 **37.** 53,600 **39.** 0.062
41. 0.000056 **43.** 0.00085 **45.** 900,000,000,000 **47.** 0.0000003 **49.** 0.00000003862 **51.** 4,600,000,000,000
53. 67,210,000,000 **55.** 5.878×10^{12} miles **57.** 5.9×10^{-7} meter **59.** 9.01×10^7 dollars
61. 8.05×10^{12} kilometers **63.** 4.39×10^{15} miles **65.** 3.624×10^8 feet **67.** 3.22×10^{36}
69. $1.387156284376542 \times 10^{15}$ **71.** 16.6334 **73.** 0.258 **75.** 2.178×10^{12}

Exercises 6.6
1. 8 yards **3.** 40 yards **5.** $33.46 **7.** 170 liters per drum **9.** 356 centimeters per piece **11.** 125 samples
13. 3°F below the record **15.** 12°F too cool **17.** 279 miles **19. (a)** 91 kilometers per hour **(b)** No
21. 30 gallons per hour **23.** $96 **25.** 1.85×10^{13} kilometers **27.** $9.97 **29.** 0.64 **31.** 11757.184 miles

Extra Practice: Examples and Exercises
Section 6.1 **1.** 5 **3.** 10,560 **5.** 5.83 **7.** 256,000 **9.** 20 **11.** 120 **13.** 9 **15.** 8 **17.** $60 **19.** $5.50
Section 6.2 **1.** 0.386 **3.** 0.065 **5.** 55,000 **7.** 0.0068, 0.0000068 **9.** 23.34, 0.02334 **11.** centimeter
13. millimeter **15.** meter **17.** 5000.3 cm **19.** 5443 m *Section 6.3* **1.** 0.233 **3.** 0.0672 **5.** 75,000 **7.** 85
9. 0.000387 **11.** 6,990,000 **13.** 900 **15.** 72,000 **17.** 19.113 L **19.** 14.082 t *Section 6.4* **1.** 11.88 m
3. 12.7 cm **5.** 2.48 miles **7.** 0.95 gal **9.** 45.48 L **11.** 411.48 cm **13.** 5.81 ft **15.** 944.88 cm **17.** 36.67°C
19. 98°F *Section 6.5* **1.** 4.45×10^2 **3.** 1.3×10^2 **5.** 9.8823×10^2 **7.** 4.56×10^{-3} **9.** 9.85×10^{-4}

11. 0.0038 **13.** 5.5 **15.** 0.000012 *Section 6.6* **1.** $8\frac{2}{3}$ yds **3.** 80 yds **5.** $35.82 **7.** 700 mL in each jar

9. 5333.33 meters **11.** 20°F **13.** 19°F too hot **15.** 19°F too hot

Review Problems Chapter 6
1. 9 **3.** 5280 **5.** 7.5 **7.** 3 **9.** 8000 **11.** 5.75 **13.** 60 **15.** 15.5 **17.** 36 **19.** 840 **21.** 560
23. 176.3 **25.** 920 **27.** 5000 **29.** 0.285 **31.** 7.93 m **33.** 35.63 m **35.** 17,000 **37.** 0.059 **39.** 196,000
41. 0.778 **43.** 0.125 **45.** 76,000 **47.** 765 **49.** 2430 **51.** 92.4 **53.** 4.58 **55.** 368.55 **57.** 54.86
59. 5.52 **61.** 9.08 **63.** 10.97 **65.** 49.6 **67.** 59 **69.** 105 **71.** 4.16×10^3 **73.** 2.18×10^5 **75.** 4.0×10^{-3}
77. 2.18×10^{-5} **79.** 5.136×10^{-1} **81.** 9,000,000 **83.** 18,900 **85.** 0.0752 **87.** 0.0000009 **89.** 0.000536
91. 8.44×10^{11} **93.** 1.342×10^{31} **95.** 1.44×10^{14} **97.** 3.12 cm **99. (a)** 200 m **(b)** 0.2 km
101. 32.6 miles farther **103.** 166.67 milliliters per jar

Quiz Sections 6.1–6.3
1. 0.75 **2.** 5000 **3.** 14.4 **4.** 10.5 **5.** 165 **6.** 0.31 **7.** 240 **8.** 8.5 **9.** 1260 **10.** 3 **11.** 4400
12. 9.5. **13.** 1800 **14.** 12,600 **15.** 5,800,000 **16.** 678,000 **17.** 236.4 **18.** 98.2 **19.** 58.2 **20.** 0.0963
21. 720 **22.** 0.05296 **23.** 4960 **24.** 980 **25.** 0.0165 **26.** 1560 **27.** 73 **28.** 0.019856 **29.** 0.007953
30. 760,000

Quiz Sections 6.4–6.6
1. 4.27 m **2.** 17.78 m **3.** 14.62 m **4.** 8.67 in. **5.** 11.16 mi **6.** 239.8 yd **7.** 16.08 L **8.** 454.8 L

9. 5.54 gal **10.** 2.89 oz **11.** 6.81 kg **12.** $\dfrac{64.4 \text{ km}}{\text{hr}}$ **13.** 15°C **14.** 122°F **15.** 1.783×10^3 **16.** 2.04×10^5

17. 8.0×10^{-5} **18.** 7.132×10^{-3} **19.** 300,000,000 **20.** 1,620,000 **21.** 0.00001367 **22.** 0.000000008
23. 6.6×10^{11} atoms **24.** 2.39×10^{15} meters **25. (a)** 21 feet **(b)** 6.41 meters **26. (a)** Carlos **(b)** 4.4 meters
27. 166.67 milliliters per jar **28.** 22.2 miles **29.** $2.25 **30.** $48

Test Chapter 6
1. 3200 **2.** 228 **3.** 84 **4.** 7 **5.** 0.75 **6.** 30 **7.** 0.273 **8.** 9200 **9.** 4.6 **10.** 0.0988 **11.** 1270
12. 9.36 **13.** 0.046 **14.** 127,000 **15.** 0.0289 **16.** 0.983 **17.** 920 **18.** 9420 **19.** 67.62 **20.** 1.63
21. 3.55 **22.** 10.03 **23.** 16.06 **24.** 9.6379×10^4 **25.** 4.0×10^{-5} **26.** 0.000000458 **27.** 6,700,000

28. (a) 20 meters **(b)** 21.8 yards **29. (a)** 15°F **(b)** Yes **30.** 1.07×10^{23} stars **31.** $\dfrac{82.5 \text{ gallons}}{\text{hour}}$

32. (a) 300 km **(b)** 14 miles

Selected Answers

Cumulative Test Chapters 1–6

1. 6028 **2.** 185,440 **3.** 69 **4.** $\dfrac{19}{42}$ **5.** $1\dfrac{3}{8}$ **6.** True **7.** $n = 6$ **8.** 209.23 grams **9.** 250% **10.** 64.8
11. 20,000 **12.** 9.5 **13.** 5000 **14.** 3.5 **15.** 300 **16.** 3700 **17.** 0.0628 **18.** 790 **19.** 0.05 **20.** 672
21. 106.12 **22.** 43.58 **23.** 25.81 **24.** 14.49 **25.** 5.79863×10^5 **26.** 7.8×10^{-4} **27.** 11.88 meters
28. 15°C is equal to 59°F. The difference is 44°F. The 15°C temperature is higher. **29.** 7 miles **30.** 1.738 centimeters

CHAPTER 7 Geometry

Pretest Chapter 7

1. 18 m **2.** 14 m **3.** 23.0 sq cm **4.** 2.4 sq cm **5.** 25.6 yd **6.** 52 ft **7.** 135 sq in. **8.** 171 sq in.
9. 97 sq m **10.** 59° **11.** 15.3 m **12.** 30 sq m **13.** 22.5 sq km **14.** 52 sq cm **15.** 28 inches **16.** 94.2 cm
17. 153.9 sq m **18.** 27.4 sq m **19.** 240 cu yd **20.** 113.0 cu ft **21.** 1846.3 cu in. **22.** 4375 cu m
23. 1130.4 cu m **24.** 120 cm **25.** 28 m **26. (a)** 3706.5 sq yd **(b)** $555.98 **27.** $870.10

Exercises 7.1

1. 15 mi **3.** 23.6 ft **5.** 49.2 ft **7.** 2.16 mm **9.** 25.46 in. **11.** 17.12 km **13.** 47.5 m **15.** 0.0338 cm
17. 36 m **19.** 4.8 mi **21.** 25.28 cm **23.** 0.0172 **25.** 94 m **27.** 180 cm **29.** 168 in.^{2} **31.** 57.76 ft^2
33. 0.288 m^2 **35.** 14,976 yd^2 **37.** 294 m^2 **39.** $157.50 **41.** $27.30 **43.** $598.22 **45.** 223.3
47. 21,842.8 **49.** 9026.56 sq ft

Exercises 7.2

1. 29.8 in. **3.** 40.2 m **5.** 113.3 cm^2 **7.** 1386 m^2 **9.** 3528 yd^2 **11.** 82 m **13.** 470 cm **15.** 54 m^2
17. 57 cm^2 **19.** 344 yd^2 **21.** 550 km^2 **23.** 718 m^2 **25.** 162.5 cm^2 **27.** 345 ft^2 **29.** $80,960.00
31. Area $= 5.2415 \times b^2$ sq units **33.** 30 ft **35.** 1800 cm **37.** 939.75 ft

Exercises 7.3

1. True **3.** True **5.** False **7.** 60° **9.** 30° **11.** 90° **13.** 135° **15.** 22.5 ft^2 **17.** 83.125 cm^2
19. 10.62 m^2 **21.** 12.25 yd^2 **23.** 25.5 in. **25.** 16.4 cm **27.** 188 yd^2 **29.** $21,060.00 **31.** 12 **33.** 393 faculty

Exercises 7.4

1. 58 in. **3.** 15 mm **5.** 22.5 yd **7.** 1.9 cm **9.** $\approx$75.4 cm **11.** $\approx$69.1 in. **13.** 78.5 yd^2 **15.** $\approx$907.5 m^2
17. $\approx$803.8 cm^2 **19.** 628 m^2 **21.** 153.9 ft^2 **23.** 11,304 mi^2 **25.** $\approx$163.3 mi^2 **27.** $\approx$189.3 m^2 **29.** $\approx$31.0 m^2
31. $1211.20 **33. (a)** $0.75, 22.1 in.^{2} **(b)** $0.67, 18.8 in.^{2} **(c)** a piece of 15-in pizza **35.** Answers vary
37. 10 revolutions **39.** 13.92 **41.** 6000 **43.** 0.70057457 m

Exercises 7.5

1. 24 m^3 **3.** 1170 mm^3 **5.** 87.9 m^3 **7.** 2260.8 m^3 **9.** 3052.1 m^3 **11.** 267.9 m^3 **13.** $\approx$718.0 m^3
15. 1004.8 in.^{3} **17.** 312.5 yd^3 **19.** 2.68×10^{11} mi^3 **21.** 21 m^3 **23.** 120 m^3 **25.** 527.5 cm^3 **27.** $\approx$1025.7 ft^3
29. 4019.2 cm^3. The sphere is larger by 167.5 cm^3 **31.** 214,500 yd^3 **33.** $9\dfrac{7}{12}$ **35.** $\dfrac{135}{16}$ or $8\dfrac{7}{16}$ **37.** 170.832816 m^3
39. 592.1 m^3 **41.** 95.6887288 ft^3

Exercises 7.6

1. 8 m **3.** 17.5 cm **5.** 1.9 yd **7.** a corresponds to f, b corresponds to e, c corresponds to d **9.** 2.2 in. **11.** 36 ft
13. 1.4 km **15.** 16.3 cm **17.** $\approx$33.3 m **19.** $\approx$11.6 yd^2 **21.** 12

Exercises 7.7

1. $9.60 **3. (a)** $\dfrac{75\ km}{hr}$ **(b)** $\dfrac{76\ km}{hr}$ **(c)** Through Woodville and Palermo **5.** 24.3 min **7.** $510.00
9. 1,507,200 gallons **11.** $74.42 **13. (a)** 40,820 km **(b)** $\dfrac{20{,}410\ km}{hr}$ **15.** 128 **17.** 0.25 **19.** $1731.70

Extra Practice: Examples and Exercises

Section 7.1 **1.** 37.2 m **3.** 7.28 cm **5.** 68 km **7.** 275 in.^{2} **9.** 144 in.^{2} **11.** $252 **13.** 480 ft **15.** $45.85
17. 886 m^2 *Section 7.2* **1.** $A = 40.8$ m^2, $P = 35.6$ m **3.** 107 m **5.** 97 yd **7.** 177 cm **9.** $A = 120$ m^2
11. $A = 780$ cm^2 **13.** $A = 93.5$ m^2 *Section 7.3* **1.** 30° **3.** 65° **5.** 50° **7.** 15 in. **9.** 28.5 m **11.** 14 ft^2
13. 19.8 cm^2 **15.** 12.96 in.^{2} **17.** 33.9 m^2 *Section 7.4* **1.** 87.92 cm **3.** 34.54 m **5.** 12.56 ft **7.** 113.04 m^2
9. 12.56 m^2 **11.** 122.46 cm^2 **13.** $A = 76.93$ m^2 **15.** $A = 19.23$ cm^2 **17.** $A = 127.17$ m^2 *Section 7.5* **1.** 140 ft^3
3. 108 yd^3 **5.** 15.312 ft^3 **7.** 221.06 cm^3 **9.** 228.91 ft^3 **11.** 1436.03 cm^3 **13.** 113.04 cm^3 **15.** 904.32 yd^3

17. 40.23 m^3 **19.** 23.55 cm^3 *Section 7.6*   **1.** 26.9 ft   **3.** $a \to f$, $b \to d$, $c \to e$ **5.** $\approx$2.1 km   **7.** $\approx$79.3 in.
Section 7.7   **1.** $11.70   **3.** $49.40   **5.** 53.39 min   **7.** $239.00   **9. (a)** 42 in.   **(b)** $183.60

Review Problems Chapter 7

1. 19.8   **3.** 23.2 yd   **5.** 16.5 cm^2 **7.** 18.5 in.^{2} **9.** 38 ft   **11.** 68 m^2 **13.** 100.4 m   **15.** 80 in.   **17.** 2700 m^2
19. 336 yd^2 **21.** 422 cm^2 **23.** 22 ft   **25.** 153°   **27.** 24.5 m^2 **29.** 45.6 cm^2 **31.** 450 m^2 **33.** 106 cm
35. 37.7 in.   **37.** 113.0 m^2 **39.** 201.0 ft^2 **41.** 226.1 in.^{2} **43.** 318.5 ft^2 **45.** 107.4 ft^2 **47.** 45 ft^3 **49.** 14,130 ft^3
51. 307.7 m^3 **53.** 1728 m^3 **55.** 3768 ft^3 **57.** 30 m   **59.** 33.8 cm   **61.** 324 cm   **63.** $92.40   **65.** $736.00
67. (a) 21,873.2 ft^3 **(b)** 17,498.6 bushels

Quiz Sections 7.1–7.4

1. 3.2 m   **2.** 15.8 cm   **3.** 96 yd^2 **4.** 225 yd^2 **5.** 25.6 m   **6.** 78 cm   **7.** 396 km^2 **8.** 550 yd^2 **9.** 7.4 in.
10. 37.7 in.   **11.** 153.9 mi^2 **12.** 13.8 yd^2 **13.** 45.3 ft^2 **14.** 38 in.^{2} **15.** 11,456 cm^2 **16.** 101.5 in.^{2}

Quiz Sections 7.5–7.7

1. 112 yd^3 **2.** 904.3 yd^3 **3.** 1780.4 in.^{3} **4.** 220 ft^3 **5.** 1186.9 in.^{3} **6.** 129.8 cm^3 **7.** 4.6   **8.** 4.8 m
9. 98 in.   **10.** 195 ft   **11. (a)** 17 yd^3 **(b)** $238.00   **12.** 9.4 hrs   **13. (a)** 2826 ft^3 **(b)** $9891.00
14. (a) 26 square yards   **(b)** Cost = $234   **15.** 94.2 cm^3 **16. (a)** 67,500 in.^{3} **(b)** 292.2 gallons

Test Chapter 7

1. 40 yd   **2.** 34 ft   **3.** 20 m   **4.** 80 m   **5.** 24.8 m   **6.** 180 yd^2 **7.** 104.0 m^2 **8.** 78 m^2 **9.** 144 m^2
10. 12 cm^2 **11.** 52.5 m^2 **12.** 37.7 in   **13.** 254.3 ft^2 **14.** 107.4 in^2 **15.** 144.3 in^2 **16.** 13.8 in^2 **17.** 840 m^3
18. 803.8 m^3 **19.** 33.5 m^3 **20.** 508.7 ft^3 **21.** 56 m^3 **22.** 43.2 cm   **23.** 46.7 ft   **24. (a)** 6456 yd^2 **(b)** $2582.40
25. (a) 113 m^3 **(b)** 72.5 kg

Cumulative Test Chapters 1–7

1. 935,760   **2.** 33,415   **3.** $\dfrac{26}{45}$ **4.** $\dfrac{4}{21}$ **5.** 56.13   **6.** 7.2272   **7.** 83   **8.** 27   **9.** 800 students   **10.** 40.7

11. 2000   **12.** 75%   **13.** 5.86 meters   **14.** 1512 inches   **15.** 54.56 miles   **16.** 50 m   **17.** 208 cm   **18.** 56.5 yd
19. 1.4 cm^2 **20.** 540 m^2 **21.** 192 m^2 **22.** 664 yd^2 **23.** 50.2 m^2 **24.** 2411.5 m^3 **25.** 3052.1 cm^3
26. 3136 cm^3 **27.** 2713.0 m^3 **28.** 33.4 m   **29.** 4.1 ft   **30. (a)** 124 yd^2 **(b)** $992.00
31. (a) 3843.63 ft^3 **(b)** 3074.69 bushels

CHAPTER 8   Statistics, Square Roots, and the Pythagorean Theorem

Pretest Chapter 8

1. Under age 18   **2.** 43%   **3.** 17%   **4.** 2200 students   **5.** 300 students   **6.** 450 people   **7.** 600 people
8. 3rd quarter of both 1989 and 1990   **9.** 4th quarter of 1989   **10.** 100 more people   **11.** 100 more people
12. August and December   **13.** December   **14.** November   **15.** 40,000 sets   **16.** 40,000 sets   **17.** 55,000 cars
18. 60,000 cars   **19.** 25,000 cars   **20.** 20,000 cars   **21.** 36 pages per day   **22.** 39 pages per day   **23.** $4.85
24. $5.35   **25.** 8   **26.** 7   **27.** 12   **28.** 12   **29.** 6.782   **30.** 9.434   **31.** 11.446   **32.** 8 ft   **33.** 13.601 ft
34. 3,606 mi   **35.** 8 feet

Exercises 8.1

1. Rent   **3.** $350   **5.** $500   **7.** $\dfrac{4}{1}$ **9.** $\dfrac{3}{10}$ **11.** Hit batters   **13.** 144   **15.** 379   **17.** $\dfrac{147}{325}$ **19.** $\dfrac{49}{24}$ **21.** 13%

23. 28%   **25.** 26,970 women   **27.** 1,650,000,000 people   **29.** 19%   **31.** 83%   **33.** 400,000,000   **35.** 42 in.^{2}

Exercises 8.2

1. 8 million people   **3.** 11 million people   **5.** 1960–1970   **7.** 40,000 men   **9.** 1990   **11.** 25,000 more
13. 1910 to 1930   **15.** 4 million dollars   **17.** 1988   **19.** One million dollars   **21.** 2 inches
23. July, August, and September   **25.** 1.5 inches more   **27.** 35

Exercises 8.3

1. 50 cars   **3.** 35 cars   **5.** 85 cars   **7.** 145 cars   **9.** 16 days   **11.** 3 days   **13.** 30 days   **15.** 37 days

17. $\dfrac{30}{47} \approx 63.8\%$ **19.** |||, 3   **21.** |||| |, 6   **23.** |||, 3   **25.** ||, 2   **27.**

29. |||, 3   **31.** ||, 2   **33.** |, 1   **35.** ||, 2
37. 7

Exercises 8.4

1. 91.8   **3.** 34   **5.** $87,000   **7.** 5 hrs   **9.** 2.59   **11.** 0.375   **13.** 23.7 mpg   **15.** 47   **17.** 1011   **19.** 0.58
21. $11,550   **23.** 31 min   **25.** $1955   **27.** 20.5 hrs   **29. (a)** $2157   **(b)** $1615
(c) The median because the mean is affected by the amount $6300   **31.** 438,645,500   **33.** 28.3 in.^{2} **35.** 110,040
37. 4850

Exercises 8.5

1. 1   **3.** 4   **5.** 5   **7.** 7   **9.** 10   **11.** 11   **13.** 13   **15.** 0   **17.** 10   **19.** 11   **21.** 14   **23.** 9   **25.** 10
27. 5   **29. (a)** Yes   **(b)** 7   **31. (a)** Yes   **(b)** 16   **33.** ≈3.873   **35.** ≈5.568   **37.** ≈9.110   **39.** ≈10.954
41. ≈13.565   **43.** 11 m   **45.** ≈5.099 m   **47.** ≈8.660 m
49. (a) 2   **(b)** 0.2   **(c)** 0.02   **(d)** Each answer is obtained from the previous answer by dividing by 10.
(e) No, because $0.004 = 4 \times 10^{-3}$, and 10^{-3} isn't a perfect square.   **51.** 22   **53.** 10.4 gals   **55.** 45.216   **57.** 18.978

Exercises 8.6

1. 5 in.   **3.** ≈8.544 yd   **5.** ≈15.199 ft   **7.** ≈3.606   **9.** ≈7.550 km   **11.** ≈11.402 m   **13.** ≈7.071 m
15. ≈6.928 yds   **17.** 17 ft   **19.** 5 miles   **21.** ≈6.9 in   **23.** ≈12.1 m   **25.** ≈8.5 m   **27.** ≈15.6 cm
29. ≈7.1 in.   **31.** 52 yd   **33.** 11.37   **35.** 3.0464   **37.** 160 in.

Extra Practice: Examples and Exercises

Section 8.1   1. Wages   **3.** $1650   **5.** $\frac{24}{119}$ **7.** 40%   **9.** 80%   **11.** 1050   **Section 8.2   1.** 3000   **3.** 9000
5. 1960–1970   **7.** 350   **9.** 150   **11.** January   **13.** 1000   **15.** 4500   **17.** 2 in.   **19.** 2 in.   **Section 8.3   1.** 45
3. 5   **5.** 50   **7.** ∥, 2   **9.** ∥, 2   **11.** ∥∥, 3   **Section 8.4   1.** 84.8   **3.** $90,000   **5.** 2.83   **7.** 23.6 mpg   **9.** 46
11. 974   **13.** 0.21   **15.** 20 students   **17.** 19 hours   **Section 8.5   1.** Yes, 11.   **3.** Yes, 1.   **5.** Yes, 12.   **7.** 18
9. 8   **11.** 5.657   **13.** 10.247   **15.** 8.888   **17.** 10 m   **19.** 8.718 m   **Section 8.6   1.** 12 km   **3.** 12.649 m
5. 11.314 m   **7.** 13.892   **9.** 6.325 miles

Review Problems Chapter 8

1. 36 computers   **3.** 20 computers   **5.** $\frac{4}{1}$ **7.** 7.5%   **9.** 15%   **11.** 57%   **13.** $192   **15.** $816

17. 6000 customers   **19.** 4th quarter 1991   **21.** 1000 customers   **23.** Yes   **25.** 400 students   **27.** 650 students
29. 100 students more   **31.** 1988–1989   **33.** 45,000 cones   **35.** 10,000 cones more   **37.** 25,000 cones more
39. The cooler the temperature, the fewer cones sold.   **41.** 50 bridges   **43.** 20 and 39   **45.** 150 bridges   **47.** ∦∥∥, 10
49. ∥∥, 3   **51.** ∥, 2   **53.** 18 times   **55.** $114   **57.** 188 women   **59.** 1353 employees   **61.** 83 students
63. $139,200   **65.** 18.6 deliveries   **67.** Median, because of the one high data item, 39   **69.** 8   **71.** 11   **73.** 15
75. 6   **77.** 12   **79.** ≈6.708   **81.** ≈8.718   **83.** ≈13.416   **85.** 5 m   **87.** 8.06 m   **89.** 9.2 ft   **91.** 3.6 ft

Quiz Sections 8.1–8.3

1. Overtime for personnel   **2.** Regular salaries of personnel   **3.** $55,000   **4.** $85,000   **5.** $\frac{17}{11}$ **6.** $\frac{17}{43}$ **7.** 14%

8. 34%   **9.** 80%   **10.** 20%   **11.** 9 cases   **12.** 42 cases   **13.** 300 students   **14.** 500 students
15. By 300 students   **16.** By 200 students   **17.** 250 houses   **18.** 400 houses   **19.** 1987   **20.** 1990
21. 2.5 million dollars   **22.** 1.5 million dollars   **23.** 1990–1991   **24.** 1989–1990   **25.** 250 deer   **26.** 550 deer
27. 1989   **28.** 1987   **29.** Yes   **30.** 1987–1988   **31.** 550 cars   **32.** 500 cars   **33.** 800 cars   **34.** 800 cars
35. 135 tickets   **36.** 480 tickets

Quiz Sections 8.4–8.6

1. 94°   **2.** 47°   **3.** 28.2 yrs   **4.** 15.5 tapes   **5.** $408.50   **6.** 230.5 flights   **7.** 86   **8.** 82   **9.** $14,925
10. $9990   **11.** $90.50   **12.** 60 hrs   **13.** 7   **14.** 5   **15.** 6   **16.** 8   **17.** 1   **18.** 0   **19.** 12   **20.** 13   **21.** 8
22. 11   **23.** 18   **24.** 15   **25.** ≈4.243   **26.** ≈4.796   **27.** ≈12.530   **28.** ≈12.689   **29.** 17 cm   **30.** 6 m
31. ≈4.796 m   **32.** ≈11.402 cm   **33.** 9.4 mi   **34.** 6.7 m

Test Chapter 8

1. 37%   **2.** 21%   **3.** 12%   **4.** 60,000 automobiles   **5.** 54,000 automobiles   **6.** 350 cars   **7.** 500 cars
8. 1st quarter 1990   **9.** 1st quarter   **10.** 50 cars more   **11.** 150 cars more   **12.** 20 yrs   **13.** 26 yrs   **14.** 12 yrs
15. Age 35   **16.** Age 65   **17.** 60,000 televisions   **18.** 25,000 televisions   **19.** 20,000 televisions
20. 60,000 televisions   **21.** 16.25   **22.** 16.5   **23.** $600   **24.** $612   **25.** 9   **26.** 8   **27.** 12   **28.** 7
29. ≈7.348   **30.** ≈10.954   **31.** ≈13.675   **32.** 9.22   **33.** 10   **34.** 5.83 cm   **35.** 9 ft

Cumulative Test Chapters 1–8

1. 20,825   **2.** 78,104   **3.** $\frac{153}{40}$ or $3\frac{33}{40}$ **4.** $\frac{108}{35}$ or $3\frac{3}{35}$ **5.** 2864.37   **6.** 72.65   **7.** 72.23   **8.** 0.6   **9.** 39 cars

10. 0.325   **11.** 350   **12.** 1.98 m   **13.** 54 ft   **14.** 28.3 in.^{2} **15.** 68 in.   **16.** 34%   **17.** 3840 students
18. 3 million dollars   **19.** 1 million dollars   **20.** 16 in.   **21.** 1950, 1960   **22.** 8 students   **23.** 16 students
24. $6.50   **25.** $4.95   **26.** 11   **27.** ≈7.550   **28.** 10.440 in.   **29.** 4.899 m   **30.** 13.9 mi

---

CHAPTER 9   Signed Numbers

Pretest Chapter 9

1. −23   **2.** −5   **3.** 2.1   **4.** −2   **5.** −$\dfrac{1}{3}$ **6.** −$\dfrac{7}{12}$ **7.** −9.4   **8.** 2.4   **9.** −5   **10.** −31   **11.** $\dfrac{8}{19}$

12. −14   **13.** 1.8   **14.** −3.8   **15.** 28   **16.** $\dfrac{19}{20}$ **17.** 10   **18.** 2   **19.** −12   **20.** −15   **21.** −26   **22.** $\dfrac{8}{9}$

23. −36   **24.** −25   **25.** −22   **26.** 1   **27.** −9   **28.** 3.5   **29.** 8   **30.** −15   **31.** $\dfrac{1}{2}$ **32.** $\dfrac{1}{3}$

Exercises 9.1

1. 14   **3.** −7   **5.** −5.7   **7.** 16.5   **9.** $\dfrac{17}{35}$ **11.** −$\dfrac{8}{9}$ **13.** 9   **15.** −5   **17.** −5   **19.** 22   **21.** −2.8

23. −$\dfrac{2}{3}$ **25.** $\dfrac{5}{9}$ **27.** −18   **29.** −0.7   **31.** $\dfrac{1}{2}$ **33.** −38   **35.** −7.6   **37.** −1   **39.** −11   **41.** −6

43. 10   **45.** −$68,000   **47.** −$5,000   **49.** $5000   **51.** 10° F   **53.** −16° F
55. the value of $a + b$ as well as $b + a$ is −5.1   **57.** evaluating $a + (b + c)$ as well as $(a + b) + c$ will yield −1

59. −$\dfrac{184}{225}$ **61.** 904.3 ft^3 **63.** 33.115

Exercises 9.2

1. −3   **3.** −14   **5.** −5   **7.** 24   **9.** −6   **11.** 3   **13.** −60   **15.** 556   **17.** −90   **19.** 40   **21.** −6.7

23. 7.1   **25.** −8.6   **27.** 8.7   **29.** 1   **31.** −$\dfrac{7}{6}$ or $1\dfrac{1}{6}$ **33.** −$\dfrac{1}{6}$ **35.** $\dfrac{3}{22}$ **37.** 15   **39.** 0   **41.** 17   **43.** −9

45. 2   **47.** 5010 ft   **49.** 55° F   **51.** $20,700   **53.** $24,000   **55.**   Bal. $50.00   **57.** −$\dfrac{1}{2}$ **59.** 5
61. 35.544   Dep. $80.00

New Bal. $130.00

Exercises 9.3

1. −14   **3.** 15   **5.** −20   **7.** 16   **9.** 60   **11.** −160   **13.** −1.5   **15.** 3.75   **17.** −$\dfrac{6}{35}$ **19.** $\dfrac{1}{23}$ **21.** −2

23. 11   **25.** 14   **27.** −2   **29.** −8   **31.** 9   **33.** −$\dfrac{5}{6}$ **35.** $\dfrac{8}{7}$ or $1\dfrac{1}{7}$ **37.** −8.2   **39.** 6.2   **41.** 112   **43.** −40

45. 16   **47.** −30   **49.** 0   **51.** 0   **53.** b is negative.   **55.** −$\dfrac{21}{64}$ **57.** 90 in.^{2} **59.** −9.656964   **61.** 15,436.61972

Exercises 9.4

1. 2   **3.** −26   **5.** 8   **7.** −12   **9.** −19   **11.** 0   **13.** −5   **15.** −12   **17.** 20   **19.** −27   **21.** −50

23. −18   **25.** −25   **27.** 1   **29.** 2   **31.** −$\dfrac{1}{2}$ **33.** −1   **35.** $\dfrac{2}{3}$ **37.** −3.6°   **39.** −$\dfrac{1}{28}$ **41.** 3.84 km

Extra Practice: Examples and Exercises

Section 9.1   **1.** 9   **3.** −$\dfrac{14}{15}$ **5.** −11.6   **7.** 14   **9.** −0.4   **11.** −14   **13.** 6   **15.** 2.4   **17.** $18,000 loss

19. 14° F   *Section 9.2*   **1.** −2   **3.** −1.6   **5.** −56   **7.** −$\dfrac{3}{20}$ **9.** $\dfrac{1}{4}$ **11.** 2   **13.** −9   **15.** −1   **17.** $48,500

19. 5080 feet   *Section 9.3*   **1.** −$\dfrac{8}{21}$ **3.** $\dfrac{3}{10}$ **5.** −$\dfrac{2}{9}$ **7.** −2   **9.** $\dfrac{13}{3}$ or $4\dfrac{1}{3}$ **11.** 1   **13.** −$\dfrac{15}{28}$ **15.** −$\dfrac{35}{8}$ or $4\dfrac{3}{8}$

17. −80   **19.** 0   *Section 9.4*   **1.** −6   **3.** −7   **5.** −33   **7.** $24\dfrac{2}{3}$ **9.** −6   **11.** 90   **13.** −$\dfrac{41}{2}$ or −$20\dfrac{1}{2}$

15. −15   **17.** $\dfrac{3}{13}$ **19.** 2

Review Problems Chapter 9

1. 11   **3.** -22 **5.** -15 **7.** -8.8 **9.** $-\dfrac{8}{15}$ **11.** 6   **13.** 11   **15.** 4   **17.** -4 **19.** -9 **21.** -15

23. 19   **25.** -1.6 **27.** $-\dfrac{1}{15}$ **29.** 13   **31.** -9 **33.** -14 **35.** 50   **37.** $\dfrac{2}{35}$ **39.** -7.8 **41.** 3   **43.** -9

45. 6   **47.** $-\dfrac{7}{10}$ **49.** 30   **51.** -112 **53.** -54 **55.** 14   **57.** 13   **59.** -11 **61.** -5 **63.** -8 **65.** -72

67. -30 **69.** 1   **71.** $\dfrac{1}{10}$ **73.** $\dfrac{2}{5}$ **75.** -2

Quiz Sections 9.1–9.2

1. -7 **2.** -11 **3.** -1.9 **4.** $-\dfrac{5}{12}$ **5.** -130 **6.** -20 **7.** $\dfrac{2}{25}$ **8.** -5.6 **9.** 10   **10.** 10   **11.** -18

12. $3°\ \text{F}$ **13.** A loss of \$70,000 or $-\$70,000$ **14.** A loss of \$25,000 or $-\$25,000$ **15.** -3 **16.** -2 **17.** $\dfrac{11}{15}$

18. -10.9 **19.** -8 **20.** 112   **21.** -2.6 **22.** $\dfrac{13}{12}$ or $1\dfrac{1}{12}$ **23.** 300   **24.** -516 **25.** $27°\ \text{F}$ **26.** $44°\ \text{F}$

27. 4330 ft   **28.** -1 **29.** -70 **30.** 20   **31.** -7 **32.** 14.1   **33.** -3.3 **34.** 33

Quiz Sections 9.3–9.4

1. 42   **2.** -36 **3.** 2   **4.** -13 **5.** -3 **6.** 4   **7.** $-\dfrac{1}{15}$ **8.** $\dfrac{3}{14}$ **9.** 3.6   **10.** -11.2 **11.** 30

12. -84 **13.** $-\dfrac{1}{10}$ **14.** $-\dfrac{3}{35}$ **15.** $\dfrac{12}{7}$ or $1\dfrac{5}{7}$ **16.** $-\dfrac{10}{9}$ or $-1\dfrac{1}{9}$ **17.** -30.1 **18.** -17.5 **19.** 16.25

20. 5.7   **21.** 24   **22.** -30 **23.** 0   **24.** -30 **25.** 30   **26.** 12   **27.** -15 **28.** -13 **29.** 2   **30.** 15

31. -25 **32.** 10   **33.** -1 **34.** $-\dfrac{1}{2}$ **35.** 1   **36.** -1

Test Chapter 9

1. -7 **2.** -46 **3.** 3.7   **4.** -3 **5.** -5 **6.** $-\dfrac{7}{6}$ or $-1\dfrac{1}{6}$ **7.** $-\dfrac{1}{2}$ **8.** 0.5   **9.** -27 **10.** 5

11. $\dfrac{19}{15}$ or $1\dfrac{4}{15}$ **12.** -11 **13.** 5.0   **14.** 3.3   **15.** $\dfrac{1}{2}$ **16.** 70   **17.** 30   **18.** -7 **19.** 5   **20.** -12 **21.** 2

22. $-\dfrac{5}{14}$ **23.** -54 **24.** -21 **25.** 30   **26.** 7   **27.** -5 **28.** -48 **29.** -35 **30.** 8   **31.** $-\dfrac{1}{2}$ **32.** $\dfrac{1}{3}$

Cumulative Test Chapters 1–9

1. 12,383   **2.** 127   **3.** $\dfrac{143}{12}$ or $11\dfrac{11}{12}$ **4.** $\dfrac{55}{12}$ or $4\dfrac{7}{12}$ **5.** 9.812   **6.** 63.46   **7.** 65.9968   **8.** 64

9. 126 defective parts   **10.** 0.304   **11.** 4000   **12.** 94,000 m   **13.** 5 yds   **14.** 78.5 m^2

15. (a) 300 students   **(b)** 1100 students   **16.** 13   **17.** -4.7 **18.** $\dfrac{5}{12}$ **19.** -11 **20.** -5 **21.** -60

22. $\dfrac{6}{5}$ or $1\dfrac{1}{5}$ **23.** 18   **24.** 4   **25.** 10   **26.** $+18$ **27.** $-\dfrac{1}{2}$ **28.** $-\dfrac{2}{3}$ **29.** -2 **30.** $-\dfrac{1}{2}$

CHAPTER 10   Algebra

Pretest Chapter 10

1. $-3x$ **2.** $-25y$ **3.** $-2a-4b$ **4.** $-3x+4y-7$ **5.** $-6a-12b-2$ **6.** $7x+3z-12w$ **7.** $30x-35y$
8. $-2x-6y+10$ **9.** $-7.5x+4.8y-6z+12$ **10.** $-12a-9b$ **11.** 4   **12.** 2   **13.** 4   **14.** 3   **15.** 22

16. $\dfrac{3}{2}$ or $1\dfrac{1}{2}$ **17.** $c-p=9$ **18.** $l=2w+11$ **19.** $a=$ height of Mt. Ararat, $a-1758=$ height of Mt. Hood

20. Let $b=$ number of degrees in angle B, let $2b=$ number of degrees in angle A, let $b-35=$ number of degrees in angle C
21. 293 miles on Monday, 349 miles on Tuesday, 256 miles on Wednesday   **22.** width $=17$ m, length $=29$ m
23. 7.75 ft, 10.25 ft   **24.** Secretary, \$14,000.   Assistant, \$26,000.   Surveyor, \$42,000.

Exercises 10.1

1. N, b, h **3.** S, r **5.** p, a, b **7.** $p = 4\,s^2$ **9.** $A = \dfrac{bh}{2}$ **11.** $V = \dfrac{4\pi r^3}{3}$ **13.** $H = 2a - 3b$ **15.** $17x$ **17.** $5x$

19. $-x$ **21.** $-21x$ **23.** $4x + 1$ **25.** $-1.1x + 6.4$ **27.** $-5.3x - 5.6$ **29.** $-32x - 3$ **31.** $2x - 3y - 4$
33. $13a - 5b - 7c$ **35.** $-3r - 8s - 5$ **37.** $8n - 29p + 4q$ **39.** $-28.5x - 4.4$ **41.** 0 **43.** $-2x - 12.9y + 6.8$
45. Associative property of addition, Commutative property of addition, Associative property of addition, Distributive property, Addition fact   **47.** $-27.4x + 15.3y - 3.4w - 27.5$ **49.** 4.8 **51.** 3 **53.** $-9.41x + 8.426y$

Exercises 10.2

1. $5x + 35$ **3.** $27x + 18$ **5.** $-2x - 2y$ **7.** $-10.5x + 21y$ **9.** $30x - 70y$ **11.** $2x + 6y - 10$
13. $-4a - 4b - 8c$ **15.** $-16a + 40b - 8c$ **17.** $-180a + 33b + 100.5$ **19.** $4a - 12b + 20c + 28$

21. $-2.6x + 17y + 10z - 24$ **23.** $x - \dfrac{3}{2}y + 2z - \dfrac{1}{4}$ **25.** $8x + y$ **27.** $21x - 53$ **29.** $-5a + 18b$

31. $14x + 12y - 28$ **33.** $8.1x + 8.1y$ **35.** $-9a + 7b - 2c$ **37.** $A = \dfrac{hB + hb}{2}$ **39.** $A = ab - ac,\ A = a(b - c)$

41. $3(2 - (-6)) = 3(8) = 24$, and $3(2 - (-6)) = 3 \cdot 2 - (3)(-6) = 6 + 18 = 24$ **43.** $-\dfrac{1}{12}x + \dfrac{1}{4}y + \dfrac{3}{2}$ **45.** 22.0 inches

47. 77 cm^2 **49.** $3.22x - 6.6332$

Exercises 10.3

1. $x = -4$ **3.** $x = 17$ **5.** $x = -1$ **7.** $x = 3$ **9.** $x = -\dfrac{4}{9}$ **11.** $x = -5$ **13.** $x = \dfrac{5}{6}$ **15.** $x = 3$ **17.** $x = 7$

19. $x = 5$ **21.** $y = -\dfrac{2}{3}$ **23.** $y = -7$ **25.** $x = -5$ **27.** Yes   **29.** No   **31.** $x = -1$ **33.** $x = 3$

35. $y = -\dfrac{7}{2}$ **37.** $x = 4$ **39. (a)** $x = 5$ **(b)** $5 = x$ **(c)** Answers vary.   **41.** $x = 0$ **43.** 7234.6 cm^3

45. $x = -0.43532$

Exercises 10.4

1. $r = c + 6$ **3.** $c = p + 29$ **5.** $c = t - \$95$ **7.** $w = t - 136$ **9.** $m = t - 19°$ **11.** $l = 2w + 7$ **13.** $s = 3f - 2$
15. $e = 3f + 1000$ **17.** $t = 800 - w$ or $w + t = 800$ **19.** $ht = 500$ **21.** $c =$ Charlie's salary, $c + 600 =$ Jim's salary
23. $j =$ Julie's mileage, $j + 386 =$ Barbara's mileage
25. $b =$ number of degrees in angle B, $b - 46° =$ number of degrees in angle A
27. $w =$ number of meters ht. of Whitney, $w + 4430 =$ number of meters ht. of Everest
29. $e =$ classes taken by Ernesto, $2e =$ number of classes taken by Wally
31. $x =$ width, $3x + 12 =$ length   **33.** $x =$ first angle, $2x =$ second angle, $x - 14 =$ third angle
35. $x - 10 =$ Mike's salary, $x - 100 =$ Bob's salary, $x - 130 =$ Fred's salary   **37.** 8 **39.** -1

Exercises 10.5

1. $x =$ length of shorter piece, shorter piece $= 5.25$ ft, longer piece $= 10.75$ ft
3. $x =$ 2nd day, 1st day $= 328.5$ mi, 2nd day $= 231.5$ mi   **5.** $x =$ truck, car $= 1960$ lbs, truck $= 3380$ lbs
7. $x =$ fall students, spring students $= 244$, summer students $= 48$, fall students $= 134$
9. $x =$ width, width $= 180$ yd, length $= 367$ yd   **11.** $x =$ width, width $= 7$ cm, length $= 19$ cm
13. $x =$ 1st side, 1st side $= 18$ m, 2nd side $= 36$ m, 3rd side $= 45$ m
15. $x =$ 1st side, 1st side $= 47$ mm, 2nd side $= 57$ mm, 3rd side $= 40$ mm
17. $x =$ angle A, $A = 28°$, $B = 84°$, $C = 68°$ **19.** $x =$ total sales, $\$40,000$ **21.** $x =$ yearly rent, $\$6000$
23. $x =$ number of people last year, 24,000 people
25. $x =$ cost of first part, 1st part $= \$8.87$, 2nd part $= \$14.74$, 3rd part $= \$49.22$ **27.** 57.75 **29.** 500
31. $W = 476.7$ yds, $L = 961.5$ yds

Extra Practice: Examples and Exercises

Section 10.1   **1.** L, r, k **3.** t, r, s **5.** s, b, c **7.** $E = MC^2$ **9.** $s = abc$ **11.** $8x$ **13.** $4x$ **15.** x
17. $-1.3a + 0.1b - 6$ **19.** $4x + y + z + 3$ *Section 10.2*   **1.** $8 + 12x$ **3.** $11x + 11$ **5.** $6.5x + 12.5$
7. $27x + 18y$ **9.** $-24t - 1$ **11.** $-42a + 14b - 7c + 35d$ **13.** $36.9s - 19.8t - 45$ **15.** $18a + 6b + 12c - 27d$

17. $21x - 78y$ **19.** $-3.5s + 30.7t$ *Section 10.3*   **1.** $x = -7$ **3.** $x = 10$ **5.** $x = -\dfrac{4}{3}$ **7.** $y = \dfrac{4}{15}$

9. $x = 2$ **11.** $x = -19$ **13.** $y = -\dfrac{2}{7}$ **15.** $y = -8$ **17.** $x = -15$ **19.** $x = -4$ *Section 10.4*   **1.** $L = A + 5$

3. $d = c - 65$ **5.** $l = e - 3$ **7.** $l = 2w + 8$ **9.** $s = 2f - 5$ **11.** $j =$ Janet's, $j + 260 =$ Sarah's
13. $b = \angle B$, $b - 23 = \angle A$ **15.** $r =$ Rosa, $r - 16 =$ Juanita   **17.** $b =$ Bob, $2b =$ Les   **19.** $x =$ width, $2x + 8 =$ length

Section 10.5 **1.** First day = 235.5 mi, Second day = 149.5 mi   **3.** First day = 333.5 mi, Second day = 261.5 mi
5. George's salary = $966, Dale's salary = $884   **7.** 548 students off campus, 663 students on campus, 493 students at home
9. May = 8300 customers, June = 16,600 customers, July = 10,800 customers   **11.** $w = 7$ cm, $l = 17$ cm
13. 1st = 19.25 m, 2nd = 38.5 m, 3rd = 44.25 m   **15.** 1st = 52 mm, 2nd = 77 mm, 3rd = 47 mm
17. $F = 40°$, $E = 80°$, $G = 60°$ **19.** $B = 23°$, $C = 69°$, $A = 88°$

Review Problems Chapter 10
1. $-5x + 2y$ **3.** $-a + 12$ **5.** $-2x - 7y$ **7.** $-x - 12y + 6$ **9.** $-2.8a + 3.4b - 2$ **11.** $-15x - 3y$
13. $2x - 6y + 8$ **15.** $-24a + 40b + 8c$ **17.** $6x + 15y - 27.5$ **19.** $-2x + 14y$ **21.** $-8a - 2b - 24$

23. $x = 5$ **25.** $x = 1$ **27.** $x = 8$ **29.** $x = 3$ **31.** $x = \dfrac{29}{5}$ **33.** $x = -2$ **35.** $y = 16$ **37.** $w = c + 3000$

39. $A = 3B$ **41.** $r =$ Roberto's salary, $r + 2050 =$ Michael's salary
43. $c =$ number of Connie's courses, $c - 6 =$ numbers of Nancy's courses   **45.** $x =$ shorter piece, 26.75 ft, 33.25 ft
47. $x =$ February customers, February = 10,550, March = 21,100, April = 13,550
49. width $= x$, width = 13 inches, length = 23 inches   **51.** $x =$ 1st side, 1st side = 32 m, 2nd side 39 m, 3rd side = 28 m

Quiz Sections 10.1–10.3
1. $-5x$ **2.** $-12y + 4$ **3.** $2a - 21b$ **4.** $-a - 8b + 6$ **5.** $-9x - 20y$ **6.** $-4x - 10y - 24$
7. $-6x - 4y + 4w$ **8.** $-19y - 14w$ **9.** $3x - 18y$ **10.** $4x + 10y$ **11.** $-4a - 8b + 12c$ **12.** $-6a - 3b + 15c$
13. $-30x - 36y + 24$ **14.** $56x - 64y - 56$ **15.** $-3x + 7.2y - 10$ **16.** $-10.4x - 14y + 28$ **17.** $8x - 16y$

18. $40x - y$ **19.** $-2x + 5y + 30$ **20.** $-10x - 26y + 28$ **21.** $x = 3$ **22.** $x = 7$ **23.** $x = -\dfrac{1}{2}$ **24.** $x = -\dfrac{1}{2}$

25. $x = 0$ **26.** $x = 0$ **27.** $x = 2$ **28.** $x = 2$ **29.** $x = -2$ **30.** $x = 8$ **31.** Yes   **32.** No

Quiz Sections 10.4–10.5
1. $h = w + 3$ **2.** $c = t + \$145$ **3.** $b = p + 367$ or $p = b - 367$ **4.** $f = s + 23$ **5.** $l = 2w + 4$ **6.** $s = 3f + 8$
7. $j =$ Jeff's salary, $j + 1800 =$ Susan's salary   **8.** $c =$ weight of chair, $c + 230 =$ weight of sofa
9. $s =$ length side, $s - 19 =$ length top   **10.** $c =$ Carlo's courses, $2c =$ Rita's courses
11. $f =$ number of cars owned by Frederick, $3f =$ number of cars owned by William
12. $B =$ degrees in angle B, $3B + 13 =$ degrees in angle A **13.** $C =$ angle C, $2C - 14° =$ angle B
14. $x =$ width, $2x - 7 =$ length   **15.** $x =$ length of 2nd side, $3x - 5 =$ length of 1st side
16. $x =$ number of times Dr. Schoenert drove, $2x =$ number of times Dr. Kaiser drove, $x + 1 =$ number of times Dr. Slater drove
17. $x =$ degrees in 1st angle, $x + 15 =$ degrees in 2nd angle, $x - 29 =$ degrees in 3rd angle
18. $x =$ cost of Frydrych's vacation, $x + 560 =$ cost of Williamson's vacation, $x - 112 =$ cost of Van Dyke's vacation
19. $x =$ short piece, short piece = 8.25 ft, long piece = 11.75 ft   **20.** Sam's salary $= x$, Sam's = $16,880, Bob's = $19,520
21. Monday's mileage $= x$, Monday = 298 miles, Tuesday = 280 miles, Wednesday = 367 miles
22. width $= x$, width = 17 m, length = 33 m   **23.** 1st side $= x$, 1st side = 12 cm, 2nd side = 15 cm, 3rd side = 18 cm
24. angle $A = x$, angle $A = 25°$, angle $B = 75°$, angle $C = 80°$ **25.** Total sales $= x$, total sales = $60,000
26. $x =$ population last year, population last year = 15,000

Test Chapter 10
1. $-5a$ **2.** $-19a$ **3.** $-13x - 2y$ **4.** $-5x - 5y - 11$ **5.** $5x - 14y + 4z$ **6.** $-4a + 7b - 15$
7. $36x - 15y$ **8.** $-4x + 8y - 12$ **9.** $-2.5x - 5y + 7.5z + 12.5$ **10.** $-11a + 14b$ **11.** $x = -3$ **12.** $x = -2$

13. $x = -3$ **14.** $x = 2$ **15.** $x = 5$ **16.** $x = -\dfrac{5}{2}$ or $-2\dfrac{1}{2}$ **17.** $s = f + 18$ **18.** $n = s - 20,000$

19. $s =$ second, $2s =$ first, $3s =$ third   **20.** $w =$ width, $2w - 5 =$ length
21. Prentice farm $= x$, Prentice farm = 47 acres, Smithfield farm = 235 acres
22. Marcia's income $= x$, Marcia = $15,200, Wanda = $14,300
23. Monday mileage $= x$,   Monday = 317 miles, Tuesday = 366 miles, Wednesday = 282 miles
24. width $= x$, width = 16 ft, length = 37 ft

Cumulative Test Chapters 1–10
1. 747   **2.** 10,815   **3.** 45,678,900   **4.** 2   **5.** $\dfrac{65}{8}$ or $8\dfrac{1}{8}$ **6.** 0.07254   **7.** 30,550.301   **8.** $n = 16$ **9.** 1596

10. 5500   **11.** 0.345 meters   **12.** 120 inches   **13.** 37.7 yards   **14.** 143 m^2 **15.** -31 **16.** 30

17. $-9a - 11b$ **18.** $-6x + 2y + 3$ **19.** $21x - 7y + 56$ **20.** $-2x - 24y$ **21.** $x = 4$ **22.** $y = -\dfrac{5}{2}$ or $-2\dfrac{1}{2}$

23. $x = 26$ **24.** $x = -3$ **25.** $p =$ weight of printer, $p + 322 =$ weight of computer
26. $f =$ students during fall, $f - 87 =$ students during summer
27. Thursday mileage $= x$, Thursday = 376 miles, Friday = 424 miles, Saturday = 281 miles

28. width $= 13\dfrac{2}{3}$ ft, length $= 35\dfrac{1}{3}$ ft

PRACTICE FINAL EXAMINATION

Chapter 1   **1.** eighty-two thousand, three hundred sixty-seven   **2.** 30,333   **3.** 173   **4.** 34,103   **5.** 4212

6. 217,745   **7.** 158   **8.** 606   **9.** 116   **10.** 32 miles per gallon   *Chapter 2*   **11.** $\frac{7}{15}$ **12.** $\frac{42}{11}$ **13.** $\frac{33}{20}$ or $1\frac{13}{20}$

14. $\frac{89}{15}$ or $5\frac{14}{15}$ **15.** $\frac{31}{14}$ or $2\frac{3}{14}$ **16.** 4   **17.** $\frac{14}{5}$ or $2\frac{4}{5}$ **18.** $\frac{22}{13}$ or $1\frac{9}{13}$ **19.** $6\frac{17}{20}$ miles   **20.** 5 packages

Chapter 3   **21.** 0.719   **22.** $\frac{43}{50}$ **23.** >   **24.** 506.38   **25.** 21.77   **26.** 0.757   **27.** 0.492   **28.** 3.69   **29.** 0.8125

30. 0.7056   *Chapter 4*   **31.** $\frac{1400 \text{ students}}{43 \text{ faculty}}$ **32.** False   **33.** $n \approx 9.4$ **34.** $n \approx 7.7$ **35.** $n = 15$ **36.** $n = 9$

37. \$3333.33   **38.** 9.75 inches   **39.** \$294.12 was withheld for federal income tax.   **40.** 1.6 pounds of butter
Chapter 5   **41.** 0.63%   **42.** 21.25%   **43.** 1.64   **44.** 17.3%   **45.** 302.4   **46.** 250   **47.** 4284   **48.** \$10,856
49. 4500 students   **50.** 34.3% increase   *Chapter 6*   **51.** 4.25 gal   **52.** 6500 lb   **53.** 192 in.   **54.** 5600 m
55. 0.0698 kg   **56.** 0.00248 L   **57.** 19.32 km   **58.** 6.3182×10^{-4} **59.** 1.264×10^{11} **60.** 1.36728 cm thick
Chapter 7   **61.** 14.4 meters   **62.** 206 cm   **63.** 5.4 square feet   **64.** 75 square meters   **65.** 113.04 square meters
66. 56.52 meters   **67.** 167.47 cubic centimeters   **68.** 205.2 cubic feet   **69.** 32.5 square meters   **70.** $n = 32.5$
Chapter 8   **71.** 8 million dollars   **72.** one million dollars   **73.** 50° F   **74.** From 1980 to 1990   **75.** 600 students
76. 1400 students   **77.** The mean is approximately 15.83. The median is 16.5   **78.** 16   **79.** 11.091   **80.** 15 feet

Chapter 9   **81.** -13 **82.** $\frac{1}{8}$ **83.** -3 **84.** -17 **85.** 24   **86.** $-\frac{8}{3}$ or $-2\frac{2}{3}$ **87.** 4   **88.** 27   **89.** 8

90. 0.5 or $\frac{1}{2}$ *Chapter 10*   **91.** $-3x - 7y$ **92.** $-7 - 4a - 17b$ **93.** $-2x + 6y + 10$ **94.** $-11x - 9y - 4$

95. $x = 2$ **96.** $x = -2$ **97.** $x = -\frac{1}{2}$ or $x = -0.5$ **98.** $x = -\frac{2}{5}$ or $x = -0.4$

99. 122 students are taking history. 110 students are taking math.   **100.** The length is 37 meters. The width is 16 meters.

APPENDIX A   The Use of a Calculator

Calculator Exercise Set
1. 11,747   **3.** 352,554   **5.** 1.294   **7.** 254   **9.** 8934.7376   **11.** 16.81   **13.** 10.03   **15.** 289.387   **17.** 378,224
19. 690,284   **21.** 1068.678   **23.** 1.58688   **25.** 1383.2   **27.** 0.7634   **29.** 16.03   **31.** \$12,562.34   **33.** 85,163.7
35. \$103,608.65   **37.** 15,782.038   **39.** 2,500344   **41.** 897   **43.** 22.5   **45.** 84.372   **47.** 4.435   **49.** 101.5673

51. 0.24685   **53.** $0.3\overline{8} + 0.692307 \approx 1.08120$ **55.** $0.25 + 0.875 + 0.2\overline{7} \approx 1.39773$ **57.** $\frac{18.61}{12.7} \approx 1.46535$

59. 196,313.1 sq. ft.   **61.** 26.9 miles per gallon   **63.** \$665.47   **65.** \$3923.31   **67.** ≈ 9.276 m/sec
69. Wally spent \$863.28, Joan spent \$1106.43 (to nearest cent), Jean spent \$243.15 more   **71.** 644   **73.** 840   **75.** 38%
77. 319.24   **79.** 3300   **81.** 64%   **83.** 69,310.64   **85.** 6420

Index